Loop Quantum Gravity

" A Theory of Quantum Space-Time "

Edited by Paul F. Kisak

Contents

Chapter 1

Loop quantum gravity

Loop quantum gravity (**LQG**) is a theory that attempts to describe the quantum properties of the universe and gravity. It is also a theory of quantum spacetime because, according to general relativity, gravity is a manifestation of the geometry of spacetime. LQG is an attempt to merge quantum mechanics and general relativity.

From the point of view of Einstein's theory, it comes as no surprise that all attempts to treat gravity simply like one more quantum force (on par with electromagnetism and the nuclear forces) have failed. According to Einstein, gravity is not a force – it is a property of space-time itself. Loop quantum gravity is an attempt to develop a quantum theory of gravity based directly on Einstein's geometrical formulation. The main output of the theory is a physical picture of space where space is granular. The granularity is a direct consequence of the quantization. It has the same nature as the granularity of the photons in the quantum theory of electromagnetism and the discrete levels of the energy of the atoms. Here, it is space itself that is discrete. In other words, there is a minimum distance possible to travel through it.

More precisely, space can be viewed as an extremely fine fabric or network "woven" of finite loops. These networks of loops are called spin networks. The evolution of a spin network over time is called a spin foam. The predicted size of this structure is the Planck length, which is approximately 10^{-35} meters. According to the theory, there is no meaning to distance at scales smaller than the Planck scale. Therefore, LQG predicts that not just matter, but space itself, has an atomic structure.

Today LQG is a vast area of research, developing in several directions, which involves about 30 research groups worldwide.[1] They all share the basic physical assumptions and the mathematical description of quantum space. The full development of the theory is being pursued in two directions: the more traditional canonical loop quantum gravity, and the newer covariant loop quantum gravity, more commonly called spin foam theory.

Research into the physical consequences of the theory is proceeding in several directions. Among these, the most well-developed is the application of LQG to cosmology, called loop quantum cosmology (LQC). LQC applies LQG ideas to the study of the early universe and the physics of the Big Bang. Its most spectacular consequence is that the evolution of the universe can be continued beyond the Big Bang. The Big Bang appears thus to be replaced by a sort of cosmic Big Bounce.

1.1 History

Main article: History of loop quantum gravity

In 1986, Abhay Ashtekar reformulated Einstein's general relativity in a language closer to that of the rest of fundamental physics. Shortly after, Ted Jacobson and Lee Smolin realized that the formal equation of quantum gravity, called the Wheeler–DeWitt equation, admitted solutions labelled by loops when rewritten in the new Ashtekar variables. Carlo Rovelli and Lee Smolin defined a nonperturbative and background-independent quantum theory of gravity in terms of these loop solutions. Jorge Pullin and Jerzy Lewandowski understood that the intersections of the loops are essential for the consistency of the theory, and the theory should be formulated in terms of intersecting loops, or graphs.

In 1994, Rovelli and Smolin showed that the quantum operators of the theory associated to area and volume have a discrete spectrum. That is, geometry is quantized. This result defines an explicit basis of states of quantum geometry, which turned out to be labelled by Roger Penrose's spin networks, which are graphs labelled by spins.

The canonical version of the dynamics was put on firm ground by Thomas Thiemann, who defined an anomaly-free Hamiltonian operator, showing the existence of a mathematically consistent background-independent theory. The covariant or spinfoam version of the dynamics developed during several decades, and crystallized in 2008, from the joint work of research groups in France, Canada, UK, Poland, and Germany, lead to the definition of a family of transition amplitudes, which in the classical limit can be shown to be related to a family of truncations of general relativity.[2] The finiteness of these amplitudes was proven in 2011.[3][4] It requires the existence of a positive cosmological constant, and this is consistent with observed acceleration in the expansion of the Universe.

1.2 General covariance and background independence

Main articles: General covariance, background-independent and diffeomorphism

In theoretical physics, general covariance is the invariance of the form of physical laws under arbitrary differentiable coordinate transformations. The essential idea is that coordinates are only artifices used in describing nature, and hence should play no role in the formulation of fundamental physical laws. A more significant requirement is the principle of general relativity that states that the laws of physics take the same form in all reference systems. This is a generalization of the principle of special relativity which states that the laws of physics take the same form in all inertial frames.

In mathematics, a diffeomorphism is an isomorphism in the category of smooth manifolds. It is an invertible function that maps one differentiable manifold to another, such that both the function and its inverse are smooth. These are the defining symmetry transformations of General Relativity since the theory is formulated only in terms of a differentiable manifold.

In general relativity, general covariance is intimately related to "diffeomorphism invariance". This symmetry is one of the defining features of the theory. However, it is a common misunderstanding that "diffeomorphism invariance" refers to the invariance of the physical predictions of a theory under arbitrary coordinate transformations; this is untrue and in fact every physical theory is invariant under coordinate transformations this way. Diffeomorphisms, as mathematicians define them, correspond to something much more radical; intuitively a way they can be envisaged is as simultaneously dragging all the physical fields (including the gravitational field) over the bare differentiable manifold while staying in the same coordinate system. Diffeomorphisms are the true symmetry transformations of general relativity, and come about from the assertion that the formulation of the theory is based on a bare differentiable manifold, but not on any prior geometry — the theory is background-independent (this is a profound shift, as all physical theories before general relativity had as part of their formulation a prior geometry). What is preserved under such transformations are the coincidences between the values the gravitational field take at such and such a "place" and the values the matter fields take there. From these relationships one can form a notion of matter being located with respect to the gravitational field, or vice versa. This is what Einstein discovered: that physical entities are located with respect to one another only and not with respect to the spacetime manifold. As Carlo Rovelli puts it: "No more fields on spacetime: just fields on fields.".[5] This is the true meaning of the saying "The stage disappears and becomes one of the actors"; space-time as a "container" over which physics takes place has no objective physical meaning and instead the gravitational interaction is represented as just one of the fields forming the world. This is known as the relationalist interpretation of space-time. The realization by Einstein that general relativity should be interpreted this way is the origin of his remark "Beyond my wildest expectations".

In LQG this aspect of general relativity is taken seriously and this symmetry is preserved by requiring that the physical states remain invariant under the generators of diffeomorphisms. The interpretation of this condition is well understood for purely spatial diffeomorphisms. However, the understanding of diffeomorphisms involving time (the Hamiltonian constraint) is more subtle because it is related to dynamics and the so-called "problem of time" in general relativity.[6] A generally accepted calculational framework to account for this constraint has yet to be found.[7][8] A plausible candidate for the quantum hamiltonian constraint is the operator introduced by Thiemann.[9]

LQG is formally background independent. The equations of LQG are not embedded in, or dependent on, space and time (except for its invariant topology). Instead, they are expected to give rise to space and time at distances which are large compared to the Planck length. The issue of background independence in LQG still has some unresolved subtleties. For example, some derivations require a fixed choice of the topology, while any consistent quantum theory of gravity should include topology change as a dynamical process.

1.3 Constraints and their Poisson bracket algebra

Main articles: Poisson bracket and Hamiltonian constraint

1.3.1 The constraints of classical canonical general relativity

Main article: Lie derivative

In the Hamiltonian formulation of ordinary classical mechanics the Poisson bracket is an important concept. A "canonical coordinate system" consists of canonical position and momentum variables that satisfy canonical Poisson-bracket relations,

$\{q_i, p_j\} = \delta_{ij}$

where the Poisson bracket is given by

$$\{f, g\} = \sum_{i=1}^{N} \left(\frac{\partial f}{\partial q_i} \frac{\partial g}{\partial p_i} - \frac{\partial f}{\partial p_i} \frac{\partial g}{\partial q_i} \right).$$

for arbitrary phase space functions $f(q_i, p_j)$ and $g(q_i, p_j)$. With the use of Poisson brackets, the Hamilton's equations can be rewritten as,

$\dot{q}_i = \{q_i, H\}$,

$\dot{p}_i = \{p_i, H\}$.

These equations describe a "flow" or orbit in phase space generated by the Hamiltonian H. Given any phase space function $F(q, p)$, we have

$\frac{d}{dt} F(q_i, p_i) = \{F, H\}.$

Let us consider constrained systems, of which General relativity is an example. In a similar way the Poisson bracket between a constraint and the phase space variables generates a flow along an orbit in (the unconstrained) phase space generated by the constraint. There are three types of constraints in Ashtekar's reformulation of classical general relativity:

$SU(2)$ Gauss gauge constraints

The Gauss constraints

$G_j(x) = 0$.

This represents an infinite number of constraints one for each value of x. These come about from re-expressing General relativity as an $\mathrm{SU}(2)$ Yang–Mills type gauge theory (Yang–Mills is a generalization of Maxwell's theory where the gauge field transforms as a vector under Gauss transformations, that is, the Gauge field is of the form $A_a^i(x)$ where i is an internal index. See Ashtekar variables). These infinite number of Gauss gauge constraints can be smeared with test fields with internal indices, $\lambda^j(x)$,

$G(\lambda) = \int d^3x G_j(x) \lambda^j(x)$.

which we demand vanish for any such function. These smeared constraints defined with respect to a suitable space of smearing functions give an equivalent description to the original constraints.

In fact Ashtekar's formulation may be thought of as ordinary $\mathrm{SU}(2)$ Yang–Mills theory together with the following special constraints, resulting from diffeomorphism invariance, and a Hamiltonian that vanishes. The dynamics of such a theory are thus very different from that of ordinary Yang–Mills theory.

Spatial diffeomorphisms constraints

The spatial diffeomorphism constraints

$C_a(x) = 0$

can be smeared by the so-called shift functions $\vec{N}(x)$ to give an equivalent set of smeared spatial diffeomorphism constraints,

$C(\vec{N}) = \int d^3x C_a(x) N^a(x)$.

These generate spatial diffeomorphisms along orbits defined by the shift function $N^a(x)$.

Hamiltonian constraints

The Hamiltonian

$H(x) = 0$

can be smeared by the so-called lapse functions $N(x)$ to give an equivalent set of smeared Hamiltonian constraints,

$H(N) = \int d^3x H(x) N(x)$.

These generate time diffeomorphisms along orbits defined by the lapse function $N(x)$.

In Ashtekar formulation the gauge field $A_a^i(x)$ is the configuration variable (the configuration variable being analogous to q in ordinary mechanics) and its conjugate momentum is the (densitized) triad (electrical field) $\tilde{E}_i^a(x)$. The constraints are certain functions of these phase space variables.

We consider the action of the constraints on arbitrary phase space functions. An important notion here is the Lie derivative, $\mathcal{L}_V$, which is basically a derivative operation that infinitesimally "shifts" functions along some orbit with tangent vector V.

1.3.2 The Poisson bracket algebra

Of particular importance is the Poisson bracket algebra formed between the (smeared) constraints themselves as it completely determines the theory. In terms of the above smeared constraints the constraint algebra amongst the Gauss' law reads,

$$\{G(\lambda), G(\mu)\} = G([\lambda, \mu])$$

where $[\lambda, \mu]^k = \lambda_i \mu_j \epsilon^{ijk}$. And so we see that the Poisson bracket of two Gauss' law is equivalent to a single Gauss' law evaluated on the commutator of the smearings. The Poisson bracket amongst spatial diffeomorphisms constraints reads

$$\{C(\vec{N}), C(\vec{M})\} = C(\mathcal{L}_{\vec{N}} \vec{M})$$

and we see that its effect is to "shift the smearing". The reason for this is that the smearing functions are not functions of the canonical variables and so the spatial diffeomorphism does not generate diffeomorphims on them. They do however generate diffeomorphims on everything else. This is equivalent to leaving everything else fixed while shifting the smearing .The action of the spatial diffeomorphism on the Gauss law is

$$\{C(\vec{N}), G(\lambda)\} = G(\mathcal{L}_{\vec{N}} \lambda) ,$$

again, it shifts the test field λ . The Gauss law has vanishing Poisson bracket with the Hamiltonian constraint. The spatial diffeomorphism constraint with a Hamiltonian gives a Hamiltonian with its smearing shifted,

$$\{C(\vec{N}), H(M)\} = H(\mathcal{L}_{\vec{N}} M) .$$

Finally, the poisson bracket of two Hamiltonians is a spatial diffeomorphism,

$$\{H(N), H(M)\} = C(K)$$

where K is some phase space function. That is, it is a sum over infinitesimal spatial diffeomorphisms constraints where the coefficients of proportionality are not constants but have non-trivial phase space dependence.

A (Poisson bracket) Lie algebra, with constraints C_I , is of the form

$$\{C_I, C_J\} = f_{IJ}^K C_K$$

where f_{IJ}^K are constants (the so-called structure constants). The above Poisson bracket algebra for General relativity does not form a true Lie algebra as we have structure functions rather than structure constants for the Poisson bracket between two Hamiltonians. This leads to difficulties.

1.3.3 Dirac observables

The constraints define a constraint surface in the original phase space. The gauge motions of the constraints apply to all phase space but have the feature that they leave the constraint surface where it is, and thus the orbit of a point in the hypersurface under gauge transformations will be an orbit entirely within it. Dirac observables are defined as phase space functions, O , that Poisson commute with all the constraints when the constraint equations are imposed,

$$\{G_j, O\}_{G_j = C_a = H = 0} = \{C_a, O\}_{G_j = C_a = H = 0} = \{H, O\}_{G_j = C_a = H = 0} = 0 ,$$

that is, they are quantities defined on the constraint surface that are invariant under the gauge transformations of the theory.

Then, solving only the constraint $G_j = 0$ and determining the Dirac observables with respect to it leads us back to the ADM phase space with constraints H, C_a . The dynamics of general relativity is generated by the constraints, it can be shown that six Einstein equations describing time evolution (really a gauge transformation) can be obtained by calculating the Poisson brackets of the three-metric and its conjugate momentum with a linear combination of the spatial diffeomorphism and Hamiltonian constraint. The vanishing of the constraints, giving the physical phase space, are the four other Einstein equations.[10]

1.4 Quantization of the constraints – the equations of quantum general relativity

1.4.1 Pre-history and Ashtekar new variables

Main articles: Frame fields in general relativity, Ashtekar variables and Self-dual Palatini action

Many of the technical problems in canonical quantum gravity revolve around the constraints. Canonical general relativity was originally formulated in terms of metric variables, but there seemed to be insurmountable mathematical difficulties in promoting the constraints to quantum operators because of their highly non-linear dependence on the canonical variables. The equations were much simplified with the introduction of Ashtekars new variables. Ashtekar variables describe canonical general relativity in terms of a new pair canonical variables closer to that of gauge theories. The first step consists of using densitized triads $\tilde{E}_i^a$ (a triad E_i^a is simply three orthogonal vector fields labeled by $i = 1, 2, 3$ and the densitized triad is defined by $\tilde{E}_i^a = \sqrt{\det(q)} E_i^a$) to encode information about the spatial metric,

$$\det(q) q^{ab} = \tilde{E}_i^a \tilde{E}_j^b \delta^{ij} .$$

(where δ^{ij} is the flat space metric, and the above equation expresses that q^{ab} , when written in terms of the basis E_i^a , is locally flat). (Formulating general relativity with triads instead of metrics was not new.) The densitized triads are not unique, and in fact one can perform a local in space rotation with respect to the internal indices i . The canonically conjugate variable is related to the extrinsic curvature by $K_a^i = K_{ab}\tilde{E}^{ai}/\sqrt{\det(q)}$. But problems similar to using the metric formulation arise when one tries to quantize the theory. Ashtekar's new insight was to introduce a new configuration variable,

$$A_a^i = \Gamma_a^i - iK_a^i$$

that behaves as a complex $SU(2)$ connection where Γ_a^i is related to the so-called spin connection via $\Gamma_a^i = \Gamma_{ajk}\epsilon^{jki}$. Here A_a^i is called the chiral spin connection. It defines a covariant derivative $\mathcal{D}_a$. It turns out that $\tilde{E}_i^a$ is the conjugate momentum of A_a^i , and together these form Ashtekar's new variables.

The expressions for the constraints in Ashtekar variables; the Gauss's law, the spatial diffeomorphism constraint and the (densitized) Hamiltonian constraint then read:

$$G^i = \mathcal{D}_a\tilde{E}_i^a = 0$$

$$C_a = \tilde{E}_i^b F_{ab}^i - A_a^i(\mathcal{D}_b\tilde{E}_i^b) = V_a - A_a^i G^i = 0\ ,$$

$$\tilde{H} = \epsilon_{ijk}\tilde{E}_i^a\tilde{E}_j^b F_{ab}^i = 0$$

respectively, where F_{ab}^i is the field strength tensor of the connection A_a^i and where V_a is referred to as the vector constraint. The above-mentioned local in space rotational invariance is the original of the $SU(2)$ gauge invariance here expressed by the Gauss law. Note that these constraints are polynomial in the fundamental variables, unlike as with the constraints in the metric formulation. This dramatic simplification seemed to open up the way to quantizing the constraints. (See the article Self-dual Palatini action for a derivation of Ashtekar's formulism).

With Ashtekar's new variables, given the configuration variable A_a^i , it is natural to consider wavefunctions $\Psi(A_a^i)$. This is the connection representation. It is analogous to ordinary quantum mechanics with configuration variable q and wavefunctions $\psi(q)$. The configuration variable gets promoted to a quantum operator via:

$$\hat{A}_a^i\Psi(A) = A_a^i\Psi(A)\ ,$$

(analogous to $\hat{q}\psi(q) = q\psi(q)$) and the triads are (functional) derivatives,

$$\hat{\tilde{E}}_i^a\Psi(A) = -i\frac{\delta\Psi(A)}{\delta A_a^i}\ .$$

(analogous to $\hat{p}\psi(q) = -i\hbar d\psi(q)/dq$). In passing over to the quantum theory the constraints become operators on a kinematic Hilbert space (the unconstrained $SU(2)$ Yang–Mills Hilbert space). Note that different ordering of the A 's and $\tilde{E}$'s when replacing the $\tilde{E}$'s with derivatives give rise to different operators - the choice made is called the factor ordering and should be chosen via physical reasoning. Formally they read

$$\hat{G}_j|\psi\rangle = 0$$

$$\hat{C}_a|\psi\rangle = 0$$

$$\hat{\tilde{H}}|\psi\rangle = 0\ .$$

There are still problems in properly defining all these equations and solving them. For example, the Hamiltonian constraint Ashtekar worked with was the densitized version instead of the original Hamiltonian, that is, he worked with $\tilde{H} = \sqrt{\det(q)}H$. There were serious difficulties in promoting this quantity to a quantum operator. Moreover, although Ashtekar variables had the virtue of simplifying the Hamiltonian, they are complex. When one quantizes the theory, it is difficult to ensure that one recovers real general relativity as opposed to complex general relativity.

1.4.2 Quantum constraints as the equations of quantum general relativity

We now move on to demonstrate an important aspect of the quantum constraints. We consider Gauss' law only. First we state the classical result that the Poisson bracket of the smeared Gauss' law $G(\lambda) = \int d^3x\lambda^j(D_aE^a)^j$ with the connections is

$$\{G(\lambda), A_a^i\} = \partial_a\lambda^i + g\epsilon^{ijk}A_a^j\lambda^k = (D_a\lambda)^i.$$

The quantum Gauss' law reads

$$\hat{G}_j\Psi(A) = -iD_a\frac{\delta\lambda\Psi[A]}{\delta A_a^j} = 0.$$

If one smears the quantum Gauss' law and study its action on the quantum state one finds that the action of the constraint on the quantum state is equivalent to shifting the argument of Ψ by an infinitesimal (in the sense of the parameter λ small) gauge transformation,

$$\left[1 + \int d^3x\lambda^j(x)\hat{G}_j\right]\Psi(A) = \Psi[A + D\lambda] = \Psi[A],$$

and the last identity comes from the fact that the constraint annihilates the state. So the constraint, as a quantum operator, is imposing the same symmetry that its vanishing imposed classically: it is telling us that the functions $\Psi[A]$ have to be gauge invariant functions of the connection. The same idea is true for the other constraints.

Therefore, the two step process in the classical theory of solving the constraints $C_I = 0$ (equivalent to solving the admissibility conditions for the initial data) and looking for the gauge orbits (solving the 'evolution' equations) is replaced by a one step process in the quantum theory, namely looking for solutions Ψ of the quantum equations $\hat{C}_I\Psi = 0$. This is because it obviously solves the constraint at the quantum level and it simultaneously looks for states that are gauge invariant because $\hat{C}_I$ is the quantum generator of gauge transformations (gauge invariant functions are constant along the gauge orbits and thus characterize them).[11] Recall that, at the classical level, solving the admissibility conditions and evolution equations was equivalent to solving all of Einstein's field equations, this underlines the central role of the quantum constraint equations in canonical quantum gravity.

1.4.3 Introduction of the loop representation

Main articles: Holonomy, Wilson loop and Knot invariant

It was in particular the inability to have good control over the space of solutions to the Gauss' law and spatial diffeomorphism constraints that led Rovelli and Smolin to consider a new representation - the loop representation in gauge theories and quantum gravity.[12]

We need the notion of a holonomy. A holonomy is a measure of how much the initial and final values of a spinor or vector differ after parallel transport around a closed loop; it is denoted

$h_\gamma[A]$.

Knowledge of the holonomies is equivalent to knowledge of the connection, up to gauge equivalence. Holonomies can also be associated with an edge; under a Gauss Law these transform as

$(h'_e)_{\alpha\beta} = U^{-1}_{\alpha\gamma}(x)(h_e)_{\gamma\sigma}U_{\sigma\beta}(y)$.

For a closed loop $x = y$ if we take the trace of this, that is, putting $\alpha = \beta$ and summing we obtain

$(h'_e)_{\alpha\alpha} = U^{-1}_{\alpha\gamma}(x)(h_e)_{\gamma\sigma}U_{\sigma\alpha}(x) = [U_{\sigma\alpha}(x)U^{-1}_{\alpha\gamma}(x)](h_e)_{\gamma\sigma} = \delta_{\sigma\gamma}(h_e)_{\gamma\sigma} = (h_e)_{\gamma\gamma}$

or

$\mathrm{Tr}\, h'_\gamma = \mathrm{Tr}\, h_\gamma$. .

The trace of an holonomy around a closed loop is written

$W_\gamma[A]$

and is called a Wilson loop. Thus Wilson loops are gauge invariant. The explicit form of the Holonomy is

$h_\gamma[A] = \mathcal{P}\exp\Big\{ - \int_{\gamma_0}^{\gamma_1} ds\dot{\gamma}^a A^i_a(\gamma(s))T_i\Big\}$

where γ is the curve along which the holonomy is evaluated, and s is a parameter along the curve, $\mathcal{P}$ denotes path ordering meaning factors for smaller values of s appear to the left, and T_i are matrices that satisfy the SU(2) algebra

$[T^i, T^j] = 2i\epsilon^{ijk}T^k$.

The Pauli matrices satisfy the above relation. It turns out that there are infinitely many more examples of sets of matrices that satisfy these relations, where each set comprises $(N+1)\times(N+1)$ matrices with $N = 1, 2, 3, \dots$, and where none of these can be thought to 'decompose' into two or more examples of lower dimension. They are called different irreducible representations of the SU(2) algebra. The most fundamental representation being the Pauli matrices. The holonomy is labelled by a half integer $N/2$ according to the irreducible representation used.

The use of Wilson loops explicitly solves the Gauss gauge constraint. To handle the spatial diffeomorphism constraint we need to go over to the loop representation. As Wilson loops form a basis we can formally expand any Gauss gauge invariant function as,

$\Psi[A] = \sum_\gamma \Psi[\gamma]W_\gamma[A]$.

This is called the loop transform. We can see the analogy with going to the momentum representation in quantum mechanics(see Position and momentum space). There one has a basis of states $\exp(ikx)$ labelled by a number k and one expands

$\psi[x] = \int dk\psi(k)\exp(ikx)$.

and works with the coefficients of the expansion $\psi(k)$.

The inverse loop transform is defined by

$\Psi[\gamma] = \int[dA]\Psi[A]W_\gamma[A]$.

This defines the loop representation. Given an operator $\hat{O}$ in the connection representation,

$\Phi[A] = \hat{O}\Psi[A] \qquad Eq\ 1$,

one should define the corresponding operator $\hat{O}'$ on $\Psi[\gamma]$ in the loop representation via,

$\Phi[\gamma] = \hat{O}'\Psi[\gamma] \qquad Eq\ 2$,

where $\Phi[\gamma]$ is defined by the usual inverse loop transform,

$\Phi[\gamma] = \int[dA]\Phi[A]W_\gamma[A] \qquad Eq\ 3.$.

A transformation formula giving the action of the operator $\hat{O}'$ on $\Psi[\gamma]$ in terms of the action of the operator $\hat{O}$ on $\Psi[A]$ is then obtained by equating the R.H.S. of Eq 2 with the R.H.S. of Eq 3 with Eq 1 substituted into Eq 3 , namely

$\hat{O}'\Psi[\gamma] = \int[dA]W_\gamma[A]\hat{O}\Psi[A]$,

or

$\hat{O}'\Psi[\gamma] = \int[dA](\hat{O}^\dagger W_\gamma[A])\Psi[A]$,

where by $\hat{O}^\dagger$ we mean the operator $\hat{O}$ but with the reverse factor ordering (remember from simple quantum mechanics where the product of operators is reversed under conjugation). We evaluate the action of this operator on the Wilson loop as a calculation in the connection representation and rearranging the result as a manipulation purely in terms of loops (one should remember that when considering the action on the Wilson loop one should choose the operator one wishes to transform with the opposite factor ordering to the one chosen for its action on wavefunctions $\Psi[A]$). This gives the physical meaning of the operator $\hat{O}'$. For example, if $\hat{O}^\dagger$ corresponded to a spatial diffeomorphism, then this can be thought of as keeping the connection field

A of $W_\gamma[A]$ where it is while performing a spatial diffeomorphism on γ instead. Therefore, the meaning of $\hat{O}'$ is a spatial diffeomorphism on γ, the argument of $\Psi[\gamma]$.

In the loop representation we can then solve the spatial diffeomorphism constraint by considering functions of loops $\Psi[\gamma]$ that are invariant under spatial diffeomorphisms of the loop γ. That is, we construct what mathematicians call knot invariants. This opened up an unexpected connection between knot theory and quantum gravity.

What about the Hamiltonian constraint? Let us go back to the connection representation. Any collection of non-intersecting Wilson loops satisfy Ashtekar's quantum Hamiltonian constraint. This can be seen from the following. With a particular ordering of terms and replacing $\tilde{E}^a_i$ by a derivative, the action of the quantum Hamiltonian constraint on a Wilson loop is

$$\hat{\tilde{H}}^\dagger W_\gamma[A] = -\epsilon_{ijk}\hat{F}^k_{ab}\frac{\delta}{\delta A^i_a}\frac{\delta}{\delta A^j_b}W_\gamma[A] .$$

When a derivative is taken it brings down the tangent vector, $\dot{\gamma}^a$, of the loop, γ. So we have something like

$$\hat{F}^i_{ab}\dot{\gamma}^a\dot{\gamma}^b .$$

However, as F^i_{ab} is anti-symmetric in the indices a and b this vanishes (this assumes that γ is not discontinuous anywhere and so the tangent vector is unique). Now let us go back to the loop representation.

We consider wavefunctions $\Psi[\gamma]$ that vanish if the loop has discontinuities and that are knot invariants. Such functions solve the Gauss law, the spatial diffeomorphism constraint and (formally) the Hamiltonian constraint. Thus we have identified an infinite set of exact (if only formal) solutions to all the equations of quantum general relativity![12] This generated a lot of interest in the approach and eventually led to LQG.

1.4.4 Geometric operators, the need for intersecting Wilson loops and spin network states

The easiest geometric quantity is the area. Let us choose coordinates so that the surface Σ is characterized by $x^3 = 0$. The area of small parallelogram of the surface Σ is the product of length of each side times $\sin\theta$ where θ is the angle between the sides. Say one edge is given by the vector $\vec{u}$ and the other by $\vec{v}$ then,

$$A = \|\vec{u}\|\|\vec{v}\|\sin\theta = \sqrt{\|\vec{u}\|^2\|\vec{v}\|^2(1-\cos^2\theta)} = \sqrt{\|\vec{u}\|^2\|\vec{v}\|^2 - (\vec{u}\cdot\vec{v})^2}$$

In the space spanned by x^1 and x^2 we have an infinitesimal parallelogram described by $\vec{u} = \vec{e}_1 dx^1$ and $\vec{v} = \vec{e}_2 dx^2$. Using $q^{(2)}_{AB} = \vec{e}_A \cdot \vec{e}_B$ (where the indices A and B run from 1 to 2), we get the area of the surface Σ to be given by

$$A_\Sigma = \int_\Sigma dx^1 dx^2 \sqrt{\det(q^{(2)})}$$

where $\det(q^{(2)}) = q_{11}q_{22} - q^2_{12}$ and is the determinant of the metric induced on Σ. The latter can be rewritten $\det(q^{(2)}) = \epsilon^{AB}\epsilon^{CD}q_{AC}q_{BD}/2$ where the indices $A \ldots D$ go from 1 to 2. This can be further rewritten as

$$\det(q^{(2)}) = \frac{\epsilon^{3ab}\epsilon^{3cd}q_{ac}q_{bc}}{2} .$$

The standard formula for an inverse matrix is

$$q^{ab} = \frac{\epsilon^{acd}\epsilon^{bef}q_{ce}q_{df}}{3!\,\det(q)}$$

Note the similarity between this and the expression for $\det(q^{(2)})$. But in Ashtekar variables we have $\tilde{E}^a_i\tilde{E}^{bi} = \det(q)q^{ab}$. Therefore,

$$A_\Sigma = \int_\Sigma dx^1 dx^2 \sqrt{\tilde{E}^3_i\tilde{E}^{3i}} .$$

According to the rules of canonical quantization we should promote the triads $\tilde{E}^3_i$ to quantum operators,

$$\hat{\tilde{E}}^3_i \sim \frac{\delta}{\delta A^i_3} .$$

It turns out that the area A_Σ can be promoted to a well defined quantum operator despite the fact that we are dealing with product of two functional derivatives and worse we have a square-root to contend with as well.[13] Putting $N = 2J$, we talk of being in the J-th representation. We note that $\sum_i T^i T^i = J(J+1)1$. This quantity is important in the final formula for the area spectrum. We simply state the result below,

$$\hat{A}_\Sigma W_\gamma[A] = 8\pi\ell^2_{\text{Planck}}\beta\sum_I\sqrt{j_I(j_I+1)}W_\gamma[A]$$

where the sum is over all edges I of the Wilson loop that pierce the surface Σ.

The formula for the volume of a region R is given by

$$V = \int_R d^3x\sqrt{\det(q)} = \frac{1}{6}\int_R dx^3\sqrt{\epsilon_{abc}\epsilon^{ijk}\tilde{E}^a_i\tilde{E}^b_j\tilde{E}^c_k} .$$

The quantization of the volume proceeds the same way as with the area. As we take the derivative, and each time we do so we bring down the tangent vector $\dot{\gamma}^a$, when the volume operator acts on non-intersecting Wilson loops the result vanishes. Quantum states with non-zero volume must therefore involve intersections. Given that the anti-symmetric summation is taken over in the formula for the volume we would need at least intersections with three non-coplanar lines. Actually it turns out that one needs at least four-valent vertices for the volume operator to be non-vanishing.

We now consider Wilson loops with intersections. We assume the real representation where the gauge group is $\mathrm{SU}(2)$. Wilson loops are an over complete basis as there are identities relating different Wilson loops. These come about from the fact that Wilson loops are based on matrices (the holonomy) and these matrices satisfy identities. Given any two $\mathrm{SU}(2)$ matrices $\mathbb{A}$ and $\mathbb{B}$ it is easy to check that,

$$\mathrm{Tr}(\mathbb{A})\,\mathrm{Tr}(\mathbb{B}) = \mathrm{Tr}(\mathbb{A}\mathbb{B}) + \mathrm{Tr}(\mathbb{A}\mathbb{B}^{-1}) \,.$$

This implies that given two loops γ and η that intersect, we will have,

$$W_\gamma[A]W_\eta[A] = W_{\gamma\circ\eta}[A] + W_{\gamma\circ\eta^{-1}}[A]$$

where by η^{-1} we mean the loop η traversed in the opposite direction and $\gamma\circ\eta$ means the loop obtained by going around the loop γ and then along η . See figure below. Given that the matrices are unitary one has that $W_\gamma[A] = W_{\gamma^{-1}}[A]$. Also given the cyclic property of the matrix traces (i.e. $Tr(\mathbb{A}\mathbb{B}) = Tr(\mathbb{B}\mathbb{A})$) one has that $W_{\gamma\circ\eta}[A] = W_{\eta\circ\gamma}[A]$. These identities can be combined with each other into further identities of increasing complexity adding more loops. These identities are the so-called Mandelstam identities. Spin networks certain are linear combinations of intersecting Wilson loops designed to address the over completeness introduced by the Mandelstam identities (for trivalent intersections they eliminate the over-compleness entirely) and actually constitute a basis for all gauge invariant functions.

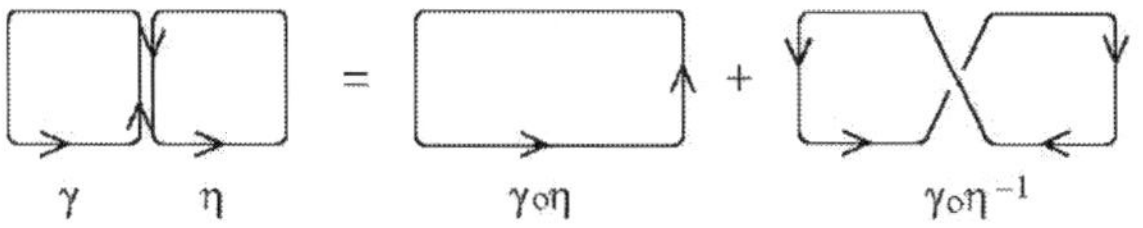

γ η = $\gamma\circ\eta$ + $\gamma\circ\eta^{-1}$

Graphical representation of the simplest non-trivial Mandestam identity relating different Wilson loops.

As mentioned above the holonomy tells you how to propagate test spin half particles. A spin network state assigns an amplitude to a set of spin half particles tracing out a path in space, merging and splitting. These are described by spin networks γ : the edges are labelled by spins together with 'intertwiners' at the vertices which are prescription for how to sum over different ways the spins are rerouted. The sum over rerouting are chosen as such to make the form of the intertwiner invariant under Gauss gauge transformations.

1.4.5 Real variables, modern analysis and LQG

Main article: Hamiltonian constraint of LQG

Let us go into more detail about the technical difficulties associated with using Ashtekar's variables:

With Ashtekar's variables one uses a complex connection and so the relevant gauge group as actually $\mathrm{SL}(2,\mathbb{C})$ and not $\mathrm{SU}(2)$. As $\mathrm{SL}(2,\mathbb{C})$ is non-compact it creates serious problems for the rigorous construction of the necessary mathematical machinery. The group $\mathrm{SU}(2)$, on the other hand, is compact and the needed constructions have been developed.

As mentioned above, because Ashtekar's variables are complex the resulting general relativity is complex. To recover the real theory, one has to impose what are known as the "reality conditions." These require that the densitized triad be real and that the real part of the Ashtekar connection equals the compatible spin connection (the compatibility condition being $\nabla_a e^I_b = 0$) determined by the desitized triad. The expression for compatible connection Γ^i_a is rather complicated and as such non-polynomial formula enters through the back door.

Before we state the next difficulty we should give a definition; a tensor density of weight W transforms like an ordinary tensor, except that in addition the W th power of the Jacobian,

$$J = \left| \frac{\partial x^a}{\partial x'^b} \right|$$

appears as a factor, i.e.

$$T'^{a\ldots}_{b\ldots} = J^W \frac{\partial x'^a}{\partial x^c} \ldots \frac{\partial x^d}{\partial x'^b} T^{c\ldots}_{d\ldots} \,.$$

It turns out that it is impossible, on general grounds, to construct a UV-finite, diffeomorphism non-violating operator corresponding to $\sqrt{\det(q)}H$. The reason is that the rescaled Hamiltonian constraint is a scalar density of weight two while it can be shown that only scalar densities of weight one have a chance to result in a well defined operator. Thus, one is forced to work with the original unrescaled, density one-valued, Hamiltonian constraint. However, this is non-polynomial and the whole virtue of the complex variables is questioned. In fact, all the solutions constructed for Ashtekar's Hamiltonian constraint only vanished for finite regularization (physics), however, this violates spatial diffeomorphism invariance.

Without the implementation and solution of the Hamiltonian constraint no progress can be made and no reliable predictions are possible.

To overcome the first problem one works with the configuration variable

$$A^i_a = \Gamma^i_a + \beta K^i_a$$

where β is real (as pointed out by Barbero, who introduced real variables some time after Ashtekar's variables[14][15]). The Guass law and the spatial diffeomorphism constraints are the same. In real Ashtekar variables the Hamiltonian is

$$H = \frac{\epsilon_{ijk} F^k_{ab} \tilde{E}^a_i \tilde{E}^b_j}{\sqrt{\det(q)}} + 2\frac{\beta^2+1}{\beta^2} \frac{(\tilde{E}^a_i \tilde{E}^b_j - \tilde{E}^a_j \tilde{E}^b_i)}{\sqrt{\det(q)}} (A^i_a - \Gamma^i_a)(A^j_b - \Gamma^j_b) = H_E + H' \,.$$

The complicated relationship between Γ^i_a and the desitized triads causes serious problems upon quantization. It is with the choice $\beta = \pm i$ that the second more complicated term is made to vanish. However, as mentioned above Γ^i_a reappears in the reality conditions. Also we still have the problem of the $1/\sqrt{\det(q)}$ factor.

Thiemann was able to make it work for real β . First he could simplify the troublesome $1/\sqrt{\det(q)}$ by using the identity

$$\{A_c^k, V\} = \frac{\epsilon_{abc}\epsilon^{ijk}\tilde{E}_i^a\tilde{E}_j^b}{\sqrt{\det(q)}}$$

where V is the volume. The A_c^k and V can be promoted to well defined operators in the loop representation and the Poisson bracket is replaced by a commutator upon quantization; this takes care of the first term. It turns out that a similar trick can be used to treat the second term. One introduces the quantity

$$K = \int d^3x K_a^i \tilde{E}_i^a$$

and notes that

$$K_a^i = \{A_a^i, K\} .$$

We are then able to write

$$A_a^i - \Gamma_a^i = \beta K_a^i = \beta\{A_a^i, K\} .$$

The reason the quantity K is easier to work with at the time of quantization is that it can be written as

$$K = -\{V, \int d^3x H_E\}$$

where we have used that the integrated densitized trace of the extrinsic curvature, K , is the "time derivative of the volume".

In the long history of canonical quantum gravity formulating the Hamiltonian constraint as a quantum operator (Wheeler–DeWitt equation) in a mathematically rigorous manner has been a formidable problem. It was in the loop representation that a mathematically well defined Hamiltonian constraint was finally formulated in 1996.[9] We leave more details of its construction to the article Hamiltonian constraint of LQG. This together with the quantum versions of the Gauss law and spatial diffeomorphism constrains written in the loop representation are the central equations of LQG (modern canonical quantum General relativity).

Finding the states that are annihilated by these constraints (the physical states), and finding the corresponding physical inner product, and observables is the main goal of the technical side of LQG.

A very important aspect of the Hamiltonian operator is that it only acts at vertices (a consequence of this is that Thiemann's Hamiltonian operator, like Ashtekar's operator, annihilates non-intersecting loops except now it is not just formal and has rigorous mathematical meaning). More precisely, its action is non-zero on at least vertices of valence three and greater and results in a linear combination of new spin networks where the original graph has been modified by the addition of lines at each vertex together and a change in the labels of the adjacent links of the vertex.

1.4.6 Implementation and solution the quantum constraints

Main articles: spectrum, dual space, Rigged Hilbert space and quantum configuration space

We solve, at least approximately, all the quantum constraint equations and for the physical inner product to make physical predictions.

Before we move on to the constraints of LQG, lets us consider certain cases. We start with a kinematic Hilbert space $\mathcal{H}_{\text{Kin}}$ as so is equipped with an inner product—the kinematic inner product $\langle\phi, \psi\rangle_{\text{Kin}}$.

i) Say we have constraints $\hat{C}_I$ whose zero eigenvalues lie in their discrete spectrum. Solutions of the first constraint, $\hat{C}_1$, correspond to a subspace of the kinematic Hilbert space, $\mathcal{H}_1 \subset \mathcal{H}_{\text{Kin}}$. There will be a projection operator P_1 mapping $\mathcal{H}_{\text{Kin}}$ onto $\mathcal{H}_1$. The kinematic inner product structure is easily employed to provide the inner product structure after solving this first constraint; the new inner product $\langle\phi, \psi\rangle_1$ is simply

$$\langle\phi, \psi\rangle_1 = \langle P\phi, P\psi\rangle_{\text{Kin}}$$

They are based on the same inner product and are states normalizable with respect to it.

ii) The zero point is not contained in the point spectrum of all the $\hat{C}_I$, there is then no non-trivial solution $\Psi \in \mathcal{H}_{\text{Kin}}$ to the system of quantum constraint equations $\hat{C}_I\Psi = 0$ for all I .

For example, the zero eigenvalue of the operator

$$\hat{C} = \left(i\frac{d}{dx} - k\right)$$

on $L_2(\mathbb{R}, dx)$ lies in the continuous spectrum $\mathbb{R}$ but the formal "eigenstate" $\exp(-ikx)$ is not normalizable in the kinematic inner product,

$$\int_{-\infty}^{\infty} dx\psi^*(x)\psi(x) = \int_{-\infty}^{\infty} dx e^{ikx}e^{-ikx} = \int_{-\infty}^{\infty} dx = \infty$$

and so does not belong to the kinematic Hilbert space $\mathcal{H}_{\text{Kin}}$. In these cases we take a dense subset $\mathcal{S}$ of $\mathcal{H}_{\text{Kin}}$ (intuitively this means either any point in $\mathcal{S}$ is either in $\mathcal{H}_{\text{Kin}}$ or arbitrarily close to a point in $\mathcal{H}_{\text{Kin}}$) with very good convergence properties and consider its dual space $\mathcal{S}'$ (intuitively these map elements of $\mathcal{S}$ onto finite complex numbers in a linear manner), then $\mathcal{S} \subset \mathcal{H}_{\text{Kin}} \subset \mathcal{S}'$ (as $\mathcal{S}'$ contains distributional functions). The constraint operator is then implemented on this larger dual space, which contains distributional functions, under the adjoint action on the operator. One looks for solutions on this larger space. This comes at the price that the solutions must be given a new Hilbert space inner product with respect to which they are normalizable (see article on rigged Hilbert space). In this case we have a generalized projection operator on the new space of states. We cannot use the above formula for the new inner product as it diverges, instead the new inner product is given by the simply modification of the above,

$\langle \phi, \psi \rangle_1 = \langle P\phi, \psi \rangle_{\text{Kin}}$.

The generalized projector P is known as a rigging map.

Implementation and solution the quantum constraints of LQG.

Let us move to LQG, additional complications will arise from that one cannot define an operator for the quantum spatial diffeomorphism constraint as the infinitesimal generator of finite diffeomorphism transformations and the fact the constraint algebra is not a Lie algebra due to the bracket between two Hamiltonian constraints.

Implementation and solution the Gauss constraint:

One does not actually need to promote the Gauss constraint to an operator since we can work directly with Gauss-gauge-invariant functions (that is, one solves the constraint classically and quantizes only the phase space reduced with respect to the Gauss constraint). The Gauss law is solved by the use of spin network states. They provide a basis for the Kinematic Hilbert space $\mathcal{H}_{\text{Kin}}$.

Implementation of the quantum spatial diffeomorphism constraint:

It turns out that one cannot define an operator for the quantum spatial diffeomorphism constraint as the infinitesimal generator of finite diffeomorphism transformations, represented on $\mathcal{H}_{\text{Kin}}$. The representation of finite diffeomorphisms is a family of unitary operators $\hat{U}_\varphi$ acting on a spin-network state ψ_γ by

$\hat{U}_\varphi \psi_\gamma := \psi_{\varphi \circ \gamma}$,

for any spatial diffeomorphism φ on Σ . To understand why one cant define an operator for the quantum spatial diffeomorphism constraint consider what is called a 1-parameter subgroup φ_t in the group of spatial diffeomorphisms, this is then represented as a 1-parameter unitary group $\hat{U}_{\varphi_t}$ on $\mathcal{H}_{\text{Kin}}$. However, $\hat{U}_{\varphi_t}$ is not weakly continuous since the subspace $\psi_{\varphi_t \circ \gamma}$ belongs to and the subspace ψ_γ belongs to are orthogonal to each other no matter how small the parameter t is. So one always has

$| < \psi_\gamma | \hat{U}_{\varphi_t} | \psi_\gamma >_{Kin} - < \psi_\gamma | \psi_\gamma >_{Kin} | = < \psi_\gamma | \psi_\gamma >_{Kin} \neq 0$,

even in the limit when t goes to zero. Therefore, the infinitesimal generator of $\hat{U}_{\varphi_t}$ does not exist.

Solution of the spatial diffeomorphism constraint.

The spatial diffeomorphism constraint has been solved. The induced inner product $< \cdot, \cdot >_{Diff}$ on $\mathcal{H}_{\text{Diff}}$ (we do not pursue the details) has a very simple description in terms of spin network states; given two spin networks s and s' , with associated spin network states ψ_s and $\psi_{s'}$, the inner product is 1 if s and s' are related to each other by a spatial diffeomorphism and zero otherwise.

We have provided a description of the implemented and complete solution of the kinematic constraints, the Gauss and spatial diffeomorphisms constraints which will be the same for any background-independent gauge field theory. The feature that distinguishes such different theories is the Hamiltonian constraint which is the only one that depends on the Lagrangian of the classical theory.

Problem arising from the Hamiltonian constraint.

Details of the implementation the quantum Hamiltonian constraint and solutions are treated in a different article Hamiltonian constraint of LQG. However, in this article we introduce an approximation scheme for the formal solution of the Hamiltonian constraint operator given in the section below on spinfoams. Here we just mention issues that arises with the Hamiltonian constraint.

The Hamiltonian constraint maps diffeomorphism invariant states onto non-diffeomorphism invariant states as so does not preserve the diffeomorphism Hilbert space $\mathcal{H}_{\text{Diff}}$. This is an unavoidable consequence of the operator algebra, in particular the commutator:

$[\hat{C}(\vec{N}), \hat{H}(M)] \propto \hat{H}(\mathcal{L}_{\vec{N}} M)$

as can be seen by applying this to $\psi_s \in \mathcal{H}_{Diff}$,

$(\hat{\vec{C}}(\vec{N}) \hat{H}(M) - \hat{H}(M) \hat{\vec{C}}(\vec{N})) \psi_s \propto \hat{H}(\mathcal{L}_{\vec{N}} M) \psi_s$

and using $\hat{\vec{C}}(\vec{N}) \psi_s = 0$ to obtain

$\hat{\vec{C}}(\vec{N}) [\hat{H}(M) \psi_s] \propto \hat{H}(\mathcal{L}_{\vec{N}} M) \psi_s \neq 0$

and so $\hat{H}(M) \psi_s$ is not in $\mathcal{H}_{Diff}$.

This means that you can't just solve the spatial diffeomorphism constraint and then the Hamiltonian constraint. This problem can be circumvented by the introduction of the master constraint, with its trivial operator algebra, one is then able in principle to construct the physical inner product from $\mathcal{H}_{\text{Diff}}$.

1.5 Spin foams

Main articles: spin network, spin foam, BF model and Barrett–Crane model

In loop quantum gravity (LQG), a spin network represents a “quantum state” of the gravitational field on a 3-dimensional hypersurface. The set of all possible spin networks (or, more accurately, "s-knots" - that is, equivalence classes of spin networks under diffeomorphisms) is countable; it constitutes a basis of LQG Hilbert space.

In physics, a spin foam is a topological structure made out of two-dimensional faces that represents one of the configurations that must be summed to obtain a Feynman's path integral (functional integration) description of quantum gravity. It is closely related to loop quantum gravity.

1.5.1 Spin foam derived from the Hamiltonian constraint operator

The Hamiltonian constraint generates 'time' evolution. Solving the Hamiltonian constraint should tell us how quantum states evolve in 'time' from an initial spin network state to a final spin network state. One approach to solving the Hamiltonian constraint starts with what is called the Dirac delta function. This is a rather singular function of the real line, denoted $\delta(x)$, that is zero everywhere except at $x = 0$ but whose integral is finite and nonzero. It can be represented as a Fourier integral,

$\delta(x) = \int e^{ikx} dk$.

One can employ the idea of the delta function to impose the condition that the Hamiltonian constraint should vanish. It is obvious that

$\prod_{x \in \Sigma} \delta(\hat{H}(x))$

is non-zero only when $\hat{H}(x) = 0$ for all x in Σ . Using this we can 'project' out solutions to the Hamiltonian constraint. With analogy to the Fourier integral given above, this (generalized) projector can formally be written as

$\int [dN] e^{i \int d^3x N(x) \hat{H}(x)}$.

Interestingly, this is formally spatially diffeomorphism-invariant. As such it can be applied at the spatially diffeomorphism-invariant level. Using this the physical inner product is formally given by

$$\left\langle \int [dN] e^{i \int d^3x N(x) \hat{H}(x)} s_{\text{int}} s_{\text{fin}} \right\rangle_{\text{Diff}}$$

where s_{int} are the initial spin network and s_{fin} is the final spin network.

The exponential can be expanded

$$\left\langle \int [dN] (1 + i \int d^3x N(x) \hat{H}(x) + \frac{i^2}{2!} [\int d^3x N(x) \hat{H}(x)] [\int d^3x' N(x') \hat{H}(x')] + \dots) s_{\text{int}}, s_{\text{fin}} \right\rangle_{\text{Diff}}$$

and each time a Hamiltonian operator acts it does so by adding a new edge at the vertex. The summation over different sequences of actions of $\hat{H}$ can be visualized as a summation over different histories of 'interaction vertices' in the 'time' evolution sending the initial spin network to the final spin network. This then naturally gives rise to the two-complex (a combinatorial set of faces that join along edges, which in turn join on vertices) underlying the spin foam description; we evolve forward an initial spin network sweeping out a surface, the action of the Hamiltonian constraint operator is to produce a new planar surface starting at the vertex. We are able to use the action of the Hamiltonian constraint on the vertex of a spin network state to associate an amplitude to each "interaction" (in analogy to Feynman diagrams). See figure below. This opens up a way of trying to directly link canonical LQG to a path integral description. Now just as a spin networks describe quantum space, each configuration contributing to these path integrals, or sums over history, describe 'quantum space-time'. Because of their resemblance to soap foams and the way they are labeled John Baez gave these 'quantum space-times' the name 'spin foams'.

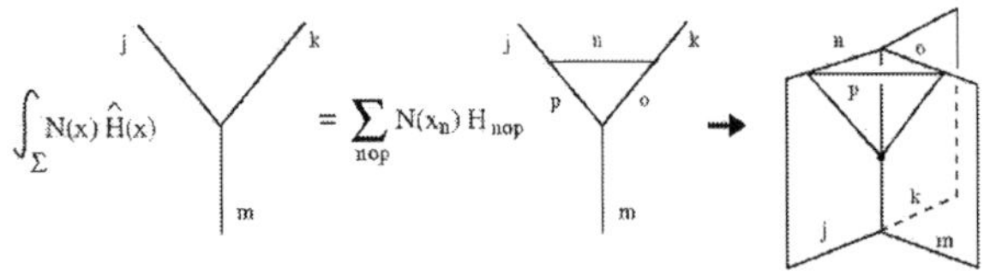

$$\int_\Sigma N(x) \hat{H}(x) \;[\text{vertex } j, k, m] = \sum_{nop} N(x_n) H_{nop} \;[\text{vertex } j, n, k, p, o, m] \rightarrow [\text{spin foam } n, o, p, j, k, m]$$

The action of the Hamiltonian constraint translated to the path integral or so-called spin foam description. A single node splits into three nodes, creating a spin foam vertex. $N(x_n)$ is the value of N at the vertex and H_{nop} are the matrix elements of the Hamiltonian constraint $\hat{H}$.

There are however severe difficulties with this particular approach, for example the Hamiltonian operator is not self-adjoint, in fact it is not even a normal operator (i.e. the operator does not commute with its adjoint) and so the spectral theorem cannot be used to define the exponential in general. The most serious problem is that the $\hat{H}(x)$'s are not mutually commuting, it can then be shown the formal quantity $\int [dN] e^{i \int d^3x N(x) \hat{H}(x)}$ cannot even define a (generalized) projector. The master constraint (see below) does not suffer from these problems and as such offers a way of connecting the canonical theory to the path integral formulation.

1.5.2 Spin foams from BF theory

It turns out there are alternative routes to formulating the path integral, however their connection to the Hamiltonian formalism is less clear. One way is to start with the BF theory. This is a simpler theory to general relativity. It has no local degrees of freedom and as such depends only on topological aspects of the fields. BF theory is what is known as a topological field theory. Surprisingly, it turns out that general relativity can be obtained from BF theory by imposing a constraint,[16] BF theory involves a field B_{ab}^{IJ} and if one chooses the field B to be the (anti-symmetric) product of two tetrads

$$B_{ab}^{IJ} = \frac{1}{2}(E_a^I E_b^J - E_b^I E_a^J)$$

(tetrads are like triads but in four spacetime dimensions), one recovers general relativity. The condition that the B field be given by the product of two tetrads is called the simplicity constraint. The spin foam dynamics of the topological field theory is well understood. Given the spin foam 'interaction' amplitudes for this simple theory, one then tries to implement the simplicity conditions to obtain a path integral for general relativity. The non-trivial task of constructing a spin foam model is then reduced to the question of how this simplicity constraint should be imposed in the quantum theory. The first attempt at this was the famous Barrett–Crane model.[17] However this model was shown to be problematic, for example there did not seem to be enough degrees of freedom to ensure the correct classical limit.[18] It has been argued that the simplicity constraint was imposed too strongly at the quantum level and should only be imposed in the sense of expectation values just as with the Lorenz gauge condition $\partial_\mu \hat{A}^\mu$ in the Gupta–Bleuler formalism of quantum electrodynamics. New models have now been put forward, sometimes motivated by imposing the simplicity conditions in a weaker sense.

Another difficulty here is that spin foams are defined on a discretization of spacetime. While this presents no problems for a topological field theory as it has no local degrees of freedom, it presents problems for GR. This is known as the problem triangularization dependence.

1.5.3 Modern formulation of spin foams

Just as imposing the classical simplicity constraint recovers general relativity from BF theory, one expects an appropriate quantum simplicity constraint will recover quantum gravity from quantum BF theory.

Much progress has been made with regard to this issue by Engle, Pereira, and Rovelli[19] and Freidel and Krasnov[20] in defining spin foam interaction amplitudes with much better behaviour.

An attempt to make contact between EPRL-FK spin foam and the canonical formulation of LQG has been made.[21]

1.5.4 Spin foam derived from the master constraint operator

See below.

1.6 The semiclassical limit

1.6.1 What is the semiclassical limit?

Main articles: Correspondence principle and classical limit

The **classical limit** or **correspondence limit** is the ability of a physical theory to approximate or "recover" classical mechanics when considered over special values of its parameters.[22] The classical limit is used with physical theories that predict non-classical behavior.

In physics, the **correspondence principle** states that the behavior of systems described by the theory of quantum mechanics (or by the old quantum theory) reproduces classical physics in the limit of large quantum numbers. In other words, it says that for large orbits and for large energies, quantum calculations must agree with classical calculations.[23]

The principle was formulated by Niels Bohr in 1920,[24] though he had previously made use of it as early as 1913 in developing his model of the atom.[25]

There are two basic requirements in establishing the semiclassical limit of any quantum theory:

i) reproduction of the Poisson brackets (of the diffeomorphism constraints in the case of general relativity). This is extremely important because, as noted above, the Poisson bracket algebra formed between the (smeared) constraints themselves completely determines the classical theory. This is analogous to establishing Ehrenfest's theorem;

ii) the specification of a complete set of classical observables whose corresponding operators (see complete set of commuting observables for the quantum mechanical definition of a complete set of observables) when acted on by appropriate semiclassical states reproduce the same classical variables with small quantum corrections (a subtle point is that states that are semiclassical for one class of observables may not be semiclassical for a different class of observables[26]).

This may be easily done, for example, in ordinary quantum mechanics for a particle but in general relativity this becomes a highly non-trivial problem as we will see below.

1.6.2 Why might LQG not have general relativity as its semiclassical limit?

Any candidate theory of quantum gravity must be able to reproduce Einstein's theory of general relativity as a classical limit of a quantum theory. This is not guaranteed because of a feature of quantum field theories which is that they have different sectors, these are analogous to the different phases that come about in the thermodynamical limit of statistical systems. Just as different phases are physically different, so are different sectors of a quantum field theory. It may turn out that LQG belongs to an unphysical sector - one in which you do not recover general relativity in the semiclassical limit (in fact there might not be any physical sector at all).

Moreover, the physical Hilbert space H_{phys} must contain enough semiclassical states to guarantee that the quantum theory one obtains can return to the classical theory when $\hbar \to 0$. In order to guarantee this one must avoid quantum anomalies at all cost, because if we do not there will be restrictions on the physical Hilbert space that have no counterpart in the classical theory, implying that the quantum theory has less degrees of freedom than the classical theory.

Theorems establishing the uniqueness of the loop representation as defined by Ashtekar et al. (i.e. a certain concrete realization of a Hilbert space and associated operators reproducing the correct loop algebra - the realization that everybody was using) have been given by two groups (Lewandowski, Okolow, Sahlmann and Thiemann;[27] and Christian Fleischhack[28]). Before this result was established it was not known whether there could be other examples of Hilbert spaces with operators invoking the same loop algebra, other realizations, not equivalent to the one that had been used so far. These uniqueness theorems imply no others exist and so if LQG does not have the correct semiclassical limit then this would mean the end of the loop representation of quantum gravity altogether.

1.6.3 Difficulties checking the semiclassical limit of LQG

There are difficulties in trying to establish LQG gives Einstein's theory of general relativity in the semiclassical limit. There are a number of particular difficulties in establishing the semiclassical limit:

1. There is no operator corresponding to infinitesimal spatial diffeomorphisms (it is not surprising that the theory has no generator of infinitesimal spatial 'translations' as it predicts spatial geometry has a discrete nature, compare to the situation in condensed matter). Instead it must be approximated by finite spatial diffeomorphisms and so the Poisson bracket structure of the classical theory is not exactly reproduced. This problem can be circumvented with the introduction of the so-called master constraint (see below)[29]

2. There is the problem of reconciling the discrete combinatorial nature of the quantum states with the continuous nature of the fields of the classical theory.

3. There are serious difficulties arising from the structure of the Poisson brackets involving the spatial diffeomorphism and Hamiltonian constraints. In particular, the algebra of (smeared) Hamiltonian constraints does not close, it is proportional to a sum over infinitesimal spatial diffeomorphisms (which, as we have just noted, does not exist in the quantum theory) where the coefficients of proportionality are not constants but have non-trivial phase space dependence – as such it does not form a Lie algebra. However, the situation is much improved by the introduction of the master constraint.[29]

4. The semiclassical machinery developed so far is only appropriate to non-graph-changing operators, however, Thiemann's Hamiltonian constraint is a graph-changing operator – the new graph it generates has degrees of freedom upon which the coherent state does not depend and so their quantum fluctuations are not suppressed. There is also the restriction, so far, that these coherent states are only defined at the Kinematic level, and now one has to lift them to the level of $\mathcal{H}_{Diff}$ and $\mathcal{H}_{Phys}$. It can be shown that Thiemann's Hamiltonian constraint is required to be graph changing in order to resolve problem 3 in some sense. The master constraint algebra however is trivial and so the requirement that it be graph changing can be lifted and indeed non-graph changing master constraint operators have been defined.

5. Formulating observables for classical general relativity is a formidable problem by itself because of its non-linear nature and space-time diffeomorphism invariance. In fact a systematic approximation scheme to calculate observables has only been recently developed.[30][31]

Difficulties in trying to examine the semiclassical limit of the theory should not be confused with it having the wrong semiclassical limit.

1.6.4 Progress in demonstrating LQG has the correct semiclassical limit

Much details here to be written up...

Concerning issue number 2 above one can consider so-called weave states. Ordinary measurements of geometric quantities are macroscopic, and planckian discreteness is smoothed out. The fabric of a T-shirt is analogous. At a distance it is a smooth curved two-dimensional surface. But a closer inspection we see that it is actually composed of thousands of one-dimensional linked threads. The image of space given in LQG is similar, consider a very large spin network formed by a very large number of nodes and links, each of Planck scale. But probed at a macroscopic scale, it appears as a three-dimensional continuous metric geometry.

As far as the editor knows problem 4 of having semiclassical machinery for non-graph changing operators is as the moment still out of reach.

To make contact with familiar low energy physics it is mandatory to have to develop approximation schemes both for the physical inner product and for Dirac observables.

The spin foam models have been intensively studied can be viewed as avenues toward approximation schemes for the physical inner product.

Markopoulou et al. adopted the idea of noiseless subsystems in an attempt to solve the problem of the low energy limit in background independent quantum gravity theories[32][33][34] The idea has even led to the intriguing possibility of matter of the standard model being identified with emergent degrees of freedom from some versions of LQG (see section below: *LQG and related research programs*).

As Wightman emphasized in the 1950s, in Minkowski QFTs the $n-$ point functions§

$$W(x_1, \ldots, x_n) = \langle 0|\phi(x_n)\ldots\phi(x_1)|0\rangle \,,$$

completely determine the theory. In particular, one can calculate the scattering amplitudes from these quantities. As explained below in the section on the *Background independent scattering amplitudes*, in the background-independent context, the $n-$ point functions refer to a state and in gravity that state can naturally encode information about a specific geometry which can then appear in the expressions of these quantities. To leading order LQG calculations have been shown to agree in an appropriate sense with the $n-$ point functions calculated in the effective low energy quantum general relativity.

1.7 Improved dynamics and the master constraint

Main articles: Hamiltonian (quantum mechanics), Hamiltonian constraint of LQG and Friedrichs extension

1.7.1 The master constraint

Thiemann's master constraint should not be confused with the master equation which has to do with random processes. The Master Constraint Programme for Loop Quantum Gravity (LQG) was proposed as a classically equivalent way to impose the infinite number of Hamiltonian constraint equations

$$H(x) = 0$$

(x being a continuous index) in terms of a single master constraint,

$$M = \int d^3x \frac{[H(x)]^2}{\sqrt{\det(q(x))}} \,.$$

which involves the square of the constraints in question. Note that $H(x)$ were infinitely many whereas the master constraint is only one. It is clear that if M vanishes then so do the infinitely many $H(x)$'s. Conversely, if all the $H(x)$'s vanish then so does M , therefore they are equivalent. The master constraint M involves an appropriate averaging over all space and so is invariant under spatial diffeomorphisms (it is invariant under spatial "shifts" as it is a summation over all such spatial "shifts" of a quantity that transforms as a scalar). Hence its Poisson bracket with the (smeared) spatial diffeomorphism constraint, $C(\vec{N})$, is simple:

$$\{M, C(\vec{N})\} = 0 \,.$$

(it is $su(2)$ invariant as well). Also, obviously as any quantity Poisson commutes with itself, and the master constraint being a single constraint, it satisfies

$$\{M, M\} = 0 \,.$$

We also have the usual algebra between spatial diffeomorphisms. This represents a dramatic simplification of the Poisson bracket structure, and raises new hope in understanding the dynamics and establishing the semiclassical limit.[35]

An initial objection to the use of the master constraint was that on first sight it did not seem to encode information about the observables; because the Mater constraint is quadratic in the constraint, when you compute its Poisson bracket with any quantity, the result is proportional to the constraint, therefore it always vanishes when the constraints are imposed and as such does not select out particular phase space functions. However, it was realized that the condition

$$\{\{M, O\}, O\}_{M=0} = 0$$

is equivalent to O being a Dirac observable. So the master constraint does capture information about the observables. Because of its significance this is known as the master equation.[35]

That the master constraint Poisson algebra is an honest Lie algebra opens up the possibility of using a certain method, known as group averaging, in order to construct solutions of the infinite number of Hamiltonian constraints, a physical inner product thereon and Dirac observables via what is known as refined algebraic quantization RAQ[36]

1.7.2 The quantum master constraint

Define the quantum master constraint (regularisation issues aside) as

$$\hat{M} := \int d^3x \left(\widehat{\frac{H}{\det(q(x))^{1/4}}} \right)^{\dagger} (x) \left(\widehat{\frac{H}{\det(q(x))^{1/4}}} \right) (x) .$$

Obviously,

$$\left(\widehat{\frac{H}{\det(q(x))^{1/4}}} \right) (x) \Psi = 0$$

for all x implies $\hat{M}\Psi = 0$. Conversely, if $\hat{M}\Psi = 0$ then

$$0 = < \Psi, \hat{M}\Psi > = \int d^3x \left\| \left(\widehat{\frac{H}{\det(q(x))^{1/4}}} \right) (x) \Psi \right\|^2 \qquad Eq\ 4$$

implies

$$\left(\widehat{\frac{H}{\det(q(x))^{1/4}}} \right) (x) \Psi = 0 .$$

What is done first is, we are able to compute the matrix elements of the would-be operator $\hat{M}$, that is, we compute the quadratic form Q_M . It turns out that as Q_M is a graph changing, diffeomorphism invariant quadratic form it cannot exist on the kinematic Hilbert space H_{Kin} , and must be defined on H_{Diff} . The fact that the master constraint operator $\hat{M}$ is densely defined on H_{Diff} , it is obvious that $\hat{M}$ is a positive and symmetric operator in H_{Diff} . Therefore, the quadratic form Q_M associated with $\hat{M}$ is closable. The closure of Q_M is the quadratic form of a unique self-adjoint operator $\hat{\overline{M}}$, called the Friedrichs extension of $\hat{M}$. We relabel $\hat{\overline{M}}$ as $\hat{M}$ for simplicity. (Note that the presence of an inner product, viz Eq 4, means there are no superfluous solutions i.e. there are no Ψ such that $\left(\widehat{\frac{H}{\det(q(x))^{1/4}}} \right) (x) \Psi \neq 0$ but for which $\hat{M}\Psi = 0$).

It is also possible to construct a quadratic form Q_{M_E} for what is called the extended master constraint (discussed below) on H_{Kin} which also involves the weighted integral of the square of the spatial diffeomorphism constraint (this is possible because Q_{M_E} is not graph changing).

The spectrum of the master constraint may not contain zero due to normal or factor ordering effects which are finite but similar in nature to the infinite vacuum energies of background-dependent quantum field theories. In this case it turns out to be physically correct to replace $\hat{M}$ with $\hat{M}' := \hat{M} - min(spec(\hat{M}))\hat{1}$ provided that the "normal ordering constant" vanishes in the classical limit, that is, $\lim_{\hbar \to 0} min(spec(\hat{M})) = 0$, so that $\hat{M}'$ is a valid quantisation of M .

1.7.3 Testing the master constraint

The constraints in their primitive form are rather singular, this was the reason for integrating them over test functions to obtain smeared constraints. However, it would appear that the equation for the master constraint, given above, is even more singular involving the product of two primitive constraints (although integrated over space). Squaring the constraint is dangerous as it could lead to worsened ultraviolent behaviour of the corresponding operator and hence the master constraint programme must be approached with due care.

In doing so the master constraint programme has been satisfactorily tested in a number of model systems with non-trivial constraint algebras, free and interacting field theories.[37][38][39][40][41] The master constraint for LQG was established as a genuine positive self-adjoint operator and the physical Hilbert space of LQG was shown to be non-empty,[42] an obvious consistency test LQG must pass to be a viable theory of quantum General relativity.

1.7.4 Applications of the master constraint

The master constraint has been employed in attempts to approximate the physical inner product and define more rigorous path integrals.[43][44][45][46]

The Consistent Discretizations approach to LQG,[47][48] is an application of the master constraint program to construct the physical Hilbert space of the canonical theory.

1.7.5 Spin foam from the master constraint

It turns out that the master constraint is easily generalized to incorporate the other constraints. It is then referred to as the extended master constraint, denoted M_E . We can define the extended master constraint which imposes both the Hamiltonian constraint and spatial diffeomorphism constraint as a single operator,

$$M_E = \int_\Sigma d^3x \frac{H(x)^2 - q^{ab} V_a(x) V_b(x)}{\sqrt{det(q)}} .$$

Setting this single constraint to zero is equivalent to $H(x) = 0$ and $V_a(x) = 0$ for all x in Σ . This constraint implements the spatial diffeomorphism and Hamiltonian constraint at the same time on the Kinematic Hilbert space. The physical inner product is then defined as

$$\langle \phi, \psi \rangle_{\text{Phys}} = \lim_{T \to \infty} \left\langle \phi, \int_{-T}^{T} dt e^{it\hat{M}_E} \psi \right\rangle$$

(as $\delta(\hat{M}_E) = \lim_{T \to \infty} \int_{-T}^{T} dt e^{it\hat{M}_E}$). A spin foam representation of this expression is obtained by splitting the t -parameter in discrete steps and writing

$e^{it\hat{M}_E} = \lim_{n\to\infty}[e^{it\hat{M}_E/n}]^n = \lim_{n\to\infty}[1 + it\hat{M}_E/n]^n$.

The spin foam description then follows from the application of $[1 + it\hat{M}_E/n]$ on a spin network resulting in a linear combination of new spin networks whose graph and labels have been modified. Obviously an approximation is made by truncating the value of n to some finite integer. An advantage of the extended master constraint is that we are working at the kinematic level and so far it is only here we have access semiclassical coherent states. Moreover, one can find none graph changing versions of this master constraint operator, which are the only type of operators appropriate for these coherent states.

1.7.6 Algebraic quantum gravity

The master constraint programme has evolved into a fully combinatorial treatment of gravity known as Algebraic Quantum Gravity (AQG).[49] The non-graph changing master constraint operator is adapted in the framework of algebraic quantum gravity. While AQG is inspired by LQG, it differs drastically from it because in AQG there is fundamentally no topology or differential structure - it is background independent in a more generalized sense and could possibly have something to say about topology change. In this new formulation of quantum gravity AQG semiclassical states always control the fluctuations of all present degrees of freedom. This makes the AQG semiclassical analysis superior over that of LQG, and progress has been made in establishing it has the correct semiclassical limit and providing contact with familiar low energy physics.[50][51] See Thiemann's book for details.

1.8 Physical applications of LQG

1.8.1 Black hole entropy

Main articles: Black hole thermodynamics, Isolated horizon and Immirzi parameter

The Immirzi parameter (also known as the Barbero-Immirzi parameter) is a numerical coefficient appearing in loop quantum gravity. It may take real or imaginary values.

Black hole thermodynamics is the area of study that seeks to reconcile the laws of thermodynamics with the existence of black hole event horizons. The no hair conjecture of general relativity states that a black hole is characterized only by its mass, its charge, and its angular momentum; hence, it has no entropy. It appears, then, that one can violate the second law of thermodynamics by dropping an object with nonzero entropy into a black hole.[52] Work by Stephen Hawking and

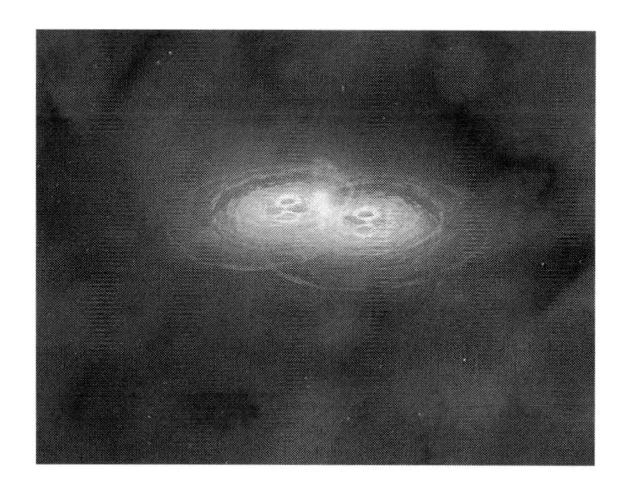

An artist depiction of two black holes merging, a process in which the laws of thermodynamics are upheld.

Jacob Bekenstein showed that one can preserve the second law of thermodynamics by assigning to each black hole a *black-hole entropy*

$$S_{\text{BH}} = \frac{k_{\text{B}} A}{4\ell_{\text{P}}^2},$$

where A is the area of the hole's event horizon, k_{B} is the Boltzmann constant, and $\ell_{\text{P}} = \sqrt{G\hbar/c^3}$ is the Planck length.[53] The fact that the black hole entropy is also the maximal entropy that can be obtained by the Bekenstein bound (wherein the Bekenstein bound becomes an equality) was the main observation that led to the holographic principle.[52]

An oversight in the application of the no-hair theorem is the assumption that the relevant degrees of freedom accounting for the entropy of the black hole must be classical in nature; what if they were purely quantum mechanical instead and had non-zero entropy? Actually, this is what is realized in the LQG derivation of black hole entropy, and can be seen as a consequence of its background-independence – the classical black hole spacetime comes about from the semiclassical limit of the quantum state of the gravitational field, but there are many quantum states that have the same semiclassical limit. Specifically, in LQG[54] it is possible to associate a quantum geometrical interpretation to the microstates: These are the quantum geometries of the horizon which are consistent with the area, A , of the black hole and the topology of the horizon (i.e. spherical). LQG offers a geometric explanation of the finiteness of the entropy and of the proportionality of the area of the horizon.[55][56] These calculations have been generalized to rotating black holes.[57]

It is possible to derive, from the covariant formulation of full quantum theory (Spinfoam) the correct relation be-

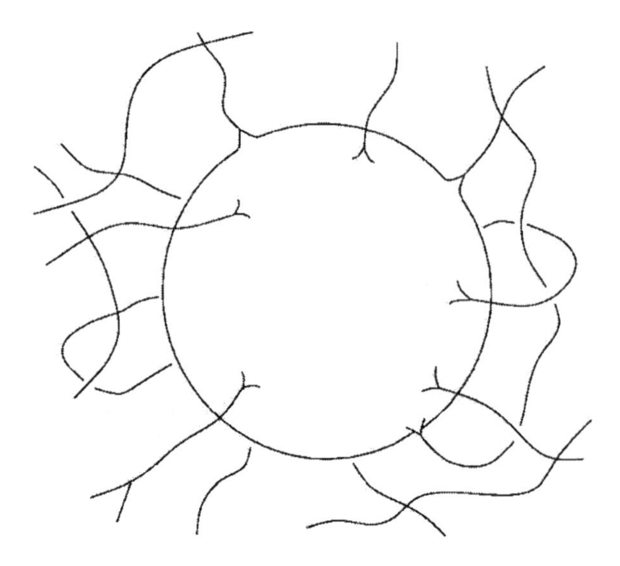

Representation of quantum geometries of the horizon. Polymer excitations in the bulk puncture the horizon, endowing it with quantized area. Intrinsically the horizon is flat except at punctures where it acquires a quantized deficit angle or quantized amount of curvature. These deficit angles add up to 4π .

tween energy and area (1st law), the Unruh temperature and the distribution that yields Hawking entropy.[58] The calculation makes use of the notion of dynamical horizon and is done for non-extremal black holes.

A recent success of the theory in this direction is the computation of the entropy of all non singular black holes directly from theory and independent of Immirzi parameter.[59] The result is the expected formula $S = A/4$, where S is the entropy and A the area of the black hole, derived by Bekenstein and Hawking on heuristic grounds. This is the only known derivation of this formula from a fundamental theory, for the case of generic non singular black holes. Older attempts at this calculation had difficulties. The problem was that although Loop quantum gravity predicted that the entropy of a black hole is proportional to the area of the event horizon, the result depended on a crucial free parameter in the theory, the above-mentioned Immirzi parameter. However, there is no known computation of the Immirzi parameter, so it had to be fixed by demanding agreement with Bekenstein and Hawking's calculation of the black hole entropy.

1.8.2 Loop quantum cosmology

Main articles: loop quantum cosmology, Big bounce and inflation (cosmology)

The popular and technical literature makes extensive references to LQG-related topic of loop quantum cosmology. LQC was mainly developed by Martin Bojowald, it was popularized Loop quantum cosmology in *Scientific American* for predicting a Big Bounce prior to the Big Bang. Loop quantum cosmology (LQC) is a symmetry-reduced model of classical general relativity quantized using methods that mimic those of loop quantum gravity (LQG) that predicts a "quantum bridge" between contracting and expanding cosmological branches.

Achievements of LQC have been the resolution of the big bang singularity, the prediction of a Big Bounce, and a natural mechanism for inflation (cosmology).

LQC models share features of LQG and so is a useful toy model. However, the results obtained are subject to the usual restriction that a truncated classical theory, then quantized, might not display the true behaviour of the full theory due to artificial suppression of degrees of freedom that might have large quantum fluctuations in the full theory. It has been argued that singularity avoidance in LQC are by mechanisms only available in these restrictive models and that singularity avoidance in the full theory can still be obtained but by a more subtle feature of LQG.[60][61]

1.8.3 Loop quantum gravity phenomenology

Quantum gravity effects are notoriously difficult to measure because the Planck length is so incredibly small. However recently physicists have started to consider the possibility of measuring quantum gravity effects mostly from astrophysical observations and gravitational wave detectors.The energy of those fluctuations at scales this small cause space-perturbations which are visible at higher scales.

1.8.4 Background independent scattering amplitudes

Loop quantum gravity is formulated in a background-independent language. No spacetime is assumed a priori, but rather it is built up by the states of theory themselves - however scattering amplitudes are derived from n -point functions (Correlation function (quantum field theory)) and these, formulated in conventional quantum field theory, are functions of points of a background space-time. The relation between the background-independent formalism and the conventional formalism of quantum field theory on a given spacetime is far from obvious, and it is far from obvious how to recover low-energy quantities from the full background-independent theory. One would like to derive the n -point functions of the theory from the background-independent formalism, in order to compare them with the standard perturbative expansion of quantum general relativity and therefore check that loop quantum gravity yields the correct low-energy limit.

A strategy for addressing this problem has been suggested;[62] the idea is to study the boundary amplitude, namely a path integral over a finite space-time region, seen as a function of the boundary value of the field.[63] In conventional quantum field theory, this boundary amplitude is well–defined[64][65] and codes the physical information of the theory; it does so in quantum gravity as well, but in a fully background–independent manner.[66] A generally covariant definition of n -point functions can then be based on the idea that the distance between physical points –arguments of the n -point function is determined by the state of the gravitational field on the boundary of the spacetime region considered.

Progress has been made in calculating background independent scattering amplitudes this way with the use of spin foams. This is a way to extract physical information from the theory. Claims to have reproduced the correct behaviour for graviton scattering amplitudes and to have recovered classical gravity have been made. "We have calculated Newton's law starting from a world with no space and no time." - Carlo Rovelli.

1.9 Gravitons, string theory, supersymmetry, extra dimensions in LQG

Main articles: graviton, string theory, supersymmetry, Kaluza–Klein theory and supergravity

Some quantum theories of gravity posit a spin-2 quantum field that is quantized, giving rise to gravitons. In string theory one generally starts with quantized excitations on top of a classically fixed background. This theory is thus described as background dependent. Particles like photons as well as changes in the spacetime geometry (gravitons) are both described as excitations on the string worldsheet. The background dependence of string theory can have important physical consequences, such as determining the number of quark generations. In contrast, loop quantum gravity, like general relativity, is manifestly background independent, eliminating the background required in string theory. Loop quantum gravity, like string theory, also aims to overcome the nonrenormalizable divergences of quantum field theories.

LQG never introduces a background and excitations living on this background, so LQG does not use gravitons as building blocks. Instead one expects that one may recover a kind of semiclassical limit or weak field limit where something like "gravitons" will show up again. In contrast, gravitons play a key role in string theory where they are among the first (massless) level of excitations of a superstring.

LQG differs from string theory in that it is formulated in 3 and 4 dimensions and without supersymmetry or Kaluza-Klein extra dimensions, while the latter requires both to be true. There is no experimental evidence to date that confirms string theory's predictions of supersymmetry and Kaluza–Klein extra dimensions. In a 2003 paper A dialog on quantum gravity,[67] Carlo Rovelli regards the fact LQG is formulated in 4 dimensions and without supersymmetry as a strength of the theory as it represents the most parsimonious explanation, consistent with current experimental results, over its rival string/M-theory. Proponents of string theory will often point to the fact that, among other things, it demonstrably reproduces the established theories of general relativity and quantum field theory in the appropriate limits, which Loop Quantum Gravity has struggled to do. In that sense string theory's connection to established physics may be considered more reliable and less speculative, at the mathematical level. Loop Quantum Gravity has nothing to say about the matter(fermions) in the universe.

Since LQG has been formulated in 4 dimensions (with and without supersymmetry), and M-theory requires supersymmetry and 11 dimensions, a direct comparison between the two has not been possible. It is possible to extend mainstream LQG formalism to higher-dimensional supergravity, general relativity with supersymmetry and Kaluza–Klein extra dimensions should experimental evidence establish their existence. It would therefore be desirable to have higher-dimensional Supergravity loop quantizations at one's disposal in order to compare these approaches. In fact a series of recent papers have been published attempting just this.[68][69][70][71][72][73][74][75] Most recently, Thiemann (and alumni) have made progress toward calculating black hole entropy for supergravity in higher dimensions. It will be interesting to compare these results to the corresponding super string calculations.[76][77]

Loop Quantum gravity like other theories of gravity remains unfalsifiable since spin foams exist at super-planck scales , it is impossible for a collider to probe those lengths in the foreseeable future.

1.10 LQG and related research programs

Main articles: noncommutative geometry, twistor theory, entropic gravity, Sundance Bilson-Thompson, Asymptotic safety in quantum gravity, Causal dynamical triangulation, group field theory and consistent discretizations

Several research groups have attempted to combine LQG with other research programs: Johannes Aastrup, Jesper M. Grimstrup et al. research combines noncommutative geometry with loop quantum gravity,[78] Laurent Freidel, Simone Speziale, et al., spinors and twistor theory with loop quantum gravity,[79] and Lee Smolin et al. with Verlinde entropic gravity and loop gravity.[80] Stephon Alexander, Antonino Marciano and Lee Smolin have attempted to explain the origins of weak force chirality in terms of Ashketar's variables, which describe gravity as chiral,[81] and LQG with Yang–Mills theory fields[82] in four dimensions. Sundance Bilson-Thompson, Hackett et al.,[83][84] has attempted to introduce standard model via LQG"s degrees of freedom as an emergent property (by employing the idea noiseless subsystems a useful notion introduced in more general situation for constrained systems by Fotini Markopoulou-Kalamara et al.[85]) LQG has also drawn philosophical comparisons with causal dynamical triangulation[86] and asymptotically safe gravity,[87] and the spinfoam with group field theory and AdS/CFT correspondence.[88] Smolin and Wen have suggested combining LQG with String-net liquid, tensors, and Smolin and Fotini Markopoulou-Kalamara Quantum Graphity. There is the consistent discretizations approach. Also, Pullin and Gambini provide a framework to connect the path integral and canonical approaches to quantum gravity. They may help reconcile the spin foam and canonical loop representation approaches. Recent research by Chris Duston and Matilde Marcolli introduces topology change via topspin networks.[89]

1.11 Problems and comparisons with alternative approaches

Main article: List of unsolved problems in physics

Some of the major unsolved problems in physics are theoretical, meaning that existing theories seem incapable of explaining a certain observed phenomenon or experimental result. The others are experimental, meaning that there is a difficulty in creating an experiment to test a proposed theory or investigate a phenomenon in greater detail.

Can quantum mechanics and general relativity be realized as a fully consistent theory (perhaps as a quantum field theory)? Is spacetime fundamentally continuous or discrete? Would a consistent theory involve a force mediated by a hypothetical graviton, or be a product of a discrete structure of spacetime itself (as in loop quantum gravity)? Are there deviations from the predictions of general relativity at very small or very large scales or in other extreme circumstances that flow from a quantum gravity theory?

The theory of LQG is one possible solution to the problem of quantum gravity, as is string theory. There are substantial differences however. For example, string theory also addresses unification, the understanding of all known forces and particles as manifestations of a single entity, by postulating extra dimensions and so-far unobserved additional particles and symmetries. Contrary to this, LQG is based only on quantum theory and general relativity and its scope is limited to understanding the quantum aspects of the gravitational interaction. On the other hand, the consequences of LQG are radical, because they fundamentally change the nature of space and time and provide a tentative but detailed physical and mathematical picture of quantum spacetime.

Presently, no semiclassical limit recovering general relativity has been shown to exist. This means it remains unproven that LQG's description of spacetime at the Planck scale has the right continuum limit (described by general relativity with possible quantum corrections). Specifically, the dynamics of the theory is encoded in the Hamiltonian constraint, but there is no candidate Hamiltonian.[90] Other technical problems include finding off-shell closure of the constraint algebra and physical inner product vector space, coupling to matter fields of Quantum field theory, fate of the renormalization of the graviton in perturbation theory that lead to ultraviolet divergence beyond 2-loops (see One-loop Feynman diagram in Feynman diagram).[90]

While there has been a recent proposal relating to observation of naked singularities,[91] and doubly special relativity as a part of a program called loop quantum cosmology, there is no experimental observation for which loop quantum gravity makes a prediction not made by the Standard Model or general relativity (a problem that plagues all current theories of quantum gravity). Because of the abovementioned lack of a semiclassical limit, LQG has not yet even reproduced the predictions made by general relativity.

An alternative criticism is that general relativity may be an effective field theory, and therefore quantization ignores the fundamental degrees of freedom.

1.12 See also

1.13 Notes

[1] Rovelli, Carlo (August 2008). "Loop Quantum Gravity" (PDF). *CERN*. Retrieved 14 September 2014.

[2] Rovelli, C. (2011). "Zakopane lectures on loop gravity". arXiv:1102.3660 [gr-qc].

[3] Muxin, H. (2011). "Cosmological constant in loop quantum gravity vertex amplitude". *Physical Review D* **84** (6): 064010.

arXiv:1105.2212. Bibcode:2011PhRvD..84f4010H. doi:10.1103/PhysRevD.84.064010.

[4] Fairbairn, W. J.; Meusburger, C. (2011). "q-Deformation of Lorentzian spin foam models". arXiv:1112.2511 [gr-qc].

[5] Rovelli, C. (2004). *Quantum Gravity*. Cambridge Monographs on Mathematical Physics. p. 71. ISBN 978-0-521-83733-0.

[6] Kauffman, S.; Smolin, L. (7 April 1997). "A Possible Solution For The Problem Of Time In Quantum Cosmology". *Edge.org*. Retrieved 2014-08-20.

[7] Smolin, L. (2006). "The Case for Background Independence". In Rickles, D.; French, S.; Saatsi, J. T. *The Structural Foundations of Quantum Gravity*. Clarendon Press. pp. 196*ff*. arXiv:hep-th/0507235. ISBN 978-0-19-926969-3.

[8] Rovelli, C. (2004). *Quantum Gravity*. Cambridge Monographs on Mathematical Physics. p. 13ff. ISBN 978-0-521-83733-0.

[9] Thiemann, T. (1996). "Anomaly-free formulation of non-perturbative, four-dimensional Lorentzian quantum gravity". *Physics Letters B* **380**: 257–264. arXiv:gr-qc/9606088. Bibcode:1996PhLB..380..257T. doi:10.1016/0370-2693(96)00532-1.

[10] Baez, J.; de Muniain, J. P. (1994). *Gauge Fields, Knots and Quantum Gravity*. Series on Knots and Everything. Vol. 4. World Scientific. Part III, chapter 4. ISBN 978-981-02-1729-7.

[11] Thiemann, T. (2003). "Lectures on Loop Quantum Gravity". *Lecture Notes in Physics* **631**: 41–135. arXiv:gr-qc/0210094. Bibcode:2003LNP...631...41T. doi:10.1007/978-3-540-45230-0_3.

[12] Rovelli, C.; Smolin, L. (1988). "Knot Theory and Quantum Gravity". *Physical Review Letters* **61** (10): 1155–1958. Bibcode:1988PhRvL..61.1155R. doi:10.1103/PhysRevLett.61.1155.

[13] Gambini, R.; Pullin, J. (2011). *A First Course in Loop Quantum Gravity*. Oxford University Press. Section 8.2. ISBN 978-0-19-959075-9.

[14] Fernando, J.; Barbero, G. (1995). "Reality Conditions and Ashtekar Variables: A Different Perspective". *Physical Review D* **51**: 5498–5506. arXiv:gr-qc/9410013. Bibcode:1995PhRvD..51.5498B. doi:10.1103/PhysRevD.51.5498.

[15] Fernando, J.; Barbero, G. (1995). "Real Ashtekar Variables for Lorentzian Signature Space-times". *Physical Review D* **51**: 5507–5520. arXiv:gr-qc/9410014. Bibcode:1995PhRvD..51.5507B. doi:10.1103/PhysRevD.51.5507.

[16] Bojowald, M.; Alejandro, P. "Spin Foam Quantization and Anomalies". arXiv:gr-qc/0303026 [gr-qc].

[17] Barrett, J.; Crane, L. (2000). "A Lorentzian signature model for quantum general relativity". *Classical and Quantum Gravity* **17**: 3101–3118. arXiv:gr-qc/9904025. Bibcode:2000CQGra..17.3101B. doi:10.1088/0264-9381/17/16/302..

[18] Rovelli, C.; Alesci, E. (2007). "The complete LQG propagator I. Difficulties with the Barrett–Crane vertex". *Physical Review D* **76**: 104012. arXiv:hep-th/0703074. Bibcode:2007PhRvD..76b4012B. doi:10.1103/PhysRevD.76.024012.

[19] Engle, J.; Pereira, R.; Rovelli, C. (2009). "Loop-Quantum-Gravity Vertex Amplitude". *Physical Review Letters* **99**: 161301. arXiv:0705.2388. Bibcode:2007PhRvL..99p1301E. doi:10.1103/physrevlett.99.161301.

[20] Freidel, L.; Krasnov, K. (2008). "A new spin foam model for 4D gravity". *Classical and Quantum Gravity* **25**: 125018. arXiv:0708.1595. Bibcode:2008CQGra..25l5018F. doi:10.1088/0264-9381/25/12/125018.

[21] Alesci, E.; Thiemann, T.; Zipfel, A. (2011). "Linking covariant and canonical LQG: new solutions to the Euclidean Scalar Constraint". arXiv:1109.1290.

[22] Bohm, D. (1989). *Quantum Theory*. Dover Publications. ISBN 978-0-486-65969-5.

[23] Tipler, P.; Llewellyn, R. (2008). *Modern Physics* (5th ed.). W. H. Freeman and Co. pp. 160–161. ISBN 978-0-7167-7550-8.

[24] Bohr, N. (1920). "Über die Serienspektra der Element". *Zeitschrift für Physik* **2** (5): 423–478. Bibcode:1920ZPhy....2..423B. doi:10.1007/BF01329978. (English translation in Bohr 1976, pp. 241–282)

[25] Jammer, M. (1989). *The Conceptual Development of Quantum Mechanics* (2nd ed.). Tomash Publishers. Section 3.2. ISBN 978-0-88318-617-6.

[26] Ashtekar, A.; Bombelli, L.; Corichi, A. (2005). "Semiclassical States for Constrained Systems". *Physical Review D* **72**: 025008. arXiv:hep-ph/0504114. Bibcode:2005PhRvD..72a5008C. doi:10.1103/PhysRevD.72.015008.

[27] Lewandowski, J.; Okołów, A.; Sahlmann, H.; Thiemann, T. (2005). "Uniqueness of Diffeomorphism Invariant States on Holonomy-Flux Algebras". *Communications in Mathematical Physics* **267**: 703–733. arXiv:gr-qc/0504147. Bibcode:2006CMaPh.267..703L. doi:10.1007/s00220-006-0100-7.

[28] Fleischhack, C. (2006). "Irreducibility of the Weyl algebra in loop quantum gravity". *Physical Review Letters* **97**: 061302. Bibcode:2006PhRvL..97f1302F. doi:10.1103/physrevlett.97.061302.

[29] Thiemann, T. (2008). *Modern Canonical General Relativity*. Cambridge Monographs on Mathematical Physics. Cambridge University Press. Section 10.6. ISBN 978-0-521-74187-3.

[30] "Partial and Complete Observables for Hamiltonian Constrained Systems". *General Relativity and Gravitation* **39**: 1891–1927. 2007. arXiv:gr-qc/0411013. Bibcode:2007GReGr..39.1891D. doi:10.1007/s10714-007-0495-2.

[31] "Partial and Complete Observables for Canonical General Relativity". *Classical and Quantum Gravity* **23**: 6155–6184. arXiv:gr-qc/0507106. Bibcode:2006CQGra..23.6155D. doi:10.1088/0264-9381/23/22/006.

[32] Dreyer, O.; Markopoulou, f.; Smolin, L. (2006). "Symmetry and entropy of black hole horizons". *Nuclear Physics B* **774**: 1–13. arXiv:hep-th/0409056. Bibcode:2006NuPhB.744....1D. doi:10.1016/j.nuclphysb.2006.02.045.

[33] Kribs, D. W.; Markopoulou, F. "Geometry from quantum particles". arXiv:gr-qc/0510052.

[34] Markopoulou, F.; Poulin, D. "Noiseless subsystems and the low energy limit of spin foam models" (unpublished).

[35] *The Phoenix Project: Master Constraint Programme for Loop Quantum Gravity*, Class.Quant.Grav.23:2211-2248,2006 or http://fr.arxiv.org/pdf/gr-qc/0305080

[36] *Modern Canonical Quantum General Relativity* by Thomas Thiemann

[37] *Testing the Master Constraint Programme for Loop Quantum Gravity I. General Framework*, Bianca Dittrich, Thomas Thiemann, Class.Quant.Grav. 23 (2006) 1025-1066.

[38] *Testing the Master Constraint Programme for Loop Quantum Gravity II. Finite Dimensional Systems*, Bianca Dittrich, Thomas Thiemann, Class.Quant.Grav. 23 (2006) 1067-1088.

[39] *Testing the Master Constraint Programme for Loop Quantum Gravity III. SL(2,R) Models*, Bianca Dittrich, Thomas Thiemann, Class.Quant.Grav. 23 (2006) 1089-1120.

[40] *Testing the Master Constraint Programme for Loop Quantum Gravity IV. Free Field Theories*, Bianca Dittrich, Thomas Thiemann, Class.Quant.Grav. 23 (2006) 1121-1142.

[41] *Testing the Master Constraint Programme for Loop Quantum Gravity V. Interacting Field Theories*, Bianca Dittrich, Thomas Thiemann, Class.Quant.Grav. 23 (2006) 1143-1162.

[42] *Quantum Spin Dynamics VIII. The Master Constraint*, Thomas Thiemann, Class.Quant.Grav. 23 (2006) 2249-2266.

[43] *Approximating the physical inner product of Loop Quantum Cosmology*, Benjamin Bahr, Thomas Thiemann, Class.Quant.Grav.24:2109-2138,2007.

[44] *On the Relation between Operator Constraint --, Master Constraint --, Reduced Phase Space --, and Path Integral Quantisation*, Muxin Han, Thomas Thiemann, Class.Quant.Grav.27:225019,2010.

[45] *On the Relation between Rigging Inner Product and Master Constraint Direct Integral Decomposition*, Muxin Han, Thomas Thiemann, J.Math.Phys.51:092501,2010.

[46] *A Path-integral for the Master Constraint of Loop Quantum Gravity*, Muxin Han, Class.Quant.Grav.27:215009,2010

[47] *Emergent diffeomorphism invariance in a discrete loop quantum gravity model*, Rodolfo Gambini, Jorge Pullin, Class.Quant.Grav.26:035002,2009

[48] Section 10.2.2 *A First Course in Loop quantum Gravity*, Rodolfo Gambinni, Jorge Pullin, Oxford University Press, first published 2011.

[49] *Algebraic Quantum Gravity (AQG) I. Conceptual Setup*, K. Giesel, T. Thiemann, Class.Quant.Grav.24:2465-2498,2007.

[50] *Algebraic Quantum Gravity (AQG) II. Semiclassical Analysis*, K. Giesel, T. Thiemann, Class.Quant.Grav.24:2499-2564,2007.

[51] *Algebraic Quantum Gravity (AQG) III. Semiclassical Perturbation Theory*, K. Giesel, T. Thiemann, Class.Quant.Grav.24:2565-2588,2007.

[52] Bousso, Raphael (2002). "The Holographic Principle". *Reviews of Modern Physics* **74** (3): 825–874. arXiv:hep-th/0203101. Bibcode:2002RvMP...74..825B. doi:10.1103/RevModPhys.74.825.

[53] Majumdar, Parthasarathi (1998). "Black Hole Entropy and Quantum Gravity" **73**: 147. arXiv:gr-qc/9807045. Bibcode:1999InJPB..73..147M.

[54] See List of loop quantum gravity researchers

[55] Rovelli, Carlo (1996). "Black Hole Entropy from Loop Quantum Gravity". *Physical Review Letters* **77** (16): 3288–3291. arXiv:gr-qc/9603063. Bibcode:1996PhRvL..77.3288R. doi:10.1103/PhysRevLett.77.3288.

[56] Ashtekar, Abhay; Baez, John; Corichi, Alejandro; Krasnov, Kirill (1998). "Quantum Geometry and Black Hole Entropy". *Physical Review Letters* **80** (5): 904–907. arXiv:gr-qc/9710007. Bibcode:1998PhRvL..80..904A. doi:10.1103/PhysRevLett.80.904.

[57] *Quantum horizons and black hole entropy: Inclusion of distortion and rotation*, Abhay Ashtekar, Jonathan Engle, Chris Van Den Broeck, Class.Quant.Grav.22:L27-L34, 2005.

[58] Bianchi, Eugenio (2012). "Entropy of Non-Extremal Black Holes from Loop Gravity". arXiv:1204.5122.

[59] http://inspirehep.net/record/940357?ln=en. http://inspirehep.net/record/1111991.

[60] *On (Cosmological) Singularity Avoidance in Loop Quantum Gravity*, Johannes Brunnemann, Thomas Thiemann, Class.Quant.Grav. 23 (2006) 1395-1428.

[61] *Unboundedness of Triad -- Like Operators in Loop Quantum Gravity*, Johannes Brunnemann, Thomas Thiemann, Class.Quant.Grav. 23 (2006) 1429-1484.

[62] L. Modesto, C. Rovelli:*Particle scattering in loop quantum gravity*, Phys Rev Lett 95 (2005) 191301

[63] R Oeckl, *A 'general boundary' formulation for quantum mechanics and quantum gravity*, Phys Lett B575 (2003) 318-324 ; *Schrodinger's cat and the clock: lessons for quantum gravity*, Class Quant Grav 20 (2003) 5371-5380l

[64] F. Conrady, C. Rovelli *Generalized Schrodinger equation in Euclidean field theory", Int J Mod Phys A 19, (2004) 1-32.*

[65] L Doplicher, *Generalized Tomonaga-Schwinger equation from the Hadamard formula*, Phys Rev D70 (2004) 064037

[66] F. Conrady, L. Doplicher, R. Oeckl, C. Rovelli, M. Testa, *Minkowski vacuum in background independent quantum gravity*, Phys Rev D69 (2004) 064019.

[67] http://arxiv.org/abs/arXiv:hep-th/0310077

[68] *New Variables for Classical and Quantum Gravity in all Dimensions I. Hamiltonian Analysis*, Norbert Bodendorfer, Thomas Thiemann, Andreas Thurn, Class. Quantum Grav. 30 (2013) 045001

[69] *New Variables for Classical and Quantum Gravity in all Dimensions II. Lagrangian Analysis*, Norbert Bodendorfer, Thomas Thiemann, Andreas Thurn, Quantum Grav. 30 (2013) 045002

[70] *New Variables for Classical and Quantum Gravity in all Dimensions III. Quantum Theory*, Norbert Bodendorfer, Thomas Thiemann, Andreas Thurn, Class. Quantum Grav. 30 (2013) 045003

[71] *New Variables for Classical and Quantum Gravity in all Dimensions IV. Matter Coupling*, Norbert Bodendorfer, Thomas Thiemann, Andreas Thurn, Class. Quantum Grav. 30 (2013) 045004

[72] *On the Implementation of the Canonical Quantum Simplicity Constraint*, Norbert Bodendorfer, Thomas Thiemann, Andreas Thurn, Class. Quantum Grav. 30 (2013) 045005

[73] *Towards Loop Quantum Supergravity (LQSG) I. Rarita-Schwinger Sector*, Norbert Bodendorfer, Thomas Thiemann, Andreas Thurn, Class. Quantum Grav. 30 (2013) 045006

[74] *Towards Loop Quantum Supergravity (LQSG) II. p-Form Sector*, Norbert Bodendorfer, Thomas Thiemann, Andreas Thurn, Class. Quantum Grav. 30 (2013) 045007

[75] *Towards Loop Quantum Supergravity (LQSG)*, Norbert Bodendorfer, Thomas Thiemann, Andreas Thurn, Phys. Lett. B 711: 205-211 (2012)

[76] *New Variables for Classical and Quantum Gravity in all Dimensions V. Isolated Horizon Boundary Degrees of Freedom*, Norbert Bodendorfer, Thomas Thiemann, Andreas Thurn, http://uk.arxiv.org/pdf/1304.2679.

[77] *Black hole entropy from loop quantum gravity in higher dimensions*, Norbert Bodendorfer http://uk.arxiv.org/pdf/1307.5029

[78] http://arxiv.org/abs/1203.6164

[79] http://arxiv.org/abs/1006.0199

[80] http://arxiv.org/abs/1001.3668

[81] http://arxiv.org/abs/1212.5246

[82] http://arxiv.org/abs/1105.3480

[83] *Quantum gravity and the standard model*, Sundance O. Bilson-Thompson, Fotini Markopoulou, Lee Smolin, Class.Quant.Grav.24:3975-3994,2007.

[84] For a precise review and outlook of this research see: *Emergent Braided Matter of Quantum Geometry*, Sundance Bilson-Thompson, Jonathan Hackett, Louis Kauffman, Yidun Wan, SIGMA 8 (2012), 014, 43 pages.

[85] *Constrained Mechanics and Noiseless Subsystems*, Tomasz Konopka, Fotini Markopoulou, arXiv:gr-qc/0601028.

[86] http://www.perimeterinstitute.ca/people/renate-loll

[87] wwnpqft.inln.cnrs.fr/pdf/Bianchi.pdf

[88] http://arxiv.org/abs/0804.0632

[89] http://arxiv.org/abs/1308.2934

[90] Nicolai, Hermann; Peeters, Kasper; Zamaklar, Marija (2005). "Loop quantum gravity: an outside view". *Classical and Quantum Gravity* **22** (19): R193–R247. arXiv:hep-th/0501114. Bibcode:2005CQGra..22R.193N. doi:10.1088/0264-9381/22/19/R01.

[91] Goswami; Joshi, Pankaj S.; Singh, Parampreet; et al. (2006). "Quantum evaporation of a naked singularity". *Physical Review Letters* **96** (3): 31302. arXiv:gr-qc/0506129. Bibcode:2006PhRvL..96c1302G. doi:10.1103/PhysRevLett.96.031302.

1.14 References

- Topical Reviews
  - Rovelli, Carlo (2011). "Zakopane lectures on loop gravity". arXiv:1102.3660.
  - Rovelli, Carlo (1998). "Loop Quantum Gravity". *Living Reviews in Relativity* **1**. Retrieved 2008-03-13.

  - Thiemann, Thomas (2003). "Lectures on Loop Quantum Gravity". *Lectures Notes in Physics*. Lecture Notes in Physics **631**: 41–135. arXiv:gr-qc/0210094. Bibcode:2003LNP...631...41T. doi:10.1007/978-3-540-45230-0_3. ISBN 978-3-540-40810-9.
  - Ashtekar, Abhay; Lewandowski, Jerzy (2004). "Background Independent Quantum Gravity: A Status Report". *Classical and Quantum Gravity* **21** (15): R53–R152. arXiv:gr-qc/0404018. Bibcode:2004CQGra..21R..53A. doi:10.1088/0264-9381/21/15/R01.
  - Carlo Rovelli and Marcus Gaul, *Loop Quantum Gravity and the Meaning of Diffeomorphism Invariance*, e-print available as gr-qc/9910079.
  - Lee Smolin, *The case for background independence*, e-print available as hep-th/0507235.
  - Alejandro Corichi, *Loop Quantum Geometry: A primer*, e-print available as .
  - Alejandro Perez, *Introduction to loop quantum gravity and spin foams*, e-print available as .
  - Hermann Nicolai and Kasper Peeters *Loop and spin foam quantum gravity: A Brief guide for beginners.*, e-print available as .

- Popular books:
  - Lee Smolin, *Three Roads to Quantum Gravity*
  - Carlo Rovelli, *Che cos'è il tempo? Che cos'è lo spazio?*, Di Renzo Editore, Roma, 2004. French translation: *Qu'est ce que le temps? Qu'est ce que l'espace?*, Bernard Gilson ed, Brussel, 2006. English translation: *What is Time? What is space?*, Di Renzo Editore, Roma, 2006.
  - Julian Barbour, *The End of Time: The Next Revolution in Our Understanding of the Universe*
  - Musser, George (2008). "The Complete Idiot's Guide to String Theory". *The Physics Teacher* (Indianapolis: Alpha) **47** (2): 368. Bibcode:2009PhTea..47Q.128H. doi:10.1119/1.3072469. ISBN 978-1-59257-702-6. – Focuses on string theory but has an extended discussion of loop gravity as well.

- Magazine articles:
  - Lee Smolin, "Atoms of Space and Time", *Scientific American*, January 2004
  - Martin Bojowald, "Following the Bouncing Universe", *Scientific American*, October 2008

- Easier introductory, expository or critical works:
  - Abhay Ashtekar, *Gravity and the quantum*, e-print available as gr-qc/0410054 (2004)
  - John C. Baez and Javier Perez de Muniain, *Gauge Fields, Knots and Quantum Gravity*, World Scientific (1994)
  - Carlo Rovelli, *A Dialog on Quantum Gravity*, e-print available as hep-th/0310077 (2003)
  - Rodolfo Gambini and Jorge Pullin, *A First Course in Loop Quantum Gravity*, Oxford (2011)
  - Carlo Rovelli and Francesca Vidotto, *Covariant Loop Quantum Gravity*, Cambridge (2014); draft available online

- More advanced introductory/expository works:
  - Carlo Rovelli, *Quantum Gravity*, Cambridge University Press (2004); draft available online
  - Thomas Thiemann, *Introduction to modern canonical quantum general relativity*, e-print available as gr-qc/0110034
  - Thomas Thiemann, *Introduction to Modern Canonical Quantum General Relativity*, Cambridge University Press (2007)
  - Abhay Ashtekar, *New Perspectives in Canonical Gravity*, Bibliopolis (1988).
  - Abhay Ashtekar, *Lectures on Non-Perturbative Canonical Gravity*, World Scientific (1991)
  - Rodolfo Gambini and Jorge Pullin, *Loops, Knots, Gauge Theories and Quantum Gravity*, Cambridge University Press (1996)
  - Hermann Nicolai, Kasper Peeters, Marija Zamaklar, *Loop quantum gravity: an outside view*, e-print available as hep-th/0501114
  - H. Nicolai and K. Peeters, *Loop and Spin Foam Quantum Gravity: A Brief Guide for Beginners*, e-print available as hep-th/0601129
  - T. Thiemann The LQG – String: Loop Quantum Gravity Quantization of String Theory (2004)

- Conference proceedings:
  - John C. Baez (ed.), *Knots and Quantum Gravity*

- Fundamental research papers:
  - Ashtekar, Abhay (1986). "New variables for classical and quantum gravity". *Physical Review Letters* **57** (18): 2244–2247. Bibcode:1986PhRvL..57.2244A. doi:10.1103/PhysRevLett.57.2244. PMID 10033673

- Ashtekar, Abhay (1987). "New Hamiltonian formulation of general relativity". *Physical Review D* **36** (6): 1587–1602. Bibcode:1987PhRvD..36.1587A. doi:10.1103/PhysRevD.36.1587
- Roger Penrose, *Angular momentum: an approach to combinatorial space-time* in *Quantum Theory and Beyond*, ed. Ted Bastin, Cambridge University Press, 1971
- Rovelli, Carlo; Smolin, Lee (1988). "Knot theory and quantum gravity". *Physical Review Letters* **61** (10): 1155–1158. Bibcode:1988PhRvL..61.1155R. doi:10.1103/PhysRevLett.61.1155.
- Rovelli, Carlo; Smolin, Lee (1990). "Loop space representation of quantum general relativity". *Nuclear Physics* **B331**: 80–152.
- Carlo Rovelli and Lee Smolin, *Discreteness of area and volume in quantum gravity*, Nucl. Phys., **B442** (1995) 593-622, e-print available as gr-qc/9411005
- Kuchař, Karel (1973). "Canonical Quantization of Gravity". In Israel, Werner. *Relativity, Astrophysics and Cosmology*. D. Reidel. pp. 237–288. ISBN 90-277-0369-8.
- Thiemann, Thomas (2006). "Loop Quantum Gravity: An Inside View". *Approaches to Fundamental Physics*. Lecture Notes in Physics **721**: 185–263. arXiv:hep-th/0608210. Bibcode:2007LNP...721..185T. doi:10.1007/978-3-540-71117-9_10. ISBN 978-3-540-71115-5.

1.15 External links

- "Loop Quantum Gravity" by Carlo Rovelli Physics World, November 2003
- Quantum Foam and Loop Quantum Gravity
- Abhay Ashtekar: Semi-Popular Articles . Some excellent popular articles suitable for beginners about space, time, GR, and LQG.
- Loop Quantum Gravity: Lee Smolin.
- Loop Quantum Gravity on arxiv.org
- A list of LQG references catered to fresh graduates
- Loop Quantum Gravity Lectures Online by Lee Smolin
- Spin networks, spin foams and loop quantum gravity
- Wired magazine, News: *Moving Beyond String Theory*
- April 2006 Scientific American Special Issue, *A Matter of Time*, has Lee Smolin LQG Article *Atoms of Space and Time*
- September 2006, The Economist, article *Looping the loop*
- Gamma-ray Large Area Space Telescope: http://glast.gsfc.nasa.gov/
- Zeno meets modern science. Article from Acta Physica Polonica B by Z.K. Silagadze.
- Did pre-big bang universe leave its mark on the sky? - According to a model based on "loop quantum gravity" theory, a parent universe that existed before ours may have left an imprint (*New Scientist*, 10 April 2008)

Chapter 2

History of loop quantum gravity

General relativity is the theory of gravitation published by Albert Einstein in **1915**. According to it, the force of gravity is a manifestation of the local geometry of spacetime. Mathematically, the theory is modelled after Bernhard Riemann's metric geometry, but the Lorentz group of spacetime symmetries (an essential ingredient of Einstein's own theory of special relativity) replaces the group of rotational symmetries of space. Loop quantum gravity inherits this geometric interpretation of gravity, and posits that a quantum theory of gravity is fundamentally a quantum theory of spacetime.

In the **1920s**, the French mathematician Élie Cartan formulated Einstein's theory in the language of bundles and connections, a generalization of Riemann's geometry to which Cartan made important contributions. The so-called Einstein–Cartan theory of gravity not only reformulated but also generalized general relativity, and allowed spacetimes with torsion as well as curvature. In Cartan's geometry of bundles, the concept of parallel transport is more fundamental than that of distance, the centerpiece of Riemannian geometry. A similar conceptual shift occurs between the invariant interval of Einstein's general relativity and the parallel transport of Einstein–Cartan theory.

In the **1960s**, physicist Roger Penrose explored the idea of space arising from a quantum combinatorial structure. His investigations resulted in the development of spin networks. Because this was a quantum theory of the rotational group and not the Lorentz group, Penrose went on to develop twistors.

In **1982**, Amitabha Sen tried to formulate a Hamiltonian formulation of general relativity based on spinorial variables, where these variables are the left and right spinorial component equivalents of Einstein–Cartan connection of general relativity. Particularly, Sen discovered a new way to write down the two constraints of ADM Hamiltonian formulation of general relativity in terms of these spinorial connections. In his form, the constraints are simply conditions that the Spinorial Weyl curvature is trace free and symmetric. He also discovered the presence of new constraints which he suggested to be interpreted as the equivalent of Gauss constraint of Yang Mills field theories. But Sen's work fell short of giving a full clear systematic theory and particularly failed to clearly discuss the conjugate momenta to the spinorial variables, its physical interpretation, and its relation to the metric (in his work he indicated this as some lambda variable).

In **1986**, physicist Abhay Ashtekar completed the project which Amitabha Sen began. He clearly identified the fundamental conjugate variables of spinorial gravity: The configuration variable is as a spinoral connection - a rule for parallel transport (technically, a connection) and the conjugate momentum variable is a coordinate frame (called a vierbein) at each point. So these variable became what we know as Ashtekar variables, a particular flavor of Einstein–Cartan theory with a complex connection.General relativity theory expressed in this way, made possible to pursue quantization of it using well-known techniques from quantum gauge field theory.

The quantization of gravity in the Ashtekar formulation was based on Wilson loops, a technique developed in the **1970s** to study the strong-interaction regime of quantum chromodynamics (QCD). It is interesting in this connection that Wilson loops were known to be ill-behaved in the case of standard quantum field theory on (flat) Minkowski space, and so did not provide a nonperturbative quantization of QCD. However, because the Ashtekar formulation was background-independent, it was possible to use Wilson loops as the basis for nonperturbative quantization of gravity.

Sen's initiation and completion by Ashtekar, finally, for the first time, in a setting where the Wheeler–DeWitt equation could be written in terms of a well-defined Hamiltonian operator on a well-defined Hilbert space, and led to construction of the first known exact solution, the so-called Chern–Simons form or Kodama state. The physical interpretation of this state remains obscure.

Around **1990**, Carlo Rovelli and Lee Smolin obtained an explicit basis of states of quantum geometry, which turned out to be labeled by Penrose's spin networks. In this context, spin networks arose as a generalization of Wilson loops necessary to deal with mutually intersecting loops. Mathematically, spin networks are related to group representation theory and can be used to construct knot invariants such as the Jones polynomial. Being closely related to topological quantum field theory and group representation theory, LQG is mostly established at the level of rigor of mathematical physics, as compared to string theory, which is established at the level of rigor of physics.

After the spin network basis was described, progress was made on the analysis of the spectra of various operators resulting in a predicted spectrum for area and volume (see below). Work on the semi-classical limit, the continuum limit, and dynamics was intense after this, but progress slower.

On the semi-classical limit front, the goal is to obtain and study analogues of the harmonic oscillator coherent states (candidates are known as weave states).

LQG was initially formulated as a quantization of the Hamiltonian ADM formalism, according to which the Einstein equations are a collection of constraints (Gauss, Diffeomorphism and Hamiltonian). The kinematics are encoded in the Gauss and Diffeomorphism constraints, whose solution is the space spanned by the spin network basis. The problem is to define the Hamiltonian constraint as a self-adjoint operator on the kinematical state space. The most promising work in this direction is Thomas Thiemann's Phoenix program.

In the recent years most of the developments in LQG has been done in the covariant formulation of the theory, called Spinfoam Theory. The present version of the covariant dynamics is due to the convergent work of different groups, but it is commonly named after a paper by Jonhatan Engle, Roberto Pereira and Carlo Rovelli in 2008. Spin foams are a framework intended to tackle the problem of dynamics and the continuum limit simultaneously. Heuristically, it would be expected that evolution between spin network states might be described by discrete combinatorial operations on the spin networks, which would then trace a two-dimensional skeleton of spacetime. This approach is related to state-sum models of statistical mechanics and topological quantum field theory such as the Turaeev–Viro model of 3D quantum gravity, and also to the Regge calculus approach to calculate the Feynman path integral of general relativity by discretizing spacetime.

2.1 Bibliography

- Topical Reviews
  - Carlo Rovelli, *Loop Quantum Gravity*, Living Reviews in Relativity **1**, (1998), 1, online article, 2001 15 August version.
  - Thomas Thiemann, *Lectures on loop quantum gravity*, e-print available as gr-qc/0210094
  - Abhay Ashtekar and Jerzy Lewandowski, *Background independent quantum gravity: a status report*, e-print available as gr-qc/0404018
  - Carlo Rovelli and Marcus Gaul, *Loop Quantum Gravity and the Meaning of Diffeomorphism Invariance*, e-print available as gr-qc/9910079.
  - Lee Smolin, *The case for background independence*, e-print available as hep-th/0507235.
- Popular books:
  - Julian Barbour, *The End of Time: The Next Revolution in Our Understanding of the Universe*
  - Lee Smolin, *Three Roads to Quantum Gravity*
  - Carlo Rovelli, *Che cos'è il tempo? Che cos'è lo spazio?*, Di Renzo Editore, Roma, 2004. French translation: *Qu'est ce que le temps? Qu'est ce que l'espace?*, Bernard Gilson ed, Brussel, 2006. English translation: *What is Time? What is space?*, Di Renzo Editore, Roma, 2006.
- Magazine articles:
  - Lee Smolin, "Atoms in Space and Time", *Scientific American*, January 2004
- Easier introductory, expository or critical works:
  - Abhay Ashtekar, *Gravity and the quantum*, e-print available as gr-qc/0410054
  - John C. Baez and Javier Perez de Muniain, *Gauge Fields, Knots and Quantum Gravity*, World Scientific (1994)
  - Carlo Rovelli, *A Dialog on Quantum Gravity*, e-print available as hep-th/0310077
- More advanced introductory/expository works:
  - Carlo Rovelli, *Quantum Gravity*, Cambridge University Press (2004); draft available online
  - Thomas Thiemann, *Introduction to modern canonical quantum general relativity*, e-print available as gr-qc/0110034
  - Abhay Ashtekar, *New Perspectives in Canonical Gravity*, Bibliopolis (1988).
  - Abhay Ashtekar, *Lectures on Non-Perturbative Canonical Gravity*, World Scientific (1991)

  - Rodolfo Gambini and Jorge Pullin, *Loops, Knots, Gauge Theories and Quantum Gravity*, Cambridge University Press (1996)
  - Hermann Nicolai, Kasper Peeters, Marija Zamaklar, *Loop quantum gravity: an outside view*, e-print available as hep-th/0501114
  - "Loop and Spin Foam Quantum Gravity: A Brief Guide for beginners arXiv:hep-th/0601129 H. Nicolai and K. Peeters.
  - Edward Witten, *Quantum Background Independence In String Theory*, e-print available as hep-th/9306122.

- Conference proceedings:
  - John C. Baez (ed.), *Knots and Quantum Gravity*

- Fundamental research papers:
  - Amitabha Sen, *Gravity as a spin system*, *Phys. Lett.* B119:89–91, December 1982.
  - Abhay Ashtekar, *New variables for classical and quantum gravity*, *Phys. Rev. Lett.*, **57**, 2244-2247, 1986
  - Abhay Ashtekar, *New Hamiltonian formulation of general relativity*, *Phys. Rev.* **D36**, 1587-1602, 1987
  - Roger Penrose, *Angular momentum: an approach to combinatorial space-time* in *Quantum Theory and Beyond*, ed. Ted Bastin, Cambridge University Press, 1971
  - Carlo Rovelli and Lee Smolin, *Knot theory and quantum gravity*, *Phys. Rev. Lett.*, **61** (1988) 1155
  - Carlo Rovelli and Lee Smolin, *Loop space representation of quantum general relativity*, *Nuclear Physics* **B331** (1990) 80-152

Chapter 3

Loop quantum cosmology

Loop quantum cosmology (LQC) is a finite, symmetry-reduced model of loop quantum gravity (LQG) that predicts a "quantum bridge" between contracting and expanding cosmological branches.

The distinguishing feature of LQC is the prominent role played by the quantum geometry effects of loop quantum gravity (LQG). In particular, quantum geometry creates a brand new repulsive force which is totally negligible at low space-time curvature but rises very rapidly in the Planck regime, overwhelming the classical gravitational attraction and thereby resolving singularities of general relativity. Once singularities are resolved, the conceptual paradigm of cosmology changes and one has to revisit many of the standard issues—e.g., the "horizon problem"—from a new perspective.

Since LQG is based on a specific quantum theory of Riemannian geometry,[1] geometric observables display a fundamental discreteness that play a key role in quantum dynamics: While predictions of LQC are very close to those of quantum geometrodynamics (QGD) away from the Planck regime, there is a dramatic difference once densities and curvatures enter the Planck scale. In LQC the big bang is replaced by a quantum bounce.

Study of LQC has led to many successes, including the emergence of a possible mechanism for cosmic inflation, resolution of gravitational singularities, as well as the development of effective semi-classical Hamiltonians.

This subfield was originally started in 1999 by Yogesh Yadav, and further developed in particular by Abhay Ashtekar and Jerzy Lewandowski. In late 2012 LQC represents a very active field in physics, with about three hundred papers on the subject published in the literature. There has also recently been work by Carlo Rovelli, et al. on relating LQC to the spinfoam-based spinfoam cosmology.

However, the results obtained in LQC are subject to the usual restriction that a truncated classical theory, then quantized, might not display the true behaviour of the full theory due to artificial suppression of degrees of freedom that might have large quantum fluctuations in the full theory. It has been argued that singularity avoidance in LQC are by mechanisms only available in these restrictive models and that singularity avoidance in the full theory can still be obtained but by a more subtle feature of LQG.[2][3]

3.1 See also

- Big Bounce
- Loop quantum gravity
- Cyclic model

3.2 References

[1] Ashtekar, Abhay (November 2008). "Loop Quantum Cosmology: An Overview". *Gen. Rel. Grav.* **41** (4): 707–741. arXiv:0812.0177. Bibcode:2009GReGr..41..707A. doi:10.1007/s10714-009-0763-4.

[2] *On (Cosmological) Singularity Avoidance in Loop Quantum Gravity*, Johannes Brunnemann, Thomas Thiemann, Class. Quant. Grav. 23 (2006) 1395-1428.

[3] *Unboundedness of Triad -- Like Operators in Loop Quantum Gravity*, Johannes Brunnemann, Thomas Thiemann, Class. Quant. Grav. 23 (2006) 1429-1484.

3.3 External links

- Loop quantum cosmology on arxiv.org
- Quantum Nature of The Big Bang in Loop Quantum Cosmology
- Gravity and the Quantum
- Loop Quantum Cosmology, Martin Bojowald

- Did our cosmos exist before the big bang?
- Abhay Ashtekar, Parampreet Singh "Loop Quantum Cosmology: A Status Report"

Chapter 4

Gravity

For other uses, see Gravity (disambiguation).
"Gravitation" and "Law of Gravity" redirect here. For other uses, see Gravitation (disambiguation) and Law of Gravity (disambiguation).
Gravity or **gravitation** is a natural phenomenon by which

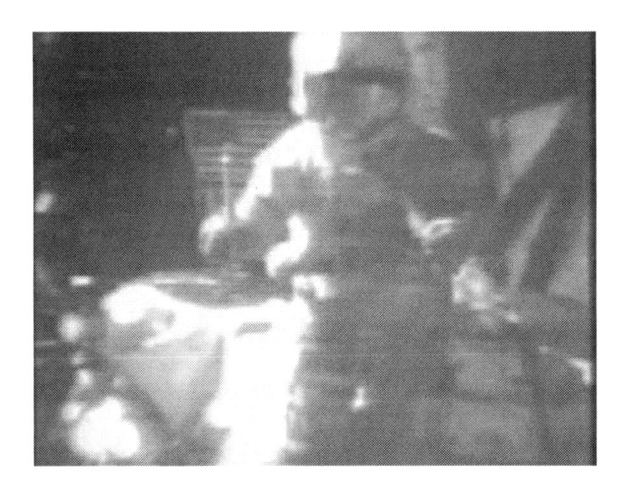

Hammer and feather drop: Apollo 15 astronaut David Scott on the Moon enacting the legend of Galileo's gravity experiment. (1.38 MB, ogg/Theora format).

all things with energy are brought toward (or *gravitate* toward) one another, including stars, planets, galaxies and even light and sub-atomic particles. Gravity is responsible for many of the structures in the Universe, by creating spheres of hydrogen — where hydrogen fuses under pressure to form stars — and grouping them into galaxies. On Earth, gravity gives weight to physical objects and causes the tides. Gravity has an infinite range, although its effects become increasingly weaker on farther objects.

Gravity is most accurately described by the general theory of relativity (proposed by Albert Einstein in 1915) which describes gravity not as a force but as a consequence of the curvature of spacetime caused by the uneven distribution of mass/energy; and resulting in gravitational time dilation, where time lapses more slowly in lower (stronger) gravitational potential. However, for most applications, gravity is well approximated by Newton's law of universal gravitation, which postulates that gravity causes a force where two bodies of mass are directly drawn (or 'attracted') to each other according to a mathematical relationship, where the attractive force is proportional to the product of their masses and inversely proportional to the square of the distance between them. This is considered to occur over an infinite range, such that all bodies (with mass) in the universe are drawn to each other no matter how far they are apart.

Gravity is the weakest of the four fundamental interactions of nature. The gravitational attraction is approximately 10^{-38} times the strength of the strong force (i.e. gravity is 38 orders of magnitude weaker), 10^{-36} times the strength of the electromagnetic force, and 10^{-29} times the strength of the weak force. As a consequence, gravity has a negligible influence on the behavior of sub-atomic particles, and plays no role in determining the internal properties of everyday matter (but see quantum gravity). On the other hand, gravity is the dominant interaction at the macroscopic scale, and is the cause of the formation, shape, and trajectory (orbit) of astronomical bodies. It is responsible for various phenomena observed on Earth and throughout the universe; for example, it causes the Earth and the other planets to orbit the Sun, the Moon to orbit the Earth, the formation of tides, and the formation and evolution of galaxies, stars and the Solar System.

In pursuit of a theory of everything, the merging of general relativity and quantum mechanics (or quantum field theory) into a more general theory of quantum gravity has become an area of research.

4.1 History of gravitational theory

Main article: History of gravitational theory

4.1.1 Scientific revolution

Modern work on gravitational theory began with the work of Galileo Galilei in the late 16th and early 17th centuries. In his famous (though possibly apocryphal[1]) experiment dropping balls from the Tower of Pisa, and later with careful measurements of balls rolling down inclines, Galileo showed that gravity accelerates all objects at the same rate. This was a major departure from Aristotle's belief that heavier objects accelerate faster.[2] Galileo postulated air resistance as the reason that lighter objects may fall more slowly in an atmosphere. Galileo's work set the stage for the formulation of Newton's theory of gravity.

4.1.2 Newton's theory of gravitation

Main article: Newton's law of universal gravitation

In 1687, English mathematician Sir Isaac Newton pub-

Sir Isaac Newton, an English physicist who lived from 1642 to 1727

lished *Principia*, which hypothesizes the inverse-square law of universal gravitation. In his own words, "I deduced that the forces which keep the planets in their orbs must [be] reciprocally as the squares of their distances from the centers about which they revolve: and thereby compared the force requisite to keep the Moon in her Orb with the force of gravity at the surface of the Earth; and found them answer pretty nearly."[3] The equation is the following:

$F = G\frac{m_1 m_2}{r^2}$

Where F is the force, m_1 and m_2 are the masses of the objects interacting, r is the distance between the centers of the masses and G is the gravitational constant.

Newton's theory enjoyed its greatest success when it was used to predict the existence of Neptune based on motions of Uranus that could not be accounted for by the actions of the other planets. Calculations by both John Couch Adams and Urbain Le Verrier predicted the general position of the planet, and Le Verrier's calculations are what led Johann Gottfried Galle to the discovery of Neptune.

A discrepancy in Mercury's orbit pointed out flaws in Newton's theory. By the end of the 19th century, it was known that its orbit showed slight perturbations that could not be accounted for entirely under Newton's theory, but all searches for another perturbing body (such as a planet orbiting the Sun even closer than Mercury) had been fruitless. The issue was resolved in 1915 by Albert Einstein's new theory of general relativity, which accounted for the small discrepancy in Mercury's orbit.

Although Newton's theory has been superseded by the Einstein's general relativity, most modern non-relativistic gravitational calculations are still made using the Newton's theory because it is simpler to work with and it gives sufficiently accurate results for most applications involving sufficiently small masses, speeds and energies.

4.1.3 Equivalence principle

The equivalence principle, explored by a succession of researchers including Galileo, Loránd Eötvös, and Einstein, expresses the idea that all objects fall in the same way, and that the effects of gravity are indistinguishable from certain aspects of acceleration and deceleration. The simplest way to test the weak equivalence principle is to drop two objects of different masses or compositions in a vacuum and see whether they hit the ground at the same time. Such experiments demonstrate that all objects fall at the same rate when other forces (such as air resistance and electromagnetic effects) are negligible. More sophisticated tests use a torsion balance of a type invented by Eötvös. Satellite experiments, for example STEP, are planned for more accurate experiments in space.[4]

Formulations of the equivalence principle include:

- The weak equivalence principle: *The trajectory of a point mass in a gravitational field depends only on its initial position and velocity, and is independent of its composition.*[5]

- The Einsteinian equivalence principle: *The outcome of any local non-gravitational experiment in a freely falling laboratory is independent of the velocity of the laboratory and its location in spacetime.*[6]

- The strong equivalence principle requiring both of the above.

4.1.4 General relativity

See also: Introduction to general relativity

In general relativity, the effects of gravitation are ascribed to spacetime curvature instead of a force. The starting point for general relativity is the equivalence principle, which equates free fall with inertial motion and describes free-falling inertial objects as being accelerated relative to non-inertial observers on the ground.[7][8] In Newtonian physics, however, no such acceleration can occur unless at least one of the objects is being operated on by a force.

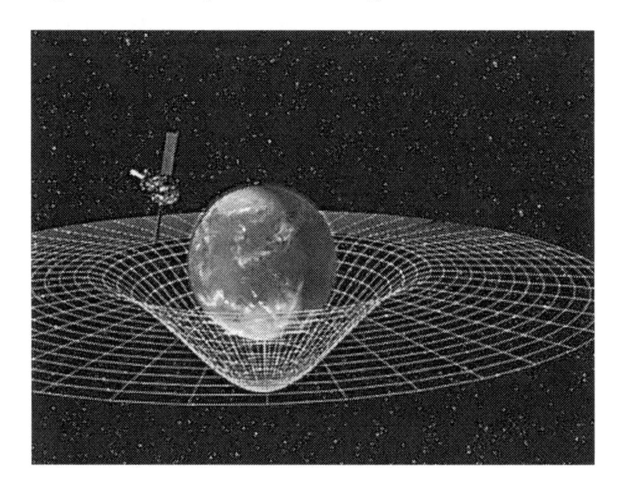

Two-dimensional analogy of spacetime distortion generated by the mass of an object. Matter changes the geometry of spacetime, this (curved) geometry being interpreted as gravity. White lines do not represent the curvature of space but instead represent the coordinate system imposed on the curved spacetime, which would be rectilinear in a flat spacetime.

Einstein proposed that spacetime is curved by matter, and that free-falling objects are moving along locally straight paths in curved spacetime. These straight paths are called geodesics. Like Newton's first law of motion, Einstein's theory states that if a force is applied on an object, it would deviate from a geodesic. For instance, we are no longer following geodesics while standing because the mechanical resistance of the Earth exerts an upward force on us, and we are non-inertial on the ground as a result. This explains why moving along the geodesics in spacetime is considered inertial.

Einstein discovered the field equations of general relativity, which relate the presence of matter and the curvature of spacetime and are named after him. The Einstein field equations are a set of 10 simultaneous, non-linear, differential equations. The solutions of the field equations are the components of the metric tensor of spacetime. A metric tensor describes a geometry of spacetime. The geodesic paths for a spacetime are calculated from the metric tensor.

Solutions

Notable solutions of the Einstein field equations include:

- The Schwarzschild solution, which describes spacetime surrounding a spherically symmetric non-rotating uncharged massive object. For compact enough objects, this solution generated a black hole with a central singularity. For radial distances from the center which are much greater than the Schwarzschild radius, the accelerations predicted by the Schwarzschild solution are practically identical to those predicted by Newton's theory of gravity.

- The Reissner-Nordström solution, in which the central object has an electrical charge. For charges with a geometrized length which are less than the geometrized length of the mass of the object, this solution produces black holes with two event horizons.

- The Kerr solution for rotating massive objects. This solution also produces black holes with multiple event horizons.

- The Kerr-Newman solution for charged, rotating massive objects. This solution also produces black holes with multiple event horizons.

- The cosmological Friedmann-Lemaître-Robertson-Walker solution, which predicts the expansion of the universe.

Tests

The tests of general relativity included the following:[9]

- General relativity accounts for the anomalous perihelion precession of Mercury.[10]

- The prediction that time runs slower at lower potentials (gravitational time dilation) has been confirmed by the Pound–Rebka experiment (1959), the Hafele–Keating experiment, and the GPS.

- The prediction of the deflection of light was first confirmed by Arthur Stanley Eddington from his observations during the Solar eclipse of May 29, 1919.[11][12] Eddington measured starlight deflections twice those predicted by Newtonian corpuscular theory, in accordance with the predictions of general relativity. However, his interpretation of the results was later disputed.[13] More recent tests using radio interferometric measurements of quasars passing behind the Sun have more accurately and consistently confirmed the deflection of light to the degree predicted by general relativity.[14] See also gravitational lens.

- The time delay of light passing close to a massive object was first identified by Irwin I. Shapiro in 1964 in interplanetary spacecraft signals.
- Gravitational radiation has been indirectly confirmed through studies of binary pulsars. On 11 February 2016, the LIGO and Virgo collaborations announced the first observation of a gravitational wave.
- Alexander Friedmann in 1922 found that Einstein equations have non-stationary solutions (even in the presence of the cosmological constant). In 1927 Georges Lemaître showed that static solutions of the Einstein equations, which are possible in the presence of the cosmological constant, are unstable, and therefore the static universe envisioned by Einstein could not exist. Later, in 1931, Einstein himself agreed with the results of Friedmann and Lemaître. Thus general relativity predicted that the Universe had to be non-static—it had to either expand or contract. The expansion of the universe discovered by Edwin Hubble in 1929 confirmed this prediction.[15]
- The theory's prediction of frame dragging was consistent with the recent Gravity Probe B results.[16]
- General relativity predicts that light should lose its energy when traveling away from massive bodies through gravitational redshift. This was verified on earth and in the solar system around 1960.

4.1.5 Gravity and quantum mechanics

Main articles: Graviton and Quantum gravity

In the decades after the discovery of general relativity, it was realized that general relativity is incompatible with quantum mechanics.[17] It is possible to describe gravity in the framework of quantum field theory like the other fundamental forces, such that the attractive force of gravity arises due to exchange of virtual gravitons, in the same way as the electromagnetic force arises from exchange of virtual photons.[18][19] This reproduces general relativity in the classical limit. However, this approach fails at short distances of the order of the Planck length,[17] where a more complete theory of quantum gravity (or a new approach to quantum mechanics) is required.

4.2 Specifics

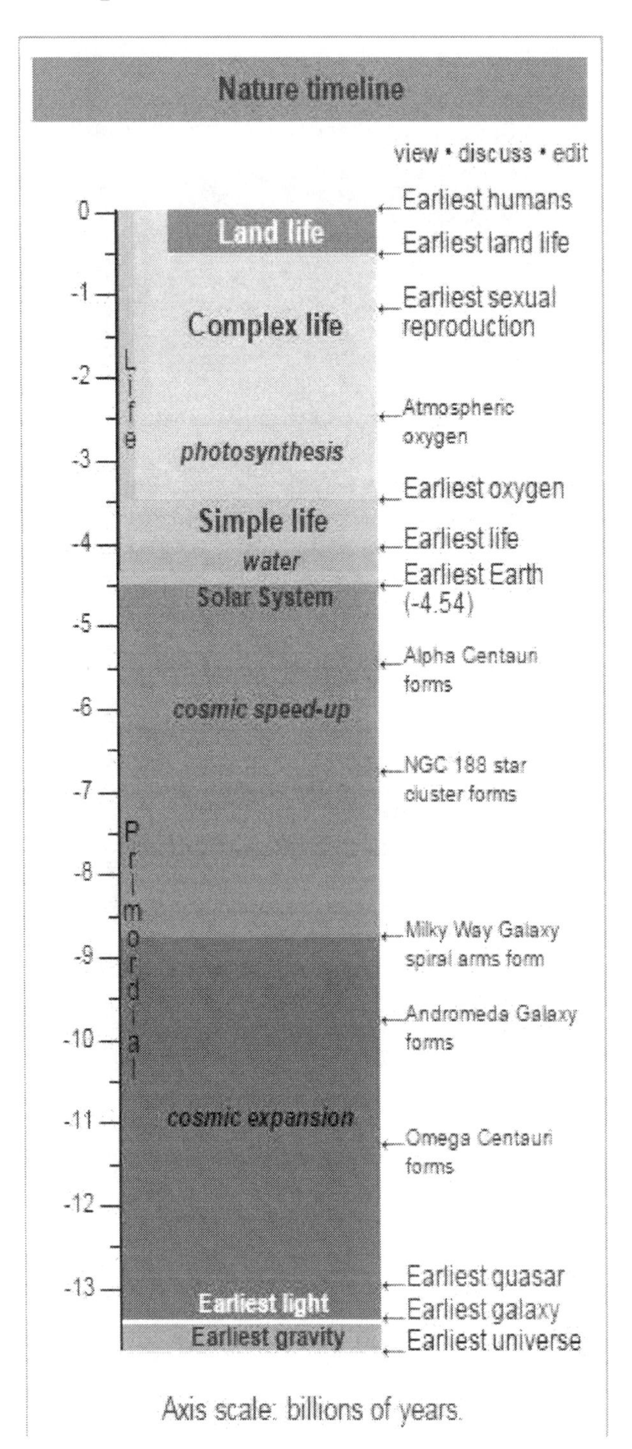

Axis scale: billions of years.

4.2.1 Earth's gravity

Main article: Earth's gravity

Every planetary body (including the Earth) is surrounded by its own gravitational field, which can be conceptualized with Newtonian physics as exerting an attractive force on all objects. Assuming a spherically symmetrical planet, the strength of this field at any given point above the surface is proportional to the planetary body's mass and inversely proportional to the square of the distance from the center of the body.

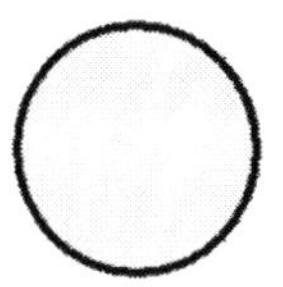

If an object with comparable mass to that of the Earth were to fall towards it, then the corresponding acceleration of the Earth would be observable.

The strength of the gravitational field is numerically equal to the acceleration of objects under its influence. The rate of acceleration of falling objects near the Earth's surface varies very slightly depending on latitude, surface features such as mountains and ridges, and perhaps unusually high or low sub-surface densities.[20] For purposes of weights and measures, a standard gravity value is defined by the International Bureau of Weights and Measures, under the International System of Units (SI).

That value, denoted g, is g = 9.80665 m/s^2 (32.1740 ft/s^2).[21][22]

The standard value of 9.80665 m/s^2 is the one originally adopted by the International Committee on Weights and Measures in 1901 for 45° latitude, even though it has been shown to be too high by about five parts in ten thousand.[23] This value has persisted in meteorology and in some standard atmospheres as the value for 45° latitude even though it applies more precisely to latitude of 45°32'33".[24]

Assuming the standardized value for g and ignoring air resistance, this means that an object falling freely near the Earth's surface increases its velocity by 9.80665 m/s (32.1740 ft/s or 22 mph) for each second of its descent. Thus, an object starting from rest will attain a velocity of 9.80665 m/s (32.1740 ft/s) after one second, approximately 19.62 m/s (64.4 ft/s) after two seconds, and so on, adding 9.80665 m/s (32.1740 ft/s) to each resulting velocity. Also, again ignoring air resistance, any and all objects, when dropped from the same height, will hit the ground at the same time.

According to Newton's 3rd Law, the Earth itself experiences a force equal in magnitude and opposite in direction to that which it exerts on a falling object. This means that the Earth also accelerates towards the object until they collide. Because the mass of the Earth is huge, however, the acceleration imparted to the Earth by this opposite force is negligible in comparison to the object's. If the object doesn't bounce after it has collided with the Earth, each of them then exerts a repulsive contact force on the other which effectively balances the attractive force of gravity and prevents further acceleration.

The force of gravity on Earth is the resultant (vector sum) of two forces: (a) The gravitational attraction in accordance with Newton's universal law of gravitation, and (b) the centrifugal force, which results from the choice of an earthbound, rotating frame of reference. At the equator, the force of gravity is the weakest due to the centrifugal force caused by the Earth's rotation. The force of gravity varies with latitude and increases from about 9.780 m/s^2 at the Equator to about 9.832 m/s^2 at the poles.

4.2.2 Equations for a falling body near the surface of the Earth

Main article: Equations for a falling body

Under an assumption of constant gravitational attraction, Newton's law of universal gravitation simplifies to $F = mg$, where m is the mass of the body and g is a constant vector with an average magnitude of 9.81 m/s^2 on Earth. This resulting force is the object's weight. The acceleration due to gravity is equal to this g. An initially stationary object which is allowed to fall freely under gravity drops a distance which is proportional to the square of the elapsed time. The image on the right, spanning half a second, was captured with a stroboscopic flash at 20 flashes per second. During the first $^1/_{20}$ of a second the ball drops one unit of distance (here, a unit is about 12 mm); by $^2/_{20}$ it has dropped at total of 4 units; by $^3/_{20}$, 9 units and so on.

Under the same constant gravity assumptions, the potential energy, Ep, of a body at height h is given by $Ep = mgh$ (or $Ep = Wh$, with W meaning weight). This expression is valid only over small distances h from the surface of the Earth. Similarly, the expression $h = \frac{v^2}{2g}$ for the maximum height reached by a vertically projected body with initial velocity v is useful for small heights and small initial velocities only.

4.2.3 Gravity and astronomy

The application of Newton's law of gravity has enabled the acquisition of much of the detailed information we have about the planets in the Solar System, the mass of the Sun, and details of quasars; even the existence of dark matter is inferred using Newton's law of gravity. Although we have not traveled to all the planets nor to the Sun, we know their masses. These masses are obtained by applying the laws of gravity to the measured characteristics of the orbit. In space an object maintains its orbit because of the force of gravity acting upon it. Planets orbit stars, stars orbit galactic centers, galaxies orbit a center of mass in clusters, and clusters orbit in superclusters. The force of gravity exerted on one object by another is directly proportional to the product of those objects' masses and inversely proportional to the square of the distance between them.

4.2.4 Gravitational radiation

Main article: Gravitational wave

According to general relativity, gravitational radiation is generated in situations where the curvature of spacetime is oscillating, such as is the case with co-orbiting objects. The gravitational radiation emitted by the Solar System is far too small to measure. However, gravitational radiation has been indirectly observed as an energy loss over time in binary pulsar systems such as PSR B1913+16. It is believed that neutron star mergers and black hole formation may create detectable amounts of gravitational radiation. Gravitational radiation observatories such as the Laser Interferometer Gravitational Wave Observatory (LIGO) have been created to study the problem. In February 2016, the Advanced LIGO team announced that they had detected gravitational waves from a black hole collision. On September 14, 2015 LIGO registered gravitational waves for the first time, as a result of the collision of two black holes 1.3 billion light-years from Earth.[26][27] This observation confirms the theoretical predictions of Einstein and others that such waves exist. The event confirms that binary black holes exist. It also opens the way for practical observation and understanding of the nature of gravity and events in the Universe including the Big Bang and what happened after it.[28][29]

4.2.5 Speed of gravity

Main article: Speed of gravity

In December 2012, a research team in China announced that it had produced measurements of the phase lag of Earth tides during full and new moons which seem to prove that the speed of gravity is equal to the speed of light.[30] This means that if the Sun suddenly disappeared, the Earth would keep orbiting it normally for 8 minutes, which is the time light takes to travel that distance. The team's findings were released in the Chinese Science Bulletin in February 2013.[31]

4.3 Anomalies and discrepancies

There are some observations that are not adequately accounted for, which may point to the need for better theories of gravity or perhaps be explained in other ways.

- **Extra-fast stars**: Stars in galaxies follow a distribution of velocities where stars on the outskirts are moving faster than they should according to the observed distributions of normal matter. Galaxies within galaxy clusters show a similar pattern. Dark matter, which would interact through gravitation but not electromagnetically, would account for the discrepancy. Various modifications to Newtonian dynamics have also been proposed.

- **Flyby anomaly**: Various spacecraft have experienced greater acceleration than expected during gravity assist maneuvers.

- **Accelerating expansion**: The metric expansion of space seems to be speeding up. Dark energy has been proposed to explain this. A recent alternative explanation is that the geometry of space is not homogeneous (due to clusters of galaxies) and that when the data are reinterpreted to take this into account, the expansion is not speeding up after all,[32] however this conclusion is disputed.[33]

- **Anomalous increase of the astronomical unit**: Recent measurements indicate that planetary orbits are widening faster than if this were solely through the Sun losing mass by radiating energy.

- **Extra energetic photons**: Photons travelling through galaxy clusters should gain energy and then lose it again on the way out. The accelerating expansion of the universe should stop the photons returning all the energy, but even taking this into account photons from the cosmic microwave background radiation gain twice as much energy as expected. This may indicate that gravity falls off *faster* than inverse-squared at certain distance scales.[34]

- **Extra massive hydrogen clouds**: The spectral lines of the Lyman-alpha forest suggest that hydrogen clouds are more clumped together at certain scales than expected and, like dark flow, may indicate that gravity falls off *slower* than inverse-squared at certain distance scales.[34]

- **Power**: Proposed extra dimensions could explain why the gravity force is so weak.[35]

4.4 Alternative theories

Main article: Alternatives to general relativity

4.4.1 Historical alternative theories

- Aristotelian theory of gravity
- Le Sage's theory of gravitation (1784) also called LeSage gravity, proposed by Georges-Louis Le Sage, based on a fluid-based explanation where a light gas fills the entire universe.
- Ritz's theory of gravitation, *Ann. Chem. Phys.* 13, 145, (1908) pp. 267–271, Weber-Gauss electrodynamics applied to gravitation. Classical advancement of perihelia.
- Nordström's theory of gravitation (1912, 1913), an early competitor of general relativity.
- Kaluza Klein theory (1921)
- Whitehead's theory of gravitation (1922), another early competitor of general relativity.

4.4.2 Modern alternative theories

- Brans–Dicke theory of gravity (1961) [36]
- Induced gravity (1967), a proposal by Andrei Sakharov according to which general relativity might arise from quantum field theories of matter
- f(R) gravity (1970)
- Horndeski theory (1974) [37]
- Supergravity (1976)
- String theory
- In the modified Newtonian dynamics (MOND) (1981), Mordehai Milgrom proposes a modification of Newton's Second Law of motion for small accelerations [38]
- The self-creation cosmology theory of gravity (1982) by G.A. Barber in which the Brans-Dicke theory is modified to allow mass creation
- Loop quantum gravity (1988) by Carlo Rovelli, Lee Smolin, and Abhay Ashtekar
- Nonsymmetric gravitational theory (NGT) (1994) by John Moffat

- Conformal gravity[39]
- Tensor–vector–scalar gravity (TeVeS) (2004), a relativistic modification of MOND by Jacob Bekenstein
- Gravity as an entropic force, gravity arising as an emergent phenomenon from the thermodynamic concept of entropy.
- In the superfluid vacuum theory the gravity and curved space-time arise as a collective excitation mode of non-relativistic background superfluid.
- Chameleon theory (2004) by Justin Khoury and Amanda Weltman.
- Pressuron theory (2013) by Olivier Minazzoli and Aurélien Hees.

4.5 See also

- Angular momentum
- Anti-gravity, the idea of neutralizing or repelling gravity
- Artificial gravity
- Birkeland current
- Cosmic gravitational wave background
- Einstein–Infeld–Hoffmann equations
- Escape velocity, the minimum velocity needed to escape from a gravity well
- g-force, a measure of acceleration
- Gauge gravitation theory
- Gauss's law for gravity
- Gravitational binding energy
- Gravitational wave
- Gravitational wave background
- Gravity assist
- Gravity gradiometry
- Gravity Recovery and Climate Experiment
- Gravity Research Foundation
- Jovian–Plutonian gravitational effect
- Kepler's third law of planetary motion
- Lagrangian point
- Micro-g environment, also called microgravity
- Mixmaster dynamics
- *n*-body problem
- Newton's laws of motion
- Pioneer anomaly
- Scalar theories of gravitation
- Speed of gravity
- Standard gravitational parameter
- Standard gravity
- Weightlessness

4.6 Footnotes

[1] Ball, Phil (June 2005). "Tall Tales". *Nature News*. doi:10.1038/news050613-10.

[2] Galileo (1638), *Two New Sciences*, First Day Salviati speaks: "If this were what Aristotle meant you would burden him with another error which would amount to a falsehood; because, since there is no such sheer height available on earth, it is clear that Aristotle could not have made the experiment; yet he wishes to give us the impression of his having performed it when he speaks of such an effect as one which we see."

[3]
- Chandrasekhar, Subrahmanyan (2003). *Newton's Principia for the common reader*. Oxford: Oxford University Press. (pp.1–2). The quotation comes from a memorandum thought to have been written about 1714. As early as 1645 Ismaël Bullialdus had argued that any force exerted by the Sun on distant objects would have to follow an inverse-square law. However, he also dismissed the idea that any such force did exist. See, for example,

Linton, Christopher M. (2004). *From Eudoxus to Einstein—A History of Mathematical Astronomy*. Cambridge: Cambridge University Press. p. 225. ISBN 978-0-521-82750-8.

[4] M.C.W.Sandford (2008). "STEP: Satellite Test of the Equivalence Principle". Rutherford Appleton Laboratory. Retrieved 2011-10-14.

[5] Paul S Wesson (2006). *Five-dimensional Physics*. World Scientific. p. 82. ISBN 981-256-661-9.

[6] Haugen, Mark P.; C. Lämmerzahl (2001). *Principles of Equivalence: Their Role in Gravitation Physics and Experiments that Test Them*. Springer. arXiv:gr-qc/0103067. ISBN 978-3-540-41236-6.

[7] "Gravity and Warped Spacetime". black-holes.org. Retrieved 2010-10-16.

[8] Dmitri Pogosyan. "Lecture 20: Black Holes—The Einstein Equivalence Principle". University of Alberta. Retrieved 2011-10-14.

[9] Pauli, Wolfgang Ernst (1958). "Part IV. General Theory of Relativity". *Theory of Relativity*. Courier Dover Publications. ISBN 978-0-486-64152-2.

[10] Max Born (1924), *Einstein's Theory of Relativity* (The 1962 Dover edition, page 348 lists a table documenting the observed and calculated values for the precession of the perihelion of Mercury, Venus, and Earth.)

[11] Dyson, F.W.; Eddington, A.S.; Davidson, C.R. (1920). "A Determination of the Deflection of Light by the Sun's Gravitational Field, from Observations Made at the Total Eclipse of May 29, 1919". *Phil. Trans. Roy. Soc. A* **220** (571–581): 291–333. Bibcode:1920RSPTA.220..291D. doi:10.1098/rsta.1920.0009.. Quote, p. 332: "Thus the results of the expeditions to Sobral and Principe can leave little doubt that a deflection of light takes place in the neighbourhood of the sun and that it is of the amount demanded by Einstein's generalised theory of relativity, as attributable to the sun's gravitational field."

[12] Weinberg, Steven (1972). *Gravitation and cosmology*. John Wiley & Sons.. Quote, p. 192: "About a dozen stars in all were studied, and yielded values 1.98 ± 0.11" and 1.61 ± 0.31", in substantial agreement with Einstein's prediction θ⊙ = 1.75"."

[13] Earman, John; Glymour, Clark (1980). "Relativity and Eclipses: The British eclipse expeditions of 1919 and their predecessors". *Historical Studies in the Physical Sciences* **11**: 49–85. doi:10.2307/27757471.

[14] Weinberg, Steven (1972). *Gravitation and cosmology*. John Wiley & Sons. p. 194.

[15] See W.Pauli, 1958, pp.219–220

[16] "NASA's Gravity Probe B Confirms Two Einstein Space-Time Theories". Nasa.gov. Retrieved 2013-07-23.

[17] Randall, Lisa (2005). *Warped Passages: Unraveling the Universe's Hidden Dimensions*. Ecco. ISBN 0-06-053108-8.

[18] Feynman, R. P.; Morinigo, F. B.; Wagner, W. G.; Hatfield, B. (1995). *Feynman lectures on gravitation*. Addison-Wesley. ISBN 0-201-62734-5.

[19] Zee, A. (2003). *Quantum Field Theory in a Nutshell*. Princeton University Press. ISBN 0-691-01019-6.

[20] "Astronomy Picture of the Day".

[21] Bureau International des Poids et Mesures (2006). "The International System of Units (SI)" (PDF) (8th ed.): 131. Retrieved 2009-11-25. Unit names are normally printed in Roman (upright) type ... Symbols for quantities are generally single letters set in an italic font, although they may be qualified by further information in subscripts or superscripts or in brackets.

[22] "SI Unit rules and style conventions". National Institute For Standards and Technology (USA). September 2004. Retrieved 2009-11-25. Variables and quantity symbols are in italic type. Unit symbols are in Roman type.

[23] List, R. J. editor, 1968, Acceleration of Gravity, *Smithsonian Meteorological Tables*, Sixth Ed. Smithsonian Institution, Washington, D.C., p. 68.

[24] U.S. Standard Atmosphere, 1976, U.S. Government Printing Office, Washington, D.C., 1976. (Linked file is very large.)

[25] "Milky Way Emerges as Sun Sets over Paranal". *www.eso.org*. European Southern Obseevatory. Retrieved 29 April 2015.

[26] Clark, Stuart (2016-02-11). "Gravitational waves: scientists announce 'we did it!' – live". *the Guardian*. Retrieved 2016-02-11.

[27] Castelvecchi, Davide; Witze, Witze (February 11, 2016). "Einstein's gravitational waves found at last". *Nature News*. doi:10.1038/nature.2016.19361. Retrieved 2016-02-11.

[28] "Scientists announce finding Gravitational Waves confirming Einstein's theory". WorldBreakingNews.

[29] "WHAT ARE GRAVITATIONAL WAVES AND WHY DO THEY MATTER?". popsci.com. Retrieved 12 February 2016.

[30] Chinese scientists find evidence for speed of gravity, astrowatch.com, 12/28/12.

[31] TANG, Ke Yun; HUA ChangCai; WEN Wu; CHI ShunLiang; YOU QingYu; YU Dan (February 2013). "Observational evidences for the speed of the gravity based on the Earth tide" (PDF). *Chinese Science Bulletin* **58** (4-5): 474–477. doi:10.1007/s11434-012-5603-3. Retrieved 12 June 2013.

[32] Dark energy may just be a cosmic illusion, *New Scientist*, issue 2646, 7 March 2008.

[33] Swiss-cheese model of the cosmos is full of holes, *New Scientist*, issue 2678, 18 October 2008.

[34] Chown, Marcus (16 March 2009). "Gravity may venture where matter fears to tread". *New Scientist*. Retrieved 4 August 2013.

[35] CERN (20 January 2012). "Extra dimensions, gravitons, and tiny black holes".

[36] Brans, C.H. (Mar 2014). "Jordan-Brans-Dicke Theory". *Scholarpedia* **9**: 31358. Bibcode:2014Schpj...931358B. doi:10.4249/scholarpedia.31358.

[37] Horndeski, G.W. (Sep 1974). "Second-Order Scalar-Tensor Field Equations in a Four-Dimensional Space". *International Journal of Theoretical Physics* **88** (10): 363–384. Bibcode:1974IJTP...10..363H. doi:10.1007/BF01807638.

[38] Milgrom, M. (Jun 2014). "The MOND paradigm of modified dynamics". *Scholarpedia* **9**: 31410. Bibcode:2014SchpJ...931410M. doi:10.4249/scholarpedia.31410.

[39] Einstein gravity from conformal gravity

4.7 References

- Halliday, David; Robert Resnick; Kenneth S. Krane (2001). *Physics v. 1*. New York: John Wiley & Sons. ISBN 0-471-32057-9.
- Serway, Raymond A.; Jewett, John W. (2004). *Physics for Scientists and Engineers* (6th ed.). Brooks/Cole. ISBN 0-534-40842-7.
- Tipler, Paul (2004). *Physics for Scientists and Engineers: Mechanics, Oscillations and Waves, Thermodynamics* (5th ed.). W. H. Freeman. ISBN 0-7167-0809-4.

4.8 Further reading

- Thorne, Kip S.; Misner, Charles W.; Wheeler, John Archibald (1973). *Gravitation*. W.H. Freeman. ISBN 0-7167-0344-0.

4.9 External links

- Hazewinkel, Michiel, ed. (2001), "Gravitation", *Encyclopedia of Mathematics*, Springer, ISBN 978-1-55608-010-4
- Hazewinkel, Michiel, ed. (2001), "Gravitation, theory of", *Encyclopedia of Mathematics*, Springer, ISBN 978-1-55608-010-4

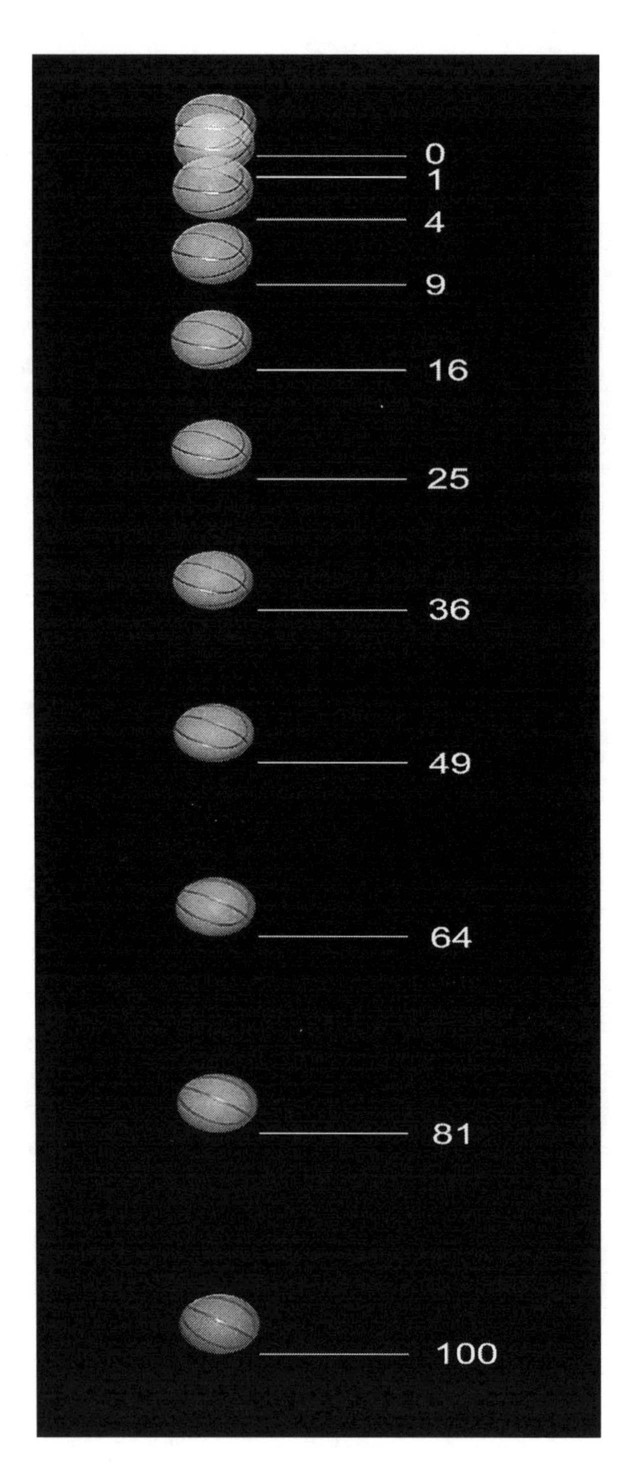

Gravity acts on stars that form our Milky Way.[25]

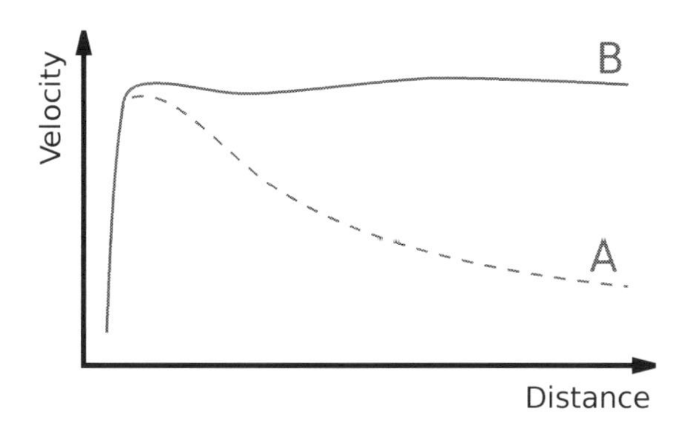

*Rotation curve of a typical spiral galaxy: predicted (**A**) and observed (**B**). The discrepancy between the curves is attributed to dark matter.*

Chapter 5

Quantum gravity

Quantum gravity (**QG**) is a field of theoretical physics that seeks to describe the force of gravity according to the principles of quantum mechanics, and where quantum effects cannot be ignored.[1]

The current understanding of gravity is based on Albert Einstein's general theory of relativity, which is formulated within the framework of classical physics. On the other hand, the nongravitational forces are described within the framework of quantum mechanics, a radically different formalism for describing physical phenomena based on the wave-like nature of matter.[2] The necessity of a quantum mechanical description of gravity follows from the fact that one cannot consistently couple a classical system to a quantum one.[3]

Although a quantum theory of gravity is needed in order to reconcile general relativity with the principles of quantum mechanics, difficulties arise when one attempts to apply the usual prescriptions of quantum field theory to the force of gravity.[4] From a technical point of view, the problem is that the theory one gets in this way is not renormalizable and therefore cannot be used to make meaningful physical predictions. As a result, theorists have taken up more radical approaches to the problem of quantum gravity, the most popular approaches being string theory and loop quantum gravity.[5] A recent development is the theory of causal fermion systems which gives quantum mechanics, general relativity, and quantum field theory as limiting cases.[6][7][8][9][10][11]

Strictly speaking, the aim of quantum gravity is only to describe the quantum behavior of the gravitational field and should not be confused with the objective of unifying all fundamental interactions into a single mathematical framework. While any substantial improvement into the present understanding of gravity would aid further work towards unification, study of quantum gravity is a field in its own right with various branches having different approaches to unification. Although some quantum gravity theories, such as string theory, try to unify gravity with the other fundamental forces, others, such as loop quantum gravity, make no such attempt; instead, they make an effort to quantize the gravitational field while it is kept separate from the other forces. A theory of quantum gravity that is also a grand unification of all known interactions is sometimes referred to as a theory of everything (TOE).

One of the difficulties of quantum gravity is that quantum gravitational effects are only expected to become apparent near the Planck scale, a scale far smaller in distance (equivalently, far larger in energy) than what is currently accessible at high energy particle accelerators. As a result, quantum gravity is a mainly theoretical enterprise, although there are speculations about how quantum gravitational effects might be observed in existing experiments.[12]

5.1 Overview

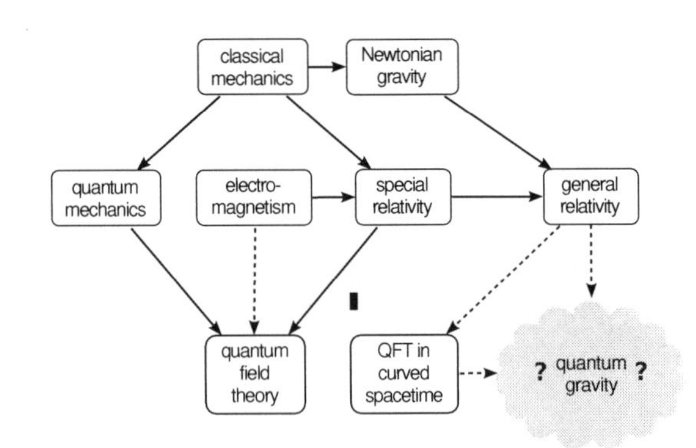

Diagram showing where quantum gravity sits in the hierarchy of physics theories

Much of the difficulty in meshing these theories at all energy scales comes from the different assumptions that these theories make on how the universe works. Quantum field theory depends on particle fields embedded in the flat space-time of special relativity. General relativity models gravity as a curvature within space-time that changes as a gravitational mass moves. Historically, the most obvious way of combining the two (such as treating gravity as simply another particle field) ran quickly into what is known as the renormalization problem. In the old-fashioned understanding of renormalization, gravity particles would attract each other and adding together all of the interactions results in many infinite values which cannot easily be cancelled out mathematically to yield sensible, finite results. This is in contrast with quantum electrodynamics where, given that the series still do not converge, the interactions sometimes evaluate to infinite results, but those are few enough in number to be removable via renormalization.

5.1.1 Effective field theories

Quantum gravity can be treated as an effective field theory. Effective quantum field theories come with some high-energy cutoff, beyond which we do not expect that the theory provides a good description of nature. The "infinities" then become large but finite quantities depending on this finite cutoff scale, and correspond to processes that involve very high energies near the fundamental cutoff. These quantities can then be absorbed into an infinite collection of coupling constants, and at energies well below the fundamental cutoff of the theory, to any desired precision; only a finite number of these coupling constants need to be measured in order to make legitimate quantum-mechanical predictions. This same logic works just as well for the highly successful theory of low-energy pions as for quantum gravity. Indeed, the first quantum-mechanical corrections to graviton-scattering and Newton's law of gravitation have been explicitly computed[13] (although they are so infinitesimally small that we may never be able to measure them). In fact, gravity is in many ways a much better quantum field theory than the Standard Model, since it appears to be valid all the way up to its cutoff at the Planck scale.

While confirming that quantum mechanics and gravity are indeed consistent at reasonable energies, it is clear that near or above the fundamental cutoff of our effective quantum theory of gravity (the cutoff is generally assumed to be of the order of the Planck scale), a new model of nature will be needed. Specifically, the problem of combining quantum mechanics and gravity becomes an issue only at very high energies, and may well require a totally new kind of model.

5.1.2 Quantum gravity theory for the highest energy scales

The general approach to deriving a quantum gravity theory that is valid at even the highest energy scales is to assume that such a theory will be simple and elegant and, accordingly, to study symmetries and other clues offered by current theories that might suggest ways to combine them into a comprehensive, unified theory. One problem with this approach is that it is unknown whether quantum gravity will actually conform to a simple and elegant theory, as it should resolve the dual conundrums of special relativity with regard to the uniformity of acceleration and gravity, and general relativity with regard to spacetime curvature.

Such a theory is required in order to understand problems involving the combination of very high energy and very small dimensions of space, such as the behavior of black holes, and the origin of the universe.

5.2 Quantum mechanics and general relativity

5.2.1 The graviton

Main article: Graviton

At present, one of the deepest problems in theoretical physics is harmonizing the theory of general relativity, which describes gravitation, and applications to large-scale structures (stars, planets, galaxies), with quantum mechanics, which describes the other three fundamental forces acting on the atomic scale. This problem must be put in the proper context, however. In particular, contrary to the popular claim that quantum mechanics and general relativity are fundamentally incompatible, one can demonstrate that the structure of general relativity essentially follows inevitably from the quantum mechanics of interacting theoretical spin-2 massless particles (called gravitons).[14][15][16][17][18]

While there is no concrete proof of the existence of gravitons, quantized theories of matter may necessitate their existence. Supporting this theory is the observation that all fundamental forces except gravity have one or more known messenger particles, leading researchers to believe that at least one most likely does exist; they have dubbed this hypothetical particle the *graviton*. The predicted find would result in the classification of the graviton as a "force particle" similar to the photon of the electromagnetic field. Many of the accepted notions of a unified theory of physics since the 1970s assume, and to some degree depend upon, the existence of the graviton. These include string theory, superstring theory, M-theory, and loop quantum gravity. Detection of gravitons is thus vital to the validation of various lines of research to unify quantum mechanics and relativity theory.

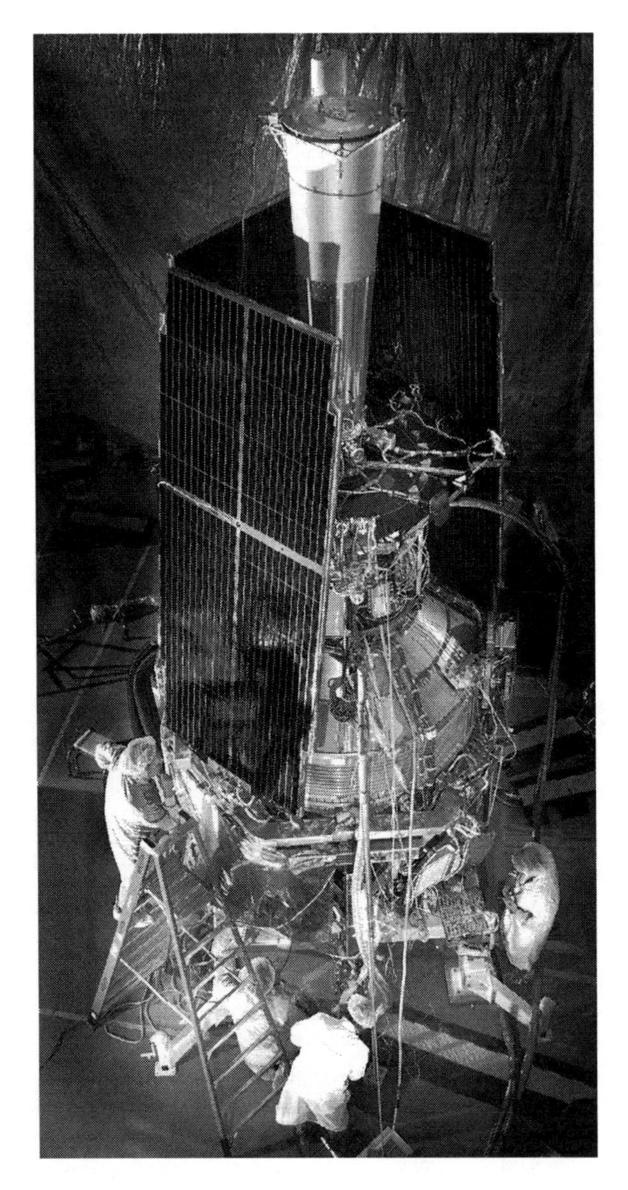

Gravity Probe B (GP-B) has measured spacetime curvature near Earth to test related models in application of Einstein's general theory of relativity.

5.2.2 The dilaton

Main article: Dilaton

The dilaton made its first appearance in Kaluza–Klein theory, a five-dimensional theory that combined gravitation and electromagnetism. Generally, it appears in string theory. More recently, however, it's become central to the lower-dimensional many-bodied gravity problem[19] based on the field theoretic approach of Roman Jackiw. The impetus arose from the fact that complete analytical solutions for the metric of a covariant N-body system have proven elusive in general relativity. To simplify the problem, the number of dimensions was lowered to $(1+1)$, i.e., one spatial dimension and one temporal dimension. This model problem, known as $R=T$ theory[20] (as opposed to the general $G=T$ theory) was amenable to exact solutions in terms of a generalization of the Lambert W function. It was also found that the field equation governing the dilaton (derived from differential geometry) was the Schrödinger equation and consequently amenable to quantization.[21]

Thus, one had a theory which combined gravity, quantization, and even the electromagnetic interaction, promising ingredients of a fundamental physical theory. It is worth noting that this outcome revealed a previously unknown and already existing *natural link* between general relativity and quantum mechanics. For some time, a generalization of this theory to $(3+1)$ dimensions was unclear. However, a recent derivation in $(3+1)$ dimensions under the right coordinate conditions yields a formulation similar to the earlier $(1+1)$ namely a dilaton field governed by the logarithmic Schrödinger equation[22] which is seen in condensed matter physics and superfluids. The field equations are indeed amenable to such a generalization (as shown with the inclusion of a one-graviton process[23]) and yield the correct Newtonian limit in d dimensions but only if a dilaton is included. Furthermore, the results become even more tantalizing in view of the apparent resemblance between the dilaton and the Higgs boson.[24] However, more experimentation is needed to resolve the relationhip between these two particles.

Since this theory can combine gravitational, electromagnetic and quantum effects, their coupling could potentially lead to a means of vindicating the theory, through cosmology and even, perhaps, experimentally. When the equation E=mC^2 was solved through the Lorentz derivative, and applied to the velocity of the electron in relationship to micro time dilation, a gravitational force was discovered as space was bent due to near C velocity of the electron and the discovery that going light speed and gravity were linked unquestionably by Einstein. It has come into question how light speed is linked to gravity, and experiment was done with atomic clocks in a centrifuge where the acceleration in the disk was causing macroscopic time dilation according to Einsteins well known works. when cesium clocks were used in the 50s on the ground and a commercial airliner, the difference in the two clocks was measured to be about 45.9 microseconds time dilation for 24 hours of flight, the clock at sea level being slower.

5.2.3 Nonrenormalizability of gravity

Further information: Renormalization

General relativity, like electromagnetism, is a classical field theory. One might expect that, as with electromagnetism, the gravitational force should also have a corresponding quantum field theory.

However, gravity is perturbatively nonrenormalizable.[25][26] For a quantum field theory to be well-defined according to this understanding of the subject, it must be asymptotically free or asymptotically safe. The theory must be characterized by a choice of *finitely many* parameters, which could, in principle, be set by experiment. For example, in quantum electrodynamics these parameters are the charge and mass of the electron, as measured at a particular energy scale.

On the other hand, in quantizing gravity there are, in perturbation theory, *infinitely many independent parameters* (counterterm coefficients) needed to define the theory. For a given choice of those parameters, one could make sense of the theory, but since it's impossible to conduct infinite experiments to fix the values of every parameter, it has been argued that one does not, in perturbation theory, have a meaningful physical theory:

- At low energies, the logic of the renormalization group tells us that, despite the unknown choices of these infinitely many parameters, quantum gravity will reduce to the usual Einstein theory of general relativity.
- On the other hand, if we could probe very high energies where quantum effects take over, then *every one* of the infinitely many unknown parameters would begin to matter, and we could make no predictions at all.

If we treat QG as an effective field theory, there is a way around this problem.

That is, the meaningful theory of quantum gravity (that makes sense and is predictive at all energy levels) inherently implies some deep principle that reduces the infinitely many unknown parameters to a finite number that can then be measured:

- One possibility is that normal perturbation theory is not a reliable guide to the renormalizability of the theory, and that there really *is* a UV fixed point for gravity. Since this is a question of non-perturbative quantum field theory, it is difficult to find a reliable answer, but some people still pursue this option.
- Another possibility is that there are new unfound symmetry principles that constrain the parameters and reduce them to a finite set. This is the route taken by string theory, where all of the excitations of the string essentially manifest themselves as new symmetries.

5.2.4 QG as an effective field theory

Main article: Effective field theory

In an effective field theory, all but the first few of the infinite set of parameters in a non-renormalizable theory are suppressed by huge energy scales and hence can be neglected when computing low-energy effects. Thus, at least in the low-energy regime, the model is indeed a predictive quantum field theory.[13] (A very similar situation occurs for the very similar effective field theory of low-energy pions.) Furthermore, many theorists agree that even the Standard Model should really be regarded as an effective field theory as well, with "nonrenormalizable" interactions suppressed by large energy scales and whose effects have consequently not been observed experimentally.

Recent work[13] has shown that by treating general relativity as an effective field theory, one can actually make legitimate predictions for quantum gravity, at least for low-energy phenomena. An example is the well-known calculation of the tiny first-order quantum-mechanical correction to the classical Newtonian gravitational potential between two masses.

5.2.5 Spacetime background dependence

Main article: Background independence

A fundamental lesson of general relativity is that there is no fixed spacetime background, as found in Newtonian mechanics and special relativity; the spacetime geometry is dynamic. While easy to grasp in principle, this is the hardest idea to understand about general relativity, and its consequences are profound and not fully explored, even at the classical level. To a certain extent, general relativity can be seen to be a relational theory,[27] in which the only physically relevant information is the relationship between different events in space-time.

On the other hand, quantum mechanics has depended since its inception on a fixed background (non-dynamic) structure. In the case of quantum mechanics, it is time that is given and not dynamic, just as in Newtonian classical mechanics. In relativistic quantum field theory, just as in classical field theory, Minkowski spacetime is the fixed background of the theory.

String theory

String theory can be seen as a generalization of quantum field theory where instead of point particles, string-like objects propagate in a fixed spacetime background, although

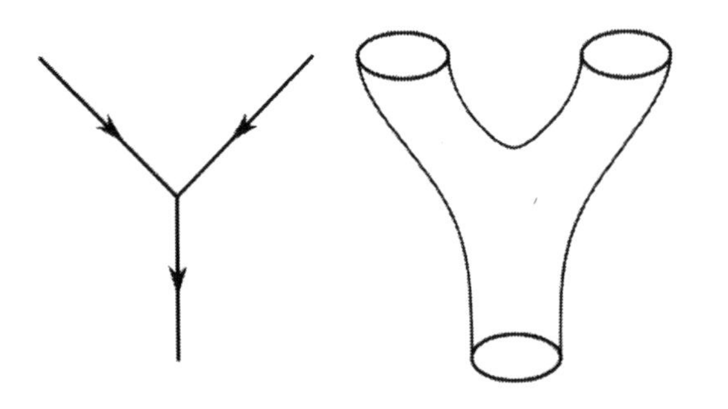

Interaction in the subatomic world: world lines of point-like particles in the Standard Model or a world sheet swept up by closed strings in string theory

the interactions among closed strings give rise to space-time in a dynamical way. Although string theory had its origins in the study of quark confinement and not of quantum gravity, it was soon discovered that the string spectrum contains the graviton, and that "condensation" of certain vibration modes of strings is equivalent to a modification of the original background. In this sense, string perturbation theory exhibits exactly the features one would expect of a perturbation theory that may exhibit a strong dependence on asymptotics (as seen, for example, in the AdS/CFT correspondence) which is a weak form of background dependence.

Background independent theories

Loop quantum gravity is the fruit of an effort to formulate a background-independent quantum theory.

Topological quantum field theory provided an example of background-independent quantum theory, but with no local degrees of freedom, and only finitely many degrees of freedom globally. This is inadequate to describe gravity in 3+1 dimensions, which has local degrees of freedom according to general relativity. In 2+1 dimensions, however, gravity is a topological field theory, and it has been successfully quantized in several different ways, including spin networks.

5.2.6 Semi-classical quantum gravity

Quantum field theory on curved (non-Minkowskian) backgrounds, while not a full quantum theory of gravity, has shown many promising early results. In an analogous way to the development of quantum electrodynamics in the early part of the 20th century (when physicists considered quantum mechanics in classical electromagnetic fields), the consideration of quantum field theory on a curved background has led to predictions such as black hole radiation.

Phenomena such as the Unruh effect, in which particles exist in certain accelerating frames but not in stationary ones, do not pose any difficulty when considered on a curved background (the Unruh effect occurs even in flat Minkowskian backgrounds). The vacuum state is the state with the least energy (and may or may not contain particles). See Quantum field theory in curved spacetime for a more complete discussion.

5.2.7 The Problem of Time

Main article: Problem of Time

A conceptual difficulty in combining quantum mechanics with general relativity arises from the contrasting role of time within these two frameworks. In quantum theories time acts as an independent background through which states evolve, with the Hamiltonian operator acting as the generator of infinitesmal translations of quantum states through time.[28] In contrast, general relativity treats time as a dynamical variable which interacts directly with matter and moreover requires the Hamiltonian constraint to vanish,[29] removing any possibility of employing a notion of time similar to that in quantum theory.

5.3 Candidate theories

There are a number of proposed quantum gravity theories.[30] Currently, there is still no complete and consistent quantum theory of gravity, and the candidate models still need to overcome major formal and conceptual problems. They also face the common problem that, as yet, there is no way to put quantum gravity predictions to experimental tests, although there is hope for this to change as future data from cosmological observations and particle physics experiments becomes available.[31][32]

5.3.1 String theory

Main article: String theory

One suggested starting point is ordinary quantum field theories which, after all, are successful in describing the other three basic fundamental forces in the context of the standard model of elementary particle physics. However, while this leads to an acceptable effective (quantum) field theory of gravity at low energies,[33] gravity turns out to be much more problematic at higher energies. For ordinary field theories such as quantum electrodynamics, a technique known as renormalization is an integral part of deriving predictions

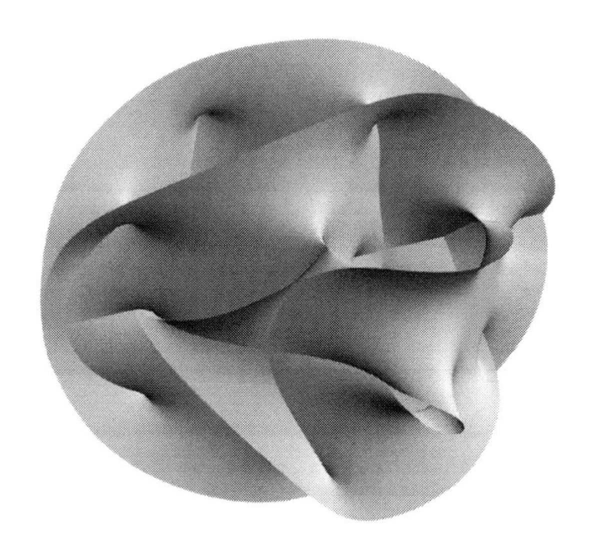

Projection of a Calabi–Yau manifold, one of the ways of compactifying the extra dimensions posited by string theory

which take into account higher-energy contributions,[34] but gravity turns out to be nonrenormalizable: at high energies, applying the recipes of ordinary quantum field theory yields models that are devoid of all predictive power.[35]

One attempt to overcome these limitations is to replace ordinary quantum field theory, which is based on the classical concept of a point particle, with a quantum theory of one-dimensional extended objects: string theory.[36] At the energies reached in current experiments, these strings are indistinguishable from point-like particles, but, crucially, different modes of oscillation of one and the same type of fundamental string appear as particles with different (electric and other) charges. In this way, string theory promises to be a unified description of all particles and interactions.[37] The theory is successful in that one mode will always correspond to a graviton, the messenger particle of gravity; however, the price of this success are unusual features such as six extra dimensions of space in addition to the usual three for space and one for time.[38]

In what is called the second superstring revolution, it was conjectured that both string theory and a unification of general relativity and supersymmetry known as supergravity[39] form part of a hypothesized eleven-dimensional model known as M-theory, which would constitute a uniquely defined and consistent theory of quantum gravity.[40][41] As presently understood, however, string theory admits a very large number (10^{500} by some estimates) of consistent vacua, comprising the so-called "string landscape". Sorting through this large family of solutions remains a major challenge.

5.3.2 Loop quantum gravity

Main article: Loop quantum gravity

Loop quantum gravity seriously considers general relativ-

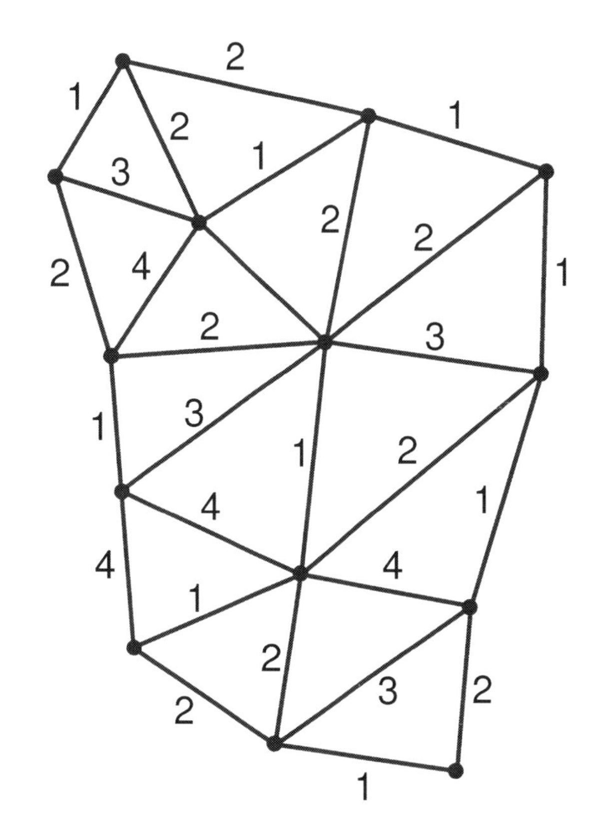

Simple spin network of the type used in loop quantum gravity

ity's insight that spacetime is a dynamical field and is therefore a quantum object. Its second idea is that the quantum discreteness that determines the particle-like behavior of other field theories (for instance, the photons of the electromagnetic field) also affects the structure of space.

The main result of loop quantum gravity is the derivation of a granular structure of space at the Planck length. This is derived from following considerations: In the case of electromagnetism, the quantum operator representing the energy of each frequency of the field has a discrete spectrum. Thus the energy of each frequency is quantized, and the quanta are the photons. In the case of gravity, the operators representing the area and the volume of each surface or space region likewise have discrete spectrum. Thus area and volume of any portion of space are also quantized, where the quanta are elementary quanta of space. It follows, then, that spacetime has an elementary quantum granular structure at the Planck scale, which cuts off the ultraviolet infinities of quantum field theory.

The quantum state of spacetime is described in the theory

by means of a mathematical structure called spin networks. Spin networks were initially introduced by Roger Penrose in abstract form, and later shown by Carlo Rovelli and Lee Smolin to derive naturally from a non-perturbative quantization of general relativity. Spin networks do not represent quantum states of a field in spacetime: they represent directly quantum states of spacetime.

The theory is based on the reformulation of general relativity known as Ashtekar variables, which represent geometric gravity using mathematical analogues of electric and magnetic fields.[42][43] In the quantum theory, space is represented by a network structure called a spin network, evolving over time in discrete steps.[44][45][46][47]

The dynamics of the theory is today constructed in several versions. One version starts with the canonical quantization of general relativity. The analogue of the Schrödinger equation is a Wheeler–DeWitt equation, which can be defined within the theory.[48] In the covariant, or spinfoam formulation of the theory, the quantum dynamics is obtained via a sum over discrete versions of spacetime, called spinfoams. These represent histories of spin networks.

5.3.3 Other approaches

There are a number of other approaches to quantum gravity. The approaches differ depending on which features of general relativity and quantum theory are accepted unchanged, and which features are modified.[49][50] Examples include:

- Asymptotic safety in quantum gravity
- Euclidean quantum gravity
- Causal dynamical triangulation[51]
- Causal fermion systems,[6][7][8][9][10][11] giving quantum mechanics, general relativity and quantum field theory as limiting cases.
- Causal sets[52]
- Covariant Feynman path integral approach
- Group field theory[53]
- Wheeler–DeWitt equation
- Geometrodynamics
- Hořava–Lifshitz gravity
- MacDowell–Mansouri action
- Path-integral based models of quantum cosmology[54]
- Regge calculus
- Shape Dynamics
- String-nets giving rise to gapless helicity ±2 excitations with no other gapless excitations[55]
- Superfluid vacuum theory a.k.a. theory of BEC vacuum
- Supergravity
- Twistor theory[56]
- Canonical quantum gravity
- E8 Theory
- Quantum holonomy theory[57]

5.4 Weinberg–Witten theorem

In quantum field theory, the Weinberg–Witten theorem places some constraints on theories of composite gravity/emergent gravity. However, recent developments attempt to show that if locality is only approximate and the holographic principle is correct, the Weinberg–Witten theorem would not be valid.

5.5 Experimental tests

As was emphasized above, quantum gravitational effects are extremely weak and therefore difficult to test. For this reason, the possibility of experimentally testing quantum gravity had not received much attention prior to the late 1990s. However, in the past decade, physicists have realized that evidence for quantum gravitational effects can guide the development of the theory. Since theoretical development has been slow, the field of phenomenological quantum gravity, which studies the possibility of experimental tests, has obtained increased attention.[58][59]

The most widely pursued possibilities for quantum gravity phenomenology include violations of Lorentz invariance, imprints of quantum gravitational effects in the cosmic microwave background (in particular its polarization), and decoherence induced by fluctuations in the space-time foam.

The BICEP2 experiment detected what was initially thought to be primordial B-mode polarization caused by gravitational waves in the early universe. If truly primordial, these waves were born as quantum fluctuations in gravity itself. Cosmologist Ken Olum (Tufts University) stated: "I think this is the only observational evidence that we have that actually shows that gravity is quantized....It's probably the only evidence of this that we will ever have."[60]

5.6 See also

5.7 References

[1] Rovelli, Carlo. "Quantum gravity - Scholarpedia". *www.scholarpedia.org*. Retrieved 2016-01-09.

[2] Griffiths, David J. (2004). *Introduction to Quantum Mechanics*. Pearson Prentice Hall. OCLC 803860989.

[3] Wald, Robert M. (1984). *General Relativity*. University of Chicago Press. p. 382. OCLC 471881415.

[4] Zee, Anthony (2010). *Quantum Field Theory in a Nutshell* (2nd ed.). Princeton University Press. p. 172. OCLC 659549695.

[5] Penrose, Roger (2007). *The road to reality : a complete guide to the laws of the universe*. Vintage. p. 1017. OCLC 716437154.

[6] F. Finster, J. Kleiner, Causal fermion systems as a candidate for a unified physical theory, arXiv:1502.03587 [math-ph] (2015)

[7] F. Finster, The Principle of the Fermionic Projector, hep-th/0001048, hep-th/0202059, hep- th/0210121, AMS/IP Studies in Advanced Mathematics, vol. **35**, American Mathematical Society, Providence, RI, 2006.

[8] F. Finster, A formulation of quantum field theory realizing a sea of interacting Dirac particles, arXiv:0911.2102 [hep-th], Lett. Math. Phys. **97** (2011), no. 2, 165–183.

[9] F. Finster, An action principle for an interacting fermion system and its analysis in the continuum limit, arXiv:0908.1542 [math-ph] (2009).

[10] F. Finster, The continuum limit of a fermion system involving neutrinos: Weak and gravitational interactions, arXiv: 1211.3351 [math-ph] (2012).

[11] F. Finster, Perturbative quantum field theory in the framework of the fermionic projector, arXiv:1310.4121 [math-ph], J. Math. Phys. **55** (2014), no. 4, 042301.

[12] Quantum effects in the early universe might have an observable effect on the structure of the present universe, for example, or gravity might play a role in the unification of the other forces. Cf. the text by Wald cited above.

[13] Donoghue (1995). "Introduction to the Effective Field Theory Description of Gravity". arXiv:gr-qc/9512024. (verify against ISBN 9789810229085)

[14] Kraichnan, R. H. (1955). "Special-Relativistic Derivation of Generally Covariant Gravitation Theory". *Physical Review* **98** (4): 1118–1122. Bibcode:1955PhRv...98.1118K. doi:10.1103/PhysRev.98.1118.

[15] Gupta, S. N. (1954). "Gravitation and Electromagnetism". *Physical Review* **96** (6): 1683–1685. Bibcode:1954PhRv...96.1683G. doi:10.1103/PhysRev.96.1683.

[16] Gupta, S. N. (1957). "Einstein's and Other Theories of Gravitation". *Reviews of Modern Physics* **29** (3): 334–336. Bibcode:1957RvMP...29..334G. doi:10.1103/RevModPhys.29.334.

[17] Gupta, S. N. (1962). "Quantum Theory of Gravitation". *Recent Developments in General Relativity*. Pergamon Press. pp. 251–258.

[18] Deser, S. (1970). "Self-Interaction and Gauge Invariance". *General Relativity and Gravitation* **1**: 9–18. arXiv:gr-qc/0411023. Bibcode:1970GReGr...1....9D. doi:10.1007/BF00759198.

[19] Ohta, Tadayuki; Mann, Robert (1996). "Canonical reduction of two-dimensional gravity for particle dynamics". *Classical and Quantum Gravity* **13** (9): 2585–2602. arXiv:gr-qc/9605004. Bibcode:1996CQGra..13.2585O. doi:10.1088/0264-9381/13/9/022.

[20] Sikkema, A E; Mann, R B (1991). "Gravitation and cosmology in (1+1) dimensions". *Classical and Quantum Gravity* **8**: 219–235. Bibcode:1991CQGra...8..219S. doi:10.1088/0264-9381/8/1/022.

[21] Farrugia; Mann; Scott (2007). "N-body Gravity and the Schroedinger Equation". *Classical and Quantum Gravity* **24** (18): 4647–4659. arXiv:gr-qc/0611144. Bibcode:2007CQGra..24.4647F. doi:10.1088/0264-9381/24/18/006.

[22] Scott, T.C.; Zhang, Xiangdong; Mann, Robert; Fee, G.J. (2016). "Canonical reduction for dilatonic gravity in 3 + 1 dimensions". *Physical Review D* **93** (8): 084017. arXiv:1605.03431. Bibcode:2016PhRvD..93h4017S. doi:10.1103/PhysRevD.93.084017.

[23] Mann, R B; Ohta, T (1997). "Exact solution for the metric and the motion of two bodies in (1+1)-dimensional gravity". *Phys. Rev. D* **55** (8): 4723–4747. arXiv:gr-qc/9611008. Bibcode:1997PhRvD..55.4723M. doi:10.1103/PhysRevD.55.4723.

[24] Bellazzini, B.; Csaki, C.; Hubisz, J.; Serra, J.; Terning, J. (2013). "A higgs-like dilaton". *Eur. Phys. J. C.* **73** (2): 2333.

[25] Feynman, R. P.; Morinigo, F. B.; Wagner, W. G.; Hatfield, B. (1995). *Feynman lectures on gravitation*. Addison-Wesley. ISBN 0-201-62734-5.

[26] Hamber, H. W. (2009). *Quantum Gravitation - The Feynman Path Integral Approach*. Springer Publishing. ISBN 978-3-540-85292-6.

[27] Smolin, Lee (2001). *Three Roads to Quantum Gravity*. Basic Books. pp. 20–25. ISBN 0-465-07835-4. Pages 220–226 are annotated references and guide for further reading.

[28] Sakurai, J. J.; Napolitano, Jim J. (2010-07-14). *Modern Quantum Mechanics* (2 ed.). Pearson. p. 68. ASIN 0805382917. ISBN 9780805382914.

[29] Novello, Mario; Bergliaffa, Santiago E. (2003-06-11). *Cosmology and Gravitation: Xth Brazilian School of Cosmology and Gravitation; 25th Anniversary (1977-2002), Mangaratiba, Rio de Janeiro, Brazil,*. Springer Science & Business Media. p. 95. ISBN 9780735401310.

[30] A timeline and overview can be found in Rovelli, Carlo (2000). "Notes for a brief history of quantum gravity". arXiv:gr-qc/0006061. (verify against ISBN 9789812777386)

[31] Ashtekar, Abhay (2007). "Loop Quantum Gravity: Four Recent Advances and a Dozen Frequently Asked Questions". *11th Marcel Grossmann Meeting on Recent Developments in Theoretical and Experimental General Relativity*. p. 126. arXiv:0705.2222. Bibcode:2008mgm..conf..126A. doi:10.1142/9789812834300_0008.

[32] Schwarz, John H. (2007). "String Theory: Progress and Problems". *Progress of Theoretical Physics Supplement* **170**: 214–226. arXiv:hep-th/0702219. Bibcode:2007PThPS.170..214S. doi:10.1143/PTPS.170.214.

[33] Donoghue, John F. (editor) (1995). "Introduction to the Effective Field Theory Description of Gravity". In Cornet, Fernando. *Effective Theories: Proceedings of the Advanced School, Almunecar, Spain, 26 June–1 July 1995*. Singapore: World Scientific. arXiv:gr-qc/9512024. ISBN 981-02-2908-9.

[34] Weinberg, Steven (1996). "Chapters 17–18". *The Quantum Theory of Fields II: Modern Applications*. Cambridge University Press. ISBN 0-521-55002-5.

[35] Goroff, Marc H.; Sagnotti, Augusto; Sagnotti, Augusto (1985). "Quantum gravity at two loops". *Physics Letters B* **160**: 81–86. Bibcode:1985PhLB..160...81G. doi:10.1016/0370-2693(85)91470-4.

[36] An accessible introduction at the undergraduate level can be found in Zwiebach, Barton (2004). *A First Course in String Theory*. Cambridge University Press. ISBN 0-521-83143-1., and more complete overviews in Polchinski, Joseph (1998). *String Theory Vol. I: An Introduction to the Bosonic String*. Cambridge University Press. ISBN 0-521-63303-6. and Polchinski, Joseph (1998b). *String Theory Vol. II: Superstring Theory and Beyond*. Cambridge University Press. ISBN 0-521-63304-4.

[37] Ibanez, L. E. (2000). "The second string (phenomenology) revolution". *Classical & Quantum Gravity* **17** (5): 1117–1128. arXiv:hep-ph/9911499. Bibcode:2000CQGra..17.1117I. doi:10.1088/0264-9381/17/5/321.

[38] For the graviton as part of the string spectrum, e.g. Green, Schwarz & Witten 1987, sec. 2.3 and 5.3; for the extra dimensions, ibid sec. 4.2.

[39] Weinberg, Steven (2000). "Chapter 31". *The Quantum Theory of Fields II: Modern Applications*. Cambridge University Press. ISBN 0-521-55002-5.

[40] Townsend, Paul K. (1996). *Four Lectures on M-Theory*. ICTP Series in Theoretical Physics. p. 385. arXiv:hep-th/9612121. Bibcode:1997hepcbconf..385T.

[41] Duff, Michael (1996). "M-Theory (the Theory Formerly Known as Strings)". *International Journal of Modern Physics A* **11** (32): 5623–5642. arXiv:hep-th/9608117. Bibcode:1996IJMPA..11.5623D. doi:10.1142/S0217751X96002583.

[42] Ashtekar, Abhay (1986). "New variables for classical and quantum gravity". *Physical Review Letters* **57** (18): 2244–2247. Bibcode:1986PhRvL..57.2244A. doi:10.1103/PhysRevLett.57.2244. PMID 10033673.

[43] Ashtekar, Abhay (1987). "New Hamiltonian formulation of general relativity". *Physical Review D* **36** (6): 1587–1602. Bibcode:1987PhRvD..36.1587A. doi:10.1103/PhysRevD.36.1587.

[44] Thiemann, Thomas (2006). "Loop Quantum Gravity: An Inside View". *Approaches to Fundamental Physics*. Lecture Notes in Physics **721**: 185. arXiv:hep-th/0608210. Bibcode:2007LNP...721..185T. doi:10.1007/978-3-540-71117-9_10. ISBN 978-3-540-71115-5.

[45] Rovelli, Carlo (1998). "Loop Quantum Gravity". *Living Reviews in Relativity* **1**. Retrieved 2008-03-13.

[46] Ashtekar, Abhay; Lewandowski, Jerzy (2004). "Background Independent Quantum Gravity: A Status Report". *Classical & Quantum Gravity* **21** (15): R53–R152. arXiv:gr-qc/0404018. Bibcode:2004CQGra..21R..53A. doi:10.1088/0264-9381/21/15/R01.

[47] Thiemann, Thomas (2003). "Lectures on Loop Quantum Gravity". *Lecture Notes in Physics*. Lecture Notes in Physics **631**: 41–135. arXiv:gr-qc/0210094. Bibcode:2003LNP...631...41T. doi:10.1007/978-3-540-45230-0_3. ISBN 978-3-540-40810-9.

[48] Rovelli, Carlo (2004). *Quantum Gravity*. Cambridge University Press. ISBN 0521715962.

[49] Isham, Christopher J. (1994). "Prima facie questions in quantum gravity". In Ehlers, Jürgen; Friedrich, Helmut. *Canonical Gravity: From Classical to Quantum*. Springer. arXiv:gr-qc/9310031. ISBN 3-540-58339-4.

[50] Sorkin, Rafael D. (1997). "Forks in the Road, on the Way to Quantum Gravity". *International Journal of Theoretical Physics* **36** (12): 2759–2781. arXiv:gr-qc/9706002. Bibcode:1997IJTP...36.2759S. doi:10.1007/BF02435709.

[51] Loll, Renate (1998). "Discrete Approaches to Quantum Gravity in Four Dimensions". *Living Reviews in Relativity* **1**: 13. arXiv:gr-qc/9805049. Bibcode:1998LRR.....1...13L. doi:10.12942/lrr-1998-13. Retrieved 2008-03-09.

[52] Sorkin, Rafael D. (2005). "Causal Sets: Discrete Gravity". In Gomberoff, Andres; Marolf, Donald. *Lectures on Quantum Gravity*. Springer. arXiv:gr-qc/0309009. ISBN 0-387-23995-2.

[53] See Daniele Oriti and references therein.

[54] Hawking, Stephen W. (1987). "Quantum cosmology". In Hawking, Stephen W.; Israel, Werner. *300 Years of Gravitation*. Cambridge University Press. pp. 631–651. ISBN 0-521-37976-8.

[55] Wen 2006

[56] See ch. 33 in Penrose 2004 and references therein.

[57] Aastrup, J. and Grimstrup, J. M. (27 Apr 2015). "Quantum Holonomy Theory" (PDF). arXiv:1504.07100.

[58] Hossenfelder, Sabine (2011). "Experimental Search for Quantum Gravity". In V. R. Frignanni. *Classical and Quantum Gravity: Theory, Analysis and Applications*. Chapter 5: Nova Publishers. ISBN 978-1-61122-957-8.

[59] Hossenfelder, Sabine (2010-10-17). V. R. Frignanni, ed. "Experimental Search for Quantum Gravity Chapter 5". *Classical and Quantum Gravity: Theory, Analysis and Applications* (Nova Publishers) **5** (2011): arXiv:1010.3420. arXiv:1010.3420. Bibcode:2010arXiv1010.3420H.

[60] Camille Carlisle. "First Direct Evidence of Big Bang Inflation". SkyandTelescope.com. Retrieved March 18, 2014.

5.8 Further reading

- Ahluwalia, D. V. (2002). "Interface of Gravitational and Quantum Realms". *Modern Physics Letters A* **17** (15–17): 1135. arXiv:gr-qc/0205121. Bibcode:2002MPLA...17.1135A. doi:10.1142/S021773230200765X.
- Ashtekar, Abhay (2005). "The winding road to quantum gravity" (PDF). *Current Science* **89**: 2064–2074.
- Carlip, Steven (2001). "Quantum Gravity: a Progress Report". *Reports on Progress in Physics* **64** (8): 885–942. arXiv:gr-qc/0108040. Bibcode:2001RPPh...64..885C. doi:10.1088/0034-4885/64/8/301.
- Herbert W. Hamber (2009). *Quantum Gravitation*. Springer Publishing. doi:10.1007/978-3-540-85293-3. ISBN 978-3-540-85292-6.
- Kiefer, Claus (2007). *Quantum Gravity*. Oxford University Press. ISBN 0-19-921252-X.
- Kiefer, Claus (2005). "Quantum Gravity: General Introduction and Recent Developments". *Annalen der Physik* **15**: 129–148. arXiv:gr-qc/0508120. Bibcode:2006AnP...518..129K. doi:10.1002/andp.200510175.
- Lämmerzahl, Claus, ed. (2003). *Quantum Gravity: From Theory to Experimental Search*. Lecture Notes in Physics. Springer. ISBN 3-540-40810-X.
- Rovelli, Carlo (2004). *Quantum Gravity*. Cambridge University Press. ISBN 0-521-83733-2.
- Quantum gravity Carlo Rovelli, Scholarpedia, 3(5):7117. doi:10.4249/scholarpedia.7117
- Trifonov, Vladimir (2008). "GR-friendly description of quantum systems". *International Journal of Theoretical Physics* **47** (2): 492–510. arXiv:math-ph/0702095. Bibcode:2008IJTP...47..492T. doi:10.1007/s10773-007-9474-3.

Chapter 6

Spin network

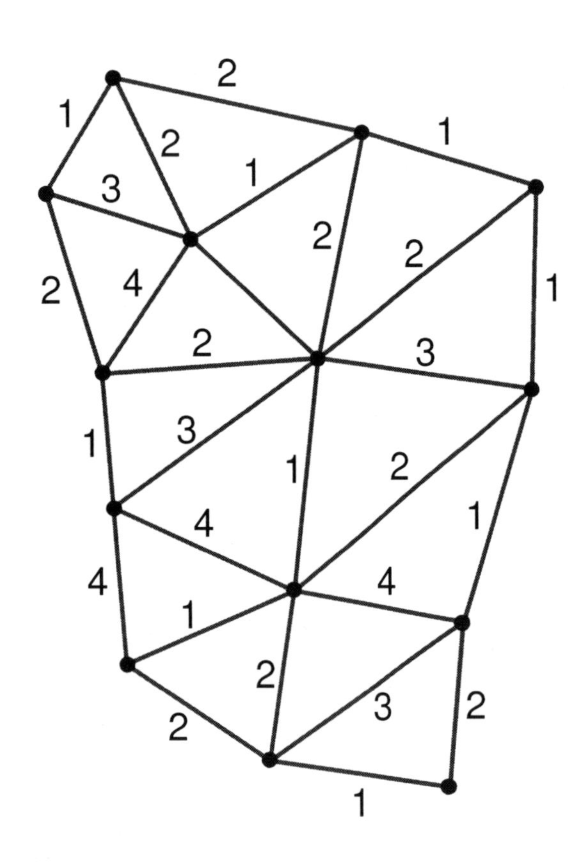

Simple spin network of the type used in loop quantum gravity

In physics, a **spin network** is a type of diagram which can be used to represent states and interactions between particles and fields in quantum mechanics. From a mathematical perspective, the diagrams are a concise way to represent multilinear functions and functions between representations of matrix groups. The diagrammatic notation often simplifies calculation because simple diagrams may be used to represent complicated functions. Roger Penrose is credited with the invention of spin networks in 1971, although similar diagrammatic techniques existed before that time.

Spin networks have been applied to the theory of quantum gravity by Carlo Rovelli, Lee Smolin, Jorge Pullin, Rodolfo Gambini and others. They can also be used to construct a particular functional on the space of connections which is invariant under local gauge transformations.

6.1 Definition

6.1.1 Penrose's original definition

A spin network, as described in Penrose 1971, is a kind of diagram in which each line segment represents the world line of a "unit" (either an elementary particle or a compound system of particles). Three line segments join at each vertex. A vertex may be interpreted as an event in which either a single unit splits into two or two units collide and join into a single unit. Diagrams whose line segments are all joined at vertices are called *closed spin networks*. Time may be viewed as going in one direction, such as from the bottom to the top of the diagram, but for closed spin networks the direction of time is irrelevant to calculations.

Each line segment is labeled with an integer called a spin number. A unit with spin number n is called an n-unit and has angular momentum $n\hbar/2$, where $\hbar$ is the reduced Planck constant. For bosons, such as photons and gluons, n is an even number. For fermions, such as electrons and quarks, n is odd.

Given any closed spin network, a non-negative integer can be calculated which is called the *norm* of the spin network. Norms can be used to calculate the probabilities of various spin values. A network whose norm is zero has zero probability of occurrence. The rules for calculating norms and probabilities are beyond the scope of this article. However, they imply that for a spin network to have nonzero norm, two requirements must be met at each vertex. Suppose a vertex joins three units with spin numbers a, b, and c. Then, these requirements are stated as:

- Triangle inequality: a must be less than or equal to b +

c, b less than or equal to $a + c$, and c less than or equal to $a + b$.

- Fermion conservation: $a + b + c$ must be an even number.

For example, $a = 3$, $b = 4$, $c = 6$ is impossible since $3 + 4 + 6 = 13$ is odd, and $a = 3$, $b = 4$, $c = 9$ is impossible since $3 + 4 < 9$. However, $a = 3$, $b = 4$, $c = 5$ is possible since $3 + 4 + 5 = 12$ is even and the triangle inequality is satisfied. Some conventions use labellings by half-integers, with the condition that the sum $a + b + c$ must be a whole number.

6.1.2 Formal definition

More formally, a **spin network** is a (directed) graph whose edges are associated with irreducible representations of a compact Lie group and whose vertices are associated with intertwiners of the edge representations adjacent to it.

A spin network, immersed into a manifold, can be used to define a functional on the space of connections on this manifold. One computes holonomies of the connection along every link (closed path) of the graph, determines representation matrices corresponding to every link, multiplies all matrices and intertwiners together, and contracts indices in a prescribed way. A remarkable feature of the resulting functional is that it is invariant under local gauge transformations.

6.2 Usage in physics

6.2.1 In the context of loop quantum gravity

In loop quantum gravity (LQG), a spin network represents a "quantum state" of the gravitational field on a 3-dimensional hypersurface. The set of all possible spin networks (or, more accurately, "s-knots" - that is, equivalence classes of spin networks under diffeomorphisms) is countable; it constitutes a basis of LQG Hilbert space.

One of the key results of loop quantum gravity is quantization of areas: the operator of the area A of a two-dimensional surface Σ should have a discrete spectrum. Every **spin network** is an eigenstate of each such operator, and the area eigenvalue equals

$$A_\Sigma = 8\pi\ell_{\text{PL}}^2\gamma\sum_i\sqrt{j_i(j_i+1)}$$

where the sum goes over all intersections i of Σ with the spin network. In this formula,

- ℓPL is the Planck Length,
- γ is the Immirzi parameter and
- ji = 0, 1/2, 1, 3/2, ... is the spin associated with the link i of the spin network. The two-dimensional area is therefore "concentrated" in the intersections with the spin network.

According to this formula, the lowest possible non-zero eigenvalue of the area operator corresponds to a link that carries spin 1/2 representation. Assuming an Immirzi parameter on the order of 1, this gives the smallest possible measurable area of ~10^{-66} cm^2.

The formula for area eigenvalues becomes somewhat more complicated if the surface is allowed to pass through the nodes (it is not yet clear if these situations are physically meaningful.)

Similar quantization applies to the volume operator. The volume of 3D submanifold that contains part of spin network is given by a sum of contributions from each node inside it. One can think that every node in a spin network is an elementary "quantum of volume" and every link is a "quantum of area" surrounding this volume.

6.2.2 More general gauge theories

Similar constructions can be made for general gauge theories with a compact Lie group G and a connection form. This is actually an exact duality over a lattice. Over a manifold however, assumptions like diffeomorphism invariance are needed to make the duality exact (smearing Wilson loops is tricky). Later, it was generalized by Robert Oeckl to representations of quantum groups in 2 and 3 dimensions using the Tannaka–Krein duality.

Michael A. Levin and Xiao-Gang Wen have also defined string-nets using tensor categories that are objects very similar to spin networks. However the exact connection with spin networks is not clear yet. String-net condensation produces topologically ordered states in condensed matter.

6.3 Usage in mathematics

In mathematics, spin networks have been used to study skein modules and character varieties, which correspond to spaces of connections.

6.4 See also

- Character variety

- Penrose graphical notation
- Spin foam
- String-net
- Trace diagram

6.5 References

Early papers:

- Sum of Wigner coefficients and their graphical representation, I. B. Levinson, "Proceed. Phys-Tech Inst. Acad Sci. Lithuanian SSR *2, 17-30 (1956)*
- Applications of negative dimensional tensors, Roger Penrose, in *Combinatorial Mathematics and its Applications*, Academic Press (1971)
- Kogut, John; Susskind, Leonard (1975). "Hamiltonian formulation of Wilson's lattice gauge theories". *Physical Review D* **11** (2): 395–408. Bibcode:1975PhRvD..11..395K. doi:10.1103/PhysRevD.11.395.
- Kogut, John B. (1983). "The lattice gauge theory approach to quantum chromodynamics". *Reviews of Modern Physics* **55** (3): 775–836. Bibcode:1983RvMP...55..775K. doi:10.1103/RevModPhys.55.775. (see the Euclidean high temperature (strong coupling) section)
- Savit, Robert (1980). "Duality in field theory and statistical systems". *Reviews of Modern Physics* **52** (2): 453–487. Bibcode:1980RvMP...52..453S. doi:10.1103/RevModPhys.52.453. (see the sections on Abelian gauge theories)

Modern papers:

- Rovelli, Carlo; Smolin, Lee (1995). "Spin networks and quantum gravity". *Phys. Rev. D* **52** (10): 5743–5759. arXiv:gr-qc/9505006. Bibcode:1995PhRvD..52.5743R. doi:10.1103/PhysRevD.52.5743.
- Pfeiffer, Hendryk; Oeckl, Robert (2002). "The dual of non-Abelian Lattice Gauge Theory". *Nuclear Physics B - Proceedings Supplements*. 106-107: 1010–1012. arXiv:hep-lat/0110034. Bibcode:2002NuPhS.106.1010P. doi:10.1016/S0920-5632(01)01913-2.
- Pfeiffer, Hendryk (2003). "Exact duality transformations for sigma models and gauge theories". *Journal of Mathematical Physics* **44** (7): 2891. arXiv:hep-lat/0205013. Bibcode:2003JMP....44.2891P. doi:10.1063/1.1580071.
- Oeckl, Robert (2003). "Generalized lattice gauge theory, spin foams and state sum invariants". *Journal of Geometry and Physics* **46** (3–4): 308–354. arXiv:hep-th/0110259. Bibcode:2003JGP....46..308O. doi:10.1016/S0393-0440(02)00148-1.
- Baez, John C. (1996). "Spin Networks in Gauge Theory". *Advances in Mathematics* **117** (2): 253–272. doi:10.1006/aima.1996.0012.
- Quantum Field Theory of Many-body Systems – from the Origin of Sound to an Origin of Light and Fermions, Xiao-Gang Wen, . (Dubbed *string-nets* here.)
- Major, Seth A. (1999). "A spin network primer". *American Journal of Physics* **67** (11): 972. arXiv:gr-qc/9905020. Bibcode:1999AmJPh..67..972M. doi:10.1119/1.19175.
- Pre-geometry and Spin Networks. An introduction. .

Books:

- Diagram Techniques in Group Theory, G. E. Stedman, Cambridge University Press, 1990
- Group Theory: Birdtracks, Lie's, and Exceptional Groups, Predrag Cvitanović, Princeton University Press, 2008, http://birdtracks.eu/

Chapter 7

Spin foam

In physics, a **spinfoam** or **spin foam** is a topological structure made out of two-dimensional faces that represents one of the configurations that must be summed by functional integration to obtain a Feynman's path integral description of quantum gravity. It is closely related to loop quantum gravity.

7.1 Spin foam in loop quantum gravity

Main article: Loop quantum gravity

Loop quantum gravity has a covariant formulation that, at present, provides the best formulation of the dynamics of the theory of quantum gravity. This is a quantum field theory where the invariance under diffeomorphisms of general relativity is implemented. The resulting path integral represents a sum over all the possible configuration of the geometry, coded in the spinfoam.

7.1.1 Spin network

A spin network is a one-dimensional graph, together with labels on its vertices and edges which encodes aspects of a spatial geometry.

A spin network is defined as a diagram (like the Feynman diagram) that makes a basis of connections between the elements of a differentiable manifold for the Hilbert spaces defined over them. Spin networks provide a representation for computations of amplitudes between two different hypersurfaces of the manifold. Any evolution of spin network provides a spin foam over a manifold of one dimension higher than the dimensions of the corresponding spin network. A spin foam may be viewed as a quantum history.

7.1.2 Spacetime

Spin networks provide a language to describe quantum geometry of space. Spin foam does the same job on spacetime.

Spacetime can be defined as as a superposition of spin foams, which is a generalized Feynman diagram where instead of a graph, a higher-dimensional complex is used. In topology this sort of space is called a 2-complex. A spin foam is a particular type of 2-complex, with labels for vertices, edges and faces. The boundary of a spin foam is a spin network, just as in the theory of manifolds, where the boundary of an n-manifold is an (n-1)-manifold.

In Loop Quantum Gravity, the present Spinfoam Theory has been inspired by the work of Ponzano-Regge model. The concept of a spin foam, although not called that at the time, was introduced in the paper "A Step Toward Pregeometry I: Ponzano-Regge Spin Networks and the Origin of Spacetime Structure in Four Dimensions" by Norman J. LaFave (gr-qc/9310036) (1993). In this paper, the concept of creating sandwiches of 4-geometry (and local time scale) from spin networks is described, along with the connection of these spin 4-geometry sandwiches to form paths of spin networks connecting given spin network boundaries (spin foams). Quantization of the structure leads to a generalized Feynman path integral over connected paths of spin networks between spin network boundaries. This paper goes beyond much of the later work by showing how 4-geometry is already present in the seemingly three dimensional spin networks, how local time scales occur, and how the field equations and conservation laws are generated by simple consistency requirements. The idea was reintroduced in [1] and later developed into the Barrett–Crane model. The formulation that is used nowadays is commonly called EPRL after the names of the authors of a series of seminal papers,[2] but the theory has also seen fundamental contributions from the work of many others, such as Laurent Freidel (FK model) and Jerzy Lewandowski (KKL model).

7.2 Definition

The partition function for a **spin foam model** is, in general,

$$Z := \sum_\Gamma w(\Gamma) \left[\sum_{j_f, i_e} \prod_f A_f(j_f) \prod_e A_e(j_f, i_e) \prod_v A_v(j_f, i_e) \right]$$

with:

- a set of 2-complexes Γ each consisting out of faces f , edges e and vertices v . Associated to each 2-complex Γ is a weight $w(\Gamma)$
- a set of irreducible representations j which label the faces and intertwiners i which label the edges.
- a vertex amplitude $A_v(j_f, i_e)$ and an edge amplitude $A_e(j_f, i_e)$
- a face amplitude $A_f(j_f)$, for which we almost always have $A_f(j_f) = \dim(j_f)$

7.3 See also

- Invariance mechanics
- Group field theory
- Loop quantum gravity
- Lorentz invariance in loop quantum gravity
- Spinfoam cosmology
- String-net

7.4 References

[1] Reisenberger, Michael P.; Rovelli, Carlo (1997). ""Sum over surfaces" form of loop quantum gravity". *Physical Review D* **56** (6): 3490. arXiv:gr-qc/9612035. Bibcode:1997PhRvD..56.3490R. doi:10.1103/PhysRevD.56.3490.

[2] Engle, Jonathan; Livine, Etera; Pereira, Roberto; Rovelli, Carlo (2008). "LQG vertex with finite Immirzi parameter". *Nuclear Physics B* **799**: 136. arXiv:0711.0146. Bibcode:2008NuPhB.799..136E. doi:10.1016/j.nuclphysb.2008.02.018.

7.5 External links

- Spin foam on arxiv.org
- John C. Baez: Spin foam models. (1997)
- Alejandro Perez: Spin Foam Models for Quantum Gravity (2003)
- Carlo Rovelli: Zakopane lectures on loop gravity (2011)

Chapter 8

Planck length

In physics, the **Planck length**, denoted ℓP, is a unit of length, equal to $1.616199(97) \times 10^{-35}$ metres. It is a base unit in the system of Planck units, developed by physicist Max Planck. The Planck length can be defined from three fundamental physical constants: the speed of light in a vacuum, the Planck constant, and the gravitational constant.

8.1 Value

The Planck length ℓP is defined as

$$\ell_{\mathrm{P}} = \sqrt{\frac{\hbar G}{c^3}} \approx 1.616\ 199(97) \times 10^{-35}\ \mathrm{m}$$

where c is the speed of light in a vacuum, G is the gravitational constant, and ħ is the reduced Planck constant. The two digits enclosed by parentheses are the estimated standard error associated with the reported numerical value.[1][2]

The Planck length is about 10^{-20} times the diameter of a proton.

8.2 Theoretical significance

There is currently no proven physical significance of the Planck length; it is, however, a topic of theoretical research. Since the Planck length is so many orders of magnitude smaller than any current instrument could possibly measure, there is no way of examining it directly. According to the generalized uncertainty principle (a concept from speculative models of quantum gravity), the Planck length is, in principle, within a factor of 10, the shortest measurable length – and no theoretically known improvement in measurement instruments could change that.

In some forms of quantum gravity, the Planck length is the length scale at which the structure of spacetime becomes dominated by quantum effects, and it is impossible to determine the difference between two locations less than one Planck length apart. The precise effects of quantum gravity are unknown; it is often guessed that spacetime might have a discrete or foamy structure at a Planck length scale.

The Planck area, equal to the square of the Planck length, plays a role in black hole entropy. The value of this entropy, in units of the Boltzmann constant, is known to be given by $\frac{A}{4\ell_{\mathrm{P}}^2}$, where A is the area of the event horizon. The Planck area is the area by which the surface of a spherical black hole increases when the black hole swallows one bit of information, as was proven by Jacob Bekenstein.[3]

If large extra dimensions exist, the measured strength of gravity may be much smaller than its true (small-scale) value. In this case the Planck length would have no fundamental physical significance, and quantum gravitational effects would appear at other scales.

In string theory, the Planck length is the order of magnitude of the oscillating strings that form elementary particles, and shorter lengths do not make physical sense.[4] The string scale l_s is related to the Planck scale by $\ell P = g_s^{1/4} l_s$, where g_s is the string coupling constant. Contrary to what the name suggests, the string coupling constant is not constant, but depends on the value of a scalar field known as the dilaton.

In loop quantum gravity, area is quantized, and the Planck area is, within a factor of 10, the smallest possible area value.

In doubly special relativity, the Planck length is observer-invariant.

The search for the laws of physics valid at the Planck length is a part of the search for the theory of everything.

8.3 Visualization

The size of the Planck length can be visualized as follows: if a particle or dot about 0.1 mm in size (which is approximately the smallest the unaided human eye can see) were magnified in size to be as large as the observable universe, then inside that universe-sized "dot", the Planck length would be roughly the size of an actual 0.1 mm dot. In other words, a 0.1 mm dot is halfway between the Planck length and the size of the observable universe on a logarithmic scale.

8.4 See also

- Fock–Lorentz symmetry
- Orders of magnitude (length)
- Planck energy
- Planck mass
- Planck epoch
- Planck scale
- Planck temperature
- Planck time

8.5 Notes and references

[1] John Baez, The Planck Length

[2] NIST, "Planck length", NIST's published CODATA constants

[3] "Phys. Rev. D 7, 2333 (1973): Black Holes and Entropy". Prd.aps.org. Retrieved 2013-10-21.

[4] Cliff Burgess; Fernando Quevedo (November 2007). "The Great Cosmic Roller-Coaster Ride". *Scientific American* (print) (Scientific American, Inc.). p. 55.

8.6 Bibliography

- Garay, Luis J. (January 1995). "Quantum gravity and minimum length". *International Journal of Modern Physics A* **10** (2): 145 ff. arXiv:gr-qc/9403008v2. Bibcode:1995IJMPA..10..145G. doi:10.1142/S0217751X95000085.

8.7 External links

- Bowley, Roger; Eaves, Laurence (2010). "Planck Length". *Sixty Symbols*. Brady Haran for the University of Nottingham.

Chapter 9

Big Bounce

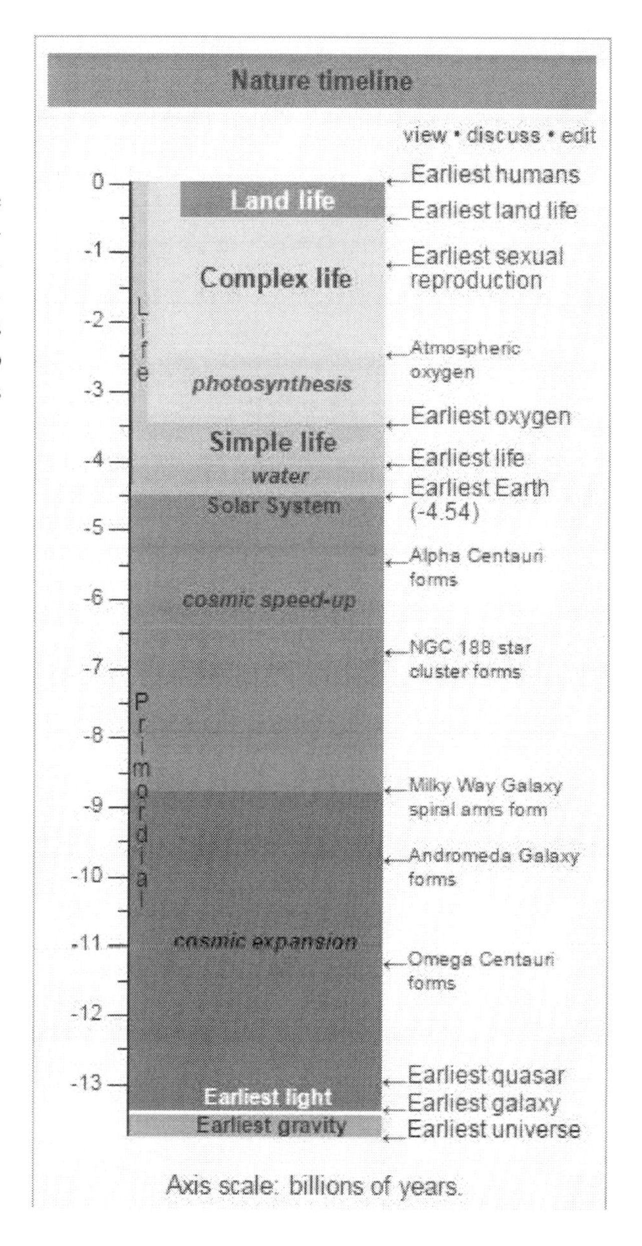

For other uses, see The Big Bounce (disambiguation).

The **Big Bounce** is a hypothetical scientific model of the formation of the known universe. It was originally suggested as a property of the *cyclic model* or *oscillatory universe* interpretation of the Big Bang where the first cosmological event was the result of the collapse of a previous universe; however, it is also a consequence of applying loop quantum gravity techniques to Big Bang cosmology and this need not be cyclic.[1]

9.1 History

Big bounce models have a venerable history and were endorsed on largely aesthetic grounds by cosmologists including Willem de Sitter, Carl Friedrich von Weizsäcker, George McVittie and George Gamow (who stressed that "from the physical point of view we must forget entirely about the precollapse period").[2] The phrase itself, however, did not appear in the scientific literature until 1987, when it was used in the title of a pair of articles (in German) in *Stern und Weltraum* by Wolfgang Priester and Hans-Joachim Blome.[3] It reappeared in 1988 in Iosif Rozental's *Big Bang, Big Bounce*, a revised English-language translation of a Russian-language book (by a different title), and in a 1991 article (in English) by Priester and Blome in *Astronomy and Astrophysics*. (The phrase apparently originated as the title of a novel by Elmore Leonard in 1969, a few years after increased public awareness of the Big Bang model in the wake of the discovery of the cosmic microwave background by Penzias and Wilson.)

9.2 Expansion and contraction

According to some oscillatory universe theorists, the Big Bang was simply the beginning of a period of expansion that followed a period of contraction. In this view, one could talk of a *Big Crunch* followed by a *Big Bang*, or more simply, a *Big Bounce*. This suggests that we could be living at any point in an infinite sequence of universes, or conversely the current universe could be the very first iteration. However, if the condition of the interval phase "between bounces", considered the 'hypothesis of the primeval atom', is taken into full contingency such enumeration may be meaningless because that condition could represent a singularity in time at each instance, if such perpetual return was absolute and undifferentiated.

The main idea behind the quantum theory of a Big Bounce is that, as density approaches infinity, the behavior of the *quantum foam* changes. All the so-called fundamental physical constants, including the speed of light in a vacuum, need not remain constant during a Big Crunch, especially in the time interval smaller than that in which measurement will ever be possible (one unit of Planck time, roughly 10^{-43} seconds) spanning or bracketing the point of inflection.

If the fundamental physical constants were determined in a quantum-mechanical manner during the Big Crunch, then their apparently inexplicable values in this universe would not be so surprising, it being understood here that a *universe* is that which exists between a Big Bang and its Big Crunch.

The Big Bounce Models, however do not explain much about that how the currently expanding universe will manage to contract. This constant and steady expansion is explained by NASA through the Metric expansion of space.

9.3 Recent developments in the theory

Martin Bojowald, an assistant professor of physics at Pennsylvania State University, published a study in July 2007 detailing work somewhat related to loop quantum gravity that claimed to mathematically solve the time before the Big Bang, which would give new weight to the oscillatory universe and Big Bounce theories.[4]

One of the main problems with the Big Bang theory is that at the moment of the Big Bang, there is a singularity of zero volume and infinite energy. This is normally interpreted as the end of the physics as we know it; in this case, of the theory of general relativity. This is why one expects quantum effects to become important and avoid the singularity.

However, research in loop quantum cosmology purported to show that a previously existing universe collapsed, not to the point of singularity, but to a point before that where the quantum effects of gravity become so strongly repulsive that the universe rebounds back out, forming a new branch. Throughout this collapse and bounce, the evolution is unitary.

Bojowald also claims that some properties of the universe that collapsed to form ours can also be determined. Some properties of the prior universe are not determinable however due to some kind of uncertainty principle.

This work is still in its early stages and very speculative. Some extensions by further scientists have been published in *Physical Review* Letters.[5]

In 2003, Peter Lynds has put forward a new cosmology model in which time is cyclic. In his theory our Universe will eventually stop expanding and then contract. Before becoming a singularity, as one would expect from Hawking's black hole theory, the universe would bounce. Lynds claims that a singularity would violate the second law of thermodynamics and this stops the universe from being bounded by singularities. The Big Crunch would be avoided with a new Big Bang. Lynds suggests the exact history of the universe would be repeated in each cycle in an eternal recurrence. Some critics argue that while the universe may be cyclic, the histories would all be variants. Lynds' theory has been dismissed by mainstream physicists for the lack of a mathematical model behind its philosophical considerations.[6]

In 2011, Nikodem Popławski showed that a nonsingular Big Bounce appears naturally in the Einstein-Cartan-Sciama-Kibble theory of gravity.[7] This theory extends general relativity by removing a constraint of the symmetry of the affine connection and regarding its antisymmetric part, the torsion tensor, as a dynamical variable. The minimal coupling between torsion and Dirac spinors generates a spin-spin interaction which is significant in fermionic matter at extremely high densities. Such an interaction averts the unphysical Big Bang singularity, replacing it with a cusp-like bounce at a finite minimum scale factor, before which the universe was contracting. This scenario also explains why the present Universe at largest scales appears spatially flat, homogeneous and isotropic, providing a physical alternative to cosmic inflation.

In 2012, a new theory of nonsingular big bounce was successfully constructed within the frame of standard Einstein gravity.[8] This theory combines the benefits of matter bounce and Ekpyrotic cosmology. Particularly, the famous BKL instability, that the homogeneous and isotropic background cosmological solution is unstable to the growth of anisotropic stress, is resolved in this theory. Moreover, curvature perturbations seeded in matter contraction are able to form a nearly scale-invariant primordial power spectrum and thus provides a consistent mechanism to explain the cosmic microwave background (CMB) observations alternative to inflation.

9.4 See also

- Abhay Ashtekar
- Anthropic principle
- Cyclic model
- Eternal return
- Martin Bojowald
- John Archibald Wheeler
- Loop quantum cosmology
- Loop quantum gravity
- Supernova

9.5 References

[1] "Penn State Researchers Look Beyond The Birth Of The Universe". *Science Daily*. May 17, 2006. Referring to Ashtekar, Abhay; Pawlowski, Tomasz; Singh, Parmpreet (2006). "Quantum Nature of the Big Bang". *Physical Review Letters* **96** (14): 141301. arXiv:gr-qc/0602086. Bibcode:2006PhRvL..96n1301A. doi:10.1103/PhysRevLett.96.141301. PMID 16712061.

[2] Kragh, Helge (1996). *Cosmology*. Princeton, NJ, USA: Princeton University Press. ISBN 0-691-00546-X.

[3] Overduin, James; Hans-Joachim Blome; Josef Hoell (June 2007). "Wolfgang Priester: from the big bounce to the Λ-dominated universe". *Naturwissenschaften* **94** (6): 417–429. arXiv:astro-ph/0608644. Bibcode:2007NW.....94..417O. doi:10.1007/s00114-006-0187-x.

[4] Bojowald, Martin (2007). "What happened before the Big Bang?". *Nature Physics* **3** (8): 523–525. Bibcode:2007NatPh...3..523B. doi:10.1038/nphys654.

[5] Ashtekar, Abhay; Corichi, Alejandro; Singh, Parampreet (2008). "Robustness of key features of loop quantum cosmology". *Physical Review D* **77** (2): 024046. arXiv:0710.3565. Bibcode:2008PhRvD..77b4046A. doi:10.1103/PhysRevD.77.024046.

[6] David Adam (14 August 2003). "The Strange story of Peter Lynds". *The Guardian*.

[7] Poplawski, N. J. (2012). "Nonsingular, big-bounce cosmology from spinor-torsion coupling". *Physical Review D* **85**: 107502. arXiv:1111.4595. Bibcode:2012PhRvD..85j7502P. doi:10.1103/PhysRevD.85.107502.

[8] Cai, Yi-Fu; Damien Easson; Robert Brandenberger (2012). "Towards a Nonsingular Bouncing Cosmology". *Journal of Cosmology and Astroparticle Physics* **08**: 020. arXiv:1206.2382. Bibcode:2012JCAP...08..020C. doi:10.1088/1475-7516/2012/08/020.

9.6 Further reading

- Magueijo, João (2003). *Faster than the Speed of Light: the Story of a Scientific Speculation*. Cambridge, MA: Perseus Publishing. ISBN 0-7382-0525-7.
- Bojowald, Martin (2008). "Follow the Bouncing Universe". *Scientific American* **299** (October 2008): 44–51. doi:10.1038/scientificamerican1008-44. PMID 18847084.

9.7 External links

- Wolfgang Priester: from the big bounce to the Lambda-dominated universe, James Overduin, 2006
- Dark Matter, Antimatter, and Time-Symmetry, Trevor Pitts, 1999
- Penn State Researchers Look Beyond The Birth Of The Universe (Penn State) May 12, 2006
- What Happened Before the Big Bang? (Penn State) July 1, 2007
- From big bang to big bounce (Pen State) NewScientist December 13, 2008
- SpringerLink - Gravitation and Cosmology, Volume 16, Number 4

Chapter 10

Accelerating expansion of the universe

The **accelerating expansion of the universe** is the observation that the universe appears to be expanding at an increasing rate. In formal terms, this means that the cosmic scale factor $a(t)$ has a positive second derivative,[1] so that the velocity at which a distant galaxy is receding from the observer is continuously increasing with time.[2]

The expansion of the universe has been accelerating since the universe entered its dark-energy-dominated era, at redshift $z \approx 0.4$ (roughly 5 billion years ago).[3] [notes 1] Within the framework of general relativity, an accelerating expansion can be accounted for by a positive value of the cosmological constant Λ, equivalent to the presence of a positive vacuum energy, dubbed "dark energy". While there are alternative possible explanations, the description assuming dark energy (positive Λ) is used in the current standard model of cosmology, known as ΛCDM ("Lambda cold dark matter").

The accelerated expansion was discovered in 1998, when two independent projects, the Supernova Cosmology Project and the High-Z Supernova Search Team simultaneously obtained results suggesting a totally unexpected acceleration in the expansion of the universe by using distant type Ia supernovae as standard candles.[4][5][6] The discovery was unexpected, cosmologists at the time expecting a deceleration in the expansion of the universe, and amounts to the realization that the universe is currently in a "dark-energy-dominated era". Three members of these two groups have subsequently been awarded Nobel Prizes for their discovery.[7] Confirmatory evidence has been found in baryon acoustic oscillations and other new results about the clustering of galaxies.

In June 2016, NASA and ESA scientists reported that the universe was found to be expanding 5% to 9% faster than thought earlier, based on studies using the Hubble Space Telescope.[8]

10.1 Background

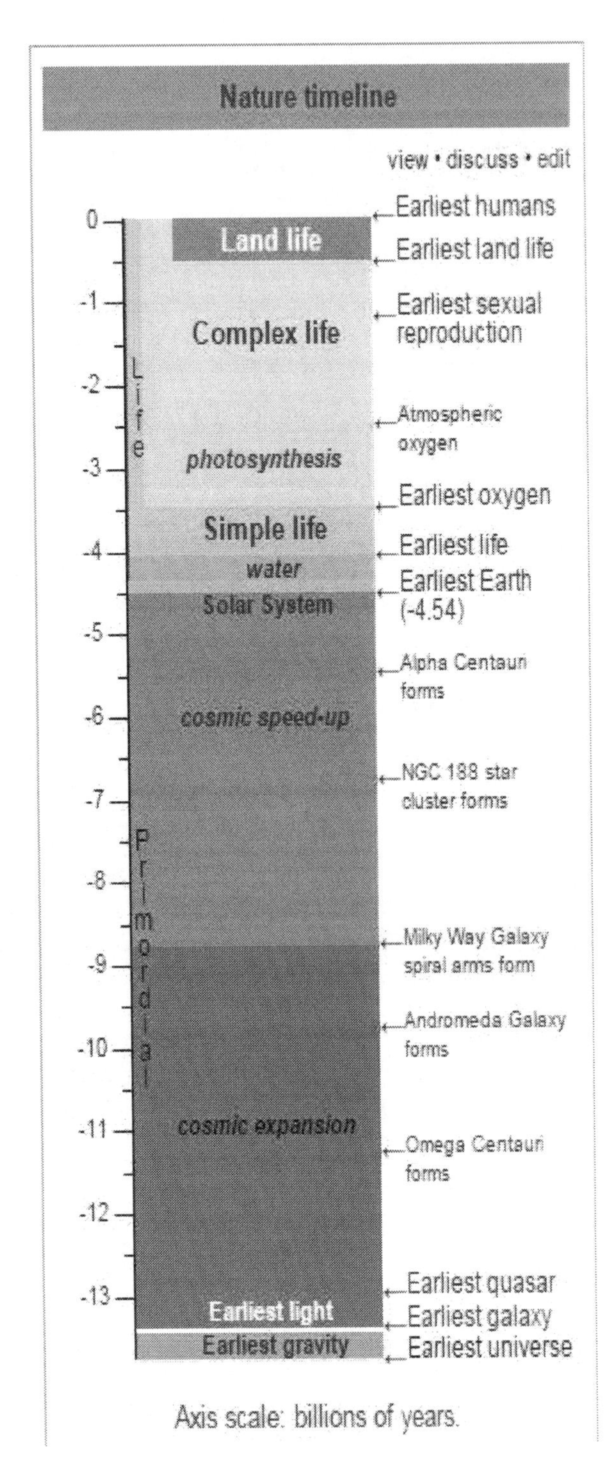

Axis scale: Billions of years.

also see {{Life timeline}}

Further information: Cosmological constant, Lambda-CDM model, Hubble's law, Friedmann–Lemaître–Robertson–Walker metric and Friedman equations

Since Hubble's discovery of the expansion of the universe in 1929,[9] the Big Bang model has become the accepted explanation for the origin of our universe. The Friedmann equation defines how the energy in the universe drives its expansion.

$$H^2 = \left(\frac{\dot{a}}{a}\right)^2 = \frac{8\pi G}{3}\rho - \frac{\mathrm{K}c^2}{R^2a^2}$$

where K represents the curvature of the universe, $a(t)$ is the scale factor, ρ is the total energy density of the universe, and H is the Hubble parameter.[10]

We define a critical density

$$\rho_c = \frac{3H^2}{8\pi G}$$

and the density parameter

$$\Omega = \frac{\rho}{\rho_c}$$

We can then rewrite the Hubble parameter as

$$H(a) = H_0\sqrt{\Omega_k a^{-2} + \Omega_m a^{-3} + \Omega_r a^{-4} + \Omega_{DE} a^{-3(1+w)}}$$

where the four currently hypothesized contributors to the energy density of the universe are curvature, matter, radiation and dark energy.[11] Each of the components decreases with the expansion of the universe (increasing scale factor), except perhaps the dark energy term. It is the values of these cosmological parameters which physicists use to determine the acceleration of the universe.

The acceleration equation describes the evolution of the scale factor with time

$$\frac{\ddot{a}}{a} = -\frac{4\pi G}{3}\left(\rho + \frac{3P}{c^2}\right)$$

where the pressure P is defined by the cosmological model chosen. (see explanatory models below)

Physicists at one time were so assured of the deceleration of the universe's expansion that they introduced a so-called deceleration parameter q_0 .[12] Current observations point towards this deceleration parameter being negative.

10.2 Evidence for acceleration

To learn about the rate of expansion of the universe we look at the magnitude-redshift relationship of astronomical objects using standard candles, or their distance-redshift relationship using standard rulers. We can also look at the growth of large-scale structure, and find that the observed values of the cosmological parameters are best described by models which include an accelerating expansion.

10.2.1 Supernova observation

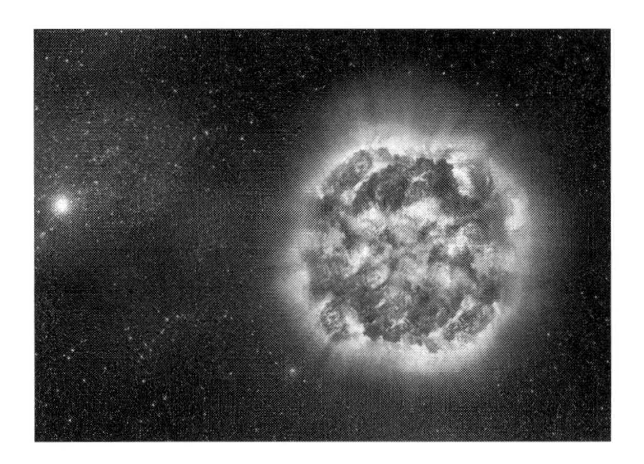

Artist's impression of a Type Ia supernova, as revealed by spectro-polarimetry observations

The first evidence for acceleration came from the observation of Type Ia supernovae, which are exploding white dwarfs that have exceeded their stability limit. Because they all have similar masses, their intrinsic luminosity is standardizable. Repeated imaging of selected areas of sky is used to discover the supernovae, then follow-up observations give their peak brightness, which is converted into a quantity known as luminosity distance (see distance measures in cosmology for details).[13] Spectral lines of their light can be used to determine their redshift.

For supernovae at redshift less than around 0.1, or light travel time less than 10 percent of the age of the universe, this gives a nearly linear distance-redshift relation due to Hubble's law. At larger distances, since the expansion rate of the universe has changed over time, the distance-redshift relation deviates from linearity, and this deviation depends on how the expansion rate has changed over time. The full calculation requires integration of the Friedmann equation, but a simple derivation can be given as follows: the redshift z directly gives the cosmic scale factor at the time the supernova exploded.

$$a(t) = \frac{1}{1+z}$$

So a supernova with a measured redshift z = 0.5 implies the universe was 1/(1+0.5) = 2/3 of its present size when the supernova exploded. In an accelerating universe, the universe was expanding more slowly in the past than it is today, which means it took a longer time to expand from 2/3 to 1.0 times its present size compared to a non-accelerating universe. This results in a larger light-travel time, larger distance and fainter supernovae, which corresponds to the actual observations. Riess found that "the distances of the high-redshift SNe Ia were, on average, 10% to 15% farther than expected in a low mass density $\Omega_M = 0.2$ universe without a cosmological constant".[14] This means that the measured high-redshift distances were too large, compared to nearby ones, for a decelerating universe.[15]

10.2.2 Baryon acoustic oscillations

Main article: Baryon acoustic oscillations

In the early universe before recombination and decoupling took place, photons and matter existed in a primordial plasma. Points of higher density in the photon-baryon plasma would contract, being compressed by gravity until the pressure became too large and they expanded again.[12] This contraction and expansion created vibrations in the plasma analogous to sound waves. Since dark matter only interacts gravitationally it stayed at the centre of the sound wave, the origin of the original overdensity. When decoupling occurred, approximately 380,000 years after the Big Bang,[16] photons separated from matter and were able to stream freely through the universe, creating the cosmic microwave background as we know it. This left shells of baryonic matter at a fixed radius from the overdensities of dark matter, a distance known as the **sound horizon**. As time passed and the universe expanded, it was at these anisotropies of matter density where galaxies started to form. So by looking at the distances at which galaxies at different redshifts tend to cluster, it is possible to determine a standard angular diameter distance and use that to compare to the distances predicted by different cosmological models.

Peaks have been found in the correlation function (the probability that two galaxies will be a certain distance apart) at $100h^{-1}$ Mpc,[11] indicating that this is the size of the sound horizon today, and by comparing this to the sound horizon at the time of decoupling (using the CMB), we can confirm that the expansion of the universe is accelerating.[17]

10.2.3 Clusters of galaxies

Measuring the mass functions of galaxy clusters, which describe the number density of the clusters above a threshold mass, also provides evidence for dark energy.[18] By comparing these mass functions at high and low redshifts to those predicted by different cosmological models, values for w and Ω_m are obtained which confirm a low matter density and a non zero amount of dark energy.[15]

10.2.4 Age of the universe

See also: Age of the universe

Given a cosmological model with certain values of the cosmological density parameters, it is possible to integrate the Friedmann equations and derive the age of the universe.

$$t_0 = \int_0^1 \frac{da}{\dot{a}}$$

By comparing this to actual measured values of the cosmological parameters, we can confirm the validity of a model which is accelerating now, and had a slower expansion in the past.[15]

10.3 Explanatory models

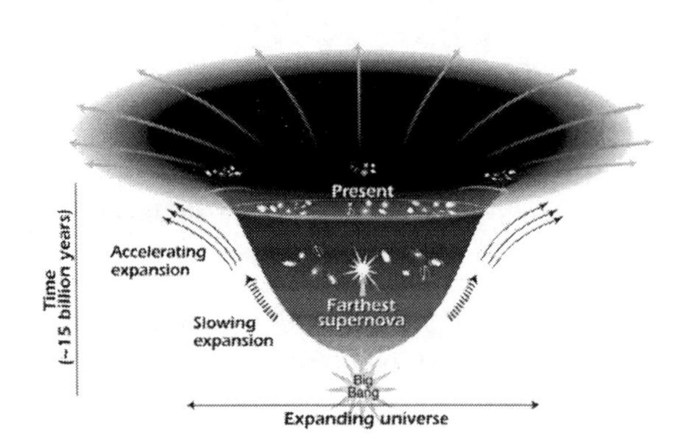

This diagram reveals changes in the rate of expansion since the universe's birth 15 billion years ago. The more shallow the curve, the faster the rate of expansion. The curve changes noticeably about 7.5 billion years ago, when objects in the universe began flying apart at a faster rate. Astronomers theorize that the faster expansion rate is due to a mysterious, dark force that is pushing galaxies apart.

The expansion of the Universe accelerating. Time flows from bottom to top

10.3.1 Dark energy

Main article: Dark energy

The most important property of dark energy is that it has negative pressure which is distributed relatively homogeneously in space.

$$P = wc^2\rho$$

where c is the speed of light, ρ is the energy density. Different theories of dark energy suggest different values of w, with $w < -1/3$ for cosmic acceleration (this leads to a positive value of $\ddot{a}$ in the acceleration equation above).

The simplest explanation for dark energy is that it is a cosmological constant or vacuum energy; in this case $w = -1$. This leads to the Lambda-CDM model, which has generally been known as the Standard Model of Cosmology from 2003 through the present, since it is the simplest model in good agreement with a variety of recent observations. Riess found that their results from supernovae observations favoured expanding models with positive cosmological constant ($\Omega_\lambda > 0$) and a current acceleration of the expansion ($q_0 < 0$).[14]

10.3.2 Phantom energy

Main article: Phantom energy

Current observations allow the possibility of a cosmological model containing a dark energy component with equation of state:

$$w < -1$$

This phantom energy density would become infinite in finite time, causing such a huge gravitational repulsion that the universe would lose all structure and end in a Big Rip.[19] For example, for $w = -3/2$ and $H_0 = 70$ km·s^{-1}·Mpc^{-1}, the time remaining before the universe ends in this "Big Rip" is 22 billion years.[20]

10.3.3 Alternative theories

Other explanations for the accelerating universe include quintessence, a proposed form of dark energy with a non-constant state equation, whose density decreases with time. Dark fluid is an alternative explanation for accelerating expansion which attempts to unite dark matter and dark energy into a single framework.[21] Alternatively, some authors have argued that the universe expansion acceleration could be due to a repulsive gravitational interaction of antimatter.[22][23][24]

Another type of model, the backreaction conjecture,[25][26] was proposed by cosmologist Syksy Räsänen:[27] the rate of expansion is not homogenous, but we are in a region where expansion is faster than the background. Inhomogeneities in the early universe cause the formation of walls and bubbles, where the inside of a bubble has less matter than on average. According to general relativity, space is less curved than on the walls, and thus appears to have more volume and a higher expansion rate. In the denser regions, the expansion is retarded by a higher gravitational attraction. Therefore, the inward collapse of the denser regions looks the same as an accelerating expansion of the bubbles, leading us to conclude that the universe is expanding at an accelerating rate.[28] The benefit is that it does not require any new physics such as dark energy. Räsänen does not consider the model likely, but without any falsification, it must remain a possibility. It would require rather large density fluctuations (20%) to work.[27]

10.4 Theories for the consequences to the universe

See also: Future of an expanding universe

As the universe expands, the density of radiation and ordinary and dark matter declines more quickly than the density of dark energy (see equation of state) and, eventually, dark energy dominates. Specifically, when the scale of the universe doubles, the density of matter is reduced by a factor of 8, but the density of dark energy is nearly unchanged (it is exactly constant if the dark energy is a cosmological constant).[12]

In models where dark energy is a cosmological constant, the universe will expand exponentially with time from now on, coming closer and closer to a de Sitter spacetime. This will eventually lead to all evidence for the Big Bang disappearing, as the cosmic microwave background is redshifted to lower intensities and longer wavelengths. Eventually its frequency will be low enough that it will be absorbed by the interstellar medium, and so be screened from any observer within the galaxy. This will occur when the universe is less than 50 times its current age, leading to the end of cosmology as we know it as the distant universe turns dark.[29]

Alternatives for the ultimate fate of the universe include the Big Rip mentioned above, a Big Bounce, Big Freeze, or Big Crunch.

10.5 See also

- Cosmological constant
- Friedmann–Lemaître–Robertson–Walker metric
- High-z Supernova Search Team
- Lambda-CDM model
- List of multiple discoveries
- Metric expansion of space
- Scale factor (cosmology)
- Supernova Cosmology Project

10.6 Notes

[1] [3]p. 6: "The Universe has gone through three distinct eras: radiation-dominated, $z \gtrsim 3000$; matter-dominated, $3000 \gtrsim z \gtrsim 0.5$; and dark-energy dominated, $z \lesssim 0.5$. The evolution of the scale factor is controlled by the dominant energy form: $a(t) \propto t2/3(1+w)$ (for constant w). During the radiation-dominated era, $a(t) \propto t1/2$; during the matter-dominated era, $a(t) \propto t2/3$; and for the dark energy-dominated era, assuming $w = -1$, asymptotically $a(t) \propto \exp(Ht)$."
p. 44: "Taken together, all the current data provide strong evidence for the existence of dark energy; they constrain the fraction of critical density contributed by dark energy, 0.76 ± 0.02, and the equation-of-state parameter, $w \approx -1 \pm 0.1$ (stat) ± 0.1 (sys), assuming that w is constant. This implies that the Universe began accelerating at redshift $z \sim 0.4$ and age $t \sim 10$ Gyr. These results are robust – data from any one method can be removed without compromising the constraints – and they are not substantially weakened by dropping the assumption of spatial flatness."

10.7 References

[1] Jones, Mark H.; Robert J. Lambourne (2004). *An Introduction to Galaxies and Cosmology*. Cambridge University Press. p. 244. ISBN 978-0-521-83738-5.

[2] Is the universe expanding faster than the speed of light? (see final paragraph)

[3] Frieman, Joshua A.; Turner, Michael S.; Huterer, Dragan (2008-01-01). "Dark Energy and the Accelerating Universe". *Annual Review of Astronomy and Astrophysics* **46** (1): 385–432. arXiv:0803.0982. Bibcode:2008ARA&A..46..385F. doi:10.1146/annurev.astro.46.060407.145243.

[4] "Nobel physics prize honours accelerating universe find". BBC News. October 4, 2011.

[5] "The Nobel Prize in Physics 2011". Nobelprize.org. Retrieved 2011-10-06.

[6] Peebles, P. J. E. & Ratra, Bharat (2003). "The cosmological constant and dark energy". *Reviews of Modern Physics* **75** (2): 559–606. arXiv:astro-ph/0207347. Bibcode:2003RvMP...75..559P. doi:10.1103/RevModPhys.75.559.

[7] *Cosmology*, Steven Weinberg, Oxford University Press, 2008

[8] Radford, Tim (3 June 2016). "Universe is expanding up to 9% faster than we thought, say scientists". *The Guardian*. Retrieved 3 June 2016.

[9] Hubble, Edwin (1929). "A relation between distance and radial velocity among extra-galactic nebulae". *PNAS* **15** (3): 168–173. Bibcode:1929PNAS...15..168H. doi:10.1073/pnas.15.3.168. PMC 522427. PMID 16577160.

[10] Nemiroff, Robert J.; Patla, Bijunath. "Adventures in Friedmann cosmology: A detailed expansion of the cosmological Friedmann equations". *American Journal of Physics* **76** (3): 265. arXiv:astro-ph/0703739. Bibcode:2008AmJPh..76..265N. doi:10.1119/1.2830536.

[11] Lapuente, P.. "Baryon Acoustic Oscillations." Dark energy: observational and theoretical approaches. Cambridge, UK: Cambridge University Press, 2010.

[12] Ryden, Barbara. "Introduction to Cosmology." Physics Today: 77. Print.

[13] Albrecht, A., Bernstein, G., Cahn, R., et al. Report of the Dark Energy TaskForce. ArXiv Astrophysics e-prints, September 2006.

[14] Riess, Adam G.; Filippenko, Alexei V.; Challis, Peter; Clocchiatti, Alejandro; Diercks, Alan; Garnavich, Peter M.; Gilliland, Ron L.; Hogan, Craig J.; Jha, Saurabh; Kirshner, Robert P.; Leibundgut, B.; Phillips, M. M.; Reiss, David; Schmidt, Brian P.; Schommer, Robert A.; Smith, R. Chris; Spyromilio, J.; Stubbs, Christopher; Suntzeff, Nicholas B.; Tonry, John. "Observational Evidence from Supernovae for an Accelerating Universe and a Cosmological Constant". *The Astronomical Journal* **116** (3): 1009–1038. arXiv:astro-ph/9805201. Bibcode:1998AJ....116.1009R. doi:10.1086/300499.

[15] Pain, Reynald. "Observational evidence of the accelerated expansion of the Universe." Comptes Rendus Physique: 521-538.

[16] Hinshaw, G. (2014). "Five-Year Wilkinson Microwave Anisotropy Probe (WMAP) Observations: Data Processing, Sky Maps, and Basic Results". *Astrophysical Journal Supplement* **180**: 225–245. arXiv:0803.0732. Bibcode:2009ApJS..180..225H. doi:10.1088/0067-0049/180/2/225.

[17] Eisenstein, Daniel J.; Zehavi, Idit; Hogg, David W.; Scoccimarro, Roman; Blanton, Michael R.; Nichol, Robert C.; Scranton, Ryan; Seo, Hee-Jong; Tegmark, Max; Zheng, Zheng; Anderson, Scott F.; Annis, Jim; Bahcall, Neta; Brinkmann, Jon; Burles, Scott; Castander, Francisco J.; Connolly, Andrew; Csabai, Istvan; Doi, Mamoru; Fukugita, Masataka; Frieman, Joshua A.; Glazebrook, Karl; Gunn, James E.; Hendry, John S.; Hennessy, Gregory; Ivezić, Zeljko; Kent, Stephen; Knapp, Gillian R.; Lin, Huan; Loh, Yeong-Shang; Lupton, Robert H.; Margon, Bruce; McKay, Timothy A.; Meiksin, Avery; Munn, Jeffery A.; Pope, Adrian; Richmond, Michael W.; Schlegel, David; Schneider, Donald P.; Shimasaku, Kazuhiro; Stoughton, Christopher; Strauss, Michael A.; SubbaRao, Mark; Szalay, Alexander S.; Szapudi, Istvan; Tucker, Douglas L.; Yanny, Brian; York, Donald G. (10 November 2005). "Detection of the Baryon Acoustic Peak in the Large-Scale Correlation Function of SDSS Luminous Red Galaxies". *The Astrophysical Journal* **633** (2): 560–574. arXiv:astro-ph/0501171. Bibcode:2005ApJ...633..560E. doi:10.1086/466512.

[18] Dekel, Avishai. Formation of structure in the Universe. New York: Cambridge University Press, 1999.

[19] Caldwell, Robert; Kamionkowski, Marc; Weinberg, Nevin (August 2003). "Phantom Energy: Dark Energy with w-1 Causes a Cosmic Doomsday". *Physical Review Letters* **91** (7): 071301. arXiv:astro-ph/0302506. Bibcode:2003PhRvL..91g1301C. doi:10.1103/PhysRevLett.91.071301. PMID 12935004.

[20] Caldwell, R.R. "A phantom menace? Cosmological consequences of a dark energy component with super-negative equation of state". *Physics Letters B* **545** (1-2): 23–29. arXiv:astro-ph/9908168. Bibcode:2002PhLB..545...23C. doi:10.1016/S0370-2693(02)02589-3.

[21] Anaelle Halle, HongSheng Zhao, Baojiu Li (2008) "Perturbations in a non-uniform dark energy fluid: equations reveal effects of modified gravity and dark matter "

[22] A. Benoit-Lévy and G. Chardin, Introducing the Dirac-Milne universe, Astronomy and Astrophysics 537, A78 (2012)

[23] D.S. Hajdukovic, Quantum vacuum and virtual gravitational dipoles: the solution to the dark energy problem?, Astrophysics and Space Science 339(1), 1−5 (2012)

[24] M. Villata, On the nature of dark energy: the lattice Universe, 2013, Astrophysics and Space Science 345, 1. Also available here

[25] "Backreaction: directions of progress". *Classical and Quantum Gravity* **28**: 164008. arXiv:1102.0408. Bibcode:2011CQGra..28p4008R. doi:10.1088/0264-9381/28/16/164008.

[26] "Backreaction in Late-Time Cosmology". *Annual Review of Nuclear and Particle Science* **62**: 57–79. arXiv:1112.5335. Bibcode:2012ARNPS..62...57B. doi:10.1146/annurev.nucl.012809.104435.

[27] https://www.newscientist.com/article/dn11498-is-dark-energy-an-illusion/

[28] "A Cosmic 'Tardis': What the Universe Has In Common with 'Doctor Who'". *Space.com*.

[29] Krauss, Lawrence M.; Scherrer, Robert J. (28 June 2007). "The return of a static universe and the end of cosmology". *General Relativity and Gravitation* **39** (10): 1545–1550. arXiv:0704.0221. Bibcode:2007GReGr..39.1545K. doi:10.1007/s10714-007-0472-9.

Chapter 11

Inflation (cosmology)

"Inflation model" and "Inflation theory" redirect here. For a general rise in the price level, see Inflation. For other uses, see Inflation (disambiguation).

In physical cosmology, **cosmic inflation**, **cosmological inflation**, or just **inflation** is a theory of exponential expansion of space in the early universe. The inflationary epoch lasted from 10^{-36} seconds after the Big Bang to sometime between 10^{-33} and 10^{-32} seconds. Following the inflationary period, the Universe continues to expand, but at a less rapid rate.[1]

Inflation theory was developed in the early 1980s. It explains the origin of the large-scale structure of the cosmos. Quantum fluctuations in the microscopic inflationary region, magnified to cosmic size, become the seeds for the growth of structure in the Universe (see galaxy formation and evolution and structure formation).[2] Many physicists also believe that inflation explains why the Universe appears to be the same in all directions (isotropic), why the cosmic microwave background radiation is distributed evenly, why the Universe is flat, and why no magnetic monopoles have been observed.

The detailed particle physics mechanism responsible for inflation is not known. The basic inflationary paradigm is accepted by most scientists, who believe a number of predictions have been confirmed by observation;[3] however, a substantial minority of scientists dissent from this position.[4][5][6] The hypothetical field thought to be responsible for inflation is called the inflaton.[7]

In 2002, three of the original architects of the theory were recognized for their major contributions; physicists Alan Guth of M.I.T., Andrei Linde of Stanford, and Paul Steinhardt of Princeton shared the prestigious Dirac Prize "for development of the concept of inflation in cosmology".[8]

11.1 Overview

Main article: Metric expansion of space

An expanding universe generally has a cosmological horizon, which, by analogy with the more familiar horizon caused by the curvature of the Earth's surface, marks the boundary of the part of the Universe that an observer can see. Light (or other radiation) emitted by objects beyond the cosmological horizon never reaches the observer, because the space in between the observer and the object is expanding too rapidly.

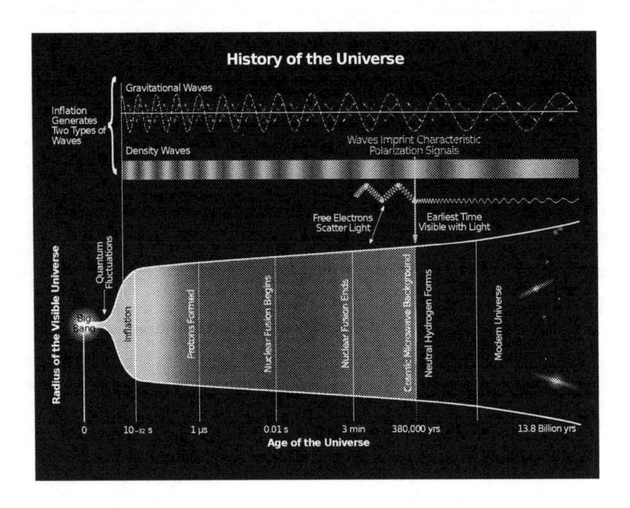

History of the Universe - gravitational waves are hypothesized to arise from cosmic inflation, a faster-than-light expansion just after the Big Bang (17 March 2014).[9][10][11]

The observable universe is one *causal patch* of a much larger unobservable universe; other parts of the Universe cannot communicate with Earth yet. These parts of the Universe are outside our current cosmological horizon. In the standard hot big bang model, without inflation, the cosmological horizon moves out, bringing new regions into view. Yet as a local observer sees such a region for the first time, it looks no different from any other region of space the local observer has already seen: its background radiation is at nearly the same temperature as the background radiation of other regions, and its space-time curvature is evolving lock-step with the others. This presents a mystery: how did these new regions know what temperature and curvature they were supposed to have? They couldn't have learned it by getting signals, because they were not previously in communication with our past light cone.[12][13]

Inflation answers this question by postulating that all the regions come from an earlier era with a big vacuum energy, or cosmological constant. A space with a cosmological constant is qualitatively different: instead of moving outward, the cosmological horizon stays put. For any one observer, the distance to the cosmological horizon is constant. With exponentially expanding space, two nearby observers are separated very quickly; so much so, that the distance between them quickly exceeds the limits of communications. The spatial slices are expanding very fast to cover huge volumes. Things are constantly moving beyond the cosmological horizon, which is a fixed distance away, and everything becomes homogeneous.

As the inflationary field slowly relaxes to the vacuum, the cosmological constant goes to zero and space begins to expand normally. The new regions that come into view during the normal expansion phase are exactly the same regions that were pushed out of the horizon during inflation, and so they are at nearly the same temperature and curvature, because they come from the same originally small patch of space.

The theory of inflation thus explains why the temperatures and curvatures of different regions are so nearly equal. It also predicts that the total curvature of a space-slice at constant global time is zero. This prediction implies that the total ordinary matter, dark matter and residual vacuum energy in the Universe have to add up to the critical density, and the evidence supports this. More strikingly, inflation allows physicists to calculate the minute differences in temperature of different regions from quantum fluctuations during the inflationary era, and many of these quantitative predictions have been confirmed.[14][15]

11.1.1 Space expands

To say that space expands exponentially means that two inertial observers are moving farther apart with accelerating velocity. In stationary coordinates for one observer, a patch of an inflating universe has the following polar metric:[16][17]

$$ds^2 = -(1 - \Lambda r^2)\, dt^2 + \frac{1}{1 - \Lambda r^2}\, dr^2 + r^2\, d\Omega^2.$$

This is just like an inside-out black hole metric—it has a zero in the dt component on a fixed radius sphere called the cosmological horizon. Objects are drawn away from the observer at $r = 0$ towards the cosmological horizon, which they cross in a finite proper time. This means that any inhomogeneities are smoothed out, just as any bumps or matter on the surface of a black hole horizon are swallowed and disappear.

Since the space–time metric has no explicit time dependence, once an observer has crossed the cosmological horizon, observers closer in take its place. This process of falling outward and replacement points closer in are always steadily replacing points further out—an exponential expansion of space–time.

This steady-state exponentially expanding spacetime is called a de Sitter space, and to sustain it there must be a cosmological constant, a vacuum energy proportional to Λ everywhere. In this case, the equation of state is $p = -\varrho$. The physical conditions from one moment to the next are stable: the rate of expansion, called the Hubble parameter, is nearly constant, and the scale factor of the Universe is proportional to e^{Ht}. Inflation is often called a period of *accelerated expansion* because the distance between two fixed observers is increasing exponentially (i.e. at an accelerating rate as they move apart), while Λ can stay approximately constant (see deceleration parameter).

11.1.2 Few inhomogeneities remain

Cosmological inflation has the important effect of smoothing out inhomogeneities, anisotropies and the curvature of space. This pushes the Universe into a very simple state, in which it is completely dominated by the inflaton field, the source of the cosmological constant, and the only significant inhomogeneities are the tiny quantum fluctuations in the inflaton. Inflation also dilutes exotic heavy particles, such as the magnetic monopoles predicted by many extensions to the Standard Model of particle physics. If the Universe was only hot enough to form such particles *before* a period of inflation, they would not be observed in nature, as they would be so rare that it is quite likely that there are none in the observable universe. Together, these effects are called the inflationary "no-hair theorem"[18] by analogy with the no hair theorem for black holes.

The "no-hair" theorem works essentially because the cosmological horizon is no different from a black-hole horizon, except for philosophical disagreements about what is on the other side. The interpretation of the no-hair theorem is that the Universe (observable and unobservable) expands by an enormous factor during inflation. In an expanding universe, energy densities generally fall, or get diluted, as the volume of the Universe increases. For example, the density of ordinary "cold" matter (dust) goes down as the inverse of the volume: when linear dimensions double, the energy density goes down by a factor of eight; the radiation energy density goes down even more rapidly as the Universe expands since the wavelength of each photon is stretched (redshifted), in addition to the photons being dispersed by the expansion. When linear dimensions are doubled, the energy density in radiation falls by a factor of sixteen (see the solution of the energy density continuity equation for an ultra-relativistic fluid). During inflation, the energy density in the inflaton field is roughly constant. However, the energy density in everything else, including inhomogeneities, curvature, anisotropies, exotic particles, and standard-model particles is falling, and through sufficient inflation these all become negligible. This leaves the Universe flat and symmetric, and (apart from the homogeneous inflaton field) mostly empty, at the moment inflation ends and reheating begins.[19]

11.1.3 Duration

A key requirement is that inflation must continue long enough to produce the present observable universe from a single, small inflationary Hubble volume. This is necessary to ensure that the Universe appears flat, homogeneous and isotropic at the largest observable scales. This requirement is generally thought to be satisfied if the Universe expanded by a factor of at least 10^{26} during inflation.[20]

11.1.4 Reheating

Inflation is a period of supercooled expansion, when the temperature drops by a factor of 100,000 or so. (The exact drop is model dependent, but in the first models it was typically from 10^{27} K down to 10^{22} K.[21]) This relatively low temperature is maintained during the inflationary phase. When inflation ends the temperature returns to the pre-inflationary temperature; this is called *reheating* or thermalization because the large potential energy of the inflaton field decays into particles and fills the Universe with Standard Model particles, including electromagnetic radiation, starting the radiation dominated phase of the Universe. Because the nature of the inflation is not known, this process is still poorly understood, although it is believed to take place through a parametric resonance.[22][23]

11.2 Motivations

Inflation resolves several problems in Big Bang cosmology that were discovered in the 1970s.[24] Inflation was first proposed by Guth while investigating the problem of why no magnetic monopoles are seen today; he found that a positive-energy false vacuum would, according to general relativity, generate an exponential expansion of space. It was very quickly realised that such an expansion would resolve many other long-standing problems. These problems arise from the observation that to look like it does *today*, the Universe would have to have started from very finely tuned, or "special" initial conditions at the Big Bang. Inflation attempts to resolve these problems by providing a dynamical mechanism that drives the Universe to this special state, thus making a universe like ours much more likely in the context of the Big Bang theory.

11.2.1 Horizon problem

Main article: Horizon problem

The horizon problem is the problem of determining why the Universe appears statistically homogeneous and isotropic in accordance with the cosmological principle.[25][26][27] For example, molecules in a canister of gas are distributed homogeneously and isotropically because they are in thermal equilibrium: gas throughout the canister has had enough time to interact to dissipate inhomogeneities and anisotropies. The situation is quite different in the big bang model without inflation, because gravitational expansion does not give the early universe enough time to equilibrate. In a big bang with only the matter and radiation known in the Standard Model, two widely separated regions of the observable universe cannot have equilibrated because they move apart from each other faster than the speed of light and thus have never come into causal contact. In the early Universe, it was not possible to send a light signal between the two regions. Because they have had no interaction, it is difficult to explain why they have the same temperature (are thermally equilibrated). Historically, proposed solutions included the *Phoenix universe* of Georges Lemaître,[28] the related oscillatory universe of Richard Chase Tolman,[29] and the Mixmaster universe of Charles Misner. Lemaître and Tolman proposed that a universe undergoing a number of cycles of contraction and expansion could come into thermal equilibrium. Their models failed, however, because of the buildup of entropy over several cycles. Misner made the (ultimately incorrect) conjecture that the Mixmaster mechanism, which made the Universe *more* chaotic, could lead to statistical homogeneity and isotropy.[26][30]

11.2.2 Flatness problem

Main article: Flatness problem

The flatness problem is sometimes called one of the Dicke

coincidences (along with the cosmological constant problem).[31][32] It became known in the 1960s that the density of matter in the Universe was comparable to the critical density necessary for a flat universe (that is, a universe whose large scale geometry is the usual Euclidean geometry, rather than a non-Euclidean hyperbolic or spherical geometry).[33]:61

Therefore, regardless of the shape of the universe the contribution of spatial curvature to the expansion of the Universe could not be much greater than the contribution of matter. But as the Universe expands, the curvature redshifts away more slowly than matter and radiation. Extrapolated into the past, this presents a fine-tuning problem because the contribution of curvature to the Universe must be exponentially small (sixteen orders of magnitude less than the density of radiation at big bang nucleosynthesis, for example). This problem is exacerbated by recent observations of the cosmic microwave background that have demonstrated that the Universe is flat to within a few percent.[34]

11.2.3 Magnetic-monopole problem

The magnetic monopole problem, sometimes called the exotic-relics problem, says that if the early universe were very hot, a large number of very heavy, stable magnetic monopoles would have been produced. This is a problem with Grand Unified Theories, which propose that at high temperatures (such as in the early universe) the electromagnetic force, strong, and weak nuclear forces are not actually fundamental forces but arise due to spontaneous symmetry breaking from a single gauge theory.[35] These theories predict a number of heavy, stable particles that have not been observed in nature. The most notorious is the magnetic monopole, a kind of stable, heavy "charge" of magnetic field.[36][37] Monopoles are predicted to be copiously produced following Grand Unified Theories at high temperature,[38][39] and they should have persisted to the present day, to such an extent that they would become the primary constituent of the Universe.[40][41] Not only is that not the case, but all searches for them have failed, placing stringent limits on the density of relic magnetic monopoles in the Universe.[42] A period of inflation that occurs below the temperature where magnetic monopoles can be produced would offer a possible resolution of this problem: monopoles would be separated from each other as the Universe around them expands, potentially lowering their observed density by many orders of magnitude. Though, as cosmologist Martin Rees has written, "Skeptics about exotic physics might not be hugely impressed by a theoretical argument to explain the absence of particles that are themselves only hypothetical. Preventive medicine can readily seem 100 percent effective against a disease that doesn't exist!"[43]

11.3 History

11.3.1 Precursors

In the early days of General Relativity, Albert Einstein introduced the cosmological constant to allow a static solution, which was a three-dimensional sphere with a uniform density of matter. Later, Willem de Sitter found a highly symmetric inflating universe, which described a universe with a cosmological constant that is otherwise empty.[44] It was discovered that Einstein's universe is unstable, and that small fluctuations cause it to collapse or turn into a de Sitter universe.

In the early 1970s Zeldovich noticed the flatness and horizon problems of Big Bang cosmology; before his work, cosmology was presumed to be symmetrical on purely philosophical grounds. In the Soviet Union, this and other considerations led Belinski and Khalatnikov to analyze the chaotic BKL singularity in General Relativity. Misner's Mixmaster universe attempted to use this chaotic behavior to solve the cosmological problems, with limited success.

In the late 1970s, Sidney Coleman applied the instanton techniques developed by Alexander Polyakov and collaborators to study the fate of the false vacuum in quantum field theory. Like a metastable phase in statistical mechanics—water below the freezing temperature or above the boiling point—a quantum field would need to nucleate a large enough bubble of the new vacuum, the new phase, in order to make a transition. Coleman found the most likely decay pathway for vacuum decay and calculated the inverse lifetime per unit volume. He eventually noted that gravitational effects would be significant, but he did not calculate these effects and did not apply the results to cosmology.

In the Soviet Union, Alexei Starobinsky noted that quantum corrections to general relativity should be important for the early universe. These generically lead to curvature-squared corrections to the Einstein–Hilbert action and a form of $f(R)$ modified gravity. The solution to Einstein's equations in the presence of curvature squared terms, when the curvatures are large, leads to an effective cosmological constant. Therefore, he proposed that the early universe went through an inflationary de Sitter era.[45] This resolved the cosmology problems and led to specific predictions for the corrections to the microwave background radiation, corrections that were then calculated in detail.

In 1978, Zeldovich noted the monopole problem, which was an unambiguous quantitative version of the horizon problem, this time in a subfield of particle physics, which led to several speculative attempts to resolve it. In 1980 Alan Guth realized that false vacuum decay in the early universe would solve the problem, leading him to propose a scalar-driven inflation. Starobinsky's and Guth's scenarios both predicted an initial deSitter phase, differing only in mechanistic details.

11.3.2 Early inflationary models

Guth proposed inflation in January 1980 to explain the nonexistence of magnetic monopoles;[46][47] it was Guth who coined the term "inflation".[48] At the same time, Starobinsky argued that quantum corrections to gravity would replace the initial singularity of the Universe with an exponentially expanding deSitter phase.[49] In October 1980, Demosthenes Kazanas suggested that exponential expansion could eliminate the particle horizon and perhaps solve the horizon problem,[50] while Sato suggested that an exponential expansion could eliminate domain walls (another kind of exotic relic).[51] In 1981 Einhorn and Sato[52] published a model similar to Guth's and showed that it would resolve the puzzle of the magnetic monopole abundance in Grand Unified Theories. Like Guth, they concluded that such a model not only required fine tuning of the cosmological constant, but also would likely lead to a much too granular universe, i.e., to large density variations resulting from bubble wall collisions.

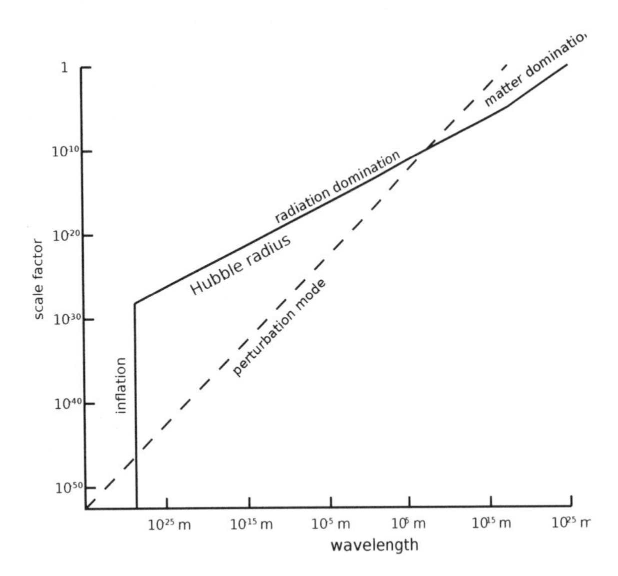

The physical size of the Hubble radius (solid line) as a function of the linear expansion (scale factor) of the universe. During cosmological inflation, the Hubble radius is constant. The physical wavelength of a perturbation mode (dashed line) is also shown. The plot illustrates how the perturbation mode grows larger than the horizon during cosmological inflation before coming back inside the horizon, which grows rapidly during radiation domination. If cosmological inflation had never happened, and radiation domination continued back until a gravitational singularity, then the mode would never have been inside the horizon in the very early universe, and no causal mechanism could have ensured that the universe was homogeneous on the scale of the perturbation mode.

Guth proposed that as the early universe cooled, it was trapped in a false vacuum with a high energy density, which is much like a cosmological constant. As the very early universe cooled it was trapped in a metastable state (it was supercooled), which it could only decay out of through the process of bubble nucleation via quantum tunneling. Bubbles of true vacuum spontaneously form in the sea of false vacuum and rapidly begin expanding at the speed of light. Guth recognized that this model was problematic because the model did not reheat properly: when the bubbles nucleated, they did not generate any radiation. Radiation could only be generated in collisions between bubble walls. But if inflation lasted long enough to solve the initial conditions problems, collisions between bubbles became exceedingly rare. In any one causal patch it is likely that only one bubble would nucleate.

11.3.3 Slow-roll inflation

The bubble collision problem was solved by Linde[53] and independently by Andreas Albrecht and Paul Steinhardt[54] in a model named *new inflation* or *slow-roll inflation* (Guth's model then became known as *old inflation*). In this model, instead of tunneling out of a false vacuum state, inflation occurred by a scalar field rolling down a potential energy hill. When the field rolls very slowly compared to the expansion of the Universe, inflation occurs. However, when the hill becomes steeper, inflation ends and reheating can occur.

11.3.4 Effects of asymmetries

Eventually, it was shown that new inflation does not produce a perfectly symmetric universe, but that quantum fluctuations in the inflaton are created. These fluctuations form the primordial seeds for all structure created in the later universe.[55] These fluctuations were first calculated by Viatcheslav Mukhanov and G. V. Chibisov in analyzing Starobinsky's similar model.[56][57][58] In the context of inflation, they were worked out independently of the work of Mukhanov and Chibisov at the three-week 1982 Nuffield Workshop on the Very Early Universe at Cambridge University.[59] The fluctuations were calculated by four groups working separately over the course of the workshop: Stephen Hawking;[60] Starobinsky;[61] Guth and So-Young Pi;[62] and Bardeen, Steinhardt and Turner.[63]

11.4 Observational status

Inflation is a mechanism for realizing the cosmological principle, which is the basis of the standard model of physical cosmology: it accounts for the homogeneity and isotropy of the observable universe. In addition, it accounts for the observed flatness and absence of magnetic monopoles. Since Guth's early work, each of these observations has received further confirmation, most impressively by the detailed observations of the cosmic microwave background made by the Wilkinson Microwave Anisotropy Probe (WMAP) spacecraft.[14] This analysis shows that the Universe is flat to within at least a few percent, and that it is homogeneous and isotropic to one part in 100,000.

In addition, inflation predicts that the structures visible in the Universe today formed through the gravitational collapse of perturbations that were formed as quantum mechanical fluctuations in the inflationary epoch. The detailed form of the spectrum of perturbations called a nearly-scale-invariant Gaussian random field (or Harrison–Zel'dovich spectrum) is very specific and has only two free parameters, the amplitude of the spectrum and the *spectral index*, which measures the slight deviation from scale invariance predicted by inflation (perfect scale invariance corresponds to the idealized de Sitter universe).[64] Inflation predicts that the observed perturbations should be in thermal equilibrium with each other (these are called *adiabatic* or *isentropic* perturbations). This structure for the perturbations has been confirmed by the WMAP spacecraft and other cosmic microwave background (CMB) experiments,[14] and galaxy surveys, especially the ongoing Sloan Digital Sky Survey.[65] These experiments have shown that the one part in 100,000 inhomogeneities observed have exactly the form predicted by theory. Moreover, there is evidence for a slight deviation from scale invariance. The *spectral index*, n_s is equal to one for a scale-invariant spectrum. The simplest inflation models predict that this quantity is between 0.92 and 0.98.[66][67][68][69] From WMAP data it can be inferred that $n_s = 0.963 \pm 0.012$,[70] implying that it differs from one at the level of two standard deviations (2σ). This is considered an important confirmation of the theory of inflation.[14]

Various inflation theories have been proposed that make radically different predictions, but they generally have much more fine tuning than should be necessary.[66][67] As a physical model, however, inflation is most valuable in that it robustly predicts the initial conditions of the Universe based on only two adjustable parameters: the spectral index (that can only change in a small range) and the amplitude of the perturbations. Except in contrived models, this is true regardless of how inflation is realized in particle physics.

Occasionally, effects are observed that appear to contradict the simplest models of inflation. The first-year WMAP data suggested that the spectrum might not be nearly scale-invariant, but might instead have a slight curvature.[71] However, the third-year data revealed that the effect was a statistical anomaly.[14] Another effect remarked upon since the first cosmic microwave background satellite, the Cosmic Background Explorer is that the amplitude of the quadrupole moment of the CMB is unexpectedly low and the other low multipoles appear to be preferentially aligned with the ecliptic plane. Some have claimed that this is a signature of non-Gaussianity and thus contradicts the simplest models of inflation. Others have suggested that the effect may be due to other new physics, foreground contamination, or even publication bias.[72]

An experimental program is underway to further test inflation with more precise CMB measurements. In particular, high precision measurements of the so-called "B-modes" of the polarization of the background radiation could provide evidence of the gravitational radiation produced by inflation, and could also show whether the energy scale of inflation predicted by the simplest models (10^{15}–10^{16} GeV) is correct.[67][68] In March 2014, it was announced that B-mode CMB polarization consistent with that predicted from inflation had been demonstrated by a South Pole experiment.[9][10][11][73][74][75] However, on 19 June 2014, lowered confidence in confirming the findings was reported;[74][76][77] on 19 September 2014, a further reduction in confidence was reported[78][79] and, on 30 January 2015, even less confidence yet was reported.[80][81]

Other potentially corroborating measurements are expected from the Planck spacecraft, although it is unclear if the signal will be visible, or if contamination from foreground sources will interfere.[82] Other forthcoming measurements, such as those of 21 centimeter radiation (radiation emitted and absorbed from neutral hydrogen before the first stars turned on), may measure the power spectrum with even greater resolution than the CMB and galaxy surveys, although it is not known if these measurements will be possible or if interference with radio sources on Earth and in the galaxy will be too great.[83]

Dark energy is broadly similar to inflation and is thought to be causing the expansion of the present-day universe to accelerate. However, the energy scale of dark energy is much lower, 10^{-12} GeV, roughly 27 orders of magnitude less than the scale of inflation.

11.5 Theoretical status

In Guth's early proposal, it was thought that the inflaton was the Higgs field, the field that explains the mass of the elementary particles.[47] It is now believed by some that the inflaton cannot be the Higgs field[84] although the recent discovery of the Higgs boson has increased the number of works considering the Higgs field as inflaton.[85] One problem of this identification is the current tension with experimental data at the electroweak scale,[86] which is currently under study at the Large Hadron Collider (LHC). Other models of inflation relied on the properties of Grand Unified Theories.[54] Since the simplest models of grand unification have failed, it is now thought by many physicists that inflation will be included in a supersymmetric theory such as string theory or a supersymmetric grand unified theory. At present, while inflation is understood principally by its detailed predictions of the initial conditions for the hot early universe, the particle physics is largely *ad hoc* modelling. As such, although predictions of inflation have been consistent with the results of observational tests, many open questions remain.

11.5.1 Fine-tuning problem

One of the most severe challenges for inflation arises from the need for fine tuning. In new inflation, the *slow-roll conditions* must be satisfied for inflation to occur. The slow-roll conditions say that the inflaton potential must be flat (compared to the large vacuum energy) and that the inflaton particles must have a small mass.[87] New inflation requires the Universe to have a scalar field with an especially flat potential and special initial conditions. However, explanations for these fine-tunings have been proposed. For example, classically scale invariant field theories, where scale invariance is broken by quantum effects, provide an explanation of the flatness of inflationary potentials, as long as the theory can be studied through perturbation theory.[88]

Andrei Linde

Linde proposed a theory known as *chaotic inflation* in which he suggested that the conditions for inflation were actually satisfied quite generically. Inflation will occur in virtually any universe that begins in a chaotic, high energy state that has a scalar field with unbounded potential energy.[89] However, in his model the inflaton field necessarily takes values larger than one Planck unit: for this reason, these are often called *large field* models and the competing new inflation models are called *small field* models. In this situation, the predictions of effective field theory are thought to be invalid, as renormalization should cause large corrections that could prevent inflation.[90] This problem has not yet been resolved and some cosmologists argue that the small field models, in which inflation can occur at a much lower energy scale, are better models.[91] While inflation depends on quantum field theory (and the semiclassical approximation to quantum gravity) in an important way, it has not been completely reconciled with these theories.

Brandenberger commented on fine-tuning in another situation.[92] The amplitude of the primordial inhomogeneities produced in inflation is directly tied to the energy scale of inflation. This scale is suggested to be around 10^{16} GeV or 10^{-3} times the Planck energy. The natural scale is naïvely the Planck scale so this small value could be seen as another form of fine-tuning (called a hierarchy problem): the energy density given by the scalar potential is down by 10^{-12} compared to the Planck density. This is not usually considered to be a critical problem, however, because the scale of inflation corresponds naturally to the scale of gauge unification.

11.5.2 Eternal inflation

Main article: Eternal inflation

In many models, the inflationary phase of the Universe's expansion lasts forever in at least some regions of the Universe. This occurs because inflating regions expand very rapidly, reproducing themselves. Unless the rate of decay to the non-inflating phase is sufficiently fast, new inflating regions are produced more rapidly than non-inflating regions. In such models most of the volume of the Universe at any given time is inflating. All models of eternal inflation produce an infinite multiverse, typically a fractal.

Although new inflation is classically rolling down the potential, quantum fluctuations can sometimes lift it to previous levels. These regions in which the inflaton fluctuates upwards expand much faster than regions in which the inflaton has a lower potential energy, and tend to dominate in terms of physical volume. This steady state, which first developed by Vilenkin,[93] is called "eternal inflation". It has been shown that any inflationary theory with an unbounded potential is eternal.[94] It is a popular conclusion among physicists that this steady state cannot continue forever into the past.[95][96][97] Inflationary spacetime, which is similar to de Sitter space, is incomplete without a contracting region. However, unlike de Sitter space, fluctuations in a contracting inflationary space collapse to form a gravitational singularity, a point where densities become infinite. Therefore, it is necessary to have a theory for the Universe's initial conditions. Linde, however, believes inflation may be past eternal.[98]

In eternal inflation, regions with inflation have an exponentially growing volume, while regions that are not inflating don't. This suggests that the volume of the inflating part of the Universe in the global picture is always unimaginably larger than the part that has stopped inflating, even though inflation eventually ends as seen by any single pre-inflationary observer. Scientists disagree about how to assign a probability distribution to this hypothetical anthropic landscape. If the probability of different regions is counted by volume, one should expect that inflation will never end or applying boundary conditions that a local observer exists to observe it, that inflation will end as late as possible. Some physicists believe this paradox can be resolved by weighting observers by their pre-inflationary volume.

11.5.3 Initial conditions

Some physicists have tried to avoid the initial conditions problem by proposing models for an eternally inflating universe with no origin.[99][100][101][102] These models propose that while the Universe, on the largest scales, expands exponentially it was, is and always will be, spatially infinite and has existed, and will exist, forever.

Other proposals attempt to describe the ex nihilo creation of the Universe based on quantum cosmology and the following inflation. Vilenkin put forth one such scenario.[93] Hartle and Hawking offered the no-boundary proposal for the initial creation of the Universe in which inflation comes about naturally.[103]

Guth described the inflationary universe as the "ultimate free lunch":[104][105] new universes, similar to our own, are continually produced in a vast inflating background. Gravitational interactions, in this case, circumvent (but do not violate) the first law of thermodynamics (energy conservation) and the second law of thermodynamics (entropy and the arrow of time problem). However, while there is consensus that this solves the initial conditions problem, some have disputed this, as it is much more likely that the Universe came about by a quantum fluctuation. Don Page was an outspoken critic of inflation because of this anomaly.[106] He stressed that the thermodynamic arrow of time necessitates low entropy initial conditions, which would be highly unlikely. According to them, rather than solving this problem, the inflation theory aggravates it – the reheating at the end of the inflation era increases entropy, making it necessary for the initial state of the Universe to be even more orderly than in other Big Bang theories with no inflation phase.

Hawking and Page later found ambiguous results when they attempted to compute the probability of inflation in the Hartle-Hawking initial state.[107] Other authors have argued that, since inflation is eternal, the probability doesn't matter as long as it is not precisely zero: once it starts, inflation perpetuates itself and quickly dominates the Universe.[4][108]:223–225 However, Albrecht and Lorenzo Sorbo argued that the probability of an inflationary cosmos, consistent with today's observations, emerging by a random fluctuation from some pre-existent state is much higher than that of a non-inflationary cosmos. This is because the "seed" amount of non-gravitational energy required for the inflationary cosmos is so much less than that for a non-inflationary alternative, which outweighs any entropic considerations.[109]

Another problem that has occasionally been mentioned is the trans-Planckian problem or trans-Planckian effects.[110] Since the energy scale of inflation and the Planck scale are relatively close, some of the quantum fluctuations that have made up the structure in our universe were smaller than the Planck length before inflation. Therefore, there ought to be corrections from Planck-scale physics, in particular the unknown quantum theory of gravity. Some disagreement remains about the magnitude of this effect: about whether it is just on the threshold of detectability or completely undetectable.[111]

11.5.4 Hybrid inflation

Another kind of inflation, called *hybrid inflation*, is an extension of new inflation. It introduces additional scalar fields, so that while one of the scalar fields is responsible for normal slow roll inflation, another triggers the end of inflation: when inflation has continued for sufficiently long, it becomes favorable to the second field to decay into a much lower energy state.[112]

In hybrid inflation, one scalar field is responsible for most of the energy density (thus determining the rate of expansion), while another is responsible for the slow roll (thus determining the period of inflation and its termination). Thus fluctuations in the former inflaton would not affect inflation termination, while fluctuations in the latter would not affect the rate of expansion. Therefore, hybrid inflation is not eternal.[113][114] When the second (slow-rolling) inflaton reaches the bottom of its potential, it changes the location of the minimum of the first inflaton's potential, which leads to a fast roll of the inflaton down its potential, leading to termination of inflation.

11.5.5 Inflation and string cosmology

The discovery of flux compactifications opened the way for reconciling inflation and string theory.[115] *Brane inflation* suggests that inflation arises from the motion of D-branes[116] in the compactified geometry, usually towards a stack of anti-D-branes. This theory, governed by the *Dirac-Born-Infeld action*, is different from ordinary inflation. The dynamics are not completely understood. It appears that special conditions are necessary since inflation occurs in tunneling between two vacua in the string landscape. The process of tunneling between two vacua is a form of old inflation, but new inflation must then occur by some other mechanism.

11.5.6 Inflation and loop quantum gravity

When investigating the effects the theory of loop quantum gravity would have on cosmology, a loop quantum cosmology model has evolved that provides a possible mechanism for cosmological inflation. Loop quantum gravity assumes a quantized spacetime. If the energy density is larger than can be held by the quantized spacetime, it is thought to bounce back.[117]

11.6 Alternatives

Other models explain some of the observations explained by inflation. However none of these "alternatives" has the same breadth of explanation and still require inflation for a more complete fit with observation. They should therefore be regarded as adjuncts to inflation, rather than as alternatives.

11.6.1 Big bounce

The big bounce hypothesis attempts to replace the cosmic singularity with a cosmic contraction and bounce, thereby explaining the initial conditions that led to the big bang.[118] The flatness and horizon problems are naturally solved in the Einstein-Cartan-Sciama-Kibble theory of gravity, without needing an exotic form of matter or free parameters.[119][120] This theory extends general relativity by removing a constraint of the symmetry of the affine connection and regarding its antisymmetric part, the torsion tensor, as a dynamical variable. The minimal coupling between torsion and Dirac spinors generates a spin-spin interaction that is significant in fermionic matter at extremely high densities. Such an interaction averts the unphysical Big Bang singularity, replacing it with a cusp-like bounce at a finite minimum scale factor, before which the Universe was contracting. The rapid expansion immediately after the Big Bounce explains why the present Universe at largest scales appears spatially flat, homogeneous and isotropic. As the density of the Universe decreases, the effects of torsion weaken and the Universe smoothly enters the radiation-dominated era.

11.6.2 String theory

String theory requires that, in addition to the three observable spatial dimensions, additional dimensions exist that are curled up or compactified (see also Kaluza–Klein theory). Extra dimensions appear as a frequent component of supergravity models and other approaches to quantum gravity. This raised the contingent question of why four space-time dimensions became large and the rest became unobservably small. An attempt to address this question, called *string gas cosmology*, was proposed by Robert Brandenberger and Cumrun Vafa.[121] This model focuses on the dynamics of the early universe considered as a hot gas of strings. Brandenberger and Vafa show that a dimension of spacetime can only expand if the strings that wind around it can efficiently annihilate each other. Each string is a one-dimensional object, and the largest number of dimensions in which two strings will generically intersect (and, presumably, annihilate) is three. Therefore, the most likely number of non-compact (large) spatial dimensions is three. Current work on this model centers on whether it can succeed in stabilizing the size of the compactified dimensions and produce the correct spectrum of primordial density perturbations.[122] Supporters admit that their model "does not solve the entropy and flatness problems of standard cosmology ..... and we can provide no explanation for why the current universe is so close to being spatially flat".[123]

11.6.3 Ekpyrotic and cyclic models

The ekpyrotic and cyclic models are also considered adjuncts to inflation. These models solve the horizon problem through an expanding epoch well *before* the Big Bang, and then generate the required spectrum of primordial density perturbations during a contracting phase leading to a Big Crunch. The Universe passes through the Big Crunch and emerges in a hot Big Bang phase. In this sense they are reminiscent of Richard Chace Tolman's oscillatory universe; in Tolman's model, however, the total age of the Universe is necessarily finite, while in these models this is not necessarily so. Whether the correct spectrum of density fluctuations can be produced, and whether the Universe can successfully navigate the Big Bang/Big Crunch transition, remains a topic of controversy and current research. Ekpyrotic models avoid the magnetic monopole problem as long as the temperature at the Big Crunch/Big Bang transition remains below the Grand Unified Scale, as this is the temperature required to produce magnetic monopoles in the first place. As things stand, there is no evidence of any 'slowing down' of the expansion, but this is not surprising as each cycle is expected to last on the order of a trillion years.

11.6.4 Varying C

Another adjunct, the varying speed of light model was offered by Jean-Pierre Petit in 1988, John Moffat in 1992 as well Albrecht and João Magueijo in 1999, instead of superluminal expansion the speed of light was 60 orders of magnitude faster than its current value solving the horizon and homogeneity problems in the early universe.

11.7 Criticisms

Since its introduction by Alan Guth in 1980, the inflationary paradigm has become widely accepted. Nevertheless, many physicists, mathematicians, and philosophers of science have voiced criticisms, claiming untestable predictions and a lack of serious empirical support.[4] In 1999, John Earman and Jesús Mosterín published a thorough critical review of inflationary cosmology, concluding, "we do not think that there are, as yet, good grounds for admitting any of the models of inflation into the standard core of cosmology."[5]

In order to work, and as pointed out by Roger Penrose from 1986 on, inflation requires extremely specific initial conditions of its own, so that the problem (or pseudo-problem) of initial conditions is not solved: "There is something fundamentally misconceived about trying to explain the uniformity of the early universe as resulting from a thermalization process. [...] For, if the thermalization is actually doing anything [...] then it represents a definite increasing of the entropy. Thus, the universe would have been even more special before the thermalization than after."[124] The problem of specific or "fine-tuned" initial conditions would not have been solved; it would have gotten worse. At a conference in 2015, Penrose said that "inflation isn't falsifiable, it's falsified. [...] BICEP did a wonderful service by bringing all the Inflation-ists out of their shell, and giving them a black eye."[6]

A recurrent criticism of inflation is that the invoked inflation field does not correspond to any known physical field, and that its potential energy curve seems to be an ad hoc contrivance to accommodate almost any data obtainable. Paul Steinhardt, one of the founding fathers of inflationary cosmology, has recently become one of its sharpest critics. He calls 'bad inflation' a period of accelerated expansion whose outcome conflicts with observations, and 'good inflation' one compatible with them: "Not only is bad inflation more likely than good inflation, but no inflation is more likely than either.... Roger Penrose considered all the possible configurations of the inflaton and gravitational fields. Some of these configurations lead to inflation ... Other configurations lead to a uniform, flat universe directly – without inflation. Obtaining a flat universe is unlikely overall. Penrose's shocking conclusion, though, was that obtaining a flat universe without inflation is much more likely than with inflation – by a factor of 10 to the googol (10 to the 100) power!"[4][108] Together with Anna Ijjas and Abraham Loeb, he wrote articles claiming that the inflationary paradigm is in trouble in view of the data from the Planck satellite.[125][126] Counter-arguments were presented by Alan Guth, David Kaiser, and Yasunori Nomura[127] and by Andrei Linde,[128] saying that "cosmic inflation is on a stronger footing than ever before".[127]

11.8 See also

- Brane cosmology
- Conservation of angular momentum
- Cosmology
- Dark flow
- Doughnut theory of the universe
- Hubble's law
- Non-minimally coupled inflation
- Nonlinear optics
- Varying speed of light
- Warm inflation

11.9 Notes

[1] "First Second of the Big Bang". *How The Universe Works 3*. 2014. Discovery Science.

[2] Tyson, Neil deGrasse and Donald Goldsmith (2004), *Origins: Fourteen Billion Years of Cosmic Evolution*, W. W. Norton & Co., pp. 84–5.

[3] Tsujikawa, Shinji (28 Apr 2003). "Introductory review of cosmic inflation": 4257. arXiv:hep-ph/0304257. Bibcode:2003hep.ph....4257T. In fact temperature anisotropies observed by the COBE satellite in 1992 exhibit nearly scale-invariant spectra as predicted by the inflationary paradigm. Recent observations of WMAP also show strong evidence for inflation.

[4] Steinhardt, Paul J. (2011). "The inflation debate: Is the theory at the heart of modern cosmology deeply flawed?" (*Scientific American*, April; pp. 18-25).

[5] Earman, John; Mosterín, Jesús (March 1999). "A Critical Look at Inflationary Cosmology". *Philosophy of Science* **66**: 1–49. doi:10.2307/188736 (inactive 2015-01-14). JSTOR 188736.

[6] Hložek, Renée (12 June 2015). "CMB@50 day three". Retrieved 15 July 2015.
This is a collation of remarks from the third day of the "Cosmic Microwave Background @50" conference held at Princeton, 10–12 June 2015.

[7] Guth, Alan H. (1997). *The Inflationary Universe: The Quest for a New Theory of Cosmic Origins*. Basic Books. pp. 233–234. ISBN 0201328402.

[8] "The Medallists: A list of past Dirac Medallists". *ictp.it*.

[9] Staff (17 March 2014). "BICEP2 2014 Results Release". *National Science Foundation*. Retrieved 18 March 2014.

[10] Clavin, Whitney (17 March 2014). "NASA Technology Views Birth of the Universe". *NASA*. Retrieved 17 March 2014.

[11] Overbye, Dennis (17 March 2014). "Space Ripples Reveal Big Bang's Smoking Gun". *The New York Times*. Retrieved 17 March 2014.

[12] Using Tiny Particles To Answer Giant Questions. Science Friday, 3 April 2009.

[13] See also Faster than light#Universal expansion.

[14] Spergel, D.N. (2006). "Three-year Wilkinson Microwave Anisotropy Probe (WMAP) observations: Implications for cosmology". WMAP... confirms the basic tenets of the inflationary paradigm...

[15] "Our Baby Universe Likely Expanded Rapidly. Study Suggests". *Space.com*.

[16] Melia, Fulvio (2007). "The Cosmic Horizon". *Monthly Notices of the Royal Astronomical Society* **382** (4): 1917–1921. arXiv:0711.4181. Bibcode:2007MNRAS.382.1917M. doi:10.1111/j.1365-2966.2007.12499.x.

[17] Melia, Fulvio; et al. (2009). "The Cosmological Spacetime". *International Journal of Modern Physics D* **18** (12): 1889–1901. arXiv:0907.5394. Bibcode:2009IJMPD..18.1889M. doi:10.1142/s0218271809015746.

[18] Kolb and Turner (1988).

[19] Barbara Sue Ryden (2003). *Introduction to cosmology*. Addison-Wesley. ISBN 978-0-8053-8912-8. Not only is inflation very effective at driving down the number density of magnetic monopoles, it is also effective at driving down the number density of every other type of particle, including photons.:202–207

[20] This is usually quoted as 60 *e*-folds of expansion, where $e^{60} \approx 10^{26}$. It is equal to the amount of expansion since reheating, which is roughly $E_{inflation}/T_0$, where $T_0 = 2.7$ K is the temperature of the cosmic microwave background today. See, *e.g.* Kolb and Turner (1998) or Liddle and Lyth (2000).

[21] Guth, *Phase transitions in the very early universe*, in *The Very Early Universe*, ISBN 0-521-31677-4 eds Hawking, Gibbon & Siklos

[22] See Kolb and Turner (1988) or Mukhanov (2005).

[23] Kofman, Lev; Linde, Andrei; Starobinsky, Alexei (1994). "Reheating after inflation". *Physical Review Letters* **73** (5): 3195–3198. arXiv:hep-th/9405187. Bibcode:1986CQGra...3..811K. doi:10.1088/0264-9381/3/5/011.

[24] Much of the historical context is explained in chapters 15–17 of Peebles (1993).

[25] Misner, Charles W.; Coley, A A; Ellis, G F R; Hancock, M (1968). "The isotropy of the universe". *Astrophysical Journal* **151** (2): 431. Bibcode:1998CQGra..15..331W. doi:10.1088/0264-9381/15/2/008.

[26] Misner, Charles; Thorne, Kip S. and Wheeler, John Archibald (1973). *Gravitation*. San Francisco: W. H. Freeman. pp. 489–490, 525–526. ISBN 0-7167-0344-0.

[27] Weinberg, Steven (1971). *Gravitation and Cosmology*. John Wiley. pp. 740, 815. ISBN 0-471-92567-5.

[28] Lemaître, Georges (1933). "The expanding universe". *Annales de la Société Scientifique de Bruxelles* **47A**: 49., English in *Gen. Rel. Grav.* **29**:641–680, 1997.

[29] R. C. Tolman (1934). *Relativity, Thermodynamics, and Cosmology*. Oxford: Clarendon Press. ISBN 0-486-65383-8. LCCN 34032023. Reissued (1987) New York: Dover ISBN 0-486-65383-8.

[30] Misner, Charles W.; Leach, P G L (1969). "Mixmaster universe". *Physical Review Letters* **22** (15): 1071–74. Bibcode:2008JPhA...41o5201A. doi:10.1088/1751-8113/41/15/155201.

[31] Dicke, Robert H. (1970). *Gravitation and the Universe*. Philadelphia: American Philosopical Society.

[32] Dicke, Robert H.; P. J. E. Peebles (1979). "The big bang cosmology – enigmas and nostrums". In ed. S. W. Hawking and W. Israel. *General Relativity: an Einstein Centenary Survey*. Cambridge University Press.

[33] Alan P. Lightman (1 January 1993). *Ancient Light: Our Changing View of the Universe*. Harvard University Press. ISBN 978-0-674-03363-4.

[34] "WMAP- Content of the Universe". *nasa.gov*.

[35] Since supersymmetric Grand Unified Theory is built into string theory, it is still a triumph for inflation that it is able to deal with these magnetic relics. See, *e.g.* Kolb and Turner (1988) and Raby, Stuart (2006). ed. Bruce Hoeneisen, ed. "Grand Unified Theories". arXiv:hep-ph/0608183.

[36] 't Hooft, Gerard (1974). "Magnetic monopoles in Unified Gauge Theories". *Nuclear Physics B* **79** (2): 276–84. Bibcode:1974NuPhB..79..276T. doi:10.1016/0550-3213(74)90486-6.

[37] Polyakov, Alexander M. (1974). "Particle spectrum in quantum field theory". *JETP Letters* **20**: 194–5. Bibcode:1974JETPL..20..194P.

[38] Guth, Alan; Tye, S. (1980). "Phase Transitions and Magnetic Monopole Production in the Very Early Universe". *Physical Review Letters* **44** (10): 631–635; Erratum *ibid.*,**44**:963, 1980. Bibcode:1980PhRvL..44..631G. doi:10.1103/PhysRevLett.44.631.

[39] Einhorn, Martin B; Stein, D. L.; Toussaint, Doug (1980). "Are Grand Unified Theories Compatible with Standard Cosmology?". *Physical Review D* **21** (12): 3295–3298. Bibcode:1980PhRvD..21.3295E. doi:10.1103/PhysRevD.21.3295.

[40] Zel'dovich, Ya.; Khlopov, M. Yu. (1978). "On the concentration of relic monopoles in the universe". *Physics Letters B* **79** (3): 239–41. Bibcode:1978PhLB...79..239Z. doi:10.1016/0370-2693(78)90232-0.

[41] Preskill, John (1979). "Cosmological production of superheavy magnetic monopoles". *Physical Review Letters* **43** (19): 1365–1368. Bibcode:1979PhRvL..43.1365P. doi:10.1103/PhysRevLett.43.1365.

[42] See, *e.g.* Yao, W.–M.; Amsler, C.; Asner, D.; Barnett, R. M.; Beringer, J.; Burchat, P. R.; Carone, C. D.; Caso, C.; Dahl, O.; d'Ambrosio, G.; De Gouvea, A.; Doser, M.; Eidelman, S.; Feng, J. L.; Gherghetta, T.; Goodman, M.; Grab, C.; Groom, D. E.; Gurtu, A.; Hagiwara, K.; Hayes, K. G.; Hernández-Rey, J. J.; Hikasa, K.; Jawahery, H.; Kolda, C.; Kwon, Y.; Mangano, M. L.; Manohar, A. V.; Masoni, A.; et al. (2006). "Review of Particle Physics". *J. Phys. G* **33** (1): 1–1232. arXiv:astro-ph/0601168. Bibcode:2006JPhG...33....1Y. doi:10.1088/0954-3899/33/1/001.

[43] Rees, Martin. (1998). *Before the Beginning* (New York: Basic Books) p. 185 ISBN 0-201-15142-1

[44] de Sitter, Willem (1917). "Einstein's theory of gravitation and its astronomical consequences. Third paper". *Monthly Notices of the Royal Astronomical Society* **78**: 3–28. Bibcode:1917MNRAS..78....3D. doi:10.1093/mnras/78.1.3.

[45] Starobinsky, A. A. (December 1979). "Spectrum Of Relict Gravitational Radiation And The Early State Of The Universe". *Journal of Experimental and Theoretical Physics Letters* **30**: 682. Bibcode:1979JETPL..30..682S.; Starobinskii, A. A. (December 1979). "Spectrum of relict gravitational radiation and the early state of the universe". *Pisma Zh. Eksp. Teor. Fiz. (Soviet Journal of Experimental and Theoretical Physics Letters)* **30**: 719. Bibcode:1979ZhPmR..30..719S.

[46] SLAC seminar, "10^{-35} seconds after the Big Bang", 23 January 1980. see Guth (1997), pg 186

[47] Guth, Alan H. (1981). "Inflationary universe: A possible solution to the horizon and flatness problems" (PDF). *Physical Review D* **23** (2): 347–356. Bibcode:1981PhRvD..23..347G. doi:10.1103/PhysRevD.23.347.

[48] Chapter 17 of Peebles (1993).

[49] Starobinsky, Alexei A. (1980). "A new type of isotropic cosmological models without singularity". *Physics Letters B* **91**: 99–102. Bibcode:1980PhLB...91...99S. doi:10.1016/0370-2693(80)90670-X.

[50] Kazanas, D. (1980). "Dynamics of the universe and spontaneous symmetry breaking". *Astrophysical Journal* **241**: L59–63. Bibcode:1980ApJ...241L..59K. doi:10.1086/183361.

[51] Sato, K. (1981). "Cosmological baryon number domain structure and the first order phase transition of a vacuum". *Physics Letters B* **33**: 66–70. Bibcode:1981PhLB...99...66S. doi:10.1016/0370-2693(81)90805-4.

[52] Einhorn, Martin B; Sato, Katsuhiko (1981). "Monopole Production In The Very Early Universe In A First Order Phase Transition". *Nuclear Physics B* **180** (3): 385–404. Bibcode:1981NuPhB.180..385E. doi:10.1016/0550-3213(81)90057-2.

[53] Linde, A (1982). "A new inflationary universe scenario: A possible solution of the horizon, flatness, homogeneity, isotropy and primordial monopole problems". *Physics Letters B* **108** (6): 389–393. Bibcode:1982PhLB..108..389L. doi:10.1016/0370-2693(82)91219-9.

[54] Albrecht, Andreas; Steinhardt, Paul (1982). "Cosmology for Grand Unified Theories with Radiatively Induced Symmetry Breaking" (PDF). *Physical Review Letters* **48** (17): 1220–1223. Bibcode:1982PhRvL..48.1220A. doi:10.1103/PhysRevLett.48.1220.

[55] J.B. Hartle (2003). *Gravity: An Introduction to Einstein's General Relativity* (1st ed.). Addison Wesley. p. 411. ISBN 0-8053-8662-9

[56] See Linde (1990) and Mukhanov (2005).

[57] Chibisov, Viatcheslav F.; Chibisov, G. V. (1981). "Quantum fluctuation and "nonsingular" universe". *JETP Letters* **33**: 532–5. Bibcode:1981JETPL..33..532M.

[58] Mukhanov, Viatcheslav F. (1982). "The vacuum energy and large scale structure of the universe". *Soviet Physics JETP* **56**: 258–65.

[59] See Guth (1997) for a popular description of the workshop, or *The Very Early Universe*, ISBN 0-521-31677-4 eds Hawking, Gibbon & Siklos for a more detailed report

[60] Hawking, S.W. (1982). "The development of irregularities in a single bubble inflationary universe". *Physics Letters B* **115** (4): 295–297. Bibcode:1982PhLB..115..295H. doi:10.1016/0370-2693(82)90373-2.

[61] Starobinsky, Alexei A. (1982). "Dynamics of phase transition in the new inflationary universe scenario and generation of perturbations". *Physics Letters B* **117** (3–4): 175–8. Bibcode:1982PhLB..117..175S. doi:10.1016/0370-2693(82)90541-X.

[62] Guth, A.H. (1982). "Fluctuations in the new inflationary universe". *Physical Review Letters* **49** (15): 1110–3. Bibcode:1982PhRvL..49.1110G. doi:10.1103/PhysRevLett.49.1110.

[63] Bardeen, James M.; Steinhardt, Paul J.; Turner, Michael S. (1983). "Spontaneous creation Of almost scale-free density perturbations in an inflationary universe". *Physical Review D* **28** (4): 679–693. Bibcode:1983PhRvD..28..679B. doi:10.1103/PhysRevD.28.679.

[64] Perturbations can be represented by Fourier modes of a wavelength. Each Fourier mode is normally distributed (usually called Gaussian) with mean zero. Different Fourier components are uncorrelated. The variance of a mode depends only on its wavelength in such a way that within any given volume each wavelength contributes an equal amount of power to the spectrum of perturbations. Since the Fourier transform is in three dimensions, this means that the variance of a mode goes as k^{-3} to compensate for the fact that within any volume, the number of modes with a given wavenumber k goes as k^3.

[65] Tegmark, M.; Eisenstein, Daniel J.; Strauss, Michael A.; Weinberg, David H.; Blanton, Michael R.; Frieman, Joshua A.; Fukugita, Masataka; Gunn, James E.; et al. (August 2006). "Cosmological constraints from the SDSS luminous red galaxies". *Physical Review D* **74** (12). arXiv:astro-ph/0608632. Bibcode:2006PhRvD..74l3507T. doi:10.1103/PhysRevD.74.123507.

[66] Steinhardt, Paul J. (2004). "Cosmological perturbations: Myths and facts". *Modern Physics Letters A* **19** (13 & 16): 967–82. Bibcode:2004MPLA...19..967S. doi:10.1142/S0217732304014252.

[67] Boyle, Latham A.; Steinhardt, PJ; Turok, N (2006). "Inflationary predictions for scalar and tensor fluctuations reconsidered". *Physical Review Letters* **96** (11): 111301. arXiv:astro-ph/0507455. Bibcode:2006PhRvL..96k1301B. doi:10.1103/PhysRevLett.96.111301. PMID 16605810.

[68] Tegmark, Max (2005). "What does inflation really predict?". *JCAP* **0504** (4): 001. arXiv:astro-ph/0410281. Bibcode:2005JCAP...04..001T. doi:10.1088/1475-7516/2005/04/001.

[69] This is known as a "red" spectrum, in analogy to redshift, because the spectrum has more power at longer wavelengths.

[70] Komatsu, E.; Smith, K. M.; Dunkley, J.; Bennett, C. L.; Gold, B.; Hinshaw, G.; Jarosik, N.; Larson, D.; et al. (January 2010). "Seven-Year Wilkinson Microwave Anisotropy Probe (WMAP) Observations: Cosmological Interpretation". *The Astrophysical Journal Supplement Series* **192** (2): 18. arXiv:1001.4538. Bibcode:2011ApJS..192...18K. doi:10.1088/0067-0049/192/2/18.

[71] Spergel, D. N.; Verde, L.; Peiris, H. V.; Komatsu, E.; Nolta, M. R.; Bennett, C. L.; Halpern, M.; Hinshaw, G.; et al. (2003). "First year Wilkinson Microwave Anisotropy Probe (WMAP) observations: determination of cosmological parameters". *Astrophysical Journal Supplement Series* **148** (1): 175–194. arXiv:astro-ph/0302209. Bibcode:2003ApJS..148..175S. doi:10.1086/377226.

[72] See cosmic microwave background#Low multipoles for details and references.

[73] Overbye, Dennis (24 March 2014). "Ripples From the Big Bang". *New York Times*. Retrieved 24 March 2014.

[74] Ade, P.A.R. (BICEP2 Collaboration); et al. (19 June 2014). "Detection of B-Mode Polarization at Degree Angular Scales by BICEP2". *Physical Review Letters* **112** (24): 241101. arXiv:1403.3985. Bibcode:2014PhRvL.112x1101A. doi:10.1103/PhysRevLett.112.241101. PMID 24996078.

[75] Woit, Peter (13 May 2014). "BICEP2 News". *Not Even Wrong*. Columbia University. Retrieved 19 January 2014.

[76] Overbye, Dennis (19 June 2014). "Astronomers Hedge on Big Bang Detection Claim". *New York Times*. Retrieved 20 June 2014.

[77] Amos, Jonathan (19 June 2014). "Cosmic inflation: Confidence lowered for Big Bang signal". *BBC News*. Retrieved 20 June 2014.

[78] Planck Collaboration Team (19 September 2014). "Planck intermediate results. XXX. The angular power spectrum of polarized dust emission at intermediate and high Galactic latitudes". *ArXiv*. arXiv:1409.5738. Bibcode:2014arXiv1409.5738P. Retrieved 22 September 2014.

[79] Overbye, Dennis (22 September 2014). "Study Confirms Criticism of Big Bang Finding". *New York Times*. Retrieved 22 September 2014.

[80] Clavin, Whitney (30 January 2015). "Gravitational Waves from Early Universe Remain Elusive". *NASA*. Retrieved 30 January 2015.

[81] Overbye, Dennis (30 January 2015). "Speck of Interstellar Dust Obscures Glimpse of Big Bang". *New York Times*. Retrieved 31 January 2015.

[82] Rosset, C.; PLANCK-HFI collaboration (2005). "Systematic effects in CMB polarization measurements". *Exploring the universe: Contents and structures of the universe (XXXIXth Rencontres de Moriond)*.

[83] Loeb, A.; Zaldarriaga, M (2004). "Measuring the small-scale power spectrum of cosmic density fluctuations through 21 cm tomography prior to the epoch of structure formation". *Physical Review Letters* **92** (21): 211301. arXiv:astro-ph/0312134. Bibcode:2004PhRvL..92u1301L. doi:10.1103/PhysRevLett.92.211301. PMID 15245272.

[84] Guth, Alan (1997). *The Inflationary Universe*. Addison–Wesley. ISBN 0-201-14942-7.

[85] Choi, Charles (Jun 29, 2012). "Could the Large Hadron Collider Discover the Particle Underlying Both Mass and Cosmic Inflation?". Scientific American. Retrieved Jun 25, 2014."The virtue of so-called Higgs inflation models

is that they might explain inflation within the current Standard Model of particle physics, which successfully describes how most known particles and forces behave. Interest in the Higgs is running hot this summer because CERN, the lab in Geneva, Switzerland, that runs the LHC, has said it will announce highly anticipated findings regarding the particle in early July."

[86] Salvio, Alberto (2013-08-09). "Higgs Inflation at NNLO after the Boson Discovery". *Phys.Lett. B727 (2013) 234-239* **727**: 234–239. arXiv:1308.2244. Bibcode:2013PhLB..727..234S. doi:10.1016/j.physletb.2013.10.042.

[87] Technically, these conditions are that the logarithmic derivative of the potential, $\epsilon = (1/2)(V'/V)^2$ and second derivative $\eta = V''/V$ are small, where V is the potential and the equations are written in reduced Planck units. See, *e.g.* Liddle and Lyth (2000), pg 42-43.

[88] Salvio, Strumia (2014-03-17). "Agravity". *JHEP 1406 (2014) 080* **2014**. arXiv:1403.4226. Bibcode:2014JHEP...06..080S. doi:10.1007/JHEP06(2014)080.

[89] Linde, Andrei D. (1983). "Chaotic inflation". *Physics Letters B* **129** (3): 171–81. Bibcode:1983PhLB..129..177L. doi:10.1016/0370-2693(83)90837-7.

[90] Technically, this is because the inflaton potential is expressed as a Taylor series in φ/mP_l, where φ is the inflaton and mP_l is the Planck mass. While for a single term, such as the mass term $m_\varphi{}^4(\varphi/mP_l)^2$, the slow roll conditions can be satisfied for φ much greater than mP_l, this is precisely the situation in effective field theory in which higher order terms would be expected to contribute and destroy the conditions for inflation. The absence of these higher order corrections can be seen as another sort of fine tuning. See *e.g.* Alabidi, Laila; Lyth, David H (2006). "Inflation models and observation". *JCAP* **0605** (5): 016. arXiv:astro-ph/0510441. Bibcode:2006JCAP...05..016A. doi:10.1088/1475-7516/2006/05/016.

[91] See, *e.g.* Lyth, David H. (1997). "What would we learn by detecting a gravitational wave signal in the cosmic microwave background anisotropy?". *Physical Review Letters* **78** (10): 1861–3. arXiv:hep-ph/9606387. Bibcode:1997PhRvL..78.1861L. doi:10.1103/PhysRevLett.78.1861.

[92] Brandenberger, Robert H. (November 2004). "Challenges for inflationary cosmology". arXiv:astro-ph/0411671.

[93] Vilenkin, Alexander (1983). "The birth of inflationary universes". *Physical Review D* **27** (12): 2848–2855. Bibcode:1983PhRvD..27.2848V. doi:10.1103/PhysRevD.27.2848.

[94] A. Linde (1986). "Eternal chaotic inflation". *Modern Physics Letters A* **1** (2): 81–85. Bibcode:1986MPLA....1...81L. doi:10.1142/S0217732386000129. A. Linde (1986). "Eternally existing self-reproducing chaotic inflationary universe" (PDF). *Physics Letters B* **175** (4): 395–400. Bibcode:1986PhLB..175..395L. doi:10.1016/0370-2693(86)90611-8.

[95] A. Borde; A. Guth; A. Vilenkin (2003). "Inflationary space-times are incomplete in past directions". *Physical Review Letters* **90** (15): 151301. arXiv:gr-qc/0110012. Bibcode:2003PhRvL..90o1301B. doi:10.1103/PhysRevLett.90.151301. PMID 12732026.

[96] A. Borde (1994). "Open and closed universes, initial singularities and inflation". *Physical Review D* **50** (6): 3692–702. arXiv:gr-qc/9403049. Bibcode:1994PhRvD..50.3692B. doi:10.1103/PhysRevD.50.3692.

[97] A. Borde; A. Vilenkin (1994). "Eternal inflation and the initial singularity". *Physical Review Letters* **72** (21): 3305–9. arXiv:gr-qc/9312022. Bibcode:1994PhRvL..72.3305B. doi:10.1103/PhysRevLett.72.3305.

[98] Linde (2005, §V).

[99] Carroll, Sean M.; Chen, Jennifer (2005). "Does inflation provide natural initial conditions for the universe?". *Gen. Rel. Grav.* **37** (10): 1671–4. arXiv:gr-qc/0505037. Bibcode:2005GReGr..37.1671C. doi:10.1007/s10714-005-0148-2.

[100] Carroll, Sean M.; Jennifer Chen (2004). "Spontaneous inflation and the origin of the arrow of time". arXiv:hep-th/0410270.

[101] Aguirre, Anthony; Gratton, Steven (2003). "Inflation without a beginning: A null boundary proposal". *Physical Review D* **67** (8): 083515. arXiv:gr-qc/0301042. Bibcode:2003PhRvD..67h3515A. doi:10.1103/PhysRevD.67.083515.

[102] Aguirre, Anthony; Gratton, Steven (2002). "Steady-State Eternal Inflation". *Physical Review D* **65** (8): 083507. arXiv:astro-ph/0111191. Bibcode:2002PhRvD..65h3507A. doi:10.1103/PhysRevD.65.083507.

[103] Hartle, J.; Hawking, S. (1983). "Wave function of the universe". *Physical Review D* **28** (12): 2960–2975. Bibcode:1983PhRvD..28.2960H. doi:10.1103/PhysRevD.28.2960.; See also Hawking (1998).

[104] Hawking (1998), p. 129.

[105] Wikiquote

[106] Page, Don N. (1983). "Inflation does not explain time asymmetry". *Nature* **304** (5921): 39–41. Bibcode:1983Natur.304...39P. doi:10.1038/304039a0.; see also Roger Penrose's book The Road to Reality: A Complete Guide to the Laws of the Universe.

[107] Hawking, S. W.; Page, Don N. (1988). "How probable is inflation?". *Nuclear Physics B* **298** (4): 789–809. Bibcode:1988NuPhB.298..789H. doi:10.1016/0550-3213(88)90008-9.

[108] Paul J. Steinhardt; Neil Turok (2007). *Endless Universe: Beyond the Big Bang*. Broadway Books. ISBN 978-0-7679-1501-4.

[109] Albrecht, Andreas; Sorbo, Lorenzo (2004). "Can the universe afford inflation?". *Physical Review D* **70** (6): 063528. arXiv:hep-th/0405270. Bibcode:2004PhRvD..70f3528A. doi:10.1103/PhysRevD.70.063528.

[110] Martin, Jerome; Brandenberger, Robert (2001). "The trans-Planckian problem of inflationary cosmology". *Physical Review D* **63** (12): 123501. arXiv:hep-th/0005209. Bibcode:2001PhRvD..63l3501M. doi:10.1103/PhysRevD.63.123501.

[111] Martin, Jerome; Ringeval, Christophe (2004). "Superimposed Oscillations in the WMAP Data?". *Physical Review D* **69** (8): 083515. arXiv:astro-ph/0310382. Bibcode:2004PhRvD..69h3515M. doi:10.1103/PhysRevD.69.083515.

[112] Robert H. Brandenberger, "A Status Review of Inflationary Cosmology", proceedings Journal-ref: BROWN-HET-1256 (2001), (available from arXiv:hep-ph/0101119v1 11 January 2001)

[113] Andrei Linde, "Prospects of Inflation", *Physica Scripta Online* (2004) (available from arXiv:hep-th/0402051)

[114] Blanco-Pillado et al., "Racetrack inflation", (2004) (available from arXiv:hep-th/0406230)

[115] Kachru, Shamit; Kallosh, Renata; Linde, Andrei; Maldacena, Juan; McAllister, Liam; Trivedi, Sandip P (2003). "Towards inflation in string theory". *JCAP* **0310** (10): 013. arXiv:hep-th/0308055. Bibcode:2003JCAP...10..013K. doi:10.1088/1475-7516/2003/10/013.

[116] G. R. Dvali, S. H. Henry Tye, *Brane inflation, Phys.Lett.* **B450**, 72-82 (1999), arXiv:hep-ph/9812483.

[117] Bojowald, Martin (October 2008). "Big Bang or Big Bounce?: New Theory on the Universe's Birth". Retrieved 2015-08-31.

[118] Itzhak Bars; Paul Steinhardt; Neil Turok (November 20, 2013). "Sailing through the big crunch-big bang transition". arXiv:1312.0739v2. In the standard big bang inflationary model, the cosmic singularity problem is left unresolved and the cosmology is geodesically incomplete. Consequently, the origin of space and time and the peculiar, exponentially fine-tuned initial conditions required to begin inflation are not explained. In a recent series of papers, we have shown how to construct the complete set of homogeneous classical cosmological solutions of the standard model coupled to gravity, in which the cosmic singularity is replaced by a bounce: the smooth transition from contraction and big crunch to big bang and expansion.

[119] Poplawski, N. J. (2010). "Cosmology with torsion: An alternative to cosmic inflation". *Physics Letters B* **694** (3): 181–185. arXiv:1007.0587. Bibcode:2010PhLB..694..181P. doi:10.1016/j.physletb.2010.09.056.

[120] Poplawski, N. (2012). "Nonsingular, big-bounce cosmology from spinor-torsion coupling". *Physical Review D* **85** (10): 107502. arXiv:1111.4595. Bibcode:2012PhRvD..85j7502P. doi:10.1103/PhysRevD.85.107502.

[121] Brandenberger, R; Vafa, C. (1989). "Superstrings in the early universe". *Nuclear Physics B* **316** (2): 391–410. Bibcode:1989NuPhB.316..391B. doi:10.1016/0550-3213(89)90037-0.

[122] Battefeld, Thorsten; Watson, Scott (2006). "String Gas Cosmology". *Reviews Modern Physics* **78** (2): 435–454. arXiv:hep-th/0510022. Bibcode:2006RvMP...78..435B. doi:10.1103/RevModPhys.78.435.

[123] Brandenberger, Robert H.; Nayeri, ALI; Patil, Subodh P.; Vafa, Cumrun (2007). "String Gas Cosmology and Structure Formation". *International Journal of Modern Physics A* **22** (21): 3621–3642. arXiv:hep-th/0608121. Bibcode:2007IJMPA..22.3621B. doi:10.1142/S0217751X07037159.

[124] Penrose, Roger (2004). *The Road to Reality: A Complete Guide to the Laws of the Universe*. London: Vintage Books, p. 755. See also Penrose, Roger (1989). "Difficulties with Inflationary Cosmology". *Annals of the New York Academy of Sciences* **271**: 249–264. Bibcode:1989NYASA.571..249P. doi:10.1111/j.1749-6632.1989.tb50513.x.

[125] Ijjas, Anna; Steinhardt, Paul J.; Loeb, Abraham. "Inflationary paradigm in trouble after Planck2013". *Physics Letters* **B723**: 261–266. arXiv:1304.2785. Bibcode:2013PhLB..723..261I. doi:10.1016/j.physletb.2013.05.023.

[126] Ijjas, Anna; Steinhardt, Paul J.; Loeb, Abraham. "Inflationary schism after Planck2013". *Physics Letters* **B736**: 142–146. Bibcode:2014PhLB..736..142I. doi:10.1016/j.physletb.2014.07.012.

[127] Guth, Alan H.; Kaiser, David I.; Nomura, Yasunori. "Inflationary paradigm after Planck 2013". *Physics Letters* **B733**: 112–119. arXiv:1312.7619. Bibcode:2014PhLB..733..112G. doi:10.1016/j.physletb.2014.03.020.

[128] Linde, Andrei. "Inflationary cosmology after Planck 2013". arXiv:1402.0526. Bibcode:2014arXiv1402.0526L.

11.10 References

- Guth, Alan (1997). *The Inflationary Universe: The Quest for a New Theory of Cosmic Origins*. Perseus. ISBN 0-201-32840-2.

- Hawking, Stephen (1998). *A Brief History of Time*. Bantam. ISBN 0-553-38016-8.
- Hawking, Stephen; Gary Gibbons (1983). *The Very Early Universe*. Cambridge University Press. ISBN 0-521-31677-4.
- Kolb, Edward; Michael Turner (1988). *The Early Universe*. Addison-Wesley. ISBN 0-201-11604-9.
- Linde, Andrei (1990). *Particle Physics and Inflationary Cosmology*. Chur, Switzerland: Harwood. arXiv:hep-th/0503203. ISBN 3-7186-0490-6.
- Linde, Andrei (2005) "Inflation and String Cosmology", *eConf* **C040802** (2004) L024; *J. Phys. Conf. Ser.* **24** (2005) 151–60; arXiv:hep-th/0503195 v1 2005-03-24.
- Liddle, Andrew; David Lyth (2000). *Cosmological Inflation and Large-Scale Structure*. Cambridge. ISBN 0-521-57598-2.
- Lyth, David H.; Riotto, Antonio (1999). "Particle physics models of inflation and the cosmological density perturbation". *Phys. Rept.* **314** (1–2): 1–146. arXiv:hep-ph/9807278. Bibcode:1999PhR...314....1L. doi:10.1016/S0370-1573(98)00128-8.
- Mukhanov, Viatcheslav (2005). *Physical Foundations of Cosmology*. Cambridge University Press. ISBN 0-521-56398-4.
- Vilenkin, Alex (2006). *Many Worlds in One: The Search for Other Universes*. Hill and Wang. ISBN 0-8090-9523-8.
- Peebles, P. J. E. (1993). *Principles of Physical Cosmology*. Princeton University Press. ISBN 0-691-01933-9.

11.11 External links

- Was Cosmic Inflation The 'Bang' Of The Big Bang?, by Alan Guth, 1997
- An Introduction to Cosmological Inflation by Andrew Liddle, 1999
- update 2004 by Andrew Liddle
- hep-ph/0309238 Laura Covi: Status of observational cosmology and inflation
- hep-th/0311040 David H. Lyth: Which is the best inflation model?
- The Growth of Inflation *Symmetry*, December 2004
- Guth's logbook showing the original idea
- WMAP Bolsters Case for Cosmic Inflation, March 2006
- NASA March 2006 WMAP press release
- Max Tegmark's *Our Mathematical Universe* (2014), "Chapter 5: Inflation"

Chapter 12

String theory

For a more accessible and less technical introduction to this topic, see Introduction to M-theory.

In physics, **string theory** is a theoretical framework in which the point-like particles of particle physics are replaced by one-dimensional objects called strings. It describes how these strings propagate through space and interact with each other. On distance scales larger than the string scale, a string looks just like an ordinary particle, with its mass, charge, and other properties determined by the vibrational state of the string. In string theory, one of the many vibrational states of the string corresponds to the graviton, a quantum mechanical particle that carries gravitational force. Thus string theory is a theory of quantum gravity.

String theory is a broad and varied subject that attempts to address a number of deep questions of fundamental physics. String theory has been applied to a variety of problems in black hole physics, early universe cosmology, nuclear physics, and condensed matter physics, and it has stimulated a number of major developments in pure mathematics. Because string theory potentially provides a unified description of gravity and particle physics, it is a candidate for a theory of everything, a self-contained mathematical model that describes all fundamental forces and forms of matter. Despite much work on these problems, it is not known to what extent string theory describes the real world or how much freedom the theory allows to choose the details.

String theory was first studied in the late 1960s as a theory of the strong nuclear force, before being abandoned in favor of quantum chromodynamics. Subsequently, it was realized that the very properties that made string theory unsuitable as a theory of nuclear physics made it a promising candidate for a quantum theory of gravity. The earliest version of string theory, bosonic string theory, incorporated only the class of particles known as bosons. It later developed into superstring theory, which posits a connection called supersymmetry between bosons and the class of particles called fermions. Five consistent versions of superstring theory were developed before it was conjectured in the mid-1990s that they were all different limiting cases of a single theory in eleven dimensions known as M-theory. In late 1997, theorists discovered an important relationship called the AdS/CFT correspondence, which relates string theory to another type of physical theory called a quantum field theory.

One of the challenges of string theory is that the full theory does not have a satisfactory definition in all circumstances. Another issue is that the theory is thought to describe an enormous landscape of possible universes, and this has complicated efforts to develop theories of particle physics based on string theory. These issues have led some in the community to criticize these approaches to physics and question the value of continued research on string theory unification.

12.1 Fundamentals

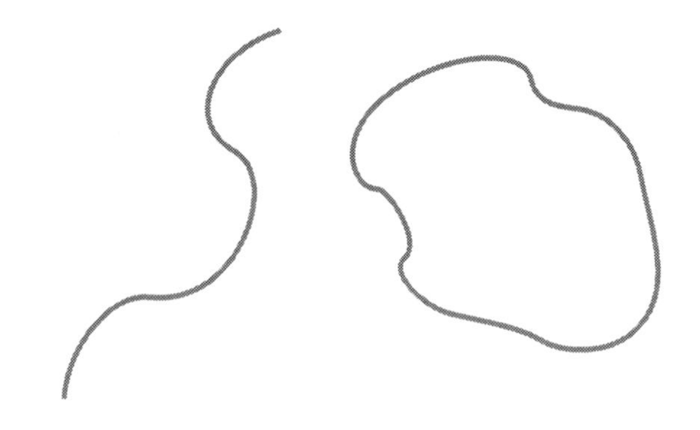

The fundamental objects of string theory are open and closed strings.

In the twentieth century, two theoretical frameworks emerged for formulating the laws of physics. One of these frameworks was Albert Einstein's general theory of relativity, a theory that explains the force of gravity and the structure of space and time. The other was quantum mechanics, a radically different formalism for describing physical phenomena using probability. By the late 1970s, these two frameworks had proven to be sufficient to explain most of the observed features of the universe, from elementary particles to atoms to the evolution of stars and the universe as a whole.[1]

In spite of these successes, there are still many problems that remain to be solved. One of the deepest problems in modern physics is the problem of quantum gravity.[1] The general theory of relativity is formulated within the framework of classical physics, whereas the other fundamental forces are described within the framework of quantum mechanics. A quantum theory of gravity is needed in order to reconcile general relativity with the principles of quantum mechanics, but difficulties arise when one attempts to apply the usual prescriptions of quantum theory to the force of gravity.[2] In addition to the problem of developing a consistent theory of quantum gravity, there are many other fundamental problems in the physics of atomic nuclei, black holes, and the early universe.[lower-alpha 1]

String theory is a theoretical framework that attempts to address these questions and many others. The starting point for string theory is the idea that the point-like particles of particle physics can also be modeled as one-dimensional objects called strings. String theory describes how strings propagate through space and interact with each other. In a given version of string theory, there is only one kind of string, which may look like a small loop or segment of ordinary string, and it can vibrate in different ways. On distance scales larger than the string scale, a string will look just like an ordinary particle, with its mass, charge, and other properties determined by the vibrational state of the string. In this way, all of the different elementary particles may be viewed as vibrating strings. In string theory, one of the vibrational states of the string gives rise to the graviton, a quantum mechanical particle that carries gravitational force. Thus string theory is a theory of quantum gravity.[3]

One of the main developments of the past several decades in string theory was the discovery of certain "dualities", mathematical transformations that identify one physical theory with another. Physicists studying string theory have discovered a number of these dualities between different versions of string theory, and this has led to the conjecture that all consistent versions of string theory are subsumed in a single framework known as M-theory.[4]

Studies of string theory have also yielded a number of results on the nature of black holes and the gravitational interaction. There are certain paradoxes that arise when one attempts to understand the quantum aspects of black holes, and work on string theory has attempted to clarify these issues. In late 1997 this line of work culminated in the discovery of the anti-de Sitter/conformal field theory correspondence or AdS/CFT.[5] This is a theoretical result which relates string theory to other physical theories which are better understood theoretically. The AdS/CFT correspondence has implications for the study of black holes and quantum gravity, and it has been applied to other subjects, including nuclear[6] and condensed matter physics.[7][8]

Since string theory incorporates all of the fundamental interactions, including gravity, many physicists hope that it fully describes our universe, making it a theory of everything. One of the goals of current research in string theory is to find a solution of the theory that reproduces the observed spectrum of elementary particles, with a small cosmological constant, containing dark matter and a plausible mechanism for cosmic inflation. While there has been progress toward these goals, it is not known to what extent string theory describes the real world or how much freedom the theory allows to choose the details.[9]

One of the challenges of string theory is that the full theory does not have a satisfactory definition in all circumstances. The scattering of strings is most straightforwardly defined using the techniques of perturbation theory, but it is not known in general how to define string theory nonperturbatively.[10] It is also not clear whether there is any principle by which string theory selects its vacuum state, the physical state that determines the properties of our universe.[11] These problems have led some in the community to criticize these approaches to the unification of physics and question the value of continued research on these problems.[12]

12.1.1 Strings

Main article: String (physics)

The application of quantum mechanics to physical objects

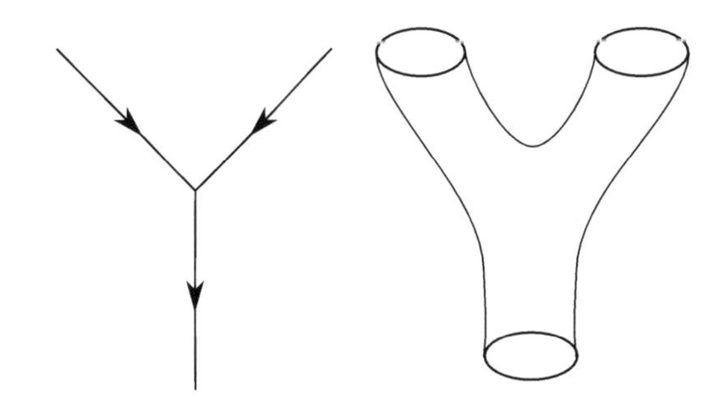

Interaction in the quantum world: worldlines of point-like particles or a worldsheet swept up by closed strings in string theory.

such as the electromagnetic field, which are extended in

space and time, is known as quantum field theory. In particle physics, quantum field theories form the basis for our understanding of elementary particles, which are modeled as excitations in the fundamental fields.[13]

In quantum field theory, one typically computes the probabilities of various physical events using the techniques of perturbation theory. Developed by Richard Feynman and others in the first half of the twentieth century, perturbative quantum field theory uses special diagrams called Feynman diagrams to organize computations. One imagines that these diagrams depict the paths of point-like particles and their interactions.[13]

The starting point for string theory is the idea that the point-like particles of quantum field theory can also be modeled as one-dimensional objects called strings.[14] The interaction of strings is most straightforwardly defined by generalizing the perturbation theory used in ordinary quantum field theory. At the level of Feynman diagrams, this means replacing the one-dimensional diagram representing the path of a point particle by a two-dimensional surface representing the motion of a string.[15] Unlike in quantum field theory, string theory does not have a full non-perturbative definition, so many of the theoretical questions that physicists would like to answer remain out of reach.[16]

In theories of particle physics based on string theory, the characteristic length scale of strings is assumed to be on the order of the Planck length, or 10^{-35} meters, the scale at which the effects of quantum gravity are believed to become significant.[15] On much larger length scales, such as the scales visible in physics laboratories, such objects would be indistinguishable from zero-dimensional point particles, and the vibrational state of the string would determine the type of particle. One of the vibrational states of a string corresponds to the graviton, a quantum mechanical particle that carries the gravitational force.[3]

The original version of string theory was bosonic string theory, but this version described only bosons, a class of particles which transmit forces between the matter particles, or fermions. Bosonic string theory was eventually superseded by theories called superstring theories. These theories describe both bosons and fermions, and they incorporate a theoretical idea called supersymmetry. This is a mathematical relation that exists in certain physical theories between the bosons and fermions. In theories with supersymmetry, each boson has a counterpart which is a fermion, and vice versa.[17]

There are several versions of superstring theory: type I, type IIA, type IIB, and two flavors of heterotic string theory ($SO(32)$ and $E_8 \times E_8$). The different theories allow different types of strings, and the particles that arise at low energies exhibit different symmetries. For example, the type I theory includes both open strings (which are segments with endpoints) and closed strings (which form closed loops), while types IIA, IIB and heterotic include only closed strings.[18]

12.1.2 Extra dimensions

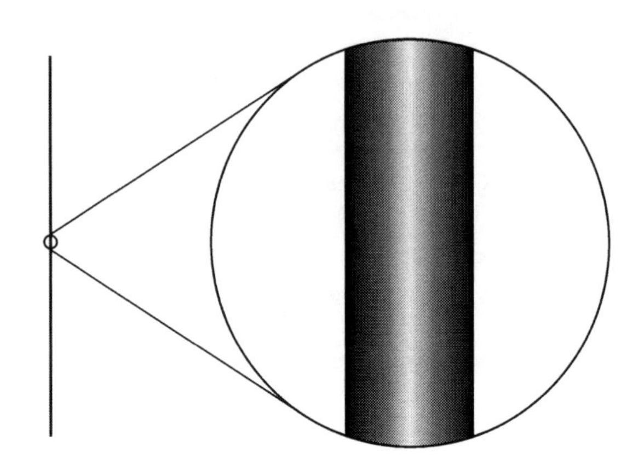

An example of compactification: At large distances, a two dimensional surface with one circular dimension looks one-dimensional.

In everyday life, there are three familiar dimensions of space: height, width and length. Einstein's general theory of relativity treats time as a dimension on par with the three spatial dimensions; in general relativity, space and time are not modeled as separate entities but are instead unified to a four-dimensional spacetime. In this framework, the phenomenon of gravity is viewed as a consequence of the geometry of spacetime.[19]

In spite of the fact that the universe is well described by four-dimensional spacetime, there are several reasons why physicists consider theories in other dimensions. In some cases, by modeling spacetime in a different number of dimensions, a theory becomes more mathematically tractable, and one can perform calculations and gain general insights more easily.[lower-alpha 2] There are also situations where theories in two or three spacetime dimensions are useful for describing phenomena in condensed matter physics.[20] Finally, there exist scenarios in which there could actually be more than four dimensions of spacetime which have nonetheless managed to escape detection.[21]

One notable feature of string theories is that these theories require extra dimensions of spacetime for their mathematical consistency. In bosonic string theory, spacetime is 26-dimensional, while in superstring theory it is ten-dimensional. In order to describe real physical phenomena using string theory, one must therefore imagine scenarios in which these extra dimensions would not be observed in experiments.[22]

Compactification is one way of modifying the number of

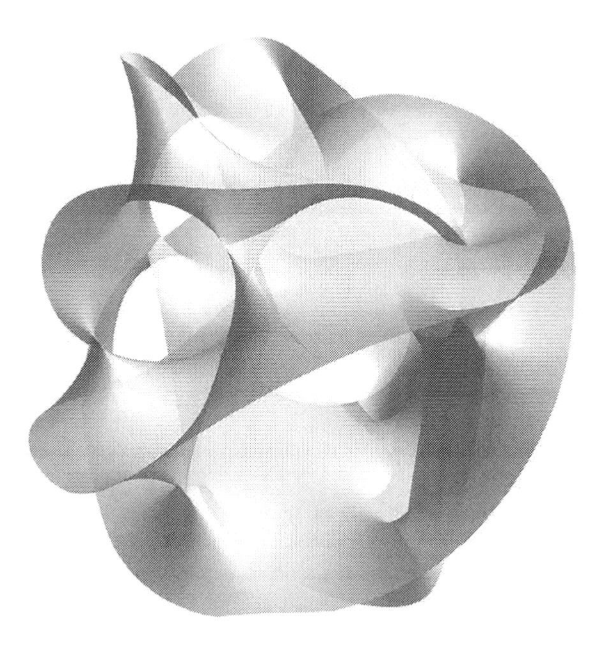

A cross section of a quintic Calabi–Yau manifold

dimensions in a physical theory. In compactification, some of the extra dimensions are assumed to "close up" on themselves to form circles.[23] In the limit where these curled up dimensions become very small, one obtains a theory in which spacetime has effectively a lower number of dimensions. A standard analogy for this is to consider a multidimensional object such as a garden hose. If the hose is viewed from a sufficient distance, it appears to have only one dimension, its length. However, as one approaches the hose, one discovers that it contains a second dimension, its circumference. Thus, an ant crawling on the surface of the hose would move in two dimensions.[24]

Compactification can be used to construct models in which spacetime is effectively four-dimensional. However, not every way of compactifying the extra dimensions produces a model with the right properties to describe nature. In a viable model of particle physics, the compact extra dimensions must be shaped like a Calabi–Yau manifold.[23] A Calabi–Yau manifold is a special space which is typically taken to be six-dimensional in applications to string theory. It is named after mathematicians Eugenio Calabi and Shing-Tung Yau.[25]

Another approach to reducing the number of dimensions is the so-called brane-world scenario. In this approach, physicists assume that the observable universe is a four-dimensional subspace of a higher dimensional space. In such models, the force-carrying bosons of particle physics arise from open strings with endpoints attached to the four-dimensional subspace, while gravity arises from closed strings propagating through the larger ambient space. This idea plays an important role in attempts to develop models of real world physics based on string theory, and it provides a natural explanation for the weakness of gravity compared to the other fundamental forces.[26]

12.1.3 Dualities

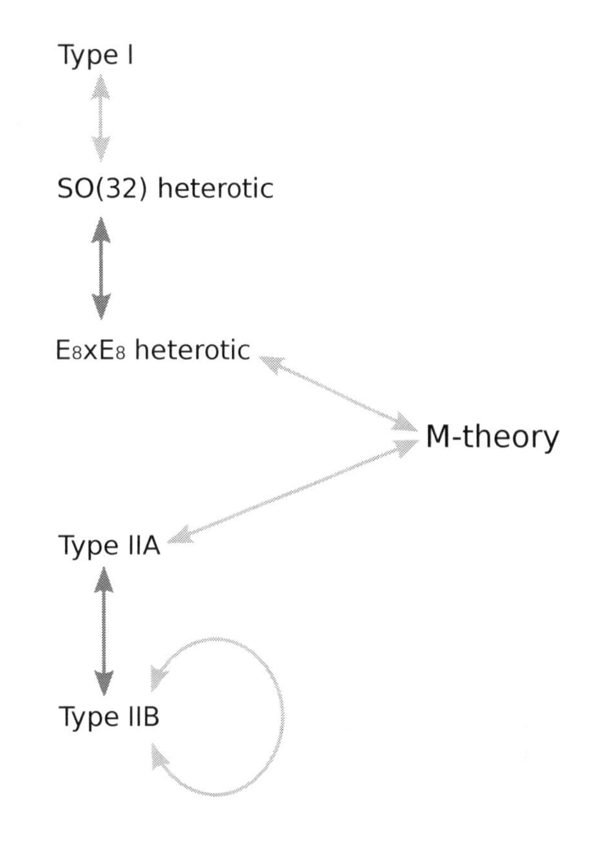

A diagram of string theory dualities. Yellow arrows indicate S-duality. Blue arrows indicate T-duality.

Main articles: S-duality and T-duality

One notable fact about string theory is that the different versions of the theory all turn out to be related in highly nontrivial ways. One of the relationships that can exist between different string theories is called S-duality. This is a relationship which says that a collection of strongly interacting particles in one theory can, in some cases, be viewed as a collection of weakly interacting particles in a completely different theory. Roughly speaking, a collection of particles is said to be strongly interacting if they combine and decay often and weakly interacting if they do so infrequently. Type I string theory turns out to be equivalent by S-duality to the $SO(32)$ heterotic string theory. Similarly, type IIB string theory is related to itself in a nontrivial way by S-duality.[27]

Another relationship between different string theories is T-duality. Here one considers strings propagating around a circular extra dimension. T-duality states that a string propagating around a circle of radius R is equivalent to a string propagating around a circle of radius $1/R$ in the sense that all observable quantities in one description are identified with quantities in the dual description. For example, a string has momentum as it propagates around a circle, and it can also wind around the circle one or more times. The number of times the string winds around a circle is called the winding number. If a string has momentum p and winding number n in one description, it will have momentum n and winding number p in the dual description. For example, type IIA string theory is equivalent to type IIB string theory via T-duality, and the two versions of heterotic string theory are also related by T-duality.[27]

In general, the term *duality* refers to a situation where two seemingly different physical systems turn out to be equivalent in a nontrivial way. Two theories related by a duality need not be string theories. For example, Montonen–Olive duality is example of an S-duality relationship between quantum field theories. The AdS/CFT correspondence is example of a duality which relates string theory to a quantum field theory. If two theories are related by a duality, it means that one theory can be transformed in some way so that it ends up looking just like the other theory. The two theories are then said to be *dual* to one another under the transformation. Put differently, the two theories are mathematically different descriptions of the same phenomena.[28]

12.1.4 Branes

Main article: Brane

In string theory and related theories, a brane is a physical object that generalizes the notion of a point particle to higher dimensions. For example, a point particle can be viewed as a brane of dimension zero, while a string can be viewed as a brane of dimension one. It is also possible to consider higher-dimensional branes. In dimension p, these are called p-branes. The word brane comes from the word "membrane" which refers to a two-dimensional brane.[29]

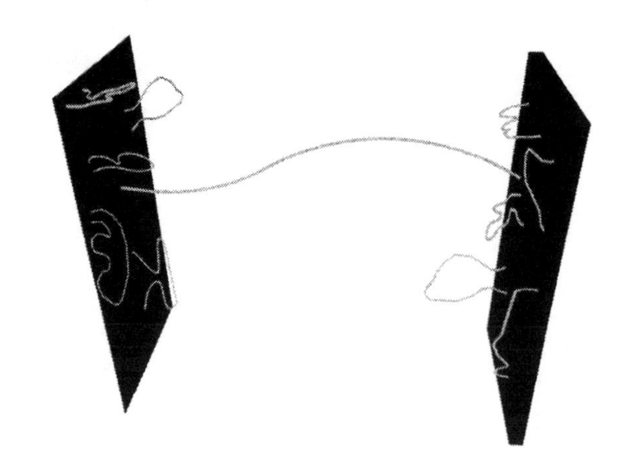

Open strings attached to a pair of D-branes

Branes are dynamical objects which can propagate through spacetime according to the rules of quantum mechanics. They have mass and can have other attributes such as charge. A p-brane sweeps out a (p+1)-dimensional volume in spacetime called its *worldvolume*. Physicists often study fields analogous to the electromagnetic field which live on the worldvolume of a brane.[29]

In string theory, D-branes are an important class of branes that arise when one considers open strings. As an open string propagates through spacetime, its endpoints are required to lie on a D-brane. The letter "D" in D-brane refers to a certain mathematical condition on the system known as the Dirichlet boundary condition. The study of D-branes in string theory has led to important results such as the AdS/CFT correspondence, which has shed light on many problems in quantum field theory.[30]

Branes are also frequently studied from a purely mathematical point of view. Mathematically, branes can be described as objects of certain categories, such as the derived category of coherent sheaves on a complex algebraic variety, or the Fukaya category of a symplectic manifold.[31] The connection between the physical notion of a brane and the mathematical notion of a category has led to important mathematical insights in the fields of algebraic and symplectic geometry[32] and representation theory.[33]

12.2 M-theory

Main article: M-theory

Prior to 1995, theorists believed that there were five consistent versions of superstring theory (type I, type IIA, type IIB, and two versions of heterotic string theory). This understanding changed in 1995 when Edward Witten suggested that the five theories were just special limiting cases of an eleven-dimensional theory called M-theory. Witten's conjecture was based on the work of a number of other physicists, including Ashoke Sen, Chris Hull, Paul Townsend, and Michael Duff. His announcement led to a flurry of research activity now known as the second superstring revolution.[34]

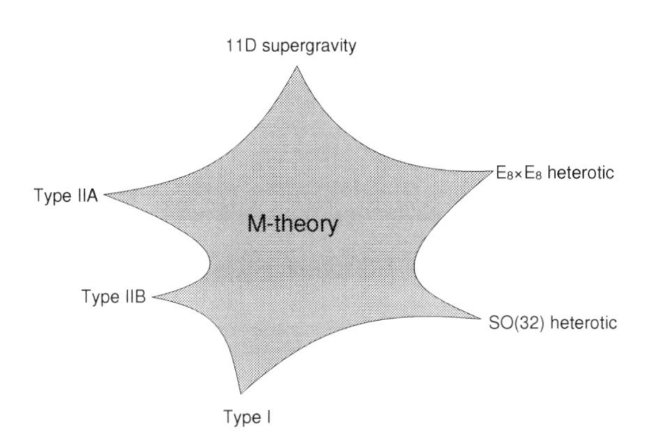

A schematic illustration of the relationship between M-theory, the five superstring theories, and eleven-dimensional supergravity. The shaded region represents a family of different physical scenarios that are possible in M-theory. In certain limiting cases corresponding to the cusps, it is natural to describe the physics using one of the six theories labeled there.

12.2.1 Unification of superstring theories

In the 1970s, many physicists became interested in supergravity theories, which combine general relativity with supersymmetry. Whereas general relativity makes sense in any number of dimensions, supergravity places an upper limit on the number of dimensions.[35] In 1978, work by Werner Nahm showed that the maximum spacetime dimension in which one can formulate a consistent supersymmetric theory is eleven.[36] In the same year, Eugene Cremmer, Bernard Julia, and Joel Scherk of the École Normale Supérieure showed that supergravity not only permits up to eleven dimensions but is in fact most elegant in this maximal number of dimensions.[37][38]

Initially, many physicists hoped that by compactifying eleven-dimensional supergravity, it might be possible to construct realistic models of our four-dimensional world. The hope was that such models would provide a unified description of the four fundamental forces of nature: electromagnetism, the strong and weak nuclear forces, and gravity. Interest in eleven-dimensional supergravity soon waned as various flaws in this scheme were discovered. One of the problems was that the laws of physics appear to distinguish between clockwise and counterclockwise, a phenomenon known as chirality. Edward Witten and others observed this chirality property cannot be readily derived by compactifying from eleven dimensions.[38]

In the first superstring revolution in 1984, many physicists turned to string theory as a unified theory of particle physics and quantum gravity. Unlike supergravity theory, string theory was able to accommodate the chirality of the standard model, and it provided a theory of gravity consistent with quantum effects.[38] Another feature of string theory that many physicists were drawn to in the 1980s and 1990s was its high degree of uniqueness. In ordinary particle theories, one can consider any collection of elementary particles whose classical behavior is described by an arbitrary Lagrangian. In string theory, the possibilities are much more constrained: by the 1990s, physicists had argued that there were only five consistent supersymmetric versions of the theory.[38]

Although there were only a handful of consistent superstring theories, it remained a mystery why there was not just one consistent formulation.[38] However, as physicists began to examine string theory more closely, they realized that these theories are related in intricate and nontrivial ways. They found that a system of strongly interacting strings can, in some cases, be viewed as a system of weakly interacting strings. This phenomenon is known as S-duality. It was studied by Ashoke Sen in the context of heterotic strings in four dimensions[39][40] and by Chris Hull and Paul Townsend in the context of the type IIB theory.[41] Theorists also found that different string theories may be related by T-duality. This duality implies that strings propagating on completely different spacetime geometries may be physically equivalent.[42]

At around the same time, as many physicists were studying the properties of strings, a small group of physicists was examining the possible applications of higher dimensional objects. In 1987, Eric Bergshoeff, Ergin Sezgin, and Paul Townsend showed that eleven-dimensional supergravity includes two-dimensional branes.[43] Intuitively, these objects look like sheets or membranes propagating through the eleven-dimensional spacetime. Shortly after this discovery, Michael Duff, Paul Howe, Takeo Inami, and Kellogg Stelle considered a particular compactification of eleven-dimensional supergravity with one of the dimensions curled up into a circle.[44] In this setting, one can imagine the membrane wrapping around the circular dimension. If the radius of the circle is sufficiently small, then this membrane looks just like a string in ten-dimensional spacetime. In fact, Duff and his collaborators showed that this construction reproduces exactly the strings appearing in type IIA superstring theory.[45]

Speaking at a string theory conference in 1995, Edward Witten made the surprising suggestion that all five superstring theories were in fact just different limiting cases of a single theory in eleven spacetime dimensions. Witten's announcement drew together all of the previous results on S- and T-duality and the appearance of higher dimensional branes in string theory.[46] In the months following Witten's announcement, hundreds of new papers appeared on the Internet confirming different parts of his proposal.[47] Today this flurry of work is known as the second superstring revolution.[48]

Initially, some physicists suggested that the new theory was a fundamental theory of membranes, but Witten was skeptical of the role of membranes in the theory. In a paper from 1996, Hořava and Witten wrote "As it has been proposed that the eleven-dimensional theory is a supermembrane theory but there are some reasons to doubt that interpretation, we will non-committally call it the M-theory, leaving to the future the relation of M to membranes."[49] In the absence of an understanding of the true meaning and structure of M-theory, Witten has suggested that the *M* should stand for "magic", "mystery", or "membrane" according to taste, and the true meaning of the title should be decided when a more fundamental formulation of the theory is known.[50]

12.2.2 Matrix theory

Main article: Matrix theory (physics)

In mathematics, a matrix is a rectangular array of numbers or other data. In physics, a matrix model is a particular kind of physical theory whose mathematical formulation involves the notion of a matrix in an important way. A matrix model describes the behavior of a set of matrices within the framework of quantum mechanics.[51]

One important example of a matrix model is the BFSS matrix model proposed by Tom Banks, Willy Fischler, Stephen Shenker, and Leonard Susskind in 1997. This theory describes the behavior of a set of nine large matrices. In their original paper, these authors showed, among other things, that the low energy limit of this matrix model is described by eleven-dimensional supergravity. These calculations led them to propose that the BFSS matrix model is exactly equivalent to M-theory. The BFSS matrix model can therefore be used as a prototype for a correct formulation of M-theory and a tool for investigating the properties of M-theory in a relatively simple setting.[51]

The development of the matrix model formulation of M-theory has led physicists to consider various connections between string theory and a branch of mathematics called noncommutative geometry. This subject is a generalization of ordinary geometry in which mathematicians define new geometric notions using tools from noncommutative algebra.[52] In a paper from 1998, Alain Connes, Michael R. Douglas, and Albert Schwarz showed that some aspects of matrix models and M-theory are described by a noncommutative quantum field theory, a special kind of physical theory in which spacetime is described mathematically using noncommutative geometry.[53] This established a link between matrix models and M-theory on the one hand, and noncommutative geometry on the other hand. It quickly led to the discovery of other important links between noncommutative geometry and various physical theories.[54][55]

12.3 Black holes

In general relativity, a black hole is defined as a region of spacetime in which the gravitational field is so strong that no particle or radiation can escape. In the currently accepted models of stellar evolution, black holes are thought to arise when massive stars undergo gravitational collapse, and many galaxies are thought to contain supermassive black holes at their centers. Black holes are also important for theoretical reasons, as they present profound challenges for theorists attempting to understand the quantum aspects of gravity. String theory has proved to be an important tool for investigating the theoretical properties of black holes because it provides a framework in which theorists can study their thermodynamics.[56]

12.3.1 Bekenstein–Hawking formula

In the branch of physics called statistical mechanics, entropy is a measure of the randomness or disorder of a physical system. This concept was studied in the 1870s by the Austrian physicist Ludwig Boltzmann, who showed that the thermodynamic properties of a gas could be derived from the combined properties of its many constituent molecules. Boltzmann argued that by averaging the behaviors of all the different molecules in a gas, one can understand macroscopic properties such as volume, temperature, and pressure. In addition, this perspective led him to give a precise definition of entropy as the natural logarithm of the number of different states of the molecules (also called *microstates*) that give rise to the same macroscopic features.[57]

In the twentieth century, physicists began to apply the same concepts to black holes. In most systems such as gases, the entropy scales with the volume. In the 1970s, the physicist Jacob Bekenstein suggested that the entropy of a black hole is instead proportional to the *surface area* of its event horizon, the boundary beyond which matter and radiation is lost to its gravitational attraction.[58] When combined with ideas of the physicist Stephen Hawking,[59] Bekenstein's work yielded a precise formula for the entropy of a black hole. The formula expresses the entropy S as

$$S = \frac{c^3 kA}{4\hbar G}$$

where c is the speed of light, k is Boltzmann's constant, $\hbar$ is the reduced Planck constant, G is Newton's constant, and A is the surface area of the event horizon.[60]

Like any physical system, a black hole has an entropy defined in terms of the number of different microstates that lead to the same macroscopic features. The Bekenstein–Hawking entropy formula gives the expected value of the entropy of a black hole, but by the 1990s, physicists still lacked a derivation of this formula by counting microstates in a theory of quantum gravity. Finding such a derivation of this formula was considered an important test of the viability of any theory of quantum gravity such as string theory.[61]

12.3.2 Derivation within string theory

In a paper from 1996, Andrew Strominger and Cumrun Vafa showed how to derive the Beckenstein–Hawking formula for certain black holes in string theory.[62] Their calculation was based on the observation that D-branes—which look like fluctuating membranes when they are weakly interacting—become dense, massive objects with event horizons when the interactions are strong. In other words, a system of strongly interacting D-branes in string theory is indistinguishable from a black hole. Strominger and Vafa analyzed such D-brane systems and calculated the number of different ways of placing D-branes in spacetime so that their combined mass and charge is equal to a given mass and charge for the resulting black hole. Their calculation reproduced the Bekenstein–Hawking formula exactly, including the factor of 1/4.[63] Subsequent work by Strominger, Vafa, and others refined the original calculations and gave the precise values of the "quantum corrections" needed to describe very small black holes.[64][65]

The black holes that Strominger and Vafa considered in their original work were quite different from real astrophysical black holes. One difference was that Strominger and Vafa considered only extremal black holes in order to make the calculation tractable. These are defined as black holes with the lowest possible mass compatible with a given charge.[66] Strominger and Vafa also restricted attention to black holes in five-dimensional spacetime with unphysical supersymmetry.[67]

Although it was originally developed in this very particular and physically unrealistic context in string theory, the entropy calculation of Strominger and Vafa has led to a qualitative understanding of how black hole entropy can be accounted for in any theory of quantum gravity. Indeed, in 1998, Strominger argued that the original result could be generalized to an arbitrary consistent theory of quantum gravity without relying on strings or supersymmetry.[68] In collaboration with several other authors in 2010, he showed that some results on black hole entropy could be extended to non-extremal astrophysical black holes.[69][70]

12.4 AdS/CFT correspondence

Main article: AdS/CFT correspondence

One approach to formulating string theory and studying its properties is provided by the anti-de Sitter/conformal field theory (AdS/CFT) correspondence. This is a theoretical result which implies that string theory is in some cases equivalent to a quantum field theory. In addition to providing insights into the mathematical structure of string theory, the AdS/CFT correspondence has shed light on many aspects of quantum field theory in regimes where traditional calculational techniques are ineffective.[6] The AdS/CFT correspondence was first proposed by Juan Maldacena in late 1997.[71] Important aspects of the correspondence were elaborated in articles by Steven Gubser, Igor Klebanov, and Alexander Markovich Polyakov,[72] and by Edward Witten.[73] By 2010, Maldacena's article had over 7000 citations, becoming the most highly cited article in the field of high energy physics.[lower-alpha 3]

12.4.1 Overview of the correspondence

In the AdS/CFT correspondence, the geometry of spacetime is described in terms of a certain vacuum solution of Einstein's equation called anti-de Sitter space.[74] In very elementary terms, anti-de Sitter space is a mathematical model of spacetime in which the notion of distance between points (the metric) is different from the notion of distance in ordinary Euclidean geometry. It is closely related to hyperbolic space, which can be viewed as a disk as illustrated on the left.[75] This image shows a tessellation of a disk by triangles and squares. One can define the distance between points of this disk in such a way that all the triangles and squares are the same size and the circular outer boundary is infinitely far from any point in the interior.[76]

One can imagine a stack of hyperbolic disks where each disk represents the state of the universe at a given time. The resulting geometric object is three-dimensional anti-de Sitter space.[75] It looks like a solid cylinder in which any cross section is a copy of the hyperbolic disk. Time runs along the vertical direction in this picture. The surface of this cylinder plays an important role in the AdS/CFT correspondence. As with the hyperbolic plane, anti-de Sitter space is curved in such a way that any point in the interior is actually infinitely far from this boundary surface.[76]

This construction describes a hypothetical universe with only two space dimensions and one time dimension, but it can be generalized to any number of dimensions. Indeed, hyperbolic space can have more than two dimensions and one can "stack up" copies of hyperbolic space to get higher-dimensional models of anti-de Sitter space.[75]

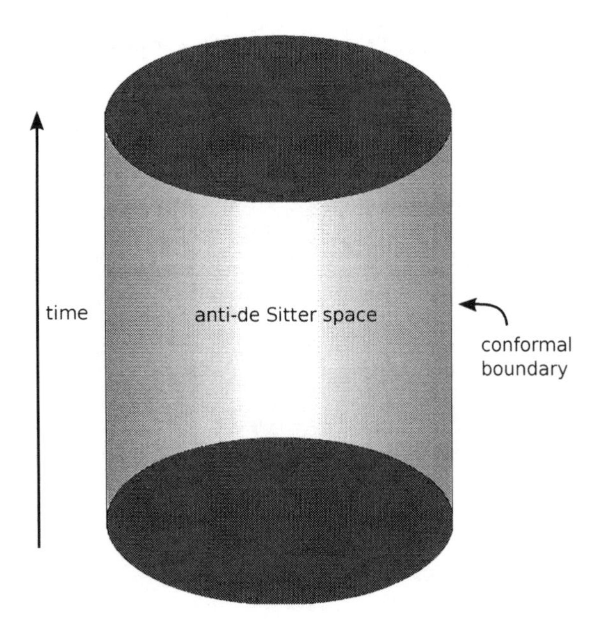

Three-dimensional anti-de Sitter space is like a stack of hyperbolic disks, each one representing the state of the universe at a given time. The resulting spacetime looks like a solid cylinder.

An important feature of anti-de Sitter space is its boundary (which looks like a cylinder in the case of three-dimensional anti-de Sitter space). One property of this boundary is that, within a small region on the surface around any given point, it looks just like Minkowski space, the model of spacetime used in nongravitational physics.[77] One can therefore consider an auxiliary theory in which "spacetime" is given by the boundary of anti-de Sitter space. This observation is the starting point for AdS/CFT correspondence, which states that the boundary of anti-de Sitter space can be regarded as the "spacetime" for a quantum field theory. The claim is that this quantum field theory is equivalent to a gravitational theory, such as string theory, in the bulk anti-de Sitter space in the sense that there is a "dictionary" for translating entities and calculations in one theory into their counterparts in the other theory. For example, a single particle in the gravitational theory might correspond to some collection of particles in the boundary theory. In addition, the predictions in the two theories are quantitatively identical so that if two particles have a 40 percent chance of colliding in the gravitational theory, then the corresponding collections in the boundary theory would also have a 40 percent chance of colliding.[78]

12.4.2 Applications to quantum gravity

The discovery of the AdS/CFT correspondence was a major advance in physicists' understanding of string theory and quantum gravity. One reason for this is that the correspondence provides a formulation of string theory in terms of quantum field theory, which is well understood by comparison. Another reason is that it provides a general framework in which physicists can study and attempt to resolve the paradoxes of black holes.[56]

In 1975, Stephen Hawking published a calculation which suggested that black holes are not completely black but emit a dim radiation due to quantum effects near the event horizon.[59] At first, Hawking's result posed a problem for theorists because it suggested that black holes destroy information. More precisely, Hawking's calculation seemed to conflict with one of the basic postulates of quantum mechanics, which states that physical systems evolve in time according to the Schrödinger equation. This property is usually referred to as unitarity of time evolution. The apparent contradiction between Hawking's calculation and the unitarity postulate of quantum mechanics came to be known as the black hole information paradox.[79]

The AdS/CFT correspondence resolves the black hole information paradox, at least to some extent, because it shows how a black hole can evolve in a manner consistent with quantum mechanics in some contexts. Indeed, one can consider black holes in the context of the AdS/CFT correspondence, and any such black hole corresponds to a configuration of particles on the boundary of anti-de Sitter space.[80] These particles obey the usual rules of quantum mechanics and in particular evolve in a unitary fashion, so the black hole must also evolve in a unitary fashion, respecting the principles of quantum mechanics.[81] In 2005, Hawking announced that the paradox had been settled in favor of information conservation by the AdS/CFT correspondence, and he suggested a concrete mechanism by which black holes might preserve information.[82]

12.4.3 Applications to quantum field theory

Main articles: AdS/QCD correspondence and AdS/CMT correspondence

In addition to its applications to theoretical problems in quantum gravity, the AdS/CFT correspondence has been applied to a variety of problems in quantum field theory. One physical system that has been studied using the AdS/CFT correspondence is the quark–gluon plasma, an exotic state of matter produced in particle accelerators. This state of matter arises for brief instants when heavy ions such as gold or lead nuclei are collided at high energies. Such collisions cause the quarks that make up atomic nuclei to deconfine at temperatures of approximately two trillion kelvins, conditions similar to those present at around 10^{-11} seconds after the Big Bang.[83]

The physics of the quark–gluon plasma is governed by a theory called quantum chromodynamics, but this the-

A magnet levitating above a high-temperature superconductor. Today some physicists are working to understand high-temperature superconductivity using the AdS/CFT correspondence.[7]

ory is mathematically intractable in problems involving the quark–gluon plasma.[lower-alpha 4] In an article appearing in 2005, Đàm Thanh Sơn and his collaborators showed that the AdS/CFT correspondence could be used to understand some aspects of the quark–gluon plasma by describing it in the language of string theory.[84] By applying the AdS/CFT correspondence, Sơn and his collaborators were able to describe the quark gluon plasma in terms of black holes in five-dimensional spacetime. The calculation showed that the ratio of two quantities associated with the quark–gluon plasma, the shear viscosity and volume density of entropy, should be approximately equal to a certain universal constant. In 2008, the predicted value of this ratio for the quark–gluon plasma was confirmed at the Relativistic Heavy Ion Collider at Brookhaven National Laboratory.[85][86]

The AdS/CFT correspondence has also been used to study aspects of condensed matter physics. Over the decades, experimental condensed matter physicists have discovered a number of exotic states of matter, including superconductors and superfluids. These states are described using the formalism of quantum field theory, but some phenomena are difficult to explain using standard field theoretic techniques. Some condensed matter theorists including Subir Sachdev hope that the AdS/CFT correspondence will make it possible to describe these systems in the language of string theory and learn more about their behavior.[85]

So far some success has been achieved in using string theory methods to describe the transition of a superfluid to an insulator. A superfluid is a system of electrically neutral atoms that flows without any friction. Such systems are often produced in the laboratory using liquid helium, but recently experimentalists have developed new ways of producing artificial superfluids by pouring trillions of cold atoms into a lattice of criss-crossing lasers. These atoms initially behave as a superfluid, but as experimentalists increase the intensity of the lasers, they become less mobile and then suddenly transition to an insulating state. During the transition, the atoms behave in an unusual way. For example, the atoms slow to a halt at a rate that depends on the temperature and on Planck's constant, the fundamental parameter of quantum mechanics, which does not enter into the description of the other phases. This behavior has recently been understood by considering a dual description where properties of the fluid are described in terms of a higher dimensional black hole.[87]

12.5 Phenomenology

Main article: String phenomenology

In addition to being an idea of considerable theoretical interest, string theory provides a framework for constructing models of real world physics that combine general relativity and particle physics. Phenomenology is the branch of theoretical physics in which physicists construct realistic models of nature from more abstract theoretical ideas. String phenomenology is the part of string theory that attempts to construct realistic or semi-realistic models based on string theory.

Partly because of theoretical and mathematical difficulties and partly because of the extremely high energies needed to test these theories experimentally, there is so far no experimental evidence that would unambiguously point to any of these models being a correct fundamental description of nature. This has led some in the community to criticize these approaches to unification and question the value of continued research on these problems.[12]

12.5.1 Particle physics

The currently accepted theory describing elementary particles and their interactions is known as the standard model of particle physics. This theory provides a unified description of three of the fundamental forces of nature: electromagnetism and the strong and weak nuclear forces. Despite its remarkable success in explaining a wide range of physical phenomena, the standard model cannot be a complete description of reality. This is because the standard model fails to incorporate the force of gravity and because of problems such as the hierarchy problem and the inability to explain the structure of fermion masses or dark matter.

String theory has been used to construct a variety of models of particle physics going beyond the standard model. Typically, such models are based on the idea of compactification. Starting with the ten- or eleven-dimensional spacetime of string or M-theory, physicists postulate a shape for the extra dimensions. By choosing this shape appropriately, they can construct models roughly similar to the standard model of particle physics, together with additional undiscovered particles.[88] One popular way of deriving realistic physics from string theory is to start with the heterotic theory in ten dimensions and assume that the six extra dimensions of spacetime are shaped like a six-dimensional Calabi–Yau manifold. Such compactifications offer many ways of extracting realistic physics from string theory. Other similar methods can be used to construct realistic or semi-realistic models of our four-dimensional world based on M-theory.[89]

12.5.2 Cosmology

Main article: String cosmology
The Big Bang theory is the prevailing cosmological model

A map of the cosmic microwave background produced by the Wilkinson Microwave Anisotropy Probe

for the universe from the earliest known periods through its subsequent large-scale evolution. Despite its success in explaining many observed features of the universe including galactic redshifts, the relative abundance of light elements such as hydrogen and helium, and the existence of a cosmic microwave background, there are several questions that remain unanswered. For example, the standard Big Bang model does not explain why the universe appears to be same in all directions, why it appears flat on very large distance scales, or why certain hypothesized particles such as magnetic monopoles are not observed in experiments.[90]

Currently, the leading candidate for a theory going beyond the Big Bang is the theory of cosmic inflation. Developed by Alan Guth and others in the 1980s, inflation postulates a period of extremely rapid accelerated expansion of the universe prior to the expansion described by the standard Big Bang theory. The theory of cosmic inflation preserves the successes of the Big Bang while providing a natural explanation for some of the mysterious features of the universe.[91] The theory has also received striking support from observations of the cosmic microwave background, the radiation that has filled the sky since around 380,000 years after the Big Bang.[92]

In the theory of inflation, the rapid initial expansion of the universe is caused by a hypothetical particle called the inflaton. The exact properties of this particle are not fixed by the theory but should ultimately be derived from a more fundamental theory such as string theory.[93] Indeed, there have been a number of attempts to identify an inflaton within the spectrum of particles described by string theory, and to study inflation using string theory. While these approaches might eventually find support in observational data such as measurements of the cosmic microwave background, the application of string theory to cosmology is still in its early stages.[94]

12.6 Connections to mathematics

In addition to influencing research in theoretical physics, string theory has stimulated a number of major developments in pure mathematics. Like many developing ideas in theoretical physics, string theory does not at present have a mathematically rigorous formulation in which all of its concepts can be defined precisely. As a result, physicists who study string theory are often guided by physical intuition to conjecture relationships between the seemingly different mathematical structures that are used to formalize different parts of the theory. These conjectures are later proved by mathematicians, and in this way, string theory serves as a source of new ideas in pure mathematics.[95]

12.6.1 Mirror symmetry

Main article: Mirror symmetry (string theory)
After Calabi–Yau manifolds had entered physics as a way to compactify extra dimensions in string theory, many physicists began studying these manifolds. In the late 1980s, several physicists noticed that given such a compactification of string theory, it is not possible to reconstruct uniquely a corresponding Calabi–Yau manifold.[96] Instead, two different versions of string theory, type IIA and type IIB, can be compactified on completely different Calabi–Yau manifolds giving rise to the same physics. In this situation, the manifolds are called mirror manifolds, and the relationship between the two physical theories is called mirror symmetry.[97]

Regardless of whether Calabi–Yau compactifications of string theory provide a correct description of nature, the existence of the mirror duality between different string theories has significant mathematical consequences. The Calabi–Yau manifolds used in string theory are of interest

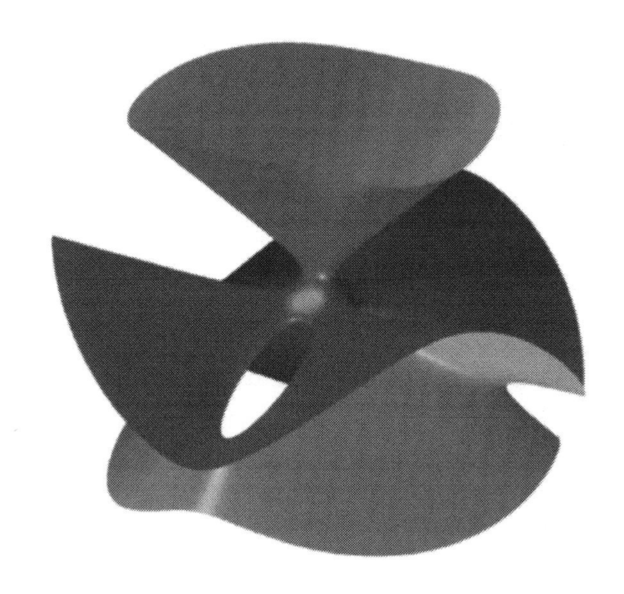

The Clebsch cubic is an example of a kind of geometric object called an algebraic variety. A classical result of enumerative geometry states that there are exactly 27 straight lines that lie entirely on this surface.

in pure mathematics, and mirror symmetry allows mathematicians to solve problems in enumerative geometry, a branch of mathematics concerned with counting the numbers of solutions to geometric questions.[31][98]

Enumerative geometry studies a class of geometric objects called algebraic varieties which are defined by the vanishing of polynomials. For example, the Clebsch cubic illustrated on the right is an algebraic variety defined using a certain polynomial of degree three in four variables. A celebrated result of nineteenth-century mathematicians Arthur Cayley and George Salmon states that there are exactly 27 straight lines that lie entirely on such a surface.[99]

Generalizing this problem, one can ask how many lines can be drawn on a quintic Calabi–Yau manifold, such as the one illustrated above, which is defined by a polynomial of degree five. This problem was solved by the nineteenth-century German mathematician Hermann Schubert, who found that there are exactly 2,875 such lines. In 1986, geometer Sheldon Katz proved that the number of curves, such as circles, that are defined by polynomials of degree two and lie entirely in the quintic is 609,250.[100]

By the year 1991, most of the classical problems of enumerative geometry had been solved and interest in enumerative geometry had begun to diminish.[101] The field was reinvigorated in May 1991 when physicists Philip Candelas, Xenia de la Ossa, Paul Green, and Linda Parks showed that mirror symmetry could be used to translate difficult mathematical questions about one Calabi–Yau manifold into easier questions about its mirror.[102] In particular, they used mirror symmetry to show that a six-dimensional Calabi–Yau manifold can contain exactly 317,206,375 curves of degree three.[101] In addition to counting degree-three curves, Candelas and his collaborators obtained a number of more general results for counting rational curves which went far beyond the results obtained by mathematicians.[103]

Originally, these results of Candelas were justified on physical grounds. However, mathematicians generally prefer rigorous proofs that do not require an appeal to physical intuition. Inspired by physicists' work on mirror symmetry, mathematicians have therefore constructed their own arguments proving the enumerative predictions of mirror symmetry.[lower-alpha 5] Today mirror symmetry is an active area of research in mathematics, and mathematicians are working to develop a more complete mathematical understanding of mirror symmetry based on physicists' intuition.[104] Major approaches to mirror symmetry include the homological mirror symmetry program of Maxim Kontsevich[32] and the SYZ conjecture of Andrew Strominger, Shing-Tung Yau, and Eric Zaslow.[105]

12.6.2 Monstrous moonshine

Main article: Monstrous moonshine

Group theory is the branch of mathematics that studies the

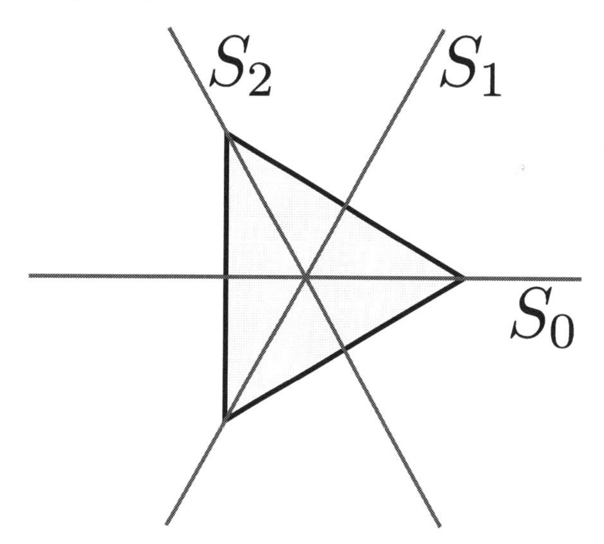

An equilateral triangle can be rotated through 120°, 240°, or 360°, or reflected in any of the three lines pictured without changing its shape.

concept of symmetry. For example, one can consider a geometric shape such as an equilateral triangle. There are various operations that one can perform on this triangle without changing its shape. One can rotate it through 120°, 240°, or 360°, or one can reflect in any of the lines labeled S_0, S_1, or S_2 in the picture. Each of these operations is called a *symmetry*, and the collection of these symmetries satisfies certain technical properties making it into what mathematicians call a group. In this particular example, the group is known as the dihedral group of order 6 because it has six elements. A general group may describe finitely many or infinitely many symmetries; if there are only finitely many symmetries, it is called a finite group.[106]

Mathematicians often strive for a classification (or list) of all mathematical objects of a given type. It is generally believed that finite groups are too diverse to admit a useful classification. A more modest but still challenging problem is to classify all finite *simple* groups. These are finite groups which may be used as building blocks for constructing arbitrary finite groups in the same way that prime numbers can be used to construct arbitrary whole numbers by taking products.[lower-alpha 6] One of the major achievements of contemporary group theory is the classification of finite simple groups, a mathematical theorem which provides a list of all possible finite simple groups.[107]

This classification theorem identifies several infinite families of groups as well as 26 additional groups which do not fit into any family. The latter groups are called the "sporadic" groups, and each one owes its existence to a remarkable combination of circumstances. The largest sporadic group, the so-called monster group, has over 10^{53} elements, more than a thousand times the number of atoms in the Earth.[108]

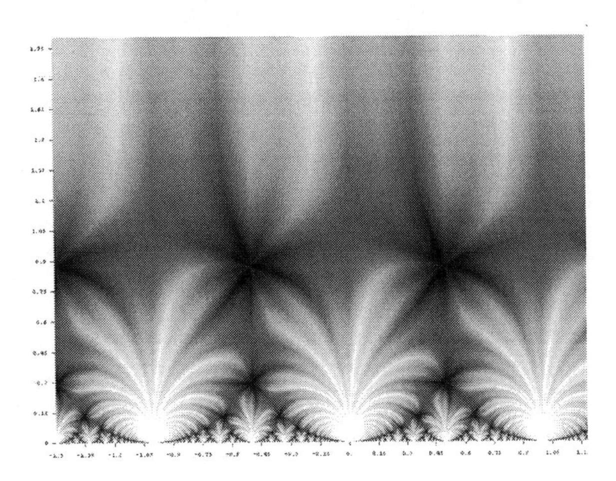

A graph of the j*-function in the complex plane*

A seemingly unrelated construction is the *j*-function of number theory. This object belongs to a special class of functions called modular functions, whose graphs form a certain kind of repeating pattern.[109] Although this function appears in a branch of mathematics which seems very different from the theory of finite groups, the two subjects turn out to be intimately related. In the late 1970s, mathematicians John McKay and John Thompson noticed that certain numbers arising in the analysis of the monster group (namely, the dimensions of its irreducible representations) are related to numbers that appear in a formula for the *j*-function (namely, the coefficients of its Fourier series).[110] This relationship was further developed by John Horton Conway and Simon Norton[111] who called it monstrous moonshine because it seemed so far fetched.[112]

In 1992, Richard Borcherds constructed a bridge between the theory of modular functions and finite groups and, in the process, explained the observations of McKay and Thompson.[113][114] Borcherds' work used ideas from string theory in an essential way, extending earlier results of Igor Frenkel, James Lepowsky, and Arne Meurman, who had realized the monster group as the symmetries of a particular version of string theory.[115] In 1998, Borcherds was awarded the Fields medal for his work.[116]

Since the 1990s, the connection between string theory and moonshine has led to further results in mathematics and physics.[108] In 2010, physicists Tohru Eguchi, Hirosi Ooguri, and Yuji Tachikawa discovered connections between a different sporadic group, the Mathieu group M_{24}, and a certain version of string theory.[117] Miranda Cheng, John Duncan, and Jeffrey A. Harvey proposed a generalization of this moonshine phenomenon called umbral moonshine,[118] and their conjecture was proved mathematically by Duncan, Michael Griffin, and Ken Ono.[119] Witten has also speculated that the version of string theory appearing in monstrous moonshine might be related to a certain simplified model of gravity in three spacetime dimensions.[120]

12.7 History

Main article: History of string theory

12.7.1 Early results

Some of the structures reintroduced by string theory arose for the first time much earlier as part of the program of classical unification started by Albert Einstein. The first person to add a fifth dimension to a theory of gravity was Gunnar Nordström in 1914, who noted that gravity in five dimensions describes both gravity and electromagnetism in four. Nordström attempted to unify electromagnetism with his theory of gravitation, which was however superseded by Einstein's general relativity in 1919. Thereafter, German mathematician Theodor Kaluza combined the fifth dimension with general relativity, and only Kaluza is usually credited with the idea. In 1926, the Swedish physicist Oskar Klein gave a physical interpretation of the unobservable extra dimension—it is wrapped into a small circle. Einstein introduced a non-symmetric metric tensor, while much later Brans and Dicke added a scalar component to gravity. These ideas would be revived within string theory, where they are demanded by consistency conditions.

Leonard Susskind

String theory was originally developed during the late 1960s and early 1970s as a never completely successful theory of hadrons, the subatomic particles like the proton and neutron that feel the strong interaction. In the 1960s, Geoffrey Chew and Steven Frautschi discovered that the mesons make families called Regge trajectories with masses related to spins in a way that was later understood by Yoichiro Nambu, Holger Bech Nielsen and Leonard Susskind to be the relationship expected from rotating strings. Chew advocated making a theory for the interactions of these trajectories that did not presume that they were composed of any fundamental particles, but would construct their interactions from self-consistency conditions on the S-matrix. The S-matrix approach was started by Werner Heisenberg in the 1940s as a way of constructing a theory that did not rely on the local notions of space and time, which Heisenberg believed break down at the nuclear scale. While the scale was off by many orders of magnitude, the approach he advocated was ideally suited for a theory of quantum gravity.

Working with experimental data, R. Dolen, D. Horn and C. Schmid developed some sum rules for hadron exchange. When a particle and antiparticle scatter, virtual particles can be exchanged in two qualitatively different ways. In the s-channel, the two particles annihilate to make temporary intermediate states that fall apart into the final state particles. In the t-channel, the particles exchange intermediate states by emission and absorption. In field theory, the two contributions add together, one giving a continuous background contribution, the other giving peaks at certain energies. In the data, it was clear that the peaks were stealing from the background—the authors interpreted this as saying that the t-channel contribution was dual to the s-channel one, meaning both described the whole amplitude and included the other.

Gabriele Veneziano

The result was widely advertised by Murray Gell-Mann, leading Gabriele Veneziano to construct a scattering am plitude that had the property of Dolen-Horn-Schmid duality, later renamed world-sheet duality. The amplitude needed poles where the particles appear, on straight line trajectories, and there is a special mathematical function whose poles are evenly spaced on half the real line— the Gamma function— which was widely used in Regge theory. By manipulating combinations of Gamma functions, Veneziano was able to find a consistent scattering amplitude with poles on straight lines, with mostly positive residues, which obeyed duality and had the appropriate Regge scaling at high energy. The amplitude could fit near-beam scattering data as well as other Regge type fits, and had a suggestive integral representation that could be used for generalization.

Over the next years, hundreds of physicists worked to complete the bootstrap program for this model, with many surprises. Veneziano himself discovered that for the scattering amplitude to describe the scattering of a particle that appears in the theory, an obvious self-consistency condition, the lightest particle must be a tachyon. Miguel Virasoro and Joel Shapiro found a different amplitude now understood to be that of closed strings, while Ziro Koba and Holger Nielsen generalized Veneziano's integral representation to multiparticle scattering. Veneziano and Sergio Fubini introduced an operator formalism for computing the scattering amplitudes that was a forerunner of world-sheet conformal theory, while Virasoro understood how to remove the poles with wrong-sign residues using a constraint on the states. Claud Lovelace calculated a loop amplitude, and noted that there is an inconsistency unless the dimension of the theory is 26. Charles Thorn, Peter Goddard and Richard Brower went on to prove that there are no wrong-sign propagating states in dimensions less than or equal to 26.

In 1969, Yoichiro Nambu, Holger Bech Nielsen, and Leonard Susskind recognized that the theory could be given a description in space and time in terms of strings. The scattering amplitudes were derived systematically from the action principle by Peter Goddard, Jeffrey Goldstone, Claudio Rebbi, and Charles Thorn, giving a space-time picture to the vertex operators introduced by Veneziano and Fubini and a geometrical interpretation to the Virasoro conditions.

In 1970, Pierre Ramond added fermions to the model, which led him to formulate a two-dimensional supersymmetry to cancel the wrong-sign states. John Schwarz and André Neveu added another sector to the fermi theory a short time later. In the fermion theories, the critical dimension was 10. Stanley Mandelstam formulated a world sheet conformal theory for both the bose and fermi case, giving a two-dimensional field theoretic path-integral to generate the operator formalism. Michio Kaku and Keiji Kikkawa gave a different formulation of the bosonic string, as a string field theory, with infinitely many particle types and with fields taking values not on points, but on loops and curves.

In 1974, Tamiaki Yoneya discovered that all the known string theories included a massless spin-two particle that obeyed the correct Ward identities to be a graviton. John Schwarz and Joel Scherk came to the same conclusion and made the bold leap to suggest that string theory was a theory of gravity, not a theory of hadrons. They reintroduced Kaluza–Klein theory as a way of making sense of the extra dimensions. At the same time, quantum chromodynamics was recognized as the correct theory of hadrons, shifting the attention of physicists and apparently leaving the bootstrap program in the dustbin of history.

String theory eventually made it out of the dustbin, but for the following decade all work on the theory was completely ignored. Still, the theory continued to develop at a steady pace thanks to the work of a handful of devotees. Ferdinando Gliozzi, Joel Scherk, and David Olive realized in 1976 that the original Ramond and Neveu Schwarz-strings were separately inconsistent and needed to be combined. The resulting theory did not have a tachyon, and was proven to have space-time supersymmetry by John Schwarz and Michael Green in 1981. The same year, Alexander Polyakov gave the theory a modern path integral formulation, and went on to develop conformal field theory extensively. In 1979, Daniel Friedan showed that the equations of motions of string theory, which are generalizations of the Einstein equations of General Relativity, emerge from the Renormalization group equations for the two-dimensional field theory. Schwarz and Green discovered T-duality, and constructed two superstring theories—IIA and IIB related by T-duality, and type I theories with open strings. The consistency conditions had been so strong, that the entire theory was nearly uniquely determined, with only a few discrete choices.

12.7.2 First superstring revolution

Edward Witten

In the early 1980s, Edward Witten discovered that most theories of quantum gravity could not accommodate chiral fermions like the neutrino. This led him, in collaboration with Luis Álvarez-Gaumé to study violations of the conservation laws in gravity theories with anomalies, concluding that type I string theories were inconsistent. Green and Schwarz discovered a contribution to the anomaly that Witten and Alvarez-Gaumé had missed, which restricted the gauge group of the type I string theory to be SO(32). In coming to understand this calculation, Edward Witten became convinced that string theory was truly a consistent theory of gravity, and he became a high-profile advocate. Following Witten's lead, between 1984 and 1986, hundreds of physicists started to work in this field, and this is sometimes called the first superstring revolution.

During this period, David Gross, Jeffrey Harvey, Emil Martinec, and Ryan Rohm discovered heterotic strings. The gauge group of these closed strings was two copies of E8, and either copy could easily and naturally include the standard model. Philip Candelas, Gary Horowitz, Andrew Strominger and Edward Witten found that the Calabi–Yau manifolds are the compactifications that preserve a realistic amount of supersymmetry, while Lance Dixon and others worked out the physical properties of orbifolds, distinctive geometrical singularities allowed in string theory. Cumrun Vafa generalized T-duality from circles to arbitrary manifolds, creating the mathematical field of mirror symmetry. Daniel Friedan, Emil Martinec and Stephen Shenker further developed the covariant quantization of the superstring using conformal field theory techniques. David Gross and Vipul Periwal discovered that string perturbation theory was divergent. Stephen Shenker showed it diverged much faster than in field theory suggesting that new nonperturbative objects were missing.

Joseph Polchinski

In the 1990s, Joseph Polchinski discovered that the theory requires higher-dimensional objects, called D-branes and identified these with the black-hole solutions of supergravity. These were understood to be the new objects suggested by the perturbative divergences, and they opened up a new field with rich mathematical structure. It quickly became clear that D-branes and other p-branes, not just strings, formed the matter content of the string theories, and the physical interpretation of the strings and branes was revealed—they are a type of black hole. Leonard Susskind had incorporated the holographic principle of Gerardus 't Hooft into string theory, identifying the long highly excited string states with ordinary thermal black hole states. As suggested by 't Hooft, the fluctuations of the black hole horizon, the world-sheet or world-volume theory, describes not only the degrees of freedom of the black hole, but all nearby objects too.

12.7.3 Second superstring revolution

In 1995, at the annual conference of string theorists at the University of Southern California (USC), Edward Witten gave a speech on string theory that in essence united the five string theories that existed at the time, and giving birth to a new 11-dimensional theory called M-theory. M-theory was also foreshadowed in the work of Paul Townsend at approximately the same time. The flurry of activity that began at this time is sometimes called the second superstring revolution.[34]

Juan Maldacena

During this period, Tom Banks, Willy Fischler, Stephen Shenker and Leonard Susskind formulated matrix theory, a full holographic description of M-theory using IIA D0 branes.[51] This was the first definition of string theory that was fully non-perturbative and a concrete mathematical realization of the holographic principle. It is an example of a gauge-gravity duality and is now understood to be a special case of the AdS/CFT correspondence. Andrew Strominger and Cumrun Vafa calculated the entropy of certain configurations of D-branes and found agreement with the semi-classical answer for extreme charged black holes.[62] Petr Hořava and Witten found the eleven-dimensional formulation of the heterotic string theories, showing that orbifolds solve the chirality problem. Witten noted that the effective description of the physics of D-branes at low energies is by a supersymmetric gauge theory, and found geometrical interpretations of mathematical structures in gauge theory that he and Nathan Seiberg had earlier discovered in terms of the location of the branes.

In 1997, Juan Maldacena noted that the low energy excitations of a theory near a black hole consist of objects close to the horizon, which for extreme charged black holes looks like an anti-de Sitter space.[71] He noted that in this limit the gauge theory describes the string excitations near the branes. So he hypothesized that string theory on a near-horizon extreme-charged black-hole geometry, an anti-deSitter space times a sphere with flux, is equally well described by the low-energy limiting gauge theory, the N = 4 supersymmetric Yang–Mills theory. This hypothesis, which is called the AdS/CFT correspondence, was further developed by Steven Gubser, Igor Klebanov and Alexander Polyakov,[72] and by Edward Witten,[73] and it is now well-accepted. It is a concrete realization of the holographic principle, which has far-reaching implications for black holes, locality and information in physics, as well as the nature of the gravitational interaction.[56] Through this relationship, string theory has been shown to be related to gauge theories like quantum chromodynamics and this has led to more quantitative understanding of the behavior of hadrons, bringing string theory back to its roots.[84]

12.8 Criticism

12.8.1 Number of solutions

Main article: String theory landscape

To construct models of particle physics based on string theory, physicists typically begin by specifying a shape for the extra dimensions of spacetime. Each of these different shapes corresponds to a different possible universe, or "vacuum state", with a different collection of particles and forces. String theory as it is currently understood has an enormous number of vacuum states, typically estimated to be around 10^{500}, and these might be sufficiently diverse to accommodate almost any phenomena that might be observed at low energies.[121]

Many critics of string theory have expressed concerns about the large number of possible universes described by string theory. In his book *Not Even Wrong*, Peter Woit, a lecturer in the mathematics department at Columbia University, has argued that the large number of different physical scenarios renders string theory vacuous as a framework for constructing models of particle physics. According to Woit,

> The possible existence of, say, 10^{500} consistent different vacuum states for superstring theory probably destroys the hope of using the theory to predict anything. If one picks among this large set just those states whose properties agree with present experimental observations, it is likely there still will be such a large number of these that one can get just about whatever value one wants for the results of any new observation.[122]

Some physicists believe this large number of solutions is actually a virtue because it may allow a natural anthropic explanation of the observed values of physical constants, in particular the small value of the cosmological constant.[122] The anthropic principle is the idea that some of the numbers appearing in the laws of physics are not fixed by any fundamental principle but must be compatible with the evolution of intelligent life. In 1987, Steven Weinberg published an article in which he argued that the cosmological constant could not have been too large, or else galaxies and intelligent life would not have been able to develop.[123] Weinberg suggested that there might be a huge number of possible consistent universes, each with a different value of the cosmological constant, and observations indicate a small value of the cosmological constant only because humans happen to live in a universe that has allowed intelligent life, and hence observers, to exist.[124]

String theorist Leonard Susskind has argued that string theory provides a natural anthropic explanation of the small value of the cosmological constant.[125] According to Susskind, the different vacuum states of string theory might be realized as different universes within a larger multiverse. The fact that the observed universe has a small cosmological constant is just a tautological consequence of the fact that a small value is required for life to exist.[126] Many prominent theorists and critics have disagreed with Susskind's conclusions.[127] According to Woit, "in this case [anthropic reasoning] is nothing more than an excuse for failure. Speculative scientific ideas fail not just when they make incorrect predictions, but also when they turn out to be vacuous and incapable of predicting anything."[128]

12.8.2 Background independence

Main article: Background independence

One of the fundamental properties of Einstein's general theory of relativity is that it is background independent, meaning that the formulation of the theory does not in any way privilege a particular spacetime geometry.[129]

One of the main criticisms of string theory from early on is that it is not manifestly background independent. In string theory, one must typically specify a fixed reference geometry for spacetime, and all other possible geometries are described as perturbations of this fixed one. In his book *The Trouble With Physics*, physicist Lee Smolin of the Perimeter Institute for Theoretical Physics claims that this is the principal weakness of string theory as a theory of quantum gravity, saying that string theory has failed to incorporate this important insight from general relativity.[130]

Others have disagreed with Smolin's characterization of string theory. In a review of Smolin's book, string theorist Joseph Polchinski writes

> [Smolin] is mistaking an aspect of the mathematical language being used for one of the physics being described. New physical theories are often discovered using a mathematical language that is not the most suitable for them... In string theory it has always been clear that the physics is background-independent even if the language being used is not, and the search for more suitable language continues. Indeed, as Smolin belatedly notes, [AdS/CFT] provides a solution to this problem, one that is unexpected and powerful.[131]

Polchinski notes that an important open problem in quantum gravity is to develop holographic descriptions of gravity which do not require the gravitational field to be asymptotically anti-de Sitter.[131]

Smolin responded that the claims about background-independence, which Polchinski presents as "clear", are in fact only an unproven hope for future results, and Smolin is skeptical about them being true at all because of fundamental reasons: "If the strong form of the AdS/CFT conjecture is shown to be correct, then a very weak, and limited form of background will have been achieved. But ... this is still a big if". Smolin points out that current results about the [AdS/CFT] conjecture rely on global super-symmetry as perturbative physics, "but the whole point of general relativity and quantum gravity is that the generic solutions are governed by no global symmetries because the geometry of spacetime is completely dynamical", which "makes it very non-trivial to show the strong form of the [AdS/CFT] conjecture, because it must extend to solutions of supergravity arbitrarily far from those with global symmetries in the bulk".[132] Smolin summarizes:

> It would be more accurate to say, "Some string theorists believe that the formulations of perturbative string theories and dualities between them that they study concretely are approximations to a deeper, background independent formulation. This missing background independent formulation is not just a different language for the theory, it is hoped to be the statement of the principles and laws that define the theory, from which everything studied so far would be derived as an approximation."[132]

12.8.3 Sociological issues

Since the superstring revolutions of the 1980s and 1990s, string theory has become the dominant paradigm of high energy theoretical physics.[133] Some string theorists have expressed the view that there does not exist an equally successful alternative theory addressing the deep questions of fundamental physics. In an interview from 1987, Nobel laureate David Gross made the following controversial comments about the reasons for the popularity of string theory:

> The most important [reason] is that there are no other good ideas around. That's what gets most people into it. When people started to get interested in string theory they didn't know anything about it. In fact, the first reaction of most people is that the theory is extremely ugly and unpleasant, at least that was the case a few years ago when the understanding of string theory was much less developed. It was difficult for people to learn about it and to be turned on. So I think the real reason why people have got attracted by it is because there is no other game in town. All other approaches of constructing grand unified theories, which were more conservative to begin with, and only gradually became more and more radical, have failed, and this game hasn't failed yet.[134]

Several other high profile theorists and commentators have expressed similar views, suggesting that there are no viable alternatives to string theory.[135]

Many critics of string theory have commented on this state of affairs. In his book criticizing string theory, Peter Woit views the status of string theory research as unhealthy and detrimental to the future of fundamental physics. He argues that the extreme popularity of string theory among theoretical physicists is partly a consequence of the financial structure of academia and the fierce competition for scarce resources.[136] In his book *The Road to Reality*, mathematical physicist Roger Penrose expresses similar views, stating "The often frantic competitiveness that this ease of communication engenders leads to 'bandwagon' effects, where researchers fear to be left behind if they do not join in."[137] Penrose also claims that the technical difficulty of modern physics forces young scientists to rely on the preferences of established researchers, rather than forging new paths of their own.[138] Lee Smolin expresses a slightly different position in his critique, claiming that string theory grew out of a tradition of particle physics which discourages speculation about the foundations of physics, while his preferred approach, loop quantum gravity, encourages more radical thinking. According to Smolin,

> String theory is a powerful, well-motivated idea and deserves much of the work that has been devoted to it. If it has so far failed, the principal reason is that its intrinsic flaws are closely tied to its strengths—and, of course, the story is unfinished, since string theory may well turn out to be part of the truth. The real question is not why we have expended so much energy on string theory but why we haven't expended nearly enough on alternative approaches.[139]

Smolin goes on to offer a number of prescriptions for how scientists might encourage a greater diversity of approaches to quantum gravity research.[140]

12.9 References

12.9.1 Notes

[1] For example, physicists are still working to understand the phenomenon of quark confinement, the paradoxes of black holes, and the origin of dark energy.

[2] For example, in the context of the AdS/CFT correspondence, theorists often formulate and study theories of gravity in unphysical numbers of spacetime dimensions.

[3] "Top Cited Articles during 2010 in hep-th". Retrieved 25 July 2013.

[4] More precisely, one cannot apply the methods of perturbative quantum field theory.

[5] Two independent mathematical proofs of mirror symmetry were given by Givental 1996, 1998 and Lian, Liu, Yau 1997, 1999, 2000.

[6] More precisely, a nontrivial group is called *simple* if its only normal subgroups are the trivial group and the group itself. The Jordan–Hölder theorem exhibits finite simple groups as the building blocks for all finite groups.

12.9.2 Citations

[1] Becker, Becker, and Schwarz 2007, p. 1

[2] Zwiebach 2009, p. 6

[3] Becker, Becker, and Schwarz 2007, pp. 2–3

[4] Becker, Becker, and Schwarz 2007, pp. 9–12

[5] Becker, Becker, and Schwarz 2007, pp. 14–15

[6] Klebanov and Maldacena 2009

[7] Merali 2011

[8] Sachdev 2013

[9] Becker, Becker, and Schwarz 2007, pp. 3, 15–16

[10] Becker, Becker, and Schwarz 2007, p. 8

[11] Becker, Becker, and Schwarz 13–14

[12] Woit 2006

[13] Zee 2010

[14] Becker, Becker, and Schwarz 2007, p. 2

[15] Becker, Becker, and Schwarz 2007, p. 6

[16] Zwiebach 2009, p. 12

[17] Becker, Becker, and Schwarz 2007, p. 4

[18] Zwiebach 2009, p. 324

[19] Wald 1984, p. 4

[20] Zee 2010, Parts V and VI

[21] Zwiebach 2009, p. 9

[22] Zwiebach 2009, p. 8

[23] Yau and Nadis 2010, Ch. 6

[24] Greene 2000, p. 186

[25] Yau and Nadis 2010, p. ix

[26] Randall and Sundrum 1999

[27] Becker, Becker, and Schwarz 2007

[28] Zwiebach 2009, p. 376

[29] Moore 2005, p. 214

[30] Moore 2005, p. 215

[31] Aspinwall et al. 2009

[32] Kontsevich 1995

[33] Kapustin and Witten 2007

[34] Duff 1998

[35] Duff 1998, p. 64

[36] Nahm 1978

[37] Cremmer, Julia, and Scherk 1978

[38] Duff 1998, p. 65

[39] Sen 1994a

[40] Sen 1994b

[41] Hull and Townsend 1995

[42] Duff 1998, p. 67

[43] Bergshoeff, Sezgin, and Townsend 1987

[44] Duff et al. 1987

[45] Duff 1998, p. 66

[46] Witten 1995

[47] Duff 1998, pp. 67–68

[48] Becker, Becker, and Schwarz 2007, p. 296

[49] Hořava and Witten 1996

[50] Duff 1996, sec. 1

[51] Banks et al. 1997

[52] Connes 1994

[53] Connes, Douglas, and Schwarz 1998

[54] Nekrasov and Schwarz 1998

[55] Seiberg and Witten 1999

[56] de Haro et al. 2013, p. 2

[57] Yau and Nadis 2010, p. 187–188

[58] Bekenstein 1973

[59] Hawking 1975

[60] Wald 1984, p. 417

[61] Yau and Nadis 2010, p. 189

[62] Strominger and Vafa 1996

[63] Yau and Nadis 2010, pp. 190–192

[64] Maldacena, Strominger, and Witten 1997

[65] Ooguri, Strominger, and Vafa 2004

[66] Yau and Nadis 2010, pp. 192–193

[67] Yau and Nadis 2010, pp. 194–195

[68] Strominger 1998

[69] Guica et al. 2009

[70] Castro, Maloney, and Strominger 2010

[71] Maldacena 1998

[72] Gubser, Klebanov, and Polyakov 1998

[73] Witten 1998

[74] Klebanov and Maldacena 2009, p. 28

[75] Maldacena 2005, p. 60

[76] Maldacena 2005, p. 61

[77] Zwiebach 2009, p. 552

[78] Maldacena 2005, pp. 61–62

[79] Susskind 2008

[80] Zwiebach 2009, p. 554

[81] Maldacena 2005, p. 63

[82] Hawking 2005

[83] Zwiebach 2009, p. 559

[84] Kovtun, Son, and Starinets 2001

[85] Merali 2011, p. 303

[86] Luzum and Romatschke 2008

[87] Sachdev 2013, p. 51

[88] Candelas et al. 1985

[89] Yau and Nadis 2010, pp. 147–150

[90] Becker, Becker, and Schwarz 2007, pp. 530–531

[91] Becker, Becker, and Schwarz 2007, p. 531

[92] Becker, Becker, and Schwarz 2007, p. 538

[93] Becker, Becker, and Schwarz 2007, p. 533

[94] Becker, Becker, and Schwarz 2007, pp. 539–543

[95] Deligne et al. 1999, p. 1

[96] Hori et al. 2003, p. xvii

[97] Aspinwall et al. 2009, p. 13

[98] Hori et al. 2003

[99] Yau and Nadis 2010, p. 167

[100] Yau and Nadis 2010, p. 166

[101] Yau and Nadis 2010, p. 169

[102] Candelas et al. 1991

[103] Yau and Nadis 2010, p. 171

[104] Hori et al. 2003, p. xix

[105] Strominger, Yau, and Zaslow 1996

[106] Dummit and Foote 2004

[107] Dummit and Foote 2004, pp. 102–103

[108] Klarreich 2015

[109] Gannon 2006, p. 2

[110] Gannon 2006, p. 4

[111] Conway and Norton 1979

[112] Gannon 2006, p. 5

[113] Gannon 2006, p. 8

[114] Borcherds 1992

[115] Frenkel, Lepowsky, and Meurman 1988

[116] Gannon 2006, p. 11

[117] Eguchi, Ooguri, and Tachikawa 2010

[118] Cheng, Duncan, and Harvey 2013

[119] Duncan, Griffin, and Ono 2015

[120] Witten 2007

[121] Woit 2006, pp. 240–242

[122] Woit 2006, p. 242

[123] Weinberg 1987

[124] Woit 2006, p. 243

[125] Susskind 2005

[126] Woit 2006, pp. 242–243

[127] Woit 2006, p. 240

[128] Woit 2006, p. 249

[129] Smolin 2006, p. 81

[130] Smolin 2006, p. 184

[131] Polchinski 2007

[132] Lee Smolin, April 2007:"Archived copy". Archived from the original on November 5, 2015. Retrieved December 31, 2015. Response to review of The Trouble with Physics by Joe Polchinski

[133] Penrose 2004, p. 1017

[134] Woit 2006, pp. 224–225

[135] Woit 2006, Ch. 16

[136] Woit 2006, p. 239

[137] Penrose 2004, p. 1018

[138] Penrose 2004, pp. 1019–1020

[139] Smolin 2006, p. 349

[140] Smolin 2006, Ch. 20

12.9.3 Bibliography

- Aspinwall, Paul; Bridgeland, Tom; Craw, Alastair; Douglas, Michael; Gross, Mark; Kapustin, Anton; Moore, Gregory; Segal, Graeme; Szendröi, Balázs; Wilson, P.M.H., eds. (2009). *Dirichlet Branes and Mirror Symmetry*. American Mathematical Society. ISBN 978-0-8218-3848-8.

- Banks, Tom; Fischler, Willy; Schenker, Stephen; Susskind, Leonard (1997). "M theory as a matrix model: A conjecture". *Physical Review D* **55** (8): 5112–5128. arXiv:hep-th/9610043. Bibcode:1997PhRvD..55.5112B. doi:10.1103/physrevd.55.5112.

- Becker, Katrin; Becker, Melanie; Schwarz, John (2007). *String theory and M-theory: A modern introduction*. Cambridge University Press. ISBN 978-0-521-86069-7.

- Bekenstein, Jacob (1973). "Black holes and entropy". *Physical Review D* **7** (8): 2333–2346. Bibcode:1973PhRvD...7.2333B. doi:10.1103/PhysRevD.7.2333.

- Bergshoeff, Eric; Sezgin, Ergin; Townsend, Paul (1987). "Supermembranes and eleven-dimensional supergravity". *Physics Letters B* **189** (1): 75–78. Bibcode:1987PhLB..189...75B. doi:10.1016/0370-2693(87)91272-X.

- Borcherds, Richard (1992). "Monstrous moonshine and Lie superalgebras". *Inventiones Mathematicae* **109** (1): 405–444. Bibcode:1992InMat.109..405B. doi:10.1007/BF01232032.

- Candelas, Philip; de la Ossa, Xenia; Green, Paul; Parks, Linda (1991). "A pair of Calabi–Yau manifolds as an exactly soluble superconformal field theory". *Nuclear Physics B* **359** (1): 21–74. Bibcode:1991NuPhB.359...21C. doi:10.1016/0550-3213(91)90292-6.
- Candelas, Philip; Horowitz, Gary; Strominger, Andrew; Witten, Edward (1985). "Vacuum configurations for superstrings". *Nuclear Physics B* **258**: 46–74. Bibcode:1985NuPhB.258...46C. doi:10.1016/0550-3213(85)90602-9.
- Castro, Alejandra; Maloney, Alexander; Strominger, Andrew (2010). "Hidden conformal symmetry of the Kerr black hole". *Physical Review D* **82** (2). arXiv:1004.0996. Bibcode:2010PhRvD..82b4008C. doi:10.1103/PhysRevD.82.024008.
- Cheng, Miranda; Duncan, John; Harvey, Jeffrey (2013). "Umbral Moonshine". arXiv:1204.2779.
- Connes, Alain (1994). *Noncommutative Geometry*. Academic Press. ISBN 978-0-12-185860-5.
- Connes, Alain; Douglas, Michael; Schwarz, Albert (1998). "Noncommutative geometry and matrix theory". *Journal of High Energy Physics*. 19981 (2): 003. arXiv:hep-th/9711162. Bibcode:1998JHEP...02..003C. doi:10.1088/1126-6708/1998/02/003.
- Conway, John; Norton, Simon (1979). "Monstrous moonshine". *Bull. London Math. Soc.* **11** (3): 308–339. doi:10.1112/blms/11.3.308.
- Cremmer, Eugene; Julia, Bernard; Scherk, Joel (1978). "Supergravity theory in eleven dimensions". *Physics Letters B* **76** (4): 409–412. Bibcode:1978PhLB...76..409C. doi:10.1016/0370-2693(78)90894-8.
- de Haro, Sebastian; Dieks, Dennis; 't Hooft, Gerard; Verlinde, Erik (2013). "Forty Years of String Theory Reflecting on the Foundations". *Foundations of Physics* **43** (1): 1–7. Bibcode:2013FoPh...43....1D. doi:10.1007/s10701-012-9691-3.
- Deligne, Pierre; Etingof, Pavel; Freed, Daniel; Jeffery, Lisa; Kazhdan, David; Morgan, John; Morrison, David; Witten, Edward, eds. (1999). *Quantum Fields and Strings: A Course for Mathematicians* **1**. American Mathematical Society. ISBN 978-0821820124.
- Duff, Michael (1996). "M-theory (the theory formerly known as strings)". *International Journal of Modern Physics A* **11** (32): 6523–41. arXiv:hep-th/9608117. Bibcode:1996IJMPA..11.5623D. doi:10.1142/S0217751X96002583.
- Duff, Michael (1998). "The theory formerly known as strings". *Scientific American* **278** (2): 64–9. doi:10.1038/scientificamerican0298-64.
- Duff, Michael; Howe, Paul; Inami, Takeo; Stelle, Kellogg (1987). "Superstrings in *D*=10 from supermembranes in *D*=11". *Nuclear Physics B* **191** (1): 70–74. Bibcode:1987PhLB..191...70D. doi:10.1016/0370-2693(87)91323-2.
- Dummit, David; Foote, Richard (2004). *Abstract Algebra*. Wiley. ISBN 978-0-471-43334-7.
- Duncan, John; Griffin, Michael; Ono, Ken (2015). "Proof of the Umbral Moonshine Conjecture". arXiv:1503.01472.
- Eguchi, Tohru; Ooguri, Hirosi; Tachikawa, Yuji (2011). "Notes on the K3 surface and the Mathieu group M_{24}". *Experimental Mathematics* **20** (1): 91–96. doi:10.1080/10586458.2011.544585.
- Frenkel, Igor; Lepowsky, James; Meurman, Arne (1988). *Vertex Operator Algebras and the Monster*. Pure and Applied Mathematics **134**. Academic Press. ISBN 0-12-267065-5.
- Gannon, Terry. *Moonshine Beyond the Monster: The Bridge Connecting Algebra, Modular Forms, and Physics*. Cambridge University Press.
- Givental, Alexander (1996). "Equivariant Gromov-Witten invariants". *International Mathematics Research Notices* **1996** (13): 613–663. doi:10.1155/S1073792896000414.
- Givental, Alexander (1998). "A mirror theorem for toric complete intersections". *Topological field theory, primitive forms and related topics*: 141–175. doi:10.1007/978-1-4612-0705-4_5. ISBN 978-1-4612-6874-1.
- Gubser, Steven; Klebanov, Igor; Polyakov, Alexander (1998). "Gauge theory correlators from non-critical string theory". *Physics Letters B* **428**: 105–114. arXiv:hep-th/9802109. Bibcode:1998PhLB..428..105G. doi:10.1016/S0370-2693(98)00377-3.
- Guica, Monica; Hartman, Thomas; Song, Wei; Strominger, Andrew (2009). "The Kerr/CFT Correspondence". *Physical Review D* **80** (12). arXiv:0809.4266. Bibcode:2009PhRvD..80l4008G. doi:10.1103/PhysRevD.80.124008.
- Hawking, Stephen (1975). "Particle creation by black holes". *Communications in Mathematical Physics* **43** (3): 199–220. Bibcode:1975CMaPh..43..199H. doi:10.1007/BF02345020.

- Hawking, Stephen (2005). "Information loss in black holes". *Physical Review D* **72** (8). arXiv:hep-th/0507171. Bibcode:2005PhRvD..72h4013H. doi:10.1103/PhysRevD.72.084013.
- Hořava, Petr; Witten, Edward (1996). "Heterotic and Type I string dynamics from eleven dimensions". *Nuclear Physics B* **460** (3): 506–524. arXiv:hep-th/9510209. Bibcode:1996NuPhB.460..506H. doi:10.1016/0550-3213(95)00621-4.
- Hori, Kentaro; Katz, Sheldon; Klemm, Albrecht; Pandharipande, Rahul; Thomas, Richard; Vafa, Cumrun; Vakil, Ravi; Zaslow, Eric, eds. (2003). *Mirror Symmetry* (PDF). American Mathematical Society. ISBN 0-8218-2955-6.
- Hull, Chris; Townsend, Paul (1995). "Unity of superstring dualities". *Nuclear Physics B* **4381** (1): 109–137. arXiv:hep-th/9410167. Bibcode:1995NuPhB.438..109H. doi:10.1016/0550-3213(94)00559-W.
- Kapustin, Anton; Witten, Edward (2007). "Electric-magnetic duality and the geometric Langlands program". *Communications in Number Theory and Physics* **1** (1): 1–236. arXiv:hep-th/0604151. Bibcode:2007CNTP....1....1K. doi:10.4310/cntp.2007.v1.n1.a1.
- Klarreich, Erica. "Mathematicians chase moonshine's shadow". *Quanta Magazine*. Retrieved March 2015.
- Klebanov, Igor; Maldacena, Juan (2009). "Solving Quantum Field Theories via Curved Spacetimes" (PDF). *Physics Today* **62**: 28–33. Bibcode:2009PhT....62a..28K. doi:10.1063/1.3074260. Archived from the original (PDF) on July 2, 2013. Retrieved May 2013.
- Kontsevich, Maxim (1995). "Homological algebra of mirror symmetry". *Proceedings of the International Congress of Mathematicians*: 120–139. arXiv:alg-geom/9411018. Bibcode:1994alg.geom.11018K.
- Kovtun, P. K.; Son, Dam T.; Starinets, A. O. (2001). "Viscosity in strongly interacting quantum field theories from black hole physics". *Physical Review Letters* **94** (11): 111601. arXiv:hep-th/0405231. Bibcode:2005PhRvL..94k1601K. doi:10.1103/PhysRevLett.94.111601. PMID 15903845.
- Lian, Bong; Liu, Kefeng; Yau, Shing-Tung (1997). "Mirror principle, I". *Asian Journal of Mathematics* **1**: 729–763. arXiv:alg-geom/9712011. Bibcode:1997alg.geom.12011L.
- Lian, Bong; Liu, Kefeng; Yau, Shing-Tung (1999a). "Mirror principle, II". *Asian Journal of Mathematics* **3**: 109–146. arXiv:math/9905006. Bibcode:1999math......5006L.
- Lian, Bong; Liu, Kefeng; Yau, Shing-Tung (1999b). "Mirror principle, III". *Asian Journal of Mathematics* **3**: 771–800. arXiv:math/9912038. Bibcode:1999math.....12038L.
- Lian, Bong; Liu, Kefeng; Yau, Shing-Tung (2000). "Mirror principle, IV". *Surveys in Differential Geometry* **7**: 475–496. arXiv:math/0007104. Bibcode:2000math......7104L. doi:10.4310/sdg.2002.v7.n1.a15.
- Luzum, Matthew; Romatschke, Paul (2008). "Conformal relativistic viscous hydrodynamics: Applications to RHIC results at √*sNN*=200 GeV". *Physical Review C* **78** (3). arXiv:0804.4015. doi:10.1103/PhysRevC.78.034915.
- Maldacena, Juan (1998). "The Large *N* limit of superconformal field theories and supergravity". *Advances in Theoretical and Mathematical Physics* **2**: 231–252. arXiv:hep-th/9711200. Bibcode:1998AdTMP...2..231M. doi:10.1063/1.59653.
- Maldacena, Juan (2005). "The Illusion of Gravity" (PDF). *Scientific American* **293** (5): 56–63. Bibcode:2005SciAm.293e..56M. doi:10.1038/scientificamerican1105-56. PMID 16318027. Archived from the original (PDF) on November 1, 2014. Retrieved July 2013.
- Maldacena, Juan; Strominger, Andrew; Witten, Edward (1997). "Black hole entropy in M-theory". *Journal of High Energy Physics* **1997** (12). arXiv:hep-th/9711053. Bibcode:1997JHEP...12..002M. doi:10.1088/1126-6708/1997/12/002.
- Merali, Zeeya (2011). "Collaborative physics: string theory finds a bench mate". *Nature* **478** (7369): 302–304. Bibcode:2011Natur.478..302M. doi:10.1038/478302a. PMID 22012369.
- Moore, Gregory (2005). "What is ... a Brane?" (PDF). *Notices of the AMS* **52**: 214. Retrieved June 2013.
- Nahm, Walter (1978). "Supersymmetries and their representations". *Nuclear Physics B* **135** (1): 149–166. Bibcode:1978NuPhB.135..149N. doi:10.1016/0550-3213(78)90218-3.

- Nekrasov, Nikita; Schwarz, Albert (1998). "Instantons on noncommutative $\mathbf{R}^4$ and (2,0) superconformal six dimensional theory". *Communications in Mathematical Physics* **198** (3): 689–703. arXiv:hep-th/9802068. Bibcode:1998CMaPh.198..689N. doi:10.1007/s002200050490.

- Ooguri, Hirosi; Strominger, Andrew; Vafa, Cumrun (2004). "Black hole attractors and the topological string". *Physical Review D* **70** (10). arXiv:hep-th/0405146. Bibcode:2004PhRvD..70j6007O. doi:10.1103/physrevd.70.106007.

- Polchinski, Joseph (2007). "All Strung Out?". *American Scientist*. Retrieved April 2015.

- Penrose, Roger (2005). *The Road to Reality: A Complete Guide to the Laws of the Universe*. Knopf. ISBN 0-679-45443-8.

- Randall, Lisa; Sundrum, Raman (1999). "An alternative to compactification". *Physical Review Letters* **83** (23): 4690–4693. arXiv:hep-th/9906064. Bibcode:1999PhRvL..83.4690R. doi:10.1103/PhysRevLett.83.4690.

- Sachdev, Subir (2013). "Strange and stringy". *Scientific American* **308** (44): 44–51. Bibcode:2012SciAm.308a..44S. doi:10.1038/scientificamerican0113-44.

- Seiberg, Nathan; Witten, Edward (1999). "String Theory and Noncommutative Geometry". *Journal of High Energy Physics* **1999** (9): 032. arXiv:hep-th/9908142. Bibcode:1999JHEP...09..032S. doi:10.1088/1126-6708/1999/09/032.

- Sen, Ashoke (1994a). "Strong-weak coupling duality in four-dimensional string theory". *International Journal of Modern Physics A* **9** (21): 3707–3750. arXiv:hep-th/9402002. Bibcode:1994IJMPA...9.3707S. doi:10.1142/S0217751X94001497.

- Sen, Ashoke (1994b). "Dyon-monopole bound states, self-dual harmonic forms on the multi-monopole moduli space, and $SL(2,\mathbf{Z})$ invariance in string theory". *Physics Letters B* **329** (2): 217–221. arXiv:hep-th/9402032. Bibcode:1994PhLB..329..217S. doi:10.1016/0370-2693(94)90763-3.

- Smolin, Lee (2006). *The Trouble with Physics: The Rise of String Theory, the Fall of a Science, and What Comes Next*. New York: Houghton Mifflin Co. ISBN 0-618-55105-0.

- Strominger, Andrew (1998). "Black hole entropy from near-horizon microstates". *Journal of High Energy Physics* **1998** (2): 009. arXiv:hep-th/9712251. Bibcode:1998JHEP...02..009S. doi:10.1088/1126-6708/1998/02/009.

- Strominger, Andrew; Vafa, Cumrun (1996). "Microscopic origin of the Bekenstein–Hawking entropy". *Physics Letters B* **379** (1): 99–104. arXiv:hep-th/9601029. Bibcode:1996PhLB..379...99S. doi:10.1016/0370-2693(96)00345-0.

- Strominger, Andrew; Yau, Shing-Tung; Zaslow, Eric (1996). "Mirror symmetry is T-duality". *Nuclear Physics B* **479** (1): 243–259. arXiv:hep-th/9606040. Bibcode:1996NuPhB.479..243S. doi:10.1016/0550-3213(96)00434-8.

- Susskind, Leonard (2005). *The Cosmic Landscape: String Theory and the Illusion of Intelligent Design*. Back Bay Books. ISBN 978-0316013338.

- Susskind, Leonard (2008). *The Black Hole War: My Battle with Stephen Hawking to Make the World Safe for Quantum Mechanics*. Little, Brown and Company. ISBN 978-0-316-01641-4.

- Wald, Robert (1984). *General Relativity*. University of Chicago Press. ISBN 978-0-226-87033-5.

- Weinberg, Steven (1987). *Anthropic bound on the cosmological constant* **59**. Physical Review Letters. p. 2607.

- Witten, Edward (1995). "String theory dynamics in various dimensions". *Nuclear Physics B* **443** (1): 85–126. arXiv:hep-th/9503124. Bibcode:1995NuPhB.443...85W. doi:10.1016/0550-3213(95)00158-O.

- Witten, Edward (1998). "Anti-de Sitter space and holography". *Advances in Theoretical and Mathematical Physics* **2**: 253–291. arXiv:hep-th/9802150. Bibcode:1998AdTMP...2..253W.

- Witten, Edward (2007). "Three-dimensional gravity revisited". arXiv:0706.3359 [hep-th].

- Woit, Peter (2006). *Not Even Wrong: The Failure of String Theory and the Search for Unity in Physical Law*. Basic Books. p. 105. ISBN 0-465-09275-6.

- Yau, Shing-Tung; Nadis, Steve (2010). *The Shape of Inner Space: String Theory and the Geometry of the Universe's Hidden Dimensions*. Basic Books. ISBN 978-0-465-02023-2.

- Zee, Anthony (2010). *Quantum Field Theory in a Nutshell* (2nd ed.). Princeton University Press. ISBN 978-0-691-14034-6.
- Zwiebach, Barton (2009). *A First Course in String Theory*. Cambridge University Press. ISBN 978-0-521-88032-9.

12.10 Further reading

12.10.1 Popularizations

General

- Greene, Brian (2003). *The Elegant Universe: Superstrings, Hidden Dimensions, and the Quest for the Ultimate Theory*. New York: W.W. Norton & Company. ISBN 0-393-05858-1.
- Greene, Brian (2004). *The Fabric of the Cosmos: Space, Time, and the Texture of Reality*. New York: Alfred A. Knopf. ISBN 0-375-41288-3.

Critical

- Penrose, Roger (2005). *The Road to Reality: A Complete Guide to the Laws of the Universe*. Knopf. ISBN 0-679-45443-8.
- Smolin, Lee (2006). *The Trouble with Physics: The Rise of String Theory, the Fall of a Science, and What Comes Next*. New York: Houghton Mifflin Co. ISBN 0-618-55105-0.
- Woit, Peter (2006). *Not Even Wrong: The Failure of String Theory And the Search for Unity in Physical Law*. London: Jonathan Cape &: New York: Basic Books. ISBN 978-0-465-09275-8.

12.10.2 Textbooks

For physicists

- Becker, Katrin; Becker, Melanie; Schwarz, John (2007). *String Theory and M-theory: A Modern Introduction*. Cambridge University Press. ISBN 978-0-521-86069-7.
- Green, Michael; Schwarz, John; Witten, Edward (2012). *Superstring theory. Vol. 1: Introduction*. Cambridge University Press. ISBN 978-1107029118.
- Green, Michael; Schwarz, John; Witten, Edward (2012). *Superstring theory. Vol. 2: Loop amplitudes, anomalies and phenomenology*. Cambridge University Press. ISBN 978-1107029132.
- Polchinski, Joseph (1998). *String Theory Vol. 1: An Introduction to the Bosonic String*. Cambridge University Press. ISBN 0-521-63303-6.
- Polchinski, Joseph (1998). *String Theory Vol. 2: Superstring Theory and Beyond*. Cambridge University Press. ISBN 0-521-63304-4.
- Zwiebach, Barton (2009). *A First Course in String Theory*. Cambridge University Press. ISBN 978-0-521-88032-9.

For mathematicians

- Deligne, Pierre; Etingof, Pavel; Freed, Daniel; Jeffery, Lisa; Kazhdan, David; Morgan, John; Morrison, David; Witten, Edward, eds. (1999). *Quantum Fields and Strings: A Course for Mathematicians, Vol. 2*. American Mathematical Society. ISBN 978-0821819883.

12.11 External links

- *The Elegant Universe*—A three-hour miniseries with Brian Greene by *NOVA* (original PBS Broadcast Dates: October 28, 8–10 p.m. and November 4, 8–9 p.m., 2003). Various images, texts, videos and animations explaining string theory.
- Not Even Wrong—A blog critical of string theory
- The Official String Theory Web Site
- Why String Theory—An introduction to string theory.

Chapter 13

Loop representation in gauge theories and quantum gravity

Attempts have been made to describe gauge theories in terms of extended objects such as Wilson loops and holonomies. The **loop representation** is a quantum hamiltonian representation of gauge theories in terms of loops. The aim of the loop representation, in the context of Yang–Mills theories is to avoid the redundancy introduced by Gauss gauge symmetries allowing to work directly in the space of physical states (Gauss gauge invariant states). The idea is well known in the context of lattice Yang–Mills theory (see lattice gauge theory). Attempts to explore the continuous loop representation was made by Gambini and Trias for canonical Yang–Mills theory, however there were difficulties as they represented singular objects. As we shall see the loop formalism goes far beyond a simple gauge invariant description, in fact it is the natural geometrical framework to treat gauge theories and quantum gravity in terms of their fundamental physical excitations.

The introduction by Ashtekar of a new set of variables (Ashtekar variables) cast general relativity in the same language as gauge theories and allowed one to apply loop techniques as a natural nonperturbative description of Einstein's theory. In canonical quantum gravity the difficulties in using the continuous loop representation are cured by the spatial diffeomorphism invariance of general relativity. The loop representation also provides a natural solution of the spatial diffeomorphism constraint, making a connection between canonical quantum gravity and knot theory. Surprisingly there were a class of loop states that provided exact (if only formal) solutions to Ashtekar's original (ill-defined) Wheeler–DeWitt equation. Hence an infinite set of exact (if only formal) solutions had been identified for all the equations of canonical quantum general gravity in this representation! This generated a lot of interest in the approach and eventually led to loop quantum gravity (LQG).

The loop representation has found application in mathematics. If topological quantum field theories are formulated in terms of loops, the resulting quantities should be what are known as knot invariants. Topological field theories only involve a finite number of degrees of freedom and so are exactly solvable. As a result, they provide concrete computable expressions that are invariants of knots. This was precisely the insight of Edward Witten[1] who noticed that computing loop dependent quantities in Chern–Simons and other three-dimensional topological quantum field theories one could come up with explicit, analytic expressions for knot invariants. For his work in this, in 1990 he was awarded the Fields Medal. He is the first and so far the only physicist to be awarded the Fields Medal, often viewed as the greatest honour in mathematics.

13.1 Gauge invariance of Maxwell's theory

The idea of gauge symmetries was introduced in Maxwell's theory. Maxwell's equations are

$\nabla \cdot \vec{E} = \frac{\rho}{\epsilon_0}$ $\quad \nabla \times \vec{B} - \epsilon_0 \mu_0 \frac{\partial \vec{E}}{\partial t} = \mu_0 \vec{J}$ $\quad \nabla \times \vec{E} + \frac{\partial \vec{B}}{\partial t} = 0$ $\quad \nabla \cdot \vec{B} = 0$

where ρ is the charge density and $\vec{J}$ the current density. The last two equations can be solved by writing fields in terms of a scalar potential, ϕ , and a vector potential, $\vec{A}$:

$\vec{E} = -\nabla \phi - \frac{\partial \vec{A}}{\partial t} \qquad \vec{B} = \nabla \times \vec{A}$.

The potentials uniquely determine the fields, but the fields do not uniquely determine the potentials - we can make the changes:

$\phi' = \phi + \frac{\partial \Lambda}{\partial t} \qquad \vec{A}' = \vec{A} - \nabla \Lambda$

without effecting the electric and magnetic fields, where $\Lambda(\vec{x}, t)$ is an arbitrary function of space-time . These are called gauge transformations. There is an elegant relativistic notation: the gauge field is

$A^\mu = (\phi, \vec{A})$

and the above gauge transformations read,

$A^{\mu\prime} = A^\mu + \partial^\mu \Lambda\ .$

The so-called field strength tensor is introduced,

$F^{\mu\nu} = \partial^\mu A^\nu - \partial^\nu A^\mu$

which is easily shown to be invariant under gauge transformations. In components,

$F^{0i} = E^i, \qquad \epsilon^{ijk} F^{jk} = B^i\ .$

Maxwell's source-free action is given by:

$S = -\frac{1}{2} \int d^4x \Big(F_{\mu\nu} F^{\mu\nu} \Big)\ .$

The ability to varying the gauge potential at different points in space and time without changing the physics is called a local invariance. Electromagnetic theory possess the simplest kind of local gauge symmetry called $U(1)$. A theory that displays local gauge invariance is called a gauge theory. In order to formulate other gauge theories we turn the above reasoning inside out. This is the subject of the next section.

13.2 The connection and gauges theories

13.2.1 The connection and Maxwell's theory

We know from quantum mechanics that if we replace the wave-function, $\psi(x)$, describing the electron field by

$\psi'(x) = \exp(i\theta)\psi(x)$

that it leaves physical predictions unchanged. We consider the imposition of local invariance on the phase of the electron field,

$\psi'(x) = \Omega\psi(x) = \exp(i\theta(x))\psi(x)$

The problem is that derivatives of $\psi(x)$ are not covariant under this transformation:

$\partial_\mu(\exp(i\theta(x))\psi(x)) = \Omega\partial_\mu\psi(x) + \partial_\mu\Omega\psi(x)\ .$

In order to cancel out the second unwanted term, one introduces a new derivative operator $\mathcal{D}_\mu$ that is covariant. To construct $\mathcal{D}_\mu$, one introduces a new field, the connection A_μ :

$\mathcal{D}_\mu = \partial_\mu + igA_\mu(x)\ .$

Then

$(\mathcal{D}_\mu\psi)' = \partial_\mu\psi' + igA'_\mu\psi' = \Omega\partial_\mu\psi + (\partial\Omega)\psi + igA'_\mu\Omega\psi$

The term $\partial_\mu\Omega$ is precisely cancelled out by requiring the connection field transforms as

$A'_\mu(x) = A_\mu(x) + \frac{i}{g}[\partial_\mu\Omega(x)]\Omega^{-1}(x) \quad Eq1.\ .$

We then have that

$(\mathcal{D}_\mu\psi)' = \Omega\mathcal{D}_\mu\psi\ .$

Note that $Eq1$ is equivalent to

$A'_\mu(x) = A_\mu(x) + \frac{1}{g}\partial_\mu\theta(x)$

which looks the same as a gauge transformation of the gauge potential of Maxwell's theory. It is possible to construct an invariant action for the connection field itself. We want an action that only has two derivatives (since actions with higher derivatives are not unitary). Define the quantity:

$F_{\mu\nu} = \frac{-i}{g}[\mathcal{D}_\mu, \mathcal{D}_\mu] = \frac{-i}{g}[\partial_\mu + igA_\mu(x), \partial_\nu + igA_\nu(x)]$

$= \frac{-i}{g}\Big([\partial_\mu, \partial_\nu] + ig(\partial_\mu A_\nu - \partial_\nu A_\mu) - g^2[A_\mu, A_\nu] \Big)$

$= \partial_\mu A_\nu - \partial_\nu A_\nu\ .$

The unique action with only two derivatives is given by:

$S = -\frac{1}{2} \int d^4x \Big(F_{\mu\nu} F^{\mu\nu} \Big)\ .$

Therefore, one can derive electromagnetic theory from arguments based solely on symmetry.

13.2.2 The connection and Yang-Mills gauge theory

We now generalize the above reasoning to general gauge groups. One begins with the generators of some Lie algebra:

$[T_i, T_j] = if^{ijk}T^k$

Let there be a fermion field that transforms as

$$\mathbf{\Psi}' \mapsto \hat{\Omega}(x)\mathbf{\Psi}(x) = \exp(i\theta^i(x)T^i)\mathbf{\Psi}(x)$$

Again the derivatives o $\mathbf{\Psi}(x)$ ıre not covariant under this transformation.We introduce a covariant derivative

$\mathcal{D}_\mu = \mathbf{I}\partial_\mu + ig\mathbf{A}_\mu(x)$

with connection field given by

$\mathbf{A}_\mu(x) = A^i_\mu(x)T^i$

We require that $\mathbf{A}_\mu(x)$ transforms as:

$\mathbf{A}'_\mu(x) = \hat{\Omega}\mathbf{A}_\mu(x)\hat{\Omega}^{-1} + \frac{i}{g}\hat{\Omega}(\partial_\mu\hat{\Omega}^{-1})$

We define the field strength operator

$$\mathbf{F}_{\mu\nu} = -\frac{i}{g}[\mathcal{D}_\mu, \mathcal{D}_\nu] = \partial_\mu \mathbf{A}_\nu - \partial_\nu \mathbf{A}_\mu + ig[\mathbf{A}_\mu, \mathbf{A}_\nu]$$

$$= (\partial_\mu A^i_\nu - \partial_\nu A^i_\mu + g f^{ijk}_\mu A^j_\mu A^k_\nu) T^i$$

As $\mathcal{D}_\mu$ is covariant, this means that the $F^i_{\mu\nu}$ tensor is also covariant:

$$\mathbf{F}_{\mu\nu} \mapsto \mathbf{F}'_{\mu\nu} = \hat{\Omega}\mathbf{F}_{\mu\nu}\hat{\Omega}^{-1}$$

Note that $\mathbf{F}_{\mu\nu}$ is only invariant under gauge transformations if $\hat{\Omega}$ is a scalar, that is, only in the case of electromagnetism.

We can now construct an invariant action out of this tensor. Again we want an action that only has two derivatives. The simplest choice is the trace of the commutator:

$$Tr(\hat{\Omega}\mathbf{F}_{\mu\nu}\hat{\Omega}^{-1}\hat{\Omega}\mathbf{F}^{\mu\nu}\hat{\Omega}^{-1}) = Tr(\mathbf{F}_{\mu\nu}\mathbf{F}^{\mu\nu})$$

The unique action with only two derivatives is given by:

$$S = -\frac{1}{2}\int d^4x Tr(\mathbf{F}_{\mu\nu}\mathbf{F}^{\mu\nu}) = -\frac{1}{2}\int d^4x Tr\left(F^i_{\mu\nu}T^j F^{\mu\nu}_j T^j\right)$$

This is the action for Yang-mills theory.

13.3 The loop representation of the Maxwell theory

We consider a change of representation in the quantum Maxwell gauge theory. The idea is to introduce a basis of states labeled by loops $\mid \gamma\rangle$ whose inner product with the connection states is given by

$$\langle A \mid \gamma\rangle = W(\gamma) = \exp\left[ie\int_\gamma dy^\alpha A_\alpha(y)\right]$$

The loop functional $W(\gamma)$ is the Wilson loop for the abelian $U(1)$ case.

13.4 The loop representation of Yang–Mills theory

We consider for simplicity (and because later we will see this is the relevant gauge group in LQG) an $SU(2)$ Yang–Mills theory in four dimensions. The field variable of the continuous theory is an $SU(2)$ connection (or gauge potential) $A^i_\mu(x)$, where i is an index in the Lie algebra of $SU(2)$. We can write for this field

$$\mathbf{A}_\mu(x) = A^i(x)\tau_i$$

where τ_i are the $su(2)$ generators, that is the Pauli matrices multiplied by $i/2$. note that unlike with Maxwell's theory, the connections $\mathbf{A}_\mu(x)$ are matrix-valued and don't commute, that is they are non-Abelian gauge theories. We must take this into account when defining the corresponding version of the holonomy for $SU(2)$ Yang–Mills theory.

We first describe the quantum theory in terms of connection variable.

13.4.1 The connection representation

In the connection representation the configuration variable is A^i_a and its conjugate momentum is the (densitized) triad $\tilde{E}^a_i$. It is most natural to consider wavefunctions $\Psi(A^i_a)$. This is known as the connection representation. The canonical variables get promoted to quantum operators:

$$\hat{A}^i{}_a\Psi[A] = A^i_a\Psi[A]$$

(analogous to the position representation $\hat{q}\psi(q) = q\psi(q)$) and the triads are functional derivatives,

$$\hat{\tilde{E}}^i_a\Psi[A] = -i\frac{\delta\Psi[A]}{\delta A^i_a}$$

(analogous to $\hat{p}\psi(q) = -i\frac{d\psi(q)}{dq}$)

13.4.2 The holonomy and Wilson loop

Let us return to the classical Yang–Mills theory. It is possible to encode the gauge invariant information of the theory in terms of `loop-like' variables.

We need the notion of a holonomy. A holonomy is a measure of how much the initial and final values of a spinor or vector differ after parallel transport around a closed loop; it is denoted

$$h_\gamma[A]$$

Knowledge of the holonomies is equivalent to knowledge of the connection, up to gauge equivalence. Holonomies can also be associated with an edge; under a Gauss Law these transform as

$$(h'_e)_{\alpha\beta} = U^{-1}_{\alpha\gamma}(x)(h_e)_{\gamma\sigma}U_{\sigma\beta}(y).$$

For a closed loop $x = y$ if we take the trace of this, that is, putting $\alpha = \beta$ and summing we obtain

$$(h'_e)_{\alpha\alpha} = U^{-1}_{\alpha\gamma}(x)(h_e)_{\gamma\sigma}U_{\sigma\alpha}(x)$$

$$= [U_{\sigma\alpha}(x)U^{-1}_{\alpha\gamma}(x)](h_e)_{\gamma\sigma} = \delta_{\sigma\gamma}(h_e)_{\gamma\sigma} = (h_e)_{\gamma\gamma}$$

or

$$\mathrm{Tr}\, h'_\gamma = \mathrm{Tr}\, h_\gamma.$$

Thus the trace of an holonomy around a closed loop is gauge invariant. It is denoted

$$W_\gamma[A]$$

and is called a Wilson loop. The explicit form of the holonomy is

$$h_\gamma[A] = \mathcal{P}\exp\Big\{-\int_{\gamma_0}^{\gamma_1} ds\dot{\gamma}^a A^i_a(\gamma(s))T_i\Big\}$$

where γ is the curve along which the holonomy is evaluated, and s is a parameter along the curve, $\mathcal{P}$ denotes path ordering meaning factors for smaller values of s appear to the left, and T_i are matrices that satisfy the $su(2)$ algebra

$$[T^i, T^j] = 2i\epsilon^{ijk}T^k.$$

The Pauli matrices satisfy the above relation. It turns out that there are infinitely many more examples of sets of matrices that satisfy these relations, where each set comprises $(N+1)\times(N+1)$ matrices with $N = 1, 2, 3, \dots$, and where none of these can be thought to 'decompose' into two or more examples of lower dimension. They are called different irreducible representations of the $su(2)$ algebra. The most fundamental representation being the Pauli matrices. The holonomy is labelled by a half integer $N/2$ according to the irreducible representation used.

13.4.3 Giles' Reconstruction theorem of gauge potentials from Wilson loops

An important theorem about Yang–Mills gauge theories is Giles' theorem, according to which if one gives the trace of the holonomy of a connection for all possible loops on a manifold one can, in principle, reconstruct all the gauge invariant information of the connection.[2] That is, Wilson loops constitute a basis of gauge invariant functions of the connection. This key result is the basis for the loop representation for gauge theories and gravity.

13.4.4 The loop transform and the loop representation

The use of Wilson loops explicitly solves the Gauss gauge constraint. As Wilson loops form a basis we can formally expand any Gauss gauge invariant function as,

$\Psi[A] = \sum_\gamma \Psi[\gamma]W_\gamma[A]$.

This is called the loop transform. We can see the analogy with going to the momentum representation in quantum mechanics. There one has a basis of states $\exp(ikx)$ labelled by a number k and one expands

$$\psi[x] = \int dk\psi(k)\exp(ikx).$$

and works with the coefficients of the expansion $\psi(k)$.

The inverse loop transform is defined by

$$\Psi[\gamma] = \int [dA]\Psi[A]W_\gamma[A].$$

This defines the loop representation. Given an operator $\hat{O}$ in the connection representation,

$$\Phi[A] = \hat{O}\Psi[A], \qquad 1\ \mathrm{Eq}$$

one should define the corresponding operator $\hat{O}'$ on $\Psi[\gamma]$ in the loop representation via,

$$\Phi[\gamma] = \hat{O}'\Psi[\gamma], \qquad 2\ \mathrm{Eq}$$

where $\Phi[\gamma]$ is defined by the usual inverse loop transform,

$$\Phi[\gamma] = \int [dA]\Phi[A]W_\gamma[A]. \qquad 3\ \mathrm{Eq}$$

A transformation formula giving the action of the operator $\hat{O}'$ on $\Psi[\gamma]$ in terms of the action of the operator $\hat{O}$ on $\Psi[A]$ is then obtained by equating the R.H.S. of Eq 2 with the R.H.S. of Eq 3 with Eq 1 substituted into Eq 3 , namely

$$\hat{O}'\Psi[\gamma] = \int [dA]W_\gamma[A]\hat{O}\Psi[A],$$

or

$$\hat{O}'\Psi[\gamma] = \int [dA](\hat{O}^\dagger W_\gamma[A])\Psi[A],$$

where by $\hat{O}^\dagger$ we mean the operator $\hat{O}$ but with the reverse factor ordering (remember from simple quantum mechanics where the product of operators is reversed under conjugation). We evaluate the action of this operator on the

Wilson loop as a calculation in the connection representation and rearranging the result as a manipulation purely in terms of loops (one should remember that when considering the action on the Wilson loop one should choose the operator one wishes to transform with the opposite factor ordering to the one chosen for its action on wavefunctions $\Psi[A]$).

13.5 The loop representation of quantum gravity

Main articles: Holonomy, Wilson loop and Knot invariant

13.5.1 Ashtekar–Barbero variables of canonical quantum gravity

The introduction of Ashtekar variables cast general relativity in the same language as gauge theories. It was in particular the inability to have good control over the space of solutions to the Gauss' law and spatial diffeomorphism constraints that led Rovelli and Smolin to consider a new representation – the loop representation.[3]

To handle the spatial diffeomorphism constraint we need to go over to the loop representation. The above reasoning gives the physical meaning of the operator $\hat{O}'$. For example, if $\hat{O}^\dagger$ corresponded to a spatial diffeomorphism, then this can be thought of as keeping the connection field A of $W_\gamma[A]$ where it is while performing a spatial diffeomorphism on γ instead. Therefore, the meaning of $\hat{O}'$ is a spatial diffeomorphism on γ , the argument of $\Psi[\gamma]$.

In the loop representation we can then solve the spatial diffeomorphism constraint by considering functions of loops $\Psi[\gamma]$ that are invariant under spatial diffeomorphisms of the loop γ . That is, we construct what mathematicians call knot invariants. This opened up an unexpected connection between knot theory and quantum gravity.

13.5.2 The loop representation and eigenfunctions of geometric quantum operators

The easiest geometric quantity is the area. Let us choose coordinates so that the surface Σ is characterized by $x^3 = 0$. The area of small parallelogram of the surface Σ is the product of length of each side times $\sin\theta$ where θ is the angle between the sides. Say one edge is given by the vector $\vec{u}$ and the other by $\vec{v}$ then,

$$A = \|\vec{u}\|\|\vec{v}\|\sin\theta = \sqrt{\|\vec{u}\|^2\|\vec{v}\|^2(1-\cos^2\theta)}$$
$$= \sqrt{\|\vec{u}\|^2\|\vec{v}\|^2 - (\vec{u}\cdot\vec{v})^2}$$

From this we get the area of the surface Σ to be given by

$$A_\Sigma = \int_\Sigma dx^1\, dx^2 \sqrt{\det\, q^{(2)}}$$

where $\det q^{(2)} = q_{11}q_{22} - q_{12}^2$ and is the determinant of the metric induced on Σ . This can be rewritten as

$$\det\, q^{(2)} = \frac{\epsilon^{3ab}\epsilon^{3cd}q_{ac}q_{bc}}{2}.$$

The standard formula for an inverse matrix is

$$q^{ab} = \frac{\epsilon^{acd}\epsilon^{bef}q_{ce}q_{df}}{3!\det(q)}$$

Note the similarity between this and the expression for $\det q^{(2)}$. But in Ashtekar variables we have $\tilde{E}_i^a\tilde{E}^{bi} = \det(q)q^{ab}$. Therefore,

$$A_\Sigma = \int_\Sigma dx^1\, dx^2 \sqrt{\tilde{E}_i^3\tilde{E}^{3i}}.$$

According to the rules of canonical quantization we should promote the triads $\tilde{E}_i^3$ to quantum operators,

$$\hat{\tilde{E}}_i^3 \sim \frac{\delta}{\delta A_3^i}.$$

It turns out that the area A_Σ can be promoted to a well defined quantum operator despite the fact that we are dealing with product of two functional derivatives and worse we have a square-root to contend with as well.[4] Putting $N = 2J$, we talk of being in the J-th representation. We note that $\sum_i T^iT^i = J(J+1)1$. This quantity is important in the final formula for the area spectrum. We simply state the result below,

$$\hat{A}_\Sigma W_\gamma[A] = 8\pi\ell_{Planck}^2\beta\sum_I\sqrt{j_I(j_I+1)}W_\gamma[A]$$

where the sum is over all edges I of the Wilson loop that pierce the surface Σ .

The formula for the volume of a region R is given by

$$V = \int_R d^3x \sqrt{\det(q)} = \frac{1}{6}\int_R dx^3 \sqrt{\epsilon_{abc}\epsilon^{ijk}\tilde{E}_i^a \tilde{E}_j^b \tilde{E}_k^c}.$$

The quantization of the volume proceeds the same way as with the area. As we take the derivative, and each time we do so we bring down the tangent vector $\dot{\gamma}^a$, when the volume operator acts on non-intersecting Wilson loops the result vanishes. Quantum states with non-zero volume must therefore involve intersections. Given that the anti-symmetric summation is taken over in the formula for the volume we would need at least intersections with three non-coplanar lines. Actually it turns out that one needs at least four-valent vertices for the volume operator to be non-vanishing.

13.5.3 Mandelstam identities: su(2) Yang–Mills

We now consider Wilson loops with intersections. We assume the real representation where the gauge group is $SU(2)$. Wilson loops are an over complete basis as there are identities relating different Wilson loops. These come about from the fact that Wilson loops are based on matrices (the holonomy) and these matrices satisfy identities, the so-called Mandelstam identities. Given any two $SU(2)$ matrices $\mathbb{A}$ and $\mathbb{B}$ it is easy to check that,

$$\mathrm{Tr}(\mathbb{A})\,\mathrm{Tr}(\mathbb{B}) = \mathrm{Tr}(\mathbb{A}\mathbb{B}) + \mathrm{Tr}(\mathbb{A}\mathbb{B}^{-1}).$$

This implies that given two loops γ and η that intersect, we will have,

$$W_\gamma[A]W_\eta[A] = W_{\gamma\circ\eta}[A] + W_{\gamma\circ\eta^{-1}}[A]$$

where by η^{-1} we mean the loop η traversed in the opposite direction and $\gamma\circ\eta$ means the loop obtained by going around the loop γ and then along η . See figure below. This is called a Mandelstam identity of the second kind. There is the Mandelstam identity of the first kind $W(\gamma_1 \circ \gamma_2) = W(\gamma_2 \circ \gamma_1)$. Spin networks are certain linear combinations of intersecting Wilson loops designed to address the over completeness introduced by the Mandelstam identities.

Graphical representation of the Mandestam identity relating different Wilson loops.

13.5.4 Spin network states

In fact spin networks constitute a basis for all gauge invariant functions which minimize the degree of over-completeness of the loop basis, and for trivalent intersections eliminate it entirely.

As mentioned above the holonomy tells you how to propagate test spin half particles. A spin network state assigns an amplitude to a set of spin half particles tracing out a path in space, merging and splitting. These are described by spin networks γ : the edges are labelled by spins together with `intertwiners' at the vertices which are prescription for how to sum over different ways the spins are rerouted. The sum over rerouting are chosen as such to make the form of the intertwiner invariant under Gauss gauge transformations.

13.5.5 Uniqueness of the loop representation in LQG

Theorems establishing the uniqueness of the loop representation as defined by Ashtekar et al. (i.e. a certain concrete realization of a Hilbert space and associated operators reproducing the correct loop algebra – the realization that everybody was using) have been given by two groups (Lewandowski, Okolow, Sahlmann and Thiemann)[5] and (Christian Fleischhack).[6] Before this result was established it was not known whether there could be other examples of Hilbert spaces with operators invoking the same loop algebra, other realizations, not equivalent to the one that had been used so far.

13.6 Knot theory and loops in topological field theory

A common method of describing a knot (or link, which are knots of several components entangled with each other) is to consider its projected image onto a plane called a knot diagram. Any given knot (or link) can be drawn in many different ways using a knot diagram. Therefore, a fundamental problem in knot theory is determining when two descriptions represent the same knot. Given a knot diagram, one tries to find a way to assign a knot invariant to it, sometimes a polynomial – called a knot polynomial. Two knot diagrams with different polynomials generated by the same procedure necessarily correspond to different knots. However, if the polynomials are the same, it may not mean that they correspond to the same knot. The better a polynomial is at distinguishing knots the more powerful it is.

In 1984, Jones [7] announced the discovery of a new link invariant, which soon led to a bewildering profusion of generalizations. He had found a new knot polynomial, the Jones polynomial. Specifically, it is an invariant of an oriented knot or link which assigns to each oriented knot or link a polynomial with integer coefficients.

In the late 1980s, Witten coined the term topological quantum field theory for a certain type of physical theory in which the expectation values of observable quantities are invariant under diffeomorphisms.

Witten [8] gave a heuristic derivation of the Jones polynomial and its generalizations from Chern–Simons theory. The basic idea is simply that the vacuum expectation values of Wilson loops in Chern–Simons theory are link invariants because of the diffeomorphism-invariance of the theory. To calculate these expectation values, however, Witten needed to use the relation between Chern–Simons theory and a conformal field theory known as the Wess–Zumino–Witten model (or the WZW model).

13.7 References

[1] E. Witten, Commun. Math. Phys **121**, 351 (1989).

[2] R. Giles, Reconstruction of gauge potentials from Wilson loops, Phys. Rev. D **24**, 2160 (1981).

[3] Rovelli, C. and Smolin, L. *Phys. Rev. Lett 61, 1155*

[4] For example see section 8.2 of *A First Course in Loop Quantum Gravity*, Gambini, R, and Pullin, J. Published by Oxford University Press 2011.

[5] Lewandowski, J., Okołów, A., Sahlmann, H., and Thiemann, T., *Uniqueness of Diffeomorphism Invariant States on Holonomy-Flux Algebras*, Commun. Math. Phys., 267, 703–733, (2005).

[6] Fleischhack, C., *Irreducibility of the Weyl algebra in loop quantum gravity*, Phys. Rev. Lett., 97, 061302, (2006).

[7] V. Jones, A polynomial invariant for knots via von Neumann algebras, reprinted in "New Developments in the Theory of Knots, *ed. T. Kohno, World Scientific, Singapore, 1989.*

[8] E. Witten, quantum field theory and the Jones polynomial, Comm. Math. Phys. **121** (1989), 351–399.

Chapter 14

Hamiltonian constraint of LQG

In the ADM formulation of general relativity one splits spacetime into spatial slices and time, the basic variables are taken to be the induced metric, $q_{ab}(x)$, on the spatial slice (the metric induced on the spatial slice by the space-time metric), and its conjugate momentum variable related to the extrinsic curvature, $K^{ab}(x)$, (this tells us how the spatial slice curves with respect to spacetime and is a measure of how the induced metric evolves in time).[1] These are the metric canonical coordinates.

Dynamics such as time-evolutions of fields are controlled by the Hamiltonian constraint.

The identity of the Hamiltonian constraint is a major open question in quantum gravity, as is extracting of physical observables from any such specific constraint.

In 1986 Abhay Ashtekar introduced a new set of canonical variables, Ashtekar variables to represent an unusual way of rewriting the metric canonical variables on the three-dimensional spatial slices in terms of a SU(2) gauge field and its complementary variable.[2] The Hamiltonian was much simplified in this reformulation. This led to the loop representation of quantum general relativity[3] and in turn loop quantum gravity.

Within the loop quantum gravity representation Thiemann was able formulate a mathematically rigorous operator as a proposal as such a constraint.[4] Although this operator defines a complete and consistent quantum theory, doubts have been raised as to the physical reality of this theory due to inconsistencies with classical general relativity (the quantum constraint algebra closes, but it is not isomorphic to the classical constraint algebra of GR, which is seen as circumstantial evidence of inconsistencies definitely not a proof of inconsistencies), and so variants have been proposed.

14.1 Classical expressions for the Hamiltonian

14.1.1 Metric formulation

The idea was to quantize the canonical variables q_{ab} and $\pi^{ab} = \sqrt{q}(K^{ab} - q^{ab}K_c^c)$, making them into operators acting on wavefunctions on the space of 3-metrics, and then to quantize the Hamiltonian (and other constraints). However, this program soon became regarded as dauntingly difficult for various reasons, one being the non-polynomial nature of the Hamiltonian constraint:

$$H = \sqrt{\det(q)}(K_{ab}K^{ab} - (K_a^a)^2 - {}^3R)$$

where ${}^3R$ is the scalar curvature of the three metric $q_{ab}(x)$. Being a non-polynomial expression in the canonical variables and their derivatives it is very difficult to promote to a quantum operator.

14.1.2 Expression using Ashtekar variables

The configuration variables of Ashtekar's variables behave like an $SU(2)$ gauge field or connection A_a^i . Its canonically conjugate momentum is $\tilde{E}_i^a$ is the densitized "electric" field or triad (densitized as $\tilde{E}_i^a = \sqrt{\det(q)}E_i^a$). What do these variables have to do with gravity? The densitized triads can be used to reconstruct the spatial metric via

$$\det(q)q^{ab} = \tilde{E}_i^a \tilde{E}_j^b \delta^{ij}$$

The densitized triads are not unique, and in fact one can perform a local in space rotation with respect to the internal indices i . This is actually the origin of the $SU(2)$ gauge invariance. The connection can be use to reconstruct the extrinsic curvature. The relation is given by

$$A_a^i = \Gamma_a^i - iK_a^i$$

where Γ_a^i is related to the spin connection, $\Gamma_a{}^j{}_i$, by $\Gamma_a^i = \Gamma_{ajk}\epsilon^{jki}$ and $K_a^i = K_{ab}\tilde{E}^{ai}/\sqrt{\det(q)}$.

In terms of Ashtekar variables, the classical expression of the constraint is given by

$$H = \frac{\epsilon_{ijk} F^k_{ab} \tilde{E}^a_i \tilde{E}^b_j}{\sqrt{\det(q)}}$$

where F^k_{ab} field strength tensor of the gauge field A^i_a . Due to the factor $1/\sqrt{\det(q)}$ this in non-polynomial in the Ashtekar's variables. Since we impose the condition

$$H = 0$$

we could consider the densitized Hamiltonian instead,

$$\tilde{H} = \sqrt{\det(q)} H = \epsilon_{ijk} F^k_{ab} \tilde{E}^a_i \tilde{E}^b_j = 0$$

This Hamiltonian is now polynomial the Ashtekar's variables. This development raised new hopes for the canonical quantum gravity programme.[5] Although Ashtekar variables had the virtue of simplifying the Hamiltonian, it has the problem that the variables become complex. When one quantizes the theory it is a difficult task ensure that one recovers real general relativity as opposed to complex general relativity. Also there were also serious difficulties promoting the densitized Hamiltonian to a quantum operator.

A way of addressing the problem of reality conditions was noting that if we took the signature to be $(+,+,+,+)$, that is Euclidean instead of Lorentzian, then one can retain the simple form of the Hamiltonian for but for real variables. One can then define what is called a generalized Wick rotation to recover the Lorentzian theory.[6] Generalized as it is a Wick transformation in phase space and has nothing to do with analytical continuation of the time parameter t .

14.1.3 Expression for real formulation of Ashtekar variables

Thomas Thiemann was able to address both the above problems.[4] He used the real connection

$$A^i_a = \Gamma^i_a + \beta K^i_a$$

In real Ashtekar variables the full Hamiltonian is

$$H = -\zeta \frac{\epsilon_{ijk} F^k_{ab} \tilde{E}^a_i \tilde{E}^b_j}{\sqrt{\det(q)}} + 2 \frac{\zeta \beta^2 - 1}{\beta^2} \frac{(\tilde{E}^a_i \tilde{E}^b_j - \tilde{E}^a_j \tilde{E}^b_i)}{\sqrt{\det(q)}} (A^i_a - \Gamma^i_a)(A^j_b - \Gamma^j_b) = H_E + H'$$

where the constant β is the Barbero-Immirzi parameter.[7] The constant ζ is -1 for Lorentzian signature and +1 for Euclidean signature. The Γ^i_a have a complicated relationship with the desitized triads and causes serious problems upon quantization. Ashtekar variables can be seen as choosing $\beta = i$ to make the second more complicated term was made to vanish (the first term is denoted H_E because for the Euclidean theory this term remains for the real choice of $\beta = \pm 1$). Also we still have the problem of the $1/\sqrt{\det(q)}$ factor.

Thiemann was able to make it work for real β . First he could simplify the troublesome $1/\sqrt{\det(q)}$ by using the identity

$$\{A^k_c, V\} = \frac{\epsilon_{abc} \epsilon^{ijk} \tilde{E}^a_i \tilde{E}^b_j}{\sqrt{\det(q)}}$$

where V is the volume,

$$V = \int d^3x \sqrt{\det(q)} = \frac{1}{6} \int d^3x \sqrt{|\tilde{E}^a_i \tilde{E}^b_j \tilde{E}^c_k \epsilon^{ijk} \epsilon_{abc}|}$$

The first term of the Hamiltonian constraint becomes

$$H_E = \{A^k_c, V\} F^k_{ab} \tilde{\epsilon}^{abc}$$

upon using Thiemann's identity. This Poisson bracket is replaced by a commutator upon quantization. It turns out that a similar trick can be used to teat the second term. Why are the Γ^i_a given by the densitized triads $\tilde{E}^a_i$? It comes about from the compatibility condition

$$D_a E^i_b = 0$$

We can solve this in much the same way as the Levi-Civita connection can be calculated from the equation $\nabla_c g_{ab} = 0$; by rotating the various indices and then adding and subtracting them (see article spin connection for more details of the derivation, although there we use slightly different notation). We then rewrite this in terms of the densitized triad using that $\det(\tilde{E}) = |\det(E)|^2$. The result is complicated and non-linear, but a homogeneous function of $\tilde{E}^a_i$ of order zero,

$$\Gamma^i_a = \frac{1}{2} \epsilon^{ijk} \tilde{E}^b_k [\tilde{E}^j_{a,b} - \tilde{E}^j_{b,a} + \tilde{E}^c_j \tilde{E}^l_a \tilde{E}^l_{c,b}] +$$

$$\frac{1}{4} \epsilon^{ijk} \tilde{E}^b_k \left[2\tilde{E}^j_a \frac{(\det(\tilde{E}))_{,b}}{\det(\tilde{E})} - \tilde{E}^j_b \frac{(\det(\tilde{E}))_{,a}}{\det(\tilde{E})} \right].$$

To circumvent the problems introduced by this complicated relationship Thiemann first defines the Gauss gauge invariant quantity

$$K = \int d^3x K_a^i \tilde{E}_i^a$$

where $K_a^i = K_{ab}\tilde{E}^{ai}/\sqrt{\det(q)}$, and notes that

$$K_a^i = \{A_a^i, K\}$$

(this is because $\{\Gamma_a^i, K\} = 0$ which comes about from the fact that βK is the generator of the canonical transformation of constant rescaling, $\tilde{E}_i^a \mapsto \tilde{E}_i^a/\beta$, and Γ_a^i is a homogeneous function of order zero). We are then able to write

$$A_a^i - \Gamma_a^i = \beta K_a^i = \beta\{A_a^i, K\}$$

and as such find an expression in terms of the configuration variable A_a^i and K for the second term of the Hamiltonian

$$H' = \epsilon^{abc}\epsilon_{ijk}\{A_a^i, K\}\{A_b^j, K\}\{A_c^k, V\}$$

Why is it easier to quantize K ? This is because it can be rewritten in terms of quantities that we already know how to quantize. Specifically K can be rewritten as

$$K = -\{V, \int d^3x H_E\}$$

where we have used that the integrated densitized trace of the extrinsic curvature is the``time derivative of the volume".

14.2 Coupling to matter

14.2.1 Coupling to scalar field

The Lagrangian for a scalar field in curved spacetime

$$L = -\int d^4x\sqrt{-\det(g)}(-g^{\mu\nu}\partial_\mu\varphi\partial_\nu\varphi - V(\varphi))$$

where μ, ν are spacetime indices. We define the conjugate momentum of the scalar field with the usual $\tilde{\pi} = \delta L/\delta\dot{\varphi}$, the Hamiltonian can be rewritten as,

$$H = \int d^3x N\left(\frac{\tilde{\pi}^2}{\sqrt{\det(q)}} + \sqrt{\det(q)}(q^{ab}\partial_a\varphi\partial_b\varphi + V(\varphi))\right) + N^a\tilde{\pi}\partial_a\varphi.$$

where N and N^a are the lapse and shift. In Ashtekar variables this reads,

$$H = \int d^3x \frac{N}{\sqrt{\det(q)}}\Big(\tilde{\pi}^2 + \tilde{E}_i^a\tilde{E}^{bi} \cdot \partial_a\varphi\partial_b\varphi + \det(q)V(\varphi)\Big) + N^a\tilde{\pi}\partial_a\varphi$$

As usual the (smeared) spatial diffeomorphisn constraint is associated with the shift function N^a and the (smeared) Hamiltonian is associated with the lapse function N . So we simply read off the spatial diffeomorphism and Hamiltonian constraint,

$$C(\vec{N})_\varphi = \int d^3x N^a\tilde{\pi}\partial_a\varphi$$

$$H(N)_\varphi = \int d^3x \frac{N}{\sqrt{\det(q)}}\Big(\tilde{\pi}^2 + \tilde{E}_i^a\tilde{E}^{bi} \cdot \partial_a\varphi\partial_b\varphi + \det(q)V(\varphi)\Big)$$

These should be added (multiplied by $8\pi G\beta$) to the spatial diffeomorphism and Hamiltonian constraint of the gravitational field, respectively. This represents the coupling of scalar matter to gravity.

14.2.2 Coupling to Fermionic field

There are problems coupling gravity to spinor fields: there are no finite-dimensional spinor representations of the general covariance group. However, there are of course spinorial representations of the Lorentz group. This fact is utilized by employing tetrad fields describing a flat tangent space at every point of spacetime. The Dirac matrices γ^I are contracted onto vierbiens,

$$\gamma^I e_I^a(x) = \gamma^a(x) \,.$$

We wish to construct a generally covariant Dirac equation. Under a flat tangent space Lorentz transformation transforms the spinor as

$$\psi \mapsto e^{i\epsilon^{IJ}(x)\sigma_{IJ}}\psi$$

We have introduced local Lorentz transformations on flat tangent space, so ϵ_{IJ} is a function of space-time. This means that the partial derivative of a spinor is no longer a genuine tensor. As usual, one introduces a connection field ω_μ^{IJ} that allows us to gauge the Lorentz group. The covariant derivative defined with the spin connection is,

$$\nabla_a\psi = (\partial_a - \tfrac{i}{4}\omega_a^{IJ}\sigma_{IJ})\psi \,,$$

and is a genuine tensor and Dirac's equation is rewritten as

$$(i\gamma^a\nabla_a - m)\psi = 0 \,.$$

The Dirac action in covariant form is

$$S_{Dirac} = \frac{1}{2}\int_{\mathcal{M}} d^4x\sqrt{-det(g)}[\overline{\Psi}\gamma^I E_I^a\nabla_a\Psi - \overline{\nabla_a\Psi}\gamma^I E_I^a\Psi]$$

where $\Psi = (\psi, \eta)$ is a Dirac bi-spinor and $\overline{\Psi} = (\Psi^*)^T\gamma^0$ its cojugate. The covariant derivative ∇_a is defined to annihilate the tetrad E_a^I.

14.2.3 Coupling to Electromagnetic field

The Lagrangian for an electromagnetic field in curved spacetime is

$$L = -\int d^4x\sqrt{-\det(g)}(g^{\mu\alpha}g^{\nu\beta}\mathcal{F}_{\mu\nu}\mathcal{F}_{\alpha\beta})$$

where

$\mathcal{F}^{\mu\nu} = \nabla^\mu\mathcal{A}^\nu - \nabla^\nu\mathcal{A}^\mu$

is the field strength tensor, in components

$\mathcal{F}^{0a} = \mathcal{E}^a$

and $\mathcal{F}^{ab} = \epsilon^{abc}B_c$

where the electric field is given by

$\mathcal{E}^a = -\nabla_a\mathcal{A}_0 - \dot{\mathcal{A}}_a$

and the magnetic field is.

$B^a = \epsilon^{abc}\nabla_b\mathcal{A}_c$.

The classical analysis with the Maxwell action followed by canonical formulation using the time gauge parametrisation results in:

$H(N, N^a, \Lambda) = \frac{1}{2}\int_\Sigma d^3xN\frac{q_{ab}}{\sqrt{det(q)}}[\tilde{\mathcal{E}}^a\tilde{\mathcal{E}}^b + B^aB^b] + N^a\mathcal{F}_{ab}\tilde{\mathcal{E}}^a + \Lambda\nabla_a\tilde{\mathcal{E}}^a$

$B^a = \epsilon^{abc}B_c \qquad \tilde{\mathcal{E}}^a = -\sqrt{q}N\mathcal{F}^{0a}$

with $\mathcal{A}_a$ and $\tilde{\mathcal{E}}^a$ being the canonical coordinates.

14.2.4 Coupling to Yang-Mills field

$H = \frac{1}{2}\int_\Sigma d^3x\frac{q_{ab}}{\sqrt{det(q)}}[\tilde{\mathcal{E}}_I^a\tilde{\mathcal{E}}_I^b + B_I^aB_I^b]$

14.2.5 Total Hamiltonian of matter coupled to gravity

The dynamics of the coupled gravity-matter system is simply defined by the adding of terms defining the matter dynamics to the gravitational hamiltonian. The full hamiltonian is described by

$H = H_{Einstein} + H_{Maxwell} + H_{Yang-Mills} + H_{Dirac} + H_{Higgs}$.

14.3 Quantum Hamiltonian constraint

In this section we discuss the quantization of the hamiltonian of pure gravity, that is in the absence of matter. The case of inclusion of matter is discussed in the next section. The constraints in their primitive form are rather singular, and so should be `smeared' by appropriate test functions. The Hamiltonian is the written as

$$H(N) = \int d^3xN\{A_c^k, V\}F_{ab}^k\epsilon^{abc}$$

For simplicity we are only considering the "Euclidean" part of the Hamiltonian constraint, extension to the full constraint can be found in the literature. There are actually many different choices for functions, and so what one then ends up with an (smeared) Hamiltonians constraints. Demanding them all to vanish is equivalent to the original description.

14.3.1 The loop representation

The Wilson loop is defined as

$$h_\gamma[A] = \mathcal{P}\exp\left\{-\int_{s_0}^{s_1} ds\dot{\gamma}^a A_a^i(\gamma(s))T_i\right\}$$

where $\mathcal{P}$ indicates a path ordering so that factors for smaller values of s appear to the left, and where the T_i satisfy the $su(2)$ algebra,

$$[T^i, T^j] = 2i\epsilon^{ijk}T^k$$

It is easy to see from this that,

$$Tr(T^iT^j) - Tr(T^jT^i) = 2i\epsilon^{ijk}Tr(T^k)$$

implies that $Tr(T^i) = 0$.

Wilson loops are not independent of each other, and in fact certain linear combinations of them called spin network states form an orthonormal basis. As spin network functions form a basis we can formally expand any Gauss gauge invariant function as,

$$\Psi[A] = \sum_\gamma \Psi[\gamma]s_\gamma[A]$$

This is called the inverse loop transform. The loop transform is given by

$$\Psi[\gamma] = \int [dA]\Psi[A]s_\gamma[A]$$

and is analogous to what one does when one goes over to the momentum representation in quantum mechanics,

$$\psi[x] = \int dk\psi(k)\exp(ikx)$$

The loop transform defines the loop representation. Given an operator $\hat{O}$ in the connection representation,

$$\Phi[A] = \hat{O}\Psi[A]$$

we define $\Phi[\gamma]$ by the loop transform,

$$\Phi[\gamma] = \int [dA]\Phi[A]s_\gamma[A]$$

This implies that one should define the corresponding operator $\hat{O}'$ on $\Psi[\gamma]$ in the loop representation as

$$\hat{O}'\Psi[\gamma] = \int [dA]s_\gamma[A]\hat{O}\Psi[A]$$

or

$$\hat{O}'\Psi[\gamma] = \int [dA](\hat{O}^\dagger s_\gamma[A])\Psi[A]$$

where by $\hat{O}^\dagger$ we mean the operator $\hat{O}$ but with the reverse factor ordering. We evaluate the action of this operator on the spin network as a calculation in the connection representation and rearranging the result as a manipulation purely in terms of loops (one should remember that when considering the action on the spin network one should choose the operator one wishes to transform with the opposite factor ordering to the one chosen for its action on wavefunctions $\Psi[A]$). This gives the physical meaning of the operator $\hat{O}'$. For example if $\hat{O}^\dagger$ were a spatial diffeomorphism, then this can be thought of as keeping the connection field A of the $s_\gamma[A]$ where it is while performing a spatial diffeomorphism on γ instead. Therefore the meaning of $\hat{O}'$ is a spatial diffeomorphism on γ , the argument of $\Psi[\gamma]$.

The holonomy operator in the loop representation is the multiplication operator,

$$\hat{h}_\gamma\Psi[\eta] = h_\gamma\Psi[\eta]$$

14.3.2 Promotion of the Hamiltonian constraint to a quantum operator

We promote the Hamiltonian constraint to a quantum operator in the loop representation. One introduces a lattice regularization procedure. we assume that space has been divided into tetrahedra Δ . One builds an expression such that the limit in which the tetrahedra shrink in size approximates the expression for the Hamiltonian constraint.

For each tetrahedron pick a vertex and call $v(\Delta)$. Let $s_i(\Delta)$ with $i = 1,2,3$ be three edges ending at $v(\Delta)$. We now construct a loop

$$\alpha_{ij} = s_i(\Delta)\cdot s_{ij}(\Delta)\cdot s_j(\Delta)^{-1}$$

by moving along $s_i(\Delta)$ then along the line joining the points s_i and s_j that are not $v(\Delta)$ (which we have denoted s_{ij}) and then returning to $v(\Delta)$ along s_j . The holonomy

$$h_\gamma[A] = \mathcal{P}\exp\left\{-\int_{s_0}^{s_1} ds\dot{\gamma}^a A_a^i(\gamma(s))T_i\right\} \approx I-(s_k^a)A_a^iT_i$$

along a line in the limit the tetraherdon shrinks approximates the connection via

$$\lim_{\Delta\to v(\Delta)} h_{s_k} = I - A_c s_k^c$$

where s_k^c is a vector in the direction of edge s_k . It can be shown that

$$\lim_{\Delta v\to(\Delta)} h_{\alpha_{ij}} = I + \frac{1}{2}F_{ab}s_i^a s_j^b$$

(this expresses the fact that the field strength tensor, or curvature, measures the holonomy around `infinitesimal loops'). We are led to trying

$$H_\Delta(N) = \sum_\Delta N(v(\Delta))\epsilon^{ijk}Tr\big(h_{\alpha_{ij}}h_{s_k}\{h_{s_k}^{-1},V\}\big)$$

where the sum is over all tetrahedra Δ . Substituting for the holonomies,

$$H_\Delta(N) = \sum_\Delta N(v(\Delta))\epsilon^{ijk}Tr\big((I+\frac{1}{2}F_{ab}s_i^a s_j^b)(I-A_c s_k^c)$$

$$\{(I+A\ \ _d s_k^d),V\}\big)$$

The identity will have vanishing Poisson bracket with the volume, so the only contribution will come from the connection. As the Poisson bracket is already proportional to s_k^c only the identity part of the holonomy h_{s_k} outside the bracket contributes. Finally we have that the holonomy around α_{ij} ; the identity term doesn't contribute as the Poisson bracket is proportional to a Pauli matrix (since $A_c = A_c^i T_i$ and the constant matrix T_i can be taken outside the Poisson bracket) and one is taking the trace. The remaining term of $h_{\alpha_{ij}}$ yields the F_{ab} . The three lengths s 's that appear combine with the summation in the limit to produce an integral.

This expression immediately can be promoted to an operator in the loop representation, both holonomies and volume promote to well defined operators there.

The triangulation is chosen to so as to be adapted to the spin network state one is acting on by choosing the vertices an lines appropriately. There will be many lines and vertices of the triangulation that do not correspond to lines and vertices of the spin network when one takes the limit. Due to the presence of the volume the Hamiltonian constraint will only contribute when there are at least three non-coplanar lines of a vertex.

Here we have only considered the action of the Hamiltonian constraint on trivalent vertices. Computing the action on higher valence vertices is more complicated. We refer the reader to the article by Borissov, De Pietri, and Rovelli.[8]

14.3.3 A finite theory

The Hamiltonian is not invariant under spatial diffeomorphisms and therefore its action can only be defined on the kinematic space. One can transfer its action to diffeomprphsm invariant states. As we will see this has implications for where precisely the new line is added. Consider a state $\langle \Psi |$ such that $\langle \Psi, s \rangle = \langle \Psi, s' \rangle$ if the spin networks s and s' are diffeomorphic to each other. Such a state is not in the kinematic space but belongs to the larger dual space of a dense subspace of the kinematic space. We then define the action of $\hat{H}(N)$ in the following way,

$$\langle \hat{H}(N)\Psi, s \rangle = \lim_{\Delta \to v} \sum_{\Delta} \langle \Psi, \hat{H}_{\Delta}(N)s \rangle$$

The position of the added line is then irrelevant. When one projects on Ψ the position of the line does not matter because one is working on the space of diffeomorphism invariant states and so the line can be moved "closer" or "further" from the vertex without changing the result.

Spatial diffeomrphism plays a crucial role in the construction. If the functions were not diffeomorphism invariant, the added line would have to be shrunk to the vertex and possible divergences could appear.

The same construction can be applied to the Hamiltonian of general relativity coupled to matter: scalar fields, Yang-Mills fields, fermions. In all cases the theory is finite, anomaly free and well defined. Gravity appears to be acting as a "fundamental regulator" of theories of matter.

14.3.4 Anomaly free

Quantum anomalies occur when the quantum constraint algebra has additional terms that don't have classical counterparts. In order to recover the correct semi classical theory these extra terms need to vanish, but this implies additional constraints and reduces the number of degrees of freedom of the theory making it unphysical. Theimann's Hamiltonian constraint can be shown to be anomaly free.

14.3.5 The kernel of the Hamiltonian constraint

The kernel is the space of states which the Hamiltonian constraint annihilates. One can outline an explicit construction of the complete and rigorous kernel of the proposed operator. They are the first with non-zero volume and which do not need non-zero cosmological constant.

The complete space of solutions to the spatial diffeomorphis $C^a(x) = 0$ for all $x \in \Sigma$ constraints has already been found long ago.[9] And even was equipped with a natural inner product induced from that of the kinematical Hilbert space $\mathcal{H}_{Kin}$ of solutions to the Gauss constraint. However, there is no chance to define the Hamiltonian constraint operators corresponding to $H(x)$ (densely) on $\mathcal{H}_{Diff}$ because the Hamiltonian constraint operators do not preserve spatial diffeomorphism invariant states. Hence one cannot simply solve the spatial diffeomorphims constraint and then the Hamiltonian constraint and so the inner product structure of $\mathcal{H}_{Diff}$ cannot be employed in the construction of the physical inner product. This problem can be circumvented with the use of the Master constraint (see below) allowing the just mentioned results to be applied to obtain the physical Hilbert space $\mathcal{H}_{Phys}$ from $\mathcal{H}_{Diff}$.

More to come here...

14.3.6 Criticisms of the Hamiltonian constraint

Recovering the constraint algebra. Classically we have

$$\{H(N), H(M)\} = C(\vec{K})$$

where

$K^a = \tilde{E}^a_i \tilde{E}^{bi}(N\partial_b M - M\partial_b N)/(\det(q))$

As we know in the loop representation a self-adjoint operator generating spatial diffeomorphims. Therefore it is not possible to implement the relation $\{H(N), H(M)\}$ for in the quantum theory with infinitesimal $\vec{C}$, it is at most possible with finite spatial dffeomoephisms.

Ultra locality of the Hamiltonian: The Hamiltonian only acts at vertices and acts by "dressing" the vertex with lines. It does not interconnect vertices nor change the valences of the lines (outside the "dressing"). The modifications that the Hamiltonian constraint operator performs at a given vertex do not propagate over the whole graph but are confined to a neighbourhood of the vertex. In fact, repeated action of the Hamiltonian generates more and more new edges ever closer to the vertex never intersecting each other. In particular there is no action at the new vertices created. This implies, for instance, that for surfaces that enclose a vertex (diffeomorphically invariantly defined) the area of such surfaces would commute with the Hamiltonian, implying no "evolution" of these areas as it is the Hamiltonian that generates "evolution". This hints at the theory "failing to propagate". However, Thiemann points out that the Hamiltonian acts every where.

There is the somewhat subtle matter that the $\hat{H}(x)$, while defined on the Hilbert space $\mathcal{H}_{Kin}$ are not explicitly known (they are known up to a spatial diffeomorphism; they exist by the axiom of choice).

These difficulties could be addressed by a new approach - the Master constraint programme.

14.4 Extension of Quantisation to Inclusion of Matter Fields

14.4.1 Fermionic matter

14.4.2 Maxwell's theory

Note that $\tilde{\mathcal{E}}^a, B^a$ are both of density weight 1. As usual, before quantisation, we need to express the constraints (and other observables) in terms of the holonomies and fluxes.

We have a common factor of $q_{ab}/\sqrt{q}$. As before, we introduce a cell decomposition and noting,

$\frac{q_{ab}}{\sqrt{q}}(x) \propto \delta_{ij}\{A^i_a(x), \sqrt{V}\}\{A^j_b(x), \sqrt{V}\}$.

14.4.3 Yang-Mills

Apart from the non-Abelian nature of the gauge field, in form, the expressions proceed in the same manner as for the Maxwell case.

14.4.4 Scalar field - Higgs field

The elementary configuration operators are analogous of the holonomy operator for connection variables and they act by multiplication as

$\hat{h}(x,\lambda)\Psi = e^{i\lambda\varphi(x)}\Psi$.

These are called point holonomies. The conjugate variable to the point holonomy which is promoted to a operator in the quantum theory, is taken to be the smeared field momentum

$P(f) = \int d^3x \pi_\varphi(x) f(x)$

where π_φ is the conjugate momentum field and $f(x)$ is a test function. Their Poisson bracket is given by

$\{h(x,\lambda), P(f)\} = i\lambda f(x) h(x,\lambda)$.

In the quantum theory one looks for a representation of the Poisson bracket as a commutator of the elementary operators,

$[\hat{h}(x,\lambda), \hat{P}(f)] = i\lambda f(x)\hat{h}(x,\lambda)$.

14.4.5 Finiteness of Theory with the Inclusion of Matter

Thiemann has illustrated how the ultraviolet diverges of ordinary quantum theory can be directly interpreted as a consequence of the approximation that disregards the quantised, discrete, nature of quantum geometry. For instance Thiemann shows how the operator for the Yang-mills hamiltonian involving E^i_a is well defined so long as we treat E as an operator, but becomes infinite as soon as we replace E with a smooth background field.

14.5 The Master constraint programme

Main article: Master constraint of LQG

14.5.1 The Master constraint

The Master Constraint Programme[10] for Loop Quantum Gravity (LQG) was proposed as a classically equivalent way to impose the infinite number of Hamiltonian constraint equations

$$H(x) = 0$$

in terms of a single Master constraint,

$$M = \int d^3x \frac{[H(x)]^2}{\sqrt{\det q(x)}}$$

which involves the square of the constraints in question. Note that $H(x)$ were infinitely many whereas the Master constraint is only one. It is clear that if M vanishes then so do the infinitely many $H(x)$'s. Conversely, if all the $H(x)$'s vanish then so does M , therefore they are equivalent.

The Master constraint M involves an appropriate averaging over all space and so is invariant under spatial diffeomorphisms (it is invariant under spatial "shifts" as it is a summation over all such spatial "shifts" of a quantity that transforms as a scalar). Hence its Poisson bracket with the (smeared) spatial diffeomorphism constraint, $C(\vec{N})$, is simple:

$$\{M, C(\vec{N})\} = 0$$

(it is $su(2)$ invariant as well). Also, obviously as any quantity Poisson commutes with itself, and the Master constraint being a single constraint, it satisfies

$$\{M, M\} = 0$$

We also have the usual algebra between spatial diffeomorphisms. This represents a dramatic simplification of the Poisson bracket structure.

14.5.2 Promotion to quantum operator

Main article: Friedrichs extension

Let us write the classical expression in the form

$$M = \int d^3x \frac{H(x)^2}{\sqrt{\det(q)(x)}} = \int d^3x (\frac{H}{[\det(q)]^{1/4}})(x) \int d^3y \delta(x,y) (\frac{H}{[\det(q)]^{1/4}})(y)$$

This expression is regulated by a one parameter function $\chi_\epsilon(x,y)$ such that $\lim_{\epsilon \to 0} \chi_\epsilon(x,y)/\epsilon^3 = \delta(x,y)$ and $\chi_\epsilon(x,x) = 1$. Define

$$V_{\epsilon,x} = \int d^3y \chi_\epsilon(x,y) \sqrt{\det(q)}(y)$$

Both terms will be similar to the expression for the Hamiltonian constraint except now it will involve $\{A, \sqrt{V_\epsilon}\}$ rather than $\{A, V\}$ which comes from the additional factor $[\det(q)]^{1/4}$. That is,

$$M = \int d^3x \epsilon^{abc} \{A_c^k, \sqrt{V}_\epsilon\} F_{ab}^k(x) \int d^3y \chi_\epsilon(x,y) \epsilon^{a'b'c'} \{A_{c'}^{k'}, \sqrt{V}_\epsilon\} F_{a'b'}^{k'}(y)$$

Thus we proceed exactly as for the Hamiltonian constraint and introduce a partition into tetrahedra, splitting both integrals into sums,

$$M = \lim_{\epsilon \to 0} \sum_{\Delta,\Delta'} \chi(v(\Delta), v(\Delta')) \overline{C_\epsilon(\Delta)} C_\epsilon(\Delta')$$

where the meaning of $C_\epsilon(\Delta)$ is similar to that of H_Δ . This is a huge simplification as $C_\epsilon(\Delta)$ can be quantized precisely as the H_Δ with a simple change in the power of the volume operator. However, it can be shown that graph-changing, spatially diffeomorphism invariant operators such as the Master constraint cannot be defined on the kinematic Hilbert space $\mathcal{H}_{Kin}$. The way out is to define $\hat{M}$ not on $\mathcal{H}_{Kin}$ but on $\mathcal{H}_{Diff}$.

What is done first is, we are able to compute the matrix elements of the would-be operator $\hat{M}$, that is, we compute the quadratic form Q_M . We would like there to be a unique, positive, self-adjoint operator $\hat{M}$ whose matrix elements reproduce Q_M . It has been shown that such an operator exists and is given by the Friedrichs extension.[11][12]

14.5.3 Solving the Master constraint and inducing the physical Hilbert space

As mentioned above one cannot simply solve the spatial diffeomorphism constraint and then the Hamiltonian constraint, inducing a physical inner product from the spatial diffeomorphism inner product, because the Hamiltonian constraint maps spatially diffeomorphism invariant states onto non-spatial diffeomorphism invariant states. However, as the Master constraint M is spatially diffeomorphism invariant it can be defined on $\mathcal{H}_{Diff}$. Therefore, we are finally able to exploit the full power of the results mentioned above in obtaining $\mathcal{H}_{Diff}$ from $\mathcal{H}_{Kin}$.[9]

14.6 External links

- Overview by Carlo Rovelli

- Thiemann's paper in Physics Letters
- Good information on LQG

14.7 References

[1] *Gravitation* by Charles W. Misner, Kip S. Thorne, John Archibald Wheeler, published by W. H. Freeman and company. New York.

[2] Ashtekar, A. (1986) *Phys. Rev. Lett.* **57**, 2244.

[3] Rovelli, C. and Smolin, L. *Phys. Rev. Lett* **61**, 1155

[4] "Anomaly-free formulation of non-perturbative, four-dimensional Lorentzian quantum gravity", T. Thiemann, Phys.Lett. B ***380*** *(1996) 257-264.*

[5] See the book *Lectures on Non-Perturbative Canonical Gravity* for more details on this and the subsequent development. First published in 1991. World Scientific Publishing Co. Pte. LtD.

[6] "Reality conditions inducing transforms for quantum gauge field theory and quantum gravity", Thomas Thiemann, *Class.Quant.Grav.* **13** (1996) 1383-1404.

[7] "Real Ashtekar Variables for Lorentzian Signature Space-times", J. Fernando, G. Barbero. *Phys.Rev.D* **51** (1995) 5507-5510

[8] Borissov, R., De Pietri R., and Rovelli, C. (1997). *Class. Quan. Grav.* **14**, 2793

[9] A. Ashtekar, J. Lewandowski, D. Marolf, J. Mour ˜ao, T. Thiemann, "Quantization for diffeomorphism invariant theories of connections with local degrees of freedom", *Journ. Math. Phys.* **36** (1995) 6456-6493, arxiv.org:gr-qc/9504018.

[10] *The Phoenix Project: Master Constraint Programme for Loop Quantum Gravity*, *Class.Quant.Grav.* **23** (2006) 2211-2248 arXiv:gr-qc/0305080

[11] *Quantum Spin Dynamics VIII. The Master Constraint*, Thomas Thiemann, *Class.Quant.Grav.* **23** (2006) 2249-2266.

[12] *Master Constraint Operator in Loop Quantum Gravity*, Muxin Han, Yongge Ma, *Phys.Lett. B* **635** (2006) 225-231.

Chapter 15

Lorentz invariance in loop quantum gravity

Lorentz invariance is a measure of universal features in hypothetical loop quantum gravity universes. The various hypothetical multiverse loop quantum gravity universe design models, could have various Lorentz invariance results.

Because loop quantum gravity model universes, space gravity theories are contenders to build and answer unification theory; the Lorentz invariance helps grade the spread of universal features throughout a proposed multiverse in time.

15.1 Grand Unified Theory

The Grand Unified Theory is the era in time in the chronology of the universe where no elementary particles existed, and the three gauge interactions of the Standard Model which define the electromagnetic, weak, and strong interactions or forces, are merged into one single force. Convention says that 3 minutes after the Big Bang, protons and neutrons began to come together to form the nuclei of simple elements.[1] Whereas, loop quantum gravity theories places the origin and the age of elementary particles and the age of Lorentz invariance, beyond 13.799 ± 0.021 billion years ago.

The permanence of our Lorentz invariance constants are based on elementary particles and their features. There are eons of time before the big bang to build the universe from black holes and older multiverses. There is a selective process that creates features in elementary particles, like accept, store and give energy. In the books of Lee Smolin about loop quantum gravity, this theory contains the evolutionary ideas of "reproduction" and "mutation" of universes, and elementary particles, so is formally analogous to models of population biology.

15.2 Earlier universes

In the early universes before the big bang, there are theories that loop quantum gravity loop quantum structures formed space. The Lorentz invariance and universal constants describe elementary particles that do not exist yet.

Fecund universes is a multiverse theory of Lee Smolin about the role of black holes. The theory has black holes and loop quantum gravity connecting early universes together. Loop quantum gravity can be pulled into black holes. In Fecund universes, each new universes, according to Lee Smolin, has slightly different laws of physics. Because these laws are only slightly different, each is assumed to be like a mutation of the early universes.

15.3 Minkowski spacetime

Loop quantum gravity (LQG) is a quantization of a classical Lagrangian field theory. It is equivalent to the usual Einstein–Cartan theory in that it leads to the same equations of motion describing general relativity with torsion. As such, it can be argued that LQG respects local Lorentz invariance.

Global Lorentz invariance is broken in LQG just like it is broken in general relativity (unless one is dealing with Minkowski spacetime, which is one particular solution of the Einstein field equations). On the other hand, there has been much talk about possible local and global violations of Lorentz invariance beyond those expected in straightforward general relativity.

Of interest in this connection would be to see whether the LQG analogue of Minkowski spacetime breaks or preserves global Lorentz invariance, and Carlo Rovelli and coworkers have recently been investigating the Minkowski state of LQG using spin foam techniques. These questions will all remain open as long as the classical limits of various LQG models (see below for the sources of variation) cannot be calculated.

15.4 Lie algebras and loop quantum gravity

Mathematically LQG is local gauge theory of the self-dual subgroup of the complexified Lorentz group, which is related to the action of the Lorentz group on Weyl spinors commonly used in elementary particle physics. This is partly a matter of mathematical convenience, as it results in a compact group SO(3) or SU(2) as gauge group, as opposed to the non-compact groups SO(3,1) or SL(2.C). The compactness of the Lie group avoids some thus-far unsolved difficulties in the quantization of gauge theories of noncompact lie groups, and is responsible for the discreteness of the area and volume spectra. The theory involving the Immirzi parameter is necessary to resolve an ambiguity in the process of complexification. These are some of the many ways in which different quantizations of the same classical theory can result in inequivalent quantum theories, or even in the impossibility to carry quantization through.

One can't distinguish between SO(3) and SU(2) or between SO(3,1) and SL(2,C) at this level: the respective Lie algebras are the same. In fact, all four groups have the same complexified Lie algebra, which makes matters even more confusing (these subtleties are usually ignored in elementary particle physics). The physical interpretation of the Lie algebra is that of infinitesimally small group transformations, and gauge bosons (such as the graviton) are Lie algebra representations, not Lie group representations. What this means for the Lorentz group is that, for sufficiently small velocity parameters, all four complexified Lie groups are indistinguishable in the absence of matter fields.

To make matters more complicated, it can be shown that a positive cosmological constant can be realized in LQG by replacing the Lorentz group with the corresponding quantum group. At the level of the Lie algebra, this corresponds to what is called q-deforming the Lie algebra, and the parameter q is related to the value of the cosmological constant. The effect of replacing a Lie algebra by a q-deformed version is that the series of its representations is truncated (in the case of the rotation group, instead of having representations labelled by all half-integral spins, one is left with all representations with total spin j less than some constant).

It is entirely possible to formulate LQG in terms of q-deformed Lie algebras instead of ordinary Lie algebras, and in the case of the Lorentz group the result would, again, be indistinguishable for sufficiently small velocity parameters.

15.5 Spin networks loop quantum gravity

In the spin-foam formalism, the Barrett–Crane model, which was for a while the most promising state-sum model of 4D Lorentzian quantum gravity, was based on representations of the noncompact groups SO(3,1) or SL(2,C), so the spin foam faces (and hence the spin network edges) were labelled by positive real numbers as opposed to the half-integer labels of SU(2) spin networks.

These and other considerations, including difficulties interpreting what it would mean to apply a Lorentz transformation to a spin network state, led Lee Smolin and others to suggest that spin network states must break Lorentz invariance. Lee Smolin and Joao Magueijo then went on to study doubly special relativity, in which not only there is a constant velocity c but also a constant distance l. They showed that there are nonlinear representations of the Lorentz Lie algebra with these properties (the usual Lorentz group being obtained from a linear representation). Doubly special relativity predicts deviations from the special relativity dispersion relation at large energies (corresponding to small wavelengths of the order of the constant length l in the doubly special theory). Giovanni Amelino-Camelia then proposed that the mystery of ultra-high-energy cosmic rays might be solved by assuming such violations of the special-relativity dispersion relation for photons.

Phenomenological (hence, not specific to LQG) constraints on anomalous dispersion relations can be obtained by considering a variety of astrophysical experimental data, of which high-energy cosmic rays are but one part. Current observations are already able to place exceedingly stringent constraints on these phenomenological parameters.

15.6 References

[1] https://lcogt.net/spacebook/early-universe/

Chapter 16

Quantum configuration space

In quantum mechanics, the Hilbert space is the space of complex-valued functions belonging to $L^2(\mathbb{R}^3, d^3x)$, where the simple $\mathbb{R}^3$ is the classical configuration space of free particle which has finite degrees of freedom, and d^3x is the Lebesgue measure on $\mathbb{R}^3$. In the quantum mechanics the domain space of the wave functions ψ is the classical configuration space $\mathbb{R}^3$.

In classical field theory, the configuration space of the field is an infinite-dimensional space. The single point denoted A in this space is represented by the set of functions $A_I(\vec{x}) \in \mathbb{R}^3$ where $\vec{x} \in \mathbb{R}^3$ and I represents an index set.

In quantum field theory, it is expected that the Hilbert space is also the L^2 space on the configuration space of the field, which is infinite dimensional, with respect to some Borel measure naturally defined. However, it is often hard to define a concrete Borel measure on the classical configuration space, since the integral theory on infinite dimensional space is involved.[1]

Thus the intuitive expectation should be modified, and the concept of quantum configuration space should be introduced as a suitable enlargement of the classical configuration space so that an infinite dimensional measure, often a cylindrical measure, can be well defined on it.

In quantum field theory, the quantum configuration space, the domain of the wave functions Ψ , is larger than the classical configuration space. While in the classical theory we can restrict ourselves to suitably smooth fields, in quantum field theory we are forced to allow distributional field configurations. In fact, in quantum field theory physically interesting measures are concentrated on distributional configurations.

That physically interesting measures are concentrated on distributional fields is the reason why in quantum theory fields arise as operator-valued distributions.[2]

The example of a scalar field can be found in the references [3][4]

16.1 References

[1] Y. Choquet-Bruhat, C. Dewitt-Morette, M. Dillard-Bleick, Analysis, Manifold, and Physics, (North-Holland Publishing Company, 1977).

[2] Conceptual Foundations of Quantum Field Theory By Tian Yu Cao

[3] A. Ashtekar and J. Lewandowski, Background independent quantum gravity: A status report, Class. Quantum Grav. 21, R53 (2004), (preprint: gr-qc/0404018).

[4] A. Ashtekar, J. Lewandowski, D. Marolf, J. Mour̃ao, and T. Thiemann, A manifestly gauge-invariant approach to quantum theories of gauge fields, (preprint: hep-th/9408108).

Chapter 17

Classical limit

The **classical limit** or **correspondence limit** is the ability of a physical theory to approximate or "recover" classical mechanics when considered over special values of its parameters.[1] The classical limit is used with physical theories that predict non-classical behavior.

17.1 Quantum theory

A heuristic postulate called the correspondence principle was introduced to quantum theory by Niels Bohr: it states that, in effect, some kind of continuity argument should apply to the classical limit of quantum systems as the value of Planck's constant normalized by the action of these systems tends to zero. Often, this is approached through "quasi-classical" techniques (cf. WKB approximation).[2]

More rigorously, the mathematical operation involved in classical limits is a group contraction, approximating physical systems where the relevant action is much larger than Planck's constant $\hbar$, so the "deformation parameter" $\hbar/S$ can be effectively taken to be zero. (cf. Weyl quantization.) Thus typically, quantum commutators (equivalently, Moyal brackets) reduce to Poisson brackets,[3] in a group contraction.

In quantum mechanics, due to Heisenberg's uncertainty principle, an electron can never be at rest; it must always have a non-zero kinetic energy, a result not found in classical mechanics. For example, if we consider something very large relative to an electron, like a baseball, the uncertainty principle predicts that it cannot really have zero kinetic energy, but the uncertainty in kinetic energy is so *small* that the baseball can effectively appear to be at rest, and hence it *appears to obey classical mechanics*. In general, if large energies and large objects (relative to the size and energy levels of an electron) are considered in quantum mechanics, the result will appear to obey classical mechanics. (The typical occupation numbers involved are huge: a macroscopic harmonic oscillator with ω=2 Hz, m=10g, and maximum amp x_0=10 cm, has $S \approx E/\omega \approx m\omega x_0^2/2 \approx 10^{-4}$Kg $m^2/s = \hbar n$, so that $n \simeq 10^{30}$. Further see coherent states.)

It is less clear, however, how the classical limit applies to chaotic systems, a field known as quantum chaos.

Quantum mechanics and classical mechanics are usually treated with entirely different formalisms: quantum theory using Hilbert space, and classical mechanics using a representation in phase space. It is possible to bring the two into a common mathematical framework in various ways. In the phase space formulation of quantum mechanics, which is statistical in nature, logical connections between quantum mechanics and classical statistical mechanics are made, enabling natural comparisons between them.[4][5]

In a crucial paper (1933), Dirac[6] explained how classical mechanics is an emergent phenomenon of quantum mechanics: destructive interference among paths with non-extremal macroscopic actions $S \gg \hbar$ obliterate amplitude contributions in the path integral he introduced, leaving the extremal action *Sclass*, thus the classical action path as the dominant contribution, an observation further elaborated by Feynman in his 1942 PhD dissertation.[7] (Further see quantum decoherence.)

17.2 Relativity and other deformations

Other familiar deformations in physics involve

- The deformation of classical Newtonian into relativistic mechanics (special relativity), with deformation parameter *v/c*; the classical limit involves small speeds, so *v/c*→0, and the systems appear to obey Newtonian mechanics.
- Similarly for the deformation of Newtonian gravity into general relativity, with deformation parameter Schwarzschild-radius/characteristic-dimension, we find that objects once again appear to obey classical mechanics (flat space), when the mass of an object times the square of the Planck length is much smaller than its size and the sizes of the problem addressed.

- Wave optics might also be regarded as a deformation of ray optics for deformation parameter λ/a.
- Likewise, thermodynamics deforms to statistical mechanics with deformation parameter $1/N$.

17.3 See also

- WKB approximation
- Quantum decoherence
- Quantum limit
- Quantum realm
- Wigner–Weyl transform
- Quantum chaos
- Fresnel integral
- Semiclassical physics

17.4 References

[1] Bohm, David (1989). *Quantum Theory*. Dover Publications. ISBN 0-486-65969-0.

[2] L.D. Landau, E.M. Lifshitz (1977). *Quantum Mechanics: Non-Relativistic Theory*. Vol. 3 (3rd ed.). Pergamon Press. ISBN 978-0-08-020940-1.

[3] Curtright, T. L.; Zachos, C. K. (2012). "Quantum Mechanics in Phase Space". *Asia Pacific Physics Newsletter* **01**: 37. doi:10.1142/S2251158X12000069.

[4] Bracken, A.; Wood, J. (2006). "Semiquantum versus semiclassical mechanics for simple nonlinear systems". *Physical Review A* **73**. arXiv:quant-ph/0511227. Bibcode:2006PhRvA..73a2104B. doi:10.1103/PhysRevA.73.012104.

[5] Conversely, in the less well-known approach presented in 1932 by Koopman and von Neumann, the dynamics of classical mechanics have been formulated in terms of an operatorial formalism in Hilbert space, a formalism used conventionally for quantum mechanics. Koopman, B. O., Neumann, J. v., Dynamical systems of continuous spectra, Proc. Natl. Acad. Sci. U.S.A., vol. 18 (1932), no. 3, pp. 255–263 (full text) . Danilo Mauro: Topics in Koopman-von Neumann Theory, arXiv:quant-ph/0301172 (2003); Bracken, A. J. (2003). "Quantum mechanics as an approximation to classical mechanics in Hilbert space". *Journal of Physics A: Mathematical and General* **36** (23): L329. doi:10.1088/0305-4470/36/23/101.

[6] Dirac, P.A.M. (1933). "The Lagrangian in quantum mechanics", *Phys. Z. der Sowjetunion* **3**: 64-71.

[7] Feynman, Richard P. (1942). Laurie M. Brown. ed. "The Principle of Least Action in Quantum Mechanics", Ph.D. Dissertation, Princeton University. (World Scientific publishers, with title "Feynman's Thesis: a New Approach to Quantum Theory", 2005.) ISBN 978-981-256-380-4.

Chapter 18

Quantum mechanics

For a more accessible and less technical introduction to this topic, see Introduction to quantum mechanics.

Quantum mechanics (**QM**; also known as **quantum physics** or **quantum theory**), including quantum field theory, is a fundamental branch of physics concerned with processes involving, for example, atoms and photons. In such processes, said to be quantized, the action has been observed to be only in integer multiples of the Planck constant. This is utterly inexplicable in classical physics.

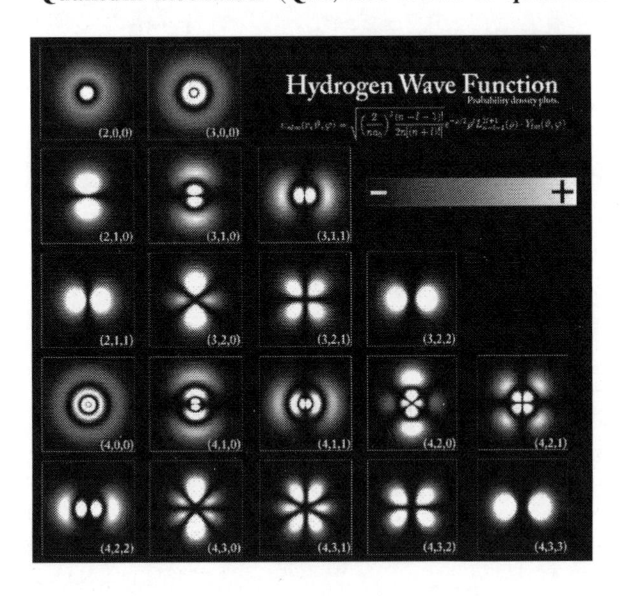

Solution to Schrödinger's equation for the hydrogen atom at different energy levels. The brighter areas represent a higher probability of finding an electron

Quantum mechanics gradually arose from Max Planck's solution in 1900 to the black-body radiation problem (reported 1859) and Albert Einstein's 1905 paper which offered a quantum-based theory to explain the photoelectric effect (reported 1887). Early quantum theory was profoundly reconceived in the mid-1920s.

The reconceived theory is formulated in various specially developed mathematical formalisms. In one of them, a mathematical function, the wave function, provides information about the probability amplitude of position, momentum, and other physical properties of a particle.

Important applications of quantum mechanical theory[1] include superconducting magnets, light-emitting diodes and the laser, the transistor and semiconductors such as the microprocessor, medical and research imaging such as magnetic resonance imaging and electron microscopy, and explanations for many biological and physical phenomena.

18.1 History

Main article: History of quantum mechanics

Scientific inquiry into the wave nature of light began in the 17th and 18th centuries, when scientists such as Robert Hooke, Christiaan Huygens and Leonhard Euler proposed a wave theory of light based on experimental observations.[2] In 1803, Thomas Young, an English polymath, performed the famous double-slit experiment that he later described in a paper titled *On the nature of light and colours*. This experiment played a major role in the general acceptance of the wave theory of light.

In 1838, Michael Faraday discovered cathode rays. These studies were followed by the 1859 statement of the black-body radiation problem by Gustav Kirchhoff, the 1877 suggestion by Ludwig Boltzmann that the energy states of a physical system can be discrete, and the 1900 quantum hypothesis of Max Planck.[3] Planck's hypothesis that energy is radiated and absorbed in discrete "quanta" (or energy elements) precisely matched the observed patterns of black-body radiation.

In 1896, Wilhelm Wien empirically determined a distribution law of black-body radiation,[4] known as Wien's law in his honor. Ludwig Boltzmann independently arrived at this result by considerations of Maxwell's equations. However, it was valid only at high frequencies and underestimated the radiance at low frequencies. Later, Planck corrected this model using Boltzmann's statistical interpretation of thermodynamics and proposed what is now called Planck's law, which led to the development of quantum mechanics.

Following Max Planck's solution in 1900 to the black-body radiation problem (reported 1859), Albert Einstein offered a quantum-based theory to explain the photoelectric effect (1905, reported 1887). Around 1900-1910, the atomic theory and the corpuscular theory of light[5] first came to be widely accepted as scientific fact; these latter theories can be viewed as quantum theories of matter and electromagnetic radiation, respectively.

Among the first to study quantum phenomena in nature were Arthur Compton, C. V. Raman, and Pieter Zeeman, each of whom has a quantum effect named after him. Robert Andrews Millikan studied the photoelectric effect experimentally, and Albert Einstein developed a theory for it. At the same time, Niels Bohr developed his theory of the atomic structure, which was later confirmed by the experiments of Henry Moseley. In 1913, Peter Debye extended Niels Bohr's theory of atomic structure, introducing elliptical orbits, a concept also introduced by Arnold Sommerfeld.[6] This phase is known as old quantum theory.

According to Planck, each energy element (E) is proportional to its frequency (ν):

$$E = h\nu$$

where h is Planck's constant.

Planck cautiously insisted that this was simply an aspect of the *processes* of absorption and emission of radiation and had nothing to do with the *physical reality* of the radiation itself.[7] In fact, he considered his quantum hypothesis a mathematical trick to get the right answer rather than a sizable discovery.[8] However, in 1905 Albert Einstein interpreted Planck's quantum hypothesis realistically and used it to explain the photoelectric effect, in which shining light on certain materials can eject electrons from the material. He won the 1921 Nobel Prize in Physics for this work.

Einstein further developed this idea to show that an electromagnetic wave such as light could also be described as a particle (later called the photon), with a discrete quantum of energy that was dependent on its frequency.[9]

The foundations of quantum mechanics were established during the first half of the 20th century by Max Planck, Niels Bohr, Werner Heisenberg, Louis de Broglie, Arthur Compton, Albert Einstein, Erwin Schrödinger, Max Born, John von Neumann, Paul Dirac, Enrico Fermi, Wolfgang Pauli, Max von Laue, Freeman Dyson, David Hilbert, Wilhelm Wien, Satyendra Nath Bose, Arnold Sommerfeld, and others. The Copenhagen interpretation of Niels Bohr became widely accepted.

In the mid-1920s, developments in quantum mechanics led to its becoming the standard formulation for atomic physics. In the summer of 1925, Bohr and Heisenberg published results that closed the old quantum theory. Out of deference to their particle-like behavior in certain processes and measurements, light quanta came to be called photons (1926). From Einstein's simple postulation was born a flurry of debating, theorizing, and testing. Thus, the entire field of quantum physics emerged, leading to its wider acceptance at the Fifth Solvay Conference in 1927.

Max Planck is considered the father of the quantum theory.

The 1927 Solvay Conference in Brussels.

It was found that subatomic particles and electromagnetic waves are neither simply particle nor wave but have certain properties of each. This originated the concept of wave–particle duality.

By 1930, quantum mechanics had been further unified and formalized by the work of David Hilbert, Paul Dirac and John von Neumann[10] with greater emphasis on measurement, the statistical nature of our knowledge of reality, and philosophical speculation about the 'observer'. It has since permeated many disciplines including quantum chemistry, quantum electronics, quantum optics, and quantum information science. Its speculative modern developments include string theory and quantum gravity theories. It also provides a useful framework for many features of the modern periodic table of elements, and describes the behaviors of atoms during chemical bonding and the flow of electrons in computer semiconductors, and therefore plays a crucial role in many modern technologies.

While quantum mechanics was constructed to describe the world of the very small, it is also needed to explain some macroscopic phenomena such as superconductors,[11] and superfluids.[12]

The word *quantum* derives from the Latin, meaning "how great" or "how much".[13] In quantum mechanics, it refers to a discrete unit assigned to certain physical quantities such as the energy of an atom at rest (see Figure 1). The discovery that particles are discrete packets of energy with wave-like properties led to the branch of physics dealing with atomic and subatomic systems which is today called quantum mechanics. It underlies the mathematical framework of many fields of physics and chemistry, including condensed matter physics, solid-state physics, atomic physics, molecular physics, computational physics, computational chemistry, quantum chemistry, particle physics, nuclear chemistry, and nuclear physics.[14] Some fundamental aspects of the theory are still actively studied.[15]

Quantum mechanics is essential to understanding the behavior of systems at atomic length scales and smaller. If the physical nature of an atom were solely described by classical mechanics, electrons would not *orbit* the nucleus, since orbiting electrons emit radiation (due to circular motion) and would eventually collide with the nucleus due to this loss of energy. This framework was unable to explain the stability of atoms. Instead, electrons remain in an uncertain, non-deterministic, *smeared*, probabilistic wave–particle orbital about the nucleus, defying the traditional assumptions of classical mechanics and electromagnetism.[16]

Quantum mechanics was initially developed to provide a better explanation and description of the atom, especially the differences in the spectra of light emitted by different isotopes of the same chemical element, as well as subatomic particles. In short, the quantum-mechanical atomic model has succeeded spectacularly in the realm where classical mechanics and electromagnetism falter.

Broadly speaking, quantum mechanics incorporates four classes of phenomena for which classical physics cannot account:

- quantization of certain physical properties
- quantum entanglement
- principle of uncertainty
- wave–particle duality

18.2 Mathematical formulations

Main article: Mathematical formulation of quantum mechanics
See also: Quantum logic

In the mathematically rigorous formulation of quantum mechanics developed by Paul Dirac,[17] David Hilbert,[18] John von Neumann,[19] and Hermann Weyl,[20] the possible states of a quantum mechanical system are symbolized[21] as unit vectors (called *state vectors*). Formally, these reside in a complex separable Hilbert space—variously called the *state space* or the *associated Hilbert space* of the system—that is well defined up to a complex number of norm 1 (the phase factor). In other words, the possible states are points in the projective space of a Hilbert space, usually called the complex projective space. The exact nature of this Hilbert space is dependent on the system—for example, the state space for position and momentum states is the space of square-integrable functions, while the state space for the spin of a single proton is just the product of two complex planes. Each observable is represented by a maximally Hermitian (precisely: by a self-adjoint) linear operator acting on the state space. Each eigenstate of an observable corresponds to an eigenvector of the operator, and the associated eigenvalue corresponds to the value of the observable in that eigenstate. If the operator's spectrum is discrete, the observable can attain only those discrete eigenvalues.

In the formalism of quantum mechanics, the state of a system at a given time is described by a complex wave function, also referred to as state vector in a complex vector space.[22] This abstract mathematical object allows for the calculation of probabilities of outcomes of concrete experiments. For example, it allows one to compute the probability of finding an electron in a particular region around the nucleus at a particular time. Contrary to classical mechanics, one can never make simultaneous predictions of conjugate variables, such as position and momentum, with accuracy. For instance, electrons may be considered (to a certain probability) to be located somewhere within a given region of space, but with their exact positions unknown. Contours of constant probability, often referred to as "clouds", may be drawn around the nucleus of an atom to conceptualize where the electron might be located with the most probability. Heisenberg's uncertainty principle quantifies the inability to precisely locate the particle given its conjugate momentum.[23]

According to one interpretation, as the result of a measurement the wave function containing the probability information for a system collapses from a given initial state to a particular eigenstate. The possible results of a measurement are the eigenvalues of the operator representing the observable—which explains the choice of *Hermitian* operators, for which all the eigenvalues are real. The probability distribution of an observable in a given state can be found by computing the spectral decomposition of the corresponding operator. Heisenberg's uncertainty principle is represented by the statement that the operators corresponding to certain observables do not commute.

The probabilistic nature of quantum mechanics thus stems from the act of measurement. This is one of the most difficult aspects of quantum systems to understand. It was the central topic in the famous Bohr–Einstein debates, in which the two scientists attempted to clarify these fundamental principles by way of thought experiments. In the decades after the formulation of quantum mechanics, the question of what constitutes a "measurement" has been extensively studied. Newer interpretations of quantum mechanics have been formulated that do away with the concept of "wave function collapse" (see, for example, the relative state interpretation). The basic idea is that when a quantum system interacts with a measuring apparatus, their respective wave functions become entangled, so that the original quantum system ceases to exist as an independent entity. For details, see the article on measurement in quantum mechanics.[24]

Generally, quantum mechanics does not assign definite values. Instead, it makes a prediction using a probability distribution; that is, it describes the probability of obtaining the possible outcomes from measuring an observable. Often these results are skewed by many causes, such as dense probability clouds. Probability clouds are approximate (but better than the Bohr model) whereby electron location is given by a probability function, the wave function eigenvalue, such that the probability is the squared modulus of the complex amplitude, or quantum state nuclear attraction.[25][26] Naturally, these probabilities will depend on the quantum state at the "instant" of the measurement. Hence, uncertainty is involved in the value. There are, however, certain states that are associated with a definite value of a particular observable. These are known as eigenstates of the observable ("eigen" can be translated from German as meaning "inherent" or "characteristic").[27]

In the everyday world, it is natural and intuitive to think of everything (every observable) as being in an eigenstate. Everything appears to have a definite position, a definite momentum, a definite energy, and a definite time of occurrence. However, quantum mechanics does not pinpoint the exact values of a particle's position and momentum (since they are conjugate pairs) or its energy and time (since they too are conjugate pairs); rather, it provides only a range of probabilities in which that particle might be given its momentum and momentum probability. Therefore, it is helpful to use different words to describe states having *uncertain* values and states having *definite* values (eigenstates). Usually, a system will not be in an eigenstate of the observable (particle) we are interested in. However, if one measures the observable, the wave function will instantaneously be an eigenstate (or "generalized" eigenstate) of that observable. This process is known as wave function collapse, a controversial and much-debated process[28] that involves expanding the system under study to include the measurement device. If one knows the corresponding wave function at the instant before the measurement, one will be able to compute the probability of the wave function collapsing into each of the possible eigenstates. For example, the free particle in the previous example will usually have a wave function that is a wave packet centered around some mean position x_0 (neither an eigenstate of position nor of momentum). When one measures the position of the particle, it is impossible to predict with certainty the result.[24] It is probable, but not certain, that it will be near x_0, where the amplitude of the wave function is large. After the measurement is performed, having obtained some result x, the wave function collapses into a position eigenstate centered at x.[29]

The time evolution of a quantum state is described by the Schrödinger equation, in which the Hamiltonian (the operator corresponding to the total energy of the system) generates the time evolution. The time evolution of wave functions is deterministic in the sense that - given a wave function at an *initial* time - it makes a definite prediction of what the wave function will be at any *later* time.[30]

During a measurement, on the other hand, the change of the initial wave function into another, later wave function is not deterministic, it is unpredictable (i.e., random). A time-evolution simulation can be seen here.[31][32]

Wave functions change as time progresses. The Schrödinger equation describes how wave functions change in time, playing a role similar to Newton's second law in classical mechanics. The Schrödinger equation, applied to the aforementioned example of the free particle, predicts that the center of a wave packet will move through space at a constant velocity (like a classical particle with no forces acting on it). However, the wave packet will also spread out as time progresses, which means that the position becomes more uncertain with time. This also has the effect of turning a position eigenstate (which can be thought of as an infinitely sharp wave packet) into a broadened wave packet that no longer represents a (definite, certain) position eigenstate.[33]

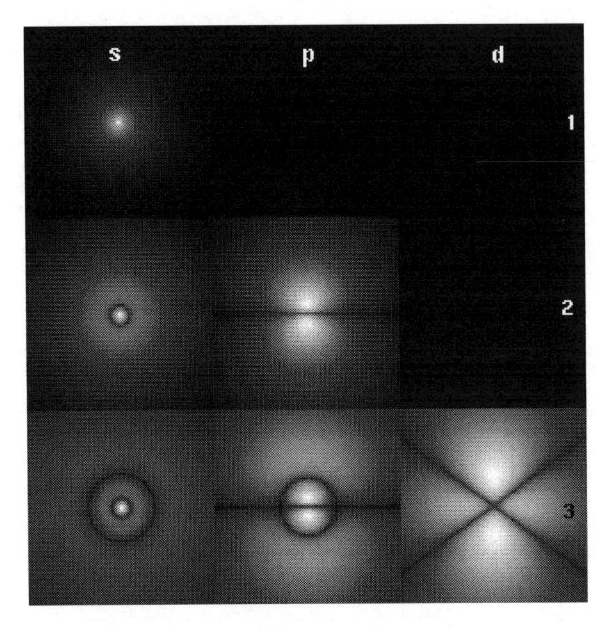

Fig. 1: Probability densities corresponding to the wave functions of an electron in a hydrogen atom possessing definite energy levels (increasing from the top of the image to the bottom: n = *1, 2, 3, ...) and angular momenta (increasing across from left to right:* s, p, d, *...). Brighter areas correspond to higher probability density in a position measurement. Such wave functions are directly comparable to Chladni's figures of acoustic modes of vibration in classical physics, and are modes of oscillation as well, possessing a sharp energy and, thus, a definite frequency. The angular momentum and energy are quantized, and take* ***only*** *discrete values like those shown (as is the case for resonant frequencies in acoustics)*

Some wave functions produce probability distributions that are constant, or independent of time—such as when in a stationary state of constant energy, time vanishes in the absolute square of the wave function. Many systems that are treated dynamically in classical mechanics are described by such "static" wave functions. For example, a single electron in an unexcited atom is pictured classically as a particle moving in a circular trajectory around the atomic nucleus, whereas in quantum mechanics it is described by a static, spherically symmetric wave function surrounding the nucleus (Fig. 1) (note, however, that only the lowest angular momentum states, labeled *s*, are spherically symmetric).[34]

The Schrödinger equation acts on the *entire* probability amplitude, not merely its absolute value. Whereas the absolute value of the probability amplitude encodes information about probabilities, its phase encodes information about the interference between quantum states. This gives rise to the "wave-like" behavior of quantum states. As it turns out, analytic solutions of the Schrödinger equation are available for only a very small number of relatively simple model Hamiltonians, of which the quantum harmonic oscillator, the particle in a box, the dihydrogen cation, and the hydrogen atom are the most important representatives. Even the helium atom—which contains just one more electron than does the hydrogen atom—has defied all attempts at a fully analytic treatment.

There exist several techniques for generating approximate solutions, however. In the important method known as perturbation theory, one uses the analytic result for a simple quantum mechanical model to generate a result for a more complicated model that is related to the simpler model by (for one example) the addition of a weak potential energy. Another method is the "semi-classical equation of motion" approach, which applies to systems for which quantum mechanics produces only weak (small) deviations from classical behavior. These deviations can then be computed based on the classical motion. This approach is particularly important in the field of quantum chaos.

18.3 Mathematically equivalent formulations of quantum mechanics

There are numerous mathematically equivalent formulations of quantum mechanics. One of the oldest and most commonly used formulations is the "transformation theory" proposed by Paul Dirac, which unifies and generalizes the two earliest formulations of quantum mechanics - matrix mechanics (invented by Werner Heisenberg) and wave mechanics (invented by Erwin Schrödinger).[35]

Especially since Werner Heisenberg was awarded the Nobel Prize in Physics in 1932 for the creation of quantum mechanics, the role of Max Born in the development of QM was overlooked until the 1954 Nobel award. The role is noted in a 2005 biography of Born, which recounts his role in the matrix formulation of quantum mechanics, and the use of probability amplitudes. Heisenberg himself acknowledges having learned matrices from Born, as published in a 1940 *festschrift* honoring Max Planck.[36] In the matrix formulation, the instantaneous state of a quantum system encodes the probabilities of its measurable properties, or "observables". Examples of observables include energy, position, momentum, and angular momentum. Observables can be either continuous (e.g., the position of a particle) or discrete (e.g., the energy of an electron bound to a hydrogen atom).[37] An alternative formulation of quantum mechanics is Feynman's path integral formulation, in which a quantum-mechanical amplitude is considered as a sum over all possible classical and non-classical paths between the initial and final states. This is the quantum-mechanical counterpart of the action principle in classical mechanics.

18.4 Interactions with other scientific theories

The rules of quantum mechanics are fundamental. They assert that the state space of a system is a Hilbert space and that observables of that system are Hermitian operators acting on that space—although they do not tell us which Hilbert space or which operators. These can be chosen appropriately in order to obtain a quantitative description of a quantum system. An important guide for making these choices is the correspondence principle, which states that the predictions of quantum mechanics reduce to those of classical mechanics when a system moves to higher energies or, equivalently, larger quantum numbers, i.e. whereas a single particle exhibits a degree of randomness, in systems incorporating millions of particles averaging takes over and, at the high energy limit, the statistical probability of random behaviour approaches zero. In other words, classical mechanics is simply a quantum mechanics of large systems. This "high energy" limit is known as the *classical* or *correspondence limit*. One can even start from an established classical model of a particular system, then attempt to guess the underlying quantum model that would give rise to the classical model in the correspondence limit.

When quantum mechanics was originally formulated, it was applied to models whose correspondence limit was non-relativistic classical mechanics. For instance, the well-known model of the quantum harmonic oscillator uses an explicitly non-relativistic expression for the kinetic energy of the oscillator, and is thus a quantum version of the classical harmonic oscillator.

Early attempts to merge quantum mechanics with special relativity involved the replacement of the Schrödinger equation with a covariant equation such as the Klein–Gordon equation or the Dirac equation. While these theories were successful in explaining many experimental results, they had certain unsatisfactory qualities stemming from their neglect of the relativistic creation and annihilation of particles. A fully relativistic quantum theory required the development of quantum field theory, which applies quantization to a field (rather than a fixed set of particles). The first complete quantum field theory, quantum electrodynamics, provides a fully quantum description of the electromagnetic interaction. The full apparatus of quantum field theory is often unnecessary for describing electrodynamic systems. A simpler approach, one that has been employed since the inception of quantum mechanics, is to treat charged particles as quantum mechanical objects being acted on by a classical electromagnetic field. For example, the elementary quantum model of the hydrogen atom describes the electric field of the hydrogen atom using a classical $-e^2/(4\pi\ \epsilon_0\ r)$ Coulomb potential. This "semi-classical" approach fails if quantum fluctuations in the electromagnetic field play an important role, such as in the emission of photons by charged particles.

Quantum field theories for the strong nuclear force and the weak nuclear force have also been developed. The quantum field theory of the strong nuclear force is called quantum chromodynamics, and describes the interactions of subnuclear particles such as quarks and gluons. The weak nuclear force and the electromagnetic force were unified, in their quantized forms, into a single quantum field theory (known as electroweak theory), by the physicists Abdus Salam, Sheldon Glashow and Steven Weinberg. These three men shared the Nobel Prize in Physics in 1979 for this work.[38]

It has proven difficult to construct quantum models of gravity, the remaining fundamental force. Semi-classical approximations are workable, and have led to predictions such as Hawking radiation. However, the formulation of a complete theory of quantum gravity is hindered by apparent incompatibilities between general relativity (the most accurate theory of gravity currently known) and some of the fundamental assumptions of quantum theory. The resolution of these incompatibilities is an area of active research, and theories such as string theory are among the possible candidates for a future theory of quantum gravity.

Classical mechanics has also been extended into the complex domain, with complex classical mechanics exhibiting behaviors similar to quantum mechanics.[39]

18.4.1 Quantum mechanics and classical physics

Predictions of quantum mechanics have been verified experimentally to an extremely high degree of accuracy.[40] According to the correspondence principle between classical and quantum mechanics, all objects obey the laws of quantum mechanics, and classical mechanics is just an approximation for large systems of objects (or a statistical quantum mechanics of a large collection of particles).[41] The laws of classical mechanics thus follow from the laws of quantum mechanics as a statistical average at the limit of large systems or large quantum numbers.[42] However, chaotic systems do not have good quantum numbers, and quantum chaos studies the relationship between classical and quantum descriptions in these systems.

Quantum coherence is an essential difference between classical and quantum theories as illustrated by the Einstein–

Podolsky–Rosen (EPR) paradox — an attack on a certain philosophical interpretation of quantum mechanics by an appeal to local realism.[43] Quantum interference involves adding together *probability amplitudes*, whereas classical "waves" infer that there is an adding together of *intensities*. For microscopic bodies, the extension of the system is much smaller than the coherence length, which gives rise to long-range entanglement and other nonlocal phenomena characteristic of quantum systems.[44] Quantum coherence is not typically evident at macroscopic scales, though an exception to this rule may occur at extremely low temperatures (i.e. approaching absolute zero) at which quantum behavior may manifest itself macroscopically.[45] This is in accordance with the following observations:

- Many macroscopic properties of a classical system are a direct consequence of the quantum behavior of its parts. For example, the stability of bulk matter (consisting of atoms and molecules which would quickly collapse under electric forces alone), the rigidity of solids, and the mechanical, thermal, chemical, optical and magnetic properties of matter are all results of the interaction of electric charges under the rules of quantum mechanics.[46]
- While the seemingly "exotic" behavior of matter posited by quantum mechanics and relativity theory become more apparent when dealing with particles of extremely small size or velocities approaching the speed of light, the laws of classical, often considered "Newtonian", physics remain accurate in predicting the behavior of the vast majority of "large" objects (on the order of the size of large molecules or bigger) at velocities much smaller than the velocity of light.[47]

18.4.2 Copenhagen interpretation of quantum versus classical kinematics

A big difference between classical and quantum mechanics is that they use very different kinematic descriptions.[48]

In Niels Bohr's mature view, quantum mechanical phenomena are required to be experiments, with complete descriptions of all the devices for the system, preparative, intermediary, and finally measuring. The descriptions are in macroscopic terms, expressed in ordinary language, supplemented with the concepts of classical mechanics.[49][50][51][52] The initial condition and the final condition of the system are respectively described by values in a configuration space, for example a position space, or some equivalent space such as a momentum space. Quantum mechanics does not admit a completely precise description, in terms of both position and momentum, of an initial condition or "state" (in the classical sense of the word) that would support a precisely deterministic and causal prediction of a final condition.[53][54] In this sense, advocated by Bohr in his mature writings, a quantum phenomenon is a process, a passage from initial to final condition, not an instantaneous "state" in the classical sense of that word.[55][56] Thus there are two kinds of processes in quantum mechanics: stationary and transitional. For a stationary process, the initial and final condition are the same. For a transition, they are different. Obviously by definition, if only the initial condition is given, the process is not determined.[53] Given its initial condition, prediction of its final condition is possible, causally but only probabilistically, because the Schrödinger equation is deterministic for wave function evolution, but the wave function describes the system only probabilistically.[57][58]

For many experiments, it is possible to think of the initial and final conditions of the system as being a particle. In some cases it appears that there are potentially several spatially distinct pathways or trajectories by which a particle might pass from initial to final condition. It is an important feature of the quantum kinematic description that it does not permit a unique definite statement of which of those pathways is actually followed. Only the initial and final conditions are definite, and, as stated in the foregoing paragraph, they are defined only as precisely as allowed by the configuration space description or its equivalent. In every case for which a quantum kinematic description is needed, there is always a compelling reason for this restriction of kinematic precision. An example of such a reason is that for a particle to be experimentally found in a definite position, it must be held motionless; for it to be experimentally found to have a definite momentum, it must have free motion; these two are logically incompatible.[59][60]

Classical kinematics does not primarily demand experimental description of its phenomena. It allows completely precise description of an instantaneous state by a value in phase space, the Cartesian product of configuration and momentum spaces. This description simply assumes or imagines a state as a physically existing entity without concern about its experimental measurability. Such a description of an initial condition, together with Newton's laws of motion, allows a precise deterministic and causal prediction of a final condition, with a definite trajectory of passage. Hamiltonian dynamics can be used for this. Classical kinematics also allows the description of a process analogous to the initial and final condition description used by quantum mechanics. Lagrangian mechanics applies to this.[61] For processes that need account to be taken of actions of a small number of Planck constants, classical kinematics is not adequate; quantum mechanics is needed.

18.4.3 General relativity and quantum mechanics

Even with the defining postulates of both Einstein's theory of general relativity and quantum theory being indisputably supported by rigorous and repeated empirical evidence, and while they do not directly contradict each other theoretically (at least with regard to their primary claims), they have proven extremely difficult to incorporate into one consistent, cohesive model.[62]

Gravity is negligible in many areas of particle physics, so that unification between general relativity and quantum mechanics is not an urgent issue in those particular applications. However, the lack of a correct theory of quantum gravity is an important issue in cosmology and the search by physicists for an elegant "Theory of Everything" (TOE). Consequently, resolving the inconsistencies between both theories has been a major goal of 20th and 21st century physics. Many prominent physicists, including Stephen Hawking, have labored for many years in the attempt to discover a theory underlying *everything*. This TOE would combine not only the different models of subatomic physics, but also derive the four fundamental forces of nature - the strong force, electromagnetism, the weak force, and gravity - from a single force or phenomenon. While Stephen Hawking was initially a believer in the Theory of Everything, after considering Gödel's Incompleteness Theorem, he has concluded that one is not obtainable, and has stated so publicly in his lecture "Gödel and the End of Physics" (2002).[63]

18.4.4 Attempts at a unified field theory

Main article: Grand unified theory

The quest to unify the fundamental forces through quantum mechanics is still ongoing. Quantum electrodynamics (or "quantum electromagnetism"), which is currently (in the perturbative regime at least) the most accurately tested physical theory in competition with general relativity,[64][65] has been successfully merged with the weak nuclear force into the electroweak force and work is currently being done to merge the electroweak and strong force into the electrostrong force. Current predictions state that at around 10^{14} GeV the three aforementioned forces are fused into a single unified field.[66] Beyond this "grand unification", it is speculated that it may be possible to merge gravity with the other three gauge symmetries, expected to occur at roughly 10^{19} GeV. However — and while special relativity is parsimoniously incorporated into quantum electrodynamics — the expanded general relativity, currently the best theory describing the gravitation force, has not been fully incorporated into quantum theory. One of those searching for a coherent TOE is Edward Witten, a theoretical physicist who formulated the M-theory, which is an attempt at describing the supersymmetrical based string theory. M-theory posits that our apparent 4-dimensional spacetime is, in reality, actually an 11-dimensional spacetime containing 10 spatial dimensions and 1 time dimension, although 7 of the spatial dimensions are - at lower energies - completely "compactified" (or infinitely curved) and not readily amenable to measurement or probing.

Another popular theory is Loop quantum gravity (LQG), a theory first proposed by Carlo Rovelli that describes the quantum properties of gravity. It is also a theory of quantum space and quantum time, because in general relativity the geometry of spacetime is a manifestation of gravity. LQG is an attempt to merge and adapt standard quantum mechanics and standard general relativity. The main output of the theory is a physical picture of space where space is granular. The granularity is a direct consequence of the quantization. It has the same nature of the granularity of the photons in the quantum theory of electromagnetism or the discrete levels of the energy of the atoms. But here it is space itself which is discrete. More precisely, space can be viewed as an extremely fine fabric or network "woven" of finite loops. These networks of loops are called spin networks. The evolution of a spin network over time is called a spin foam. The predicted size of this structure is the Planck length, which is approximately 1.616×10^{-35} m. According to theory, there is no meaning to length shorter than this (cf. Planck scale energy). Therefore, LQG predicts that not just matter, but also space itself, has an atomic structure.

18.5 Philosophical implications

Main article: Interpretations of quantum mechanics

Since its inception, the many counter-intuitive aspects and results of quantum mechanics have provoked strong philosophical debates and many interpretations. Even fundamental issues, such as Max Born's basic rules concerning probability amplitudes and probability distributions, took decades to be appreciated by society and many leading scientists. Richard Feynman once said, "I think I can safely say that nobody understands quantum mechanics."[67] According to Steven Weinberg, "There is now in my opinion no entirely satisfactory interpretation of quantum mechanics."[68]

The Copenhagen interpretation — due largely to Niels Bohr and Werner Heisenberg — remains most widely accepted amongst physicists, some 75 years after its enunciation. According to this interpretation, the probabilistic nature of quantum mechanics is not a *temporary* feature which will eventually be replaced by a deterministic theory, but instead must be considered a *final* renunciation of the classical idea of "causality." It is also believed therein that any well-defined application of the quantum mechanical formalism must always make reference to the experimental arrangement, due to the conjugate nature of evidence obtained under different experimental situations.

Albert Einstein, himself one of the founders of quantum theory, did not accept some of the more philosophical or metaphysical interpretations of quantum mechanics, such as rejection of determinism and of causality. He is famously quoted as saying, in response to this aspect, "God does not play with dice".[69] He rejected the concept that the state of a physical system depends on the experimental arrangement for its measurement. He held that a state of nature occurs in its own right, regardless of whether or how it might be observed. In that view, he is supported by the currently accepted definition of a quantum state, which remains invariant under arbitrary choice of configuration space for its representation, that is to say, manner of observation. He also held that underlying quantum mechanics there should be a theory that thoroughly and directly expresses the rule against action at a distance; in other words, he insisted on the principle of locality. He considered, but rejected on theoretical grounds, a particular proposal for hidden variables to obviate the indeterminism or acausality of quantum mechanical measurement. He considered that quantum mechanics was a currently valid but not a permanently definitive theory for quantum phenomena. He thought its future replacement would require profound conceptual advances, and would not come quickly or easily. The Bohr-Einstein debates provide a vibrant critique of the Copenhagen Interpretation from an epistemological point of view. In arguing for his views, he produced a series of objections, the most famous of which has become known as the Einstein–Podolsky–Rosen paradox.

John Bell showed that this "EPR" paradox led to experimentally testable differences between quantum mechanics and theories that rely on added hidden variables. Experiments have been performed confirming the accuracy of quantum mechanics, thereby demonstrating that quantum mechanics cannot be improved upon by addition of hidden variables.[70] Alain Aspect's initial experiments in 1982, and many subsequent experiments since, have definitively verified quantum entanglement.

Entanglement, as demonstrated in Bell-type experiments, does not, however, violate causality, since no transfer of information happens. Quantum entanglement forms the basis of quantum cryptography, which is proposed for use in high-security commercial applications in banking and government.

The Everett many-worlds interpretation, formulated in 1956, holds that *all* the possibilities described by quantum theory *simultaneously* occur in a multiverse composed of mostly independent parallel universes.[71] This is not accomplished by introducing some "new axiom" to quantum mechanics, but on the contrary, by *removing* the axiom of the collapse of the wave packet. *All* of the possible consistent states of the measured system and the measuring apparatus (including the observer) are present in a *real* physical - not just formally mathematical, as in other interpretations - quantum superposition. Such a superposition of consistent state combinations of different systems is called an entangled state. While the multiverse is deterministic, we perceive non-deterministic behavior governed by probabilities, because we can only observe the universe (i.e., the consistent state contribution to the aforementioned superposition) that we, as observers, inhabit. Everett's interpretation is perfectly consistent with John Bell's experiments and makes them intuitively understandable. However, according to the theory of quantum decoherence, these "parallel universes" will never be accessible to us. The inaccessibility can be understood as follows: once a measurement is done, the measured system becomes entangled with *both* the physicist who measured it *and* a huge number of other particles, some of which are photons flying away at the speed of light towards the other end of the universe. In order to prove that the wave function did not collapse, one would have to bring *all* these particles back and measure them again, together with the system that was originally measured. Not only is this completely impractical, but even if one *could* theoretically do this, it would have to destroy any evidence that the original measurement took place (including the physicist's memory). In light of these Bell tests, Cramer (1986) formulated his transactional interpretation.[72] Relational quantum mechanics appeared in the late 1990s as the modern derivative of the Copenhagen Interpretation.

18.6 Applications

Quantum mechanics has had enormous[73] success in explaining many of the features of our universe. Quantum mechanics is often the only tool available that can reveal the individual behaviors of the subatomic particles that make up all forms of matter (electrons, protons, neutrons, photons, and others). Quantum mechanics has strongly influenced string theories, candidates for a Theory of Everything (see reductionism).

Quantum mechanics is also critically important for understanding how individual atoms combine covalently to form molecules. The application of quantum mechanics to chemistry is known as quantum chemistry. Relativistic quantum mechanics can, in principle, mathematically describe most of chemistry. Quantum mechanics can also provide quantitative insight into ionic and covalent bonding processes by explicitly showing which molecules are energetically favorable to which others and the magnitudes of the energies involved.[74] Furthermore, most of the calculations performed in modern computational chemistry rely on quantum mechanics.

In many aspects modern technology operates at a scale where quantum effects are significant.

18.6.1 Electronics

Many modern electronic devices are designed using quantum mechanics. Examples include the laser, the transistor (and thus the microchip), the electron microscope, and magnetic resonance imaging (MRI). The study of semiconductors led to the invention of the diode and the transistor, which are indispensable parts of modern electronics systems, computer and telecommunication devices. Another application is the light emitting diode which is a high-efficiency source of light.

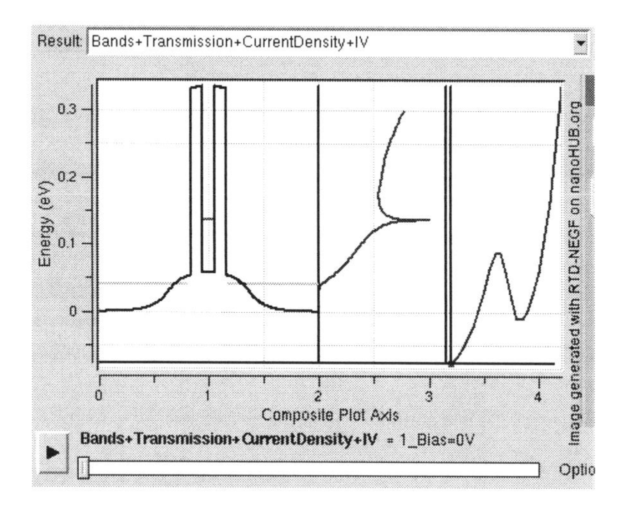

A working mechanism of a resonant tunneling diode device, based on the phenomenon of quantum tunneling through potential barriers.(Left: band diagram; Center: transmission coefficient; Right: current-voltage characteristics) As shown in the band diagram(left), although there are two barriers, electrons still tunnel through via the confined states between two barriers(center), conducting current.

Many electronic devices operate under effect of Quantum tunneling. It even exists in the simple light switch. The switch would not work if electrons could not quantum tunnel through the layer of oxidation on the metal contact surfaces. Flash memory chips found in USB drives use quantum tunneling to erase their memory cells. Some negative differential resistance devices also utilizes quantum tunneling effect, such as resonant tunneling diode .Unlike classical diodes, its current is carried by resonant tunneling through two potential barriers (see right figure). Its negative resistance behavior can only be understood with quantum mechanics: As the confined state moves close to Fermi level, tunnel current increases. As it moves away, current decreases. Quantum mechanics is vital to understanding and designing such electronic devices.

18.6.2 Cryptography

Researchers are currently seeking robust methods of directly manipulating quantum states. Efforts are being made to more fully develop quantum cryptography, which will theoretically allow guaranteed secure transmission of information.

18.6.3 Quantum computing

A more distant goal is the development of quantum computers, which are expected to perform certain computational tasks exponentially faster than classical computers. Instead of using classical bits, quantum computers use qubits, which can be in superpositions of states. Another active research topic is quantum teleportation, which deals with techniques to transmit quantum information over arbitrary distances.

18.6.4 Macroscale quantum effects

While quantum mechanics primarily applies to the smaller atomic regimes of matter and energy, some systems exhibit quantum mechanical effects on a large scale. Superfluidity, the frictionless flow of a liquid at temperatures near absolute zero, is one well-known example. So is the closely related phenomenon of superconductivity, the frictionless flow of an electron gas in a conducting material (an electric current) at sufficiently low temperatures.

18.6.5 Quantum theory

Quantum theory also provides accurate descriptions for many previously unexplained phenomena, such as black-body radiation and the stability of the orbitals of electrons in atoms. It has also given insight into the workings of many different biological systems, including smell receptors and protein structures.[75] Recent work on photosynthesis has provided evidence that quantum correlations play an essential role in this fundamental process of plants and many other organisms.[76] Even so, classical physics can often provide good approximations to results otherwise obtained by quantum physics, typically in circumstances with large numbers of particles or large quantum numbers. Since classical formulas are much simpler and easier to compute than quantum formulas, classical approximations are used and preferred when the system is large enough to render the effects of quantum mechanics insignificant.

18.7 Examples

18.7.1 Free particle

For example, consider a free particle. In quantum mechanics, there is wave–particle duality, so the properties of the particle can be described as the properties of a wave. Therefore, its quantum state can be represented as a wave of arbitrary shape and extending over space as a wave function. The position and momentum of the particle are observables. The Uncertainty Principle states that both the position and the momentum cannot simultaneously be measured with complete precision. However, one *can* measure the position (alone) of a moving free particle, creating an eigenstate of position with a wave function that is very large (a Dirac delta) at a particular position *x*, and zero everywhere else. If one performs a position measurement on such a wave function, the resultant *x* will be obtained with 100% probability (i.e., with full certainty, or complete precision). This is called an eigenstate of position—or, stated in mathematical terms, a *generalized position eigenstate (eigendistribution)*. If the particle is in an eigenstate of position, then its momentum is completely unknown. On the other hand, if the particle is in an eigenstate of momentum, then its position is completely unknown.[77] In an eigenstate of momentum having a plane wave form, it can be shown that the wavelength is equal to *h/p*, where *h* is Planck's constant and *p* is the momentum of the eigenstate.[78]

18.7.2 Step potential

Main article: Solution of Schrödinger equation for a step potential

The potential in this case is given by:

$$V(x) = \begin{cases} 0, & x < 0, \\ V_0, & x \geq 0. \end{cases}$$

The solutions are superpositions of left- and right-moving waves:

$$\psi_1(x) = \frac{1}{\sqrt{k_1}} \left(A_{\rightarrow} e^{ik_1 x} + A_{\leftarrow} e^{-ik_1 x} \right) \quad x < 0$$

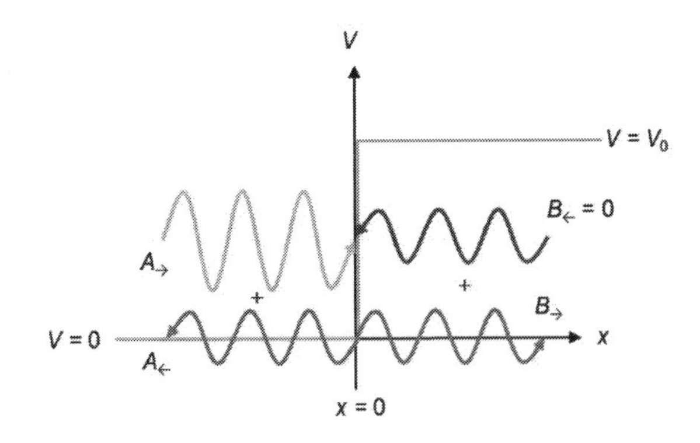

Scattering at a finite potential step of height V_0, *shown in green. The amplitudes and direction of left- and right-moving waves are indicated. Yellow is the incident wave, blue are reflected and transmitted waves, red does not occur.* $E > V_0$ *for this figure.*

$$\psi_2(x) = \frac{1}{\sqrt{k_2}} \left(B_{\rightarrow} e^{ik_2 x} + B_{\leftarrow} e^{-ik_2 x} \right) \quad x > 0$$

where the wave vectors are related to the energy via

$$k_1 = \sqrt{2mE/\hbar^2}$$

$$k_2 = \sqrt{2m(E - V_0)/\hbar^2}$$

with coefficients A and B determined from the boundary conditions and by imposing a continuous derivative on the solution.

Each term of the solution can be interpreted as an incident, reflected, or transmitted component of the wave, allowing the calculation of transmission and reflection coefficients. Notably, in contrast to classical mechanics, incident particles with energies greater than the potential step are partially reflected.

18.7.3 Rectangular potential barrier

Main article: Rectangular potential barrier

This is a model for the quantum tunneling effect which plays an important role in the performance of modern technologies such as flash memory and scanning tunneling microscopy. Quantum tunneling is central to physical phenomena involved in superlattices.

18.7.4 Particle in a box

Main article: Particle in a box

The particle in a one-dimensional potential energy box is the most mathematically simple example where restraints

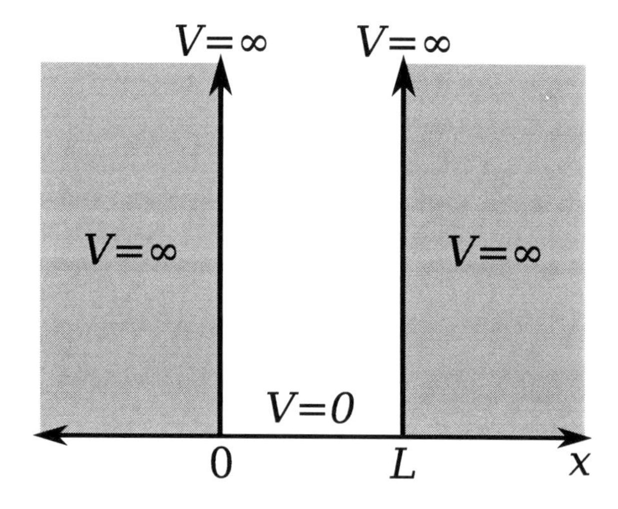

1-dimensional potential energy box (or infinite potential well)

lead to the quantization of energy levels. The box is defined as having zero potential energy everywhere *inside* a certain region, and infinite potential energy everywhere *outside* that region. For the one-dimensional case in the x direction, the time-independent Schrödinger equation may be written[79]

$$-\frac{\hbar^2}{2m}\frac{d^2\psi}{dx^2} = E\psi.$$

With the differential operator defined by

$$\hat{p}_x = -i\hbar\frac{d}{dx}$$

the previous equation is evocative of the classic kinetic energy analogue,

$$\frac{1}{2m}\hat{p}_x^2 = E,$$

with state ψ in this case having energy E coincident with the kinetic energy of the particle.

The general solutions of the Schrödinger equation for the particle in a box are

$$\psi(x) = Ae^{ikx} + Be^{-ikx} \qquad E = \frac{\hbar^2 k^2}{2m}$$

or, from Euler's formula,

$$\psi(x) = C\sin kx + D\cos kx.$$

The infinite potential walls of the box determine the values of C, D, and k at $x = 0$ and $x = L$ where ψ must be zero. Thus, at $x = 0$,

$$\psi(0) = 0 = C\sin 0 + D\cos 0 = D$$

and $D = 0$. At $x = L$,

$$\psi(L) = 0 = C\sin kL.$$

in which C cannot be zero as this would conflict with the Born interpretation. Therefore, since $\sin(kL) = 0$, kL must be an integer multiple of π,

$$k = \frac{n\pi}{L} \qquad n = 1, 2, 3, \ldots.$$

The quantization of energy levels follows from this constraint on k, since

$$E = \frac{\hbar^2\pi^2 n^2}{2mL^2} = \frac{n^2 h^2}{8mL^2}.$$

18.7.5 Finite potential well

Main article: Finite potential well

A finite potential well is the generalization of the infinite potential well problem to potential wells having finite depth.

The finite potential well problem is mathematically more complicated than the infinite particle-in-a-box problem as the wave function is not pinned to zero at the walls of the well. Instead, the wave function must satisfy more complicated mathematical boundary conditions as it is nonzero in regions outside the well.

18.7.6 Harmonic oscillator

Main article: Quantum harmonic oscillator

As in the classical case, the potential for the quantum harmonic oscillator is given by

$$V(x) = \frac{1}{2}m\omega^2 x^2$$

This problem can either be treated by directly solving the Schrödinger equation, which is not trivial, or by using the more elegant "ladder method" first proposed by Paul Dirac. The eigenstates are given by

$$\psi_n(x) = \sqrt{\frac{1}{2^n\, n!}}\cdot\left(\frac{m\omega}{\pi\hbar}\right)^{1/4}\cdot e^{-\frac{m\omega x^2}{2\hbar}}\cdot H_n\left(\sqrt{\frac{m\omega}{\hbar}}x\right),$$

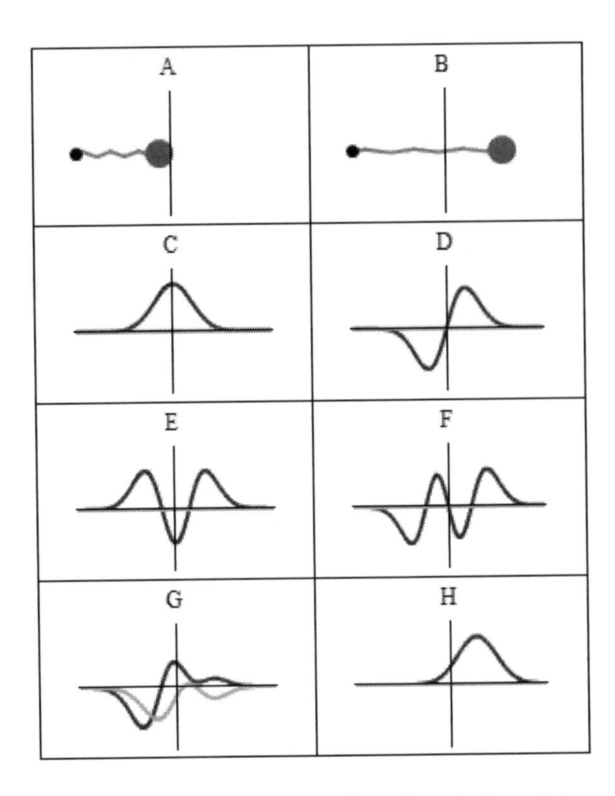

Some trajectories of a harmonic oscillator (i.e. a ball attached to a spring) in classical mechanics (A-B) and quantum mechanics (C-H). In quantum mechanics, the position of the ball is represented by a wave (called the wave function), with the real part shown in blue and the imaginary part shown in red. Some of the trajectories (such as C,D,E,and F) are standing waves (or "stationary states"). Each standing-wave frequency is proportional to a possible energy level of the oscillator. This "energy quantization" does not occur in classical physics, where the oscillator can have any *energy.*

$n = 0, 1, 2, \ldots.$

where *Hn* are the Hermite polynomials,

$$H_n(x) = (-1)^n e^{x^2} \frac{d^n}{dx^n} \left(e^{-x^2}\right)$$

and the corresponding energy levels are

$$E_n = \hbar\omega \left(n + \frac{1}{2}\right)$$

This is another example illustrating the quantization of energy for bound states.

18.8 See also

- Angular momentum diagrams (quantum mechanics)
- EPR paradox
- Fractional quantum mechanics
- List of quantum-mechanical systems with analytical solutions
- Macroscopic quantum phenomena
- Phase space formulation
- Regularization (physics)
- Spherical basis

18.9 Notes

[1] Matson, John. "What Is Quantum Mechanics Good for?". Scientific American. Retrieved 18 May 2016.

[2] Max Born & Emil Wolf, Principles of Optics, 1999, Cambridge University Press

[3] Mehra, J.; Rechenberg, H. (1982). *The historical development of quantum theory*. New York: Springer-Verlag. ISBN 0387906428.

[4] Kragh, Helge (2002). *Quantum Generations: A History of Physics in the Twentieth Century*. Princeton University Press. p. 58. ISBN 0-691-09552-3. Extract of page 58

[5] Ben-Menahem, Ari (2009). *Historical Encyclopedia of Natural and Mathematical Sciences, Volume 1*. Springer. p. 3678. ISBN 3540688315. Extract of page 3678

[6] E Arunan (2010). "Peter Debye" (PDF). *Resonance (journal)* (Indian Academy of Sciences) **15** (12).

[7] Kuhn, T. S. (1978). *Black-body theory and the quantum discontinuity 1894-1912*. Oxford: Clarendon Press. ISBN 0195023838.

[8] Kragh, Helge (1 December 2000), *Max Planck: the reluctant revolutionary*, PhysicsWorld.com

[9] Einstein, A. (1905). "Über einen die Erzeugung und Verwandlung des Lichtes betreffenden heuristischen Gesichtspunkt" [On a heuristic point of view concerning the production and transformation of light]. *Annalen der Physik* **17** (6): 132–148. Bibcode:1905AnP...322..132E. doi:10.1002/andp.19053220607. Reprinted in *The collected papers of Albert Einstein*, John Stachel, editor, Princeton University Press, 1989, Vol. 2, pp. 149-166, in German; see also *Einstein's early work on the quantum hypothesis*, ibid. pp. 134-148.

[10] van Hove, Leon (1958). "Von Neumann's contributions to quantum mechanics" (PDF). *Bulletin of the American Mathematical Society* **64**: Part2:95–99. doi:10.1090/s0002-9904-1958-10206-2.

[11] *The Feynman Lectures on Physics* **III** 21-4 "...it was long believed that the wave function of the Schrödinger equation would never have a macroscopic representation analogous to the macroscopic representation of the amplitude for photons. On the other hand, it is now realized that the phenomena of superconductivity presents us with just this situation. accessdate=2015-11-24

[12] Richard Packard (2006) "Berkeley Experiments on Superfluid Macroscopic Quantum Effects" accessdate=2015-11-24

[13] "Quantum - Definition and More from the Free Merriam-Webster Dictionary". Merriam-webster.com. Retrieved 2012-08-18.

[14] https://web.archive.org/web/20091007133943/http://mooni.fccj.org/%7Eethall/quantum/quant.htm. Archived from the original on October 7, 2009. Retrieved May 23, 2009. Missing or empty |title= (help)

[15] "ysfine.com". *ysfine.com*. Retrieved 11 September 2015.

[16] Oocities.com at the Wayback Machine (archived October 26, 2009)

[17] P.A.M. Dirac, *The Principles of Quantum Mechanics*, Clarendon Press, Oxford, 1930.

[18] D. Hilbert *Lectures on Quantum Theory*, 1915–1927

[19] J. von Neumann, *Mathematische Grundlagen der Quantenmechanik*, Springer, Berlin, 1932 (English translation: *Mathematical Foundations of Quantum Mechanics*, Princeton University Press, 1955).

[20] H.Weyl "The Theory of Groups and Quantum Mechanics", 1931 (original title: "Gruppentheorie und Quantenmechanik").

[21] Dirac, P.A.M. (1958). *The Principles of Quantum Mechanics*, 4th edition, Oxford University Press, Oxford UK, p. ix: "For this reason I have chosen the symbolic method, introducing the representatives later merely as an aid to practical calculation."

[22] Greiner, Walter; Müller, Berndt (1994). *Quantum Mechanics Symmetries, Second edition*. Springer-Verlag. p. 52. ISBN 3-540-58080-8., Chapter 1, p. 52

[23] "Heisenberg - Quantum Mechanics, 1925–1927: The Uncertainty Relations". Aip.org. Retrieved 2012-08-18.

[24] Greenstein, George; Zajonc, Arthur (2006). *The Quantum Challenge: Modern Research on the Foundations of Quantum Mechanics, Second edition*. Jones and Bartlett Publishers, Inc. p. 215. ISBN 0-7637-2470-X., Chapter 8, p. 215

[25] "[Abstract] Visualization of Uncertain Particle Movement". Actapress.com. Retrieved 2012-08-18.

[26] Hirshleifer, Jack (2001). *The Dark Side of the Force: Economic Foundations of Conflict Theory*. Campbridge University Press. p. 265. ISBN 0-521-80412-4., Chapter , p.

[27] "dict.cc dictionary :: eigen :: German-English translation". *dict.cc*. Retrieved 11 September 2015.

[28] "Topics: Wave-Function Collapse". Phy.olemiss.edu. 2012-07-27. Retrieved 2012-08-18.

[29] "Collapse of the wave-function". Farside.ph.utexas.edu. Retrieved 2012-08-18.

[30] "Determinism and Naive Realism : philosophy". Reddit.com. 2009-06-01. Retrieved 2012-08-18.

[31] Michael Trott. "Time-Evolution of a Wavepacket in a Square Well — Wolfram Demonstrations Project". Demonstrations.wolfram.com. Retrieved 2010-10-15.

[32] Michael Trott. "Time Evolution of a Wavepacket In a Square Well". Demonstrations.wolfram.com. Retrieved 2010-10-15.

[33] Mathews, Piravonu Mathews; Venkatesan, K. (1976). *A Textbook of Quantum Mechanics*. Tata McGraw-Hill. p. 36. ISBN 0-07-096510-2., Chapter 2, p. 36

[34] "Wave Functions and the Schrödinger Equation" (PDF). Retrieved 2010-10-15.

[35] (PDF) http://th-www.if.uj.edu.pl/acta/vol19/pdf/v19p0683.pdf. Retrieved June 4, 2009. Missing or empty |title= (help)

[36] Nancy Thorndike Greenspan, "The End of the Certain World: The Life and Science of Max Born" (Basic Books, 2005), pp. 124-8 and 285-6.

[37] http://ocw.usu.edu/physics/classical-mechanics/pdf_lectures/06.pdf

[38] "The Nobel Prize in Physics 1979". Nobel Foundation. Retrieved 2010-02-16.

[39] Carl M. Bender, Daniel W. Hook, Karta Kooner (2009-12-31). "Complex Elliptic Pendulum". arXiv:1001.0131 [hep-th].

[40] See, for example, Precision tests of QED. The relativistic refinement of quantum mechanics known as quantum electrodynamics (QED) has been shown to agree with experiment to within 1 part in 10^8 for some atomic properties.

[41] Tipler, Paul; Llewellyn, Ralph (2008). *Modern Physics* (5 ed.). W. H. Freeman and Company. pp. 160–161. ISBN 978-0-7167-7550-8.

[42] "Quantum mechanics course iwhatisquantummechanics". Scribd.com. 2008-09-14. Retrieved 2012-08-18.

[43] A. Einstein, B. Podolsky, and N. Rosen, *Can quantum-mechanical description of physical reality be considered complete?* Phys. Rev. **47** 777 (1935).

[44] "Between classical and quantum�" (PDF). Retrieved 2012-08-19. replacement character in |title= at position 30 (help)

[45] (see macroscopic quantum phenomena, Bose–Einstein condensate, and Quantum machine)

[46] "Atomic Properties". Academic.brooklyn.cuny.edu. Retrieved 2012-08-18.

[47] http://assets.cambridge.org/97805218/29526/excerpt/9780521829526_excerpt.pdf

[48] Born, M., Heisenberg, W., Jordan, P. (1926). *Z. Phys.* **35**: 557–615. Translated as 'On quantum mechanics II', pp. 321–385 in Van der Waerden, B.L. (1967), *Sources of Quantum Mechanics*, North-Holland, Amsterdam, "The basic difference between the theory proposed here and that used hitherto ... lies in the characteristic kinematics ...", p. 385.

[49] Dirac, P.A.M. (1930/1958). *The Principles of Quantum Mechanics*, fourth edition, Oxford University Press, Oxford UK, p. 5: "A question about what will happen to a particular photon under certain conditions is not really very precise. To make it precise one must imagine some experiment performed having a bearing on the question, and enquire what will be the result of the experiment. Only questions about the results of experiments have a real significance and it is only such questions that theoretical physics has to consider."

[50] Bohr, N. (1939). The Causality Problem in Atomic Physics, in *New Theories in Physics, Conference organized in collaboration with the International Union of Physics and the Polish Intellectual Co-operation Committee, Warsaw, May 30th – June 3rd 1938*, International Institute of Intellectual Co-operation, Paris, 1939, pp. 11–30, reprinted in *Niels Bohr, Collected Works*, volume 7 (1933 – 1958) edited by J. Kalckar, Elsevier, Amsterdam, ISBN 0-444-89892-1, pp. 303–322. "The essential lesson of the analysis of measurements in quantum theory is thus the emphasis on the necessity, in the account of the phenomena, of taking the whole experimental arrangement into consideration, in complete conformity with the fact that all unambiguous interpretation of the quantum mechanical formalism involves the fixation of the external conditions, defining the initial state of the atomic system and the character of the possible predictions as regards subsequent observable properties of that system. Any measurement in quantum theory can in fact only refer either to a fixation of the initial state or to the test of such predictions, and it is first the combination of both kinds which constitutes a well-defined phenomenon."

[51] Bohr, N. (1948). On the notions of complementarity and causality, *Dialectica* **2**: 312–319. "As a more appropriate way of expression, one may advocate limitation of the use of the word *phenomenon* to refer to observations obtained under specified circumstances, including an account of the whole experiment."

[52] Ludwig, G. (1987). *An Axiomatic Basis for Quantum Mechanics*, volume 2, *Quantum Mechanics and Macrosystems*, translated by K. Just, Springer, Berlin, ISBN 978-3-642-71899-1, Chapter XIII, Special Structures in Preparation and Registration Devices, §1, Measurement chains, p. 132.

[53] Heisenberg, W. (1927). Über den anschaulichen Inhalt der quantentheoretischen Kinematik und Mechanik, *Z. Phys.* **43**: 172–198. Translation as 'The actual content of quantum theoretical kinematics and mechanics' here , "But in the rigorous formulation of the law of causality, — "If we know the present precisely, we can calculate the future" — it is not the conclusion that is faulty, but the premise."

[54] Green, H.S. (1965). *Matrix Mechanics*, with a foreword by Max Born, P. Noordhoff Ltd, Groningen. "It is not possible, therefore, to provide 'initial conditions' for the prediction of the behaviour of atomic systems, in the way contemplated by classical physics. This is accepted by quantum theory, not merely as an experimental difficulty, but as a fundamental law of nature", p. 32.

[55] Rosenfeld, L. (1957). Misunderstandings about the foundations of quantum theory, pp. 41–45 in *Observation and Interpretation*, edited by S. Körner, Butterworths, London. "A phenomenon is therefore a process (endowed with the characteristic quantal wholeness) involving a definite type of interaction between the system and the apparatus."

[56] Dirac, P.A.M. (1973). Development of the physicist's conception of nature, pp. 1–55 in *The Physicist's Conception of Nature*, edited by J. Mehra, D. Reidel, Dordrecht, ISBN 90-277-0345-0, p. 5: "That led Heisenberg to his really masterful step forward, resulting in the new quantum mechanics. His idea was to build up a theory entirely in terms of quantities referring to two states."

[57] Born, M. (1927). Physical aspects of quantum mechanics, *Nature* **119**: 354–357, "These probabilities are thus dynamically determined. But what the system actually does is not determined ..."

[58] Messiah, A. (1961). *Quantum Mechanics*, volume 1, translated by G.M. Temmer from the French *Mécanique Quantique*, North-Holland, Amsterdam, p. 157.

[59] Bohr, N. (1928). The Quantum postulate and the recent development of atomic theory, *Nature* **121**: 580–590.

[60] Heisenberg, W. (1930). *The Physical Principles of the Quantum Theory*, translated by C. Eckart and F.C. Hoyt, University of Chicago Press.

[61] Goldstein, H. (1950). *Classical Mechanics*, Addison-Wesley, ISBN 0-201-02510-8.

[62] "There is as yet no logically consistent and complete relativistic quantum field theory.", p. 4. — V. B. Berestetskii, E. M. Lifshitz, L P Pitaevskii (1971). J. B. Sykes, J. S. Bell (translators). *Relativistic Quantum Theory* **4, part I**. *Course of Theoretical Physics (Landau and Lifshitz)* ISBN 0-08-016025-5

[63] "Stephen Hawking; Gödel and the end of physics". *cam.ac.uk*. Retrieved 11 September 2015.

[64] "The Nature of Space and Time". *google.com*. Retrieved 11 September 2015.

[65] Tatsumi Aoyama, Masashi Hayakawa, Toichiro Kinoshita, Makiko Nio (2012). "Tenth-Order QED Contribution to the Electron g-2 and an Improved Value of the Fine Structure Constant". *Physical Review Letters* **109** (11): 111807. arXiv:1205.5368v2. Bibcode:2012PhRvL.109k1807A. doi:10.1103/PhysRevLett.109.111807.

[66] Parker, B. (1993). *Overcoming some of the problems*. pp. 259–279.

[67] The Character of Physical Law (1965) Ch. 6; also quoted in The New Quantum Universe (2003), by Tony Hey and Patrick Walters

[68] Weinberg, S. "Collapse of the State Vector", Phys. Rev. A 85, 062116 (2012).

[69] Harrison, Edward (16 March 2000). *Cosmology: The Science of the Universe*. Cambridge University Press. p. 239. ISBN 978-0-521-66148-5.

[70] "Action at a Distance in Quantum Mechanics (Stanford Encyclopedia of Philosophy)". Plato.stanford.edu. 2007-01-26. Retrieved 2012-08-18.

[71] "Everett's Relative-State Formulation of Quantum Mechanics (Stanford Encyclopedia of Philosophy)". Plato.stanford.edu. Retrieved 2012-08-18.

[72] The Transactional Interpretation of Quantum Mechanics by John Cramer. *Reviews of Modern Physics* 58, 647-688, July (1986)

[73] See, for example, the Feynman Lectures on Physics for some of the technological applications which use quantum mechanics, e.g., transistors (vol **III**, pp. 14–11 ff), integrated circuits, which are follow-on technology in solid-state physics (vol **II**, pp. 8–6), and lasers (vol **III**, pp. 9–13).

[74] *Introduction to Quantum Mechanics with Applications to Chemistry - Linus Pauling, E. Bright Wilson*. 1985-03-01. ISBN 9780486648712. Retrieved 2012-08-18.

[75] Anderson, Mark (2009-01-13). "Is Quantum Mechanics Controlling Your Thoughts? | Subatomic Particles". DISCOVER Magazine. Retrieved 2012-08-18.

[76] "Quantum mechanics boosts photosynthesis". physicsworld.com. Retrieved 2010-10-23.

[77] Davies, P. C. W.; Betts, David S. (1984). *Quantum Mechanics, Second edition*. Chapman and Hall. p. 79. ISBN 0-7487-4446-0., Chapter 6, p. 79

[78] Baofu, Peter (2007-12-31). *The Future of Complexity: Conceiving a Better Way to Understand Order and Chaos*. ISBN 9789812708991. Retrieved 2012-08-18.

[79] Derivation of particle in a box, chemistry.tidalswan.com

18.10 References

The following titles, all by working physicists, attempt to communicate quantum theory to lay people, using a minimum of technical apparatus.

- Chester, Marvin (1987) *Primer of Quantum Mechanics*. John Wiley. ISBN 0-486-42878-8
- Cox, Brian; Forshaw, Jeff (2011). *The Quantum Universe: Everything That Can Happen Does Happen:*. Allen Lane. ISBN 1-84614-432-9.
- Richard Feynman, 1985. *QED: The Strange Theory of Light and Matter*, Princeton University Press. ISBN 0-691-08388-6. Four elementary lectures on quantum electrodynamics and quantum field theory, yet containing many insights for the expert.
- Ghirardi, GianCarlo, 2004. *Sneaking a Look at God's Cards*, Gerald Malsbary, trans. Princeton Univ. Press. The most technical of the works cited here. Passages using algebra, trigonometry, and bra–ket notation can be passed over on a first reading.
- N. David Mermin, 1990, "Spooky actions at a distance: mysteries of the QT" in his *Boojums all the way through*. Cambridge University Press: 110-76.
- Victor Stenger, 2000. *Timeless Reality: Symmetry, Simplicity, and Multiple Universes*. Buffalo NY: Prometheus Books. Chpts. 5-8. Includes cosmological and philosophical considerations.

More technical:

- Bryce DeWitt, R. Neill Graham, eds., 1973. *The Many-Worlds Interpretation of Quantum Mechanics*, Princeton Series in Physics, Princeton University Press. ISBN 0-691-08131-X
- Dirac, P. A. M. (1930). *The Principles of Quantum Mechanics*. ISBN 0-19-852011-5. The beginning chapters make up a very clear and comprehensible introduction.
- Hugh Everett, 1957, "Relative State Formulation of Quantum Mechanics", *Reviews of Modern Physics* 29: 454-62.
- Feynman, Richard P.; Leighton, Robert B.; Sands, Matthew (1965). *The Feynman Lectures on Physics* **1–3**. Addison-Wesley. ISBN 0-7382-0008-5.
- Griffiths, David J. (2004). *Introduction to Quantum Mechanics (2nd ed.)*. Prentice Hall. ISBN 0-13-111892-7. OCLC 40251748. A standard undergraduate text.

- Max Jammer, 1966. *The Conceptual Development of Quantum Mechanics*. McGraw Hill.
- Hagen Kleinert, 2004. *Path Integrals in Quantum Mechanics, Statistics, Polymer Physics, and Financial Markets*, 3rd ed. Singapore: World Scientific. Draft of 4th edition.
- Gunther Ludwig, 1968. *Wave Mechanics*. London: Pergamon Press. ISBN 0-08-203204-1
- George Mackey (2004). *The mathematical foundations of quantum mechanics*. Dover Publications. ISBN 0-486-43517-2.
- Albert Messiah, 1966. *Quantum Mechanics* (Vol. I), English translation from French by G. M. Temmer. North Holland, John Wiley & Sons. Cf. chpt. IV, section III.
- Omnès, Roland (1999). *Understanding Quantum Mechanics*. Princeton University Press. ISBN 0-691-00435-8. OCLC 39849482.
- Scerri, Eric R., 2006. *The Periodic Table: Its Story and Its Significance*. Oxford University Press. Considers the extent to which chemistry and the periodic system have been reduced to quantum mechanics. ISBN 0-19-530573-6
- Transnational College of Lex (1996). *What is Quantum Mechanics? A Physics Adventure*. Language Research Foundation, Boston. ISBN 0-9643504-1-6. OCLC 34661512.
- von Neumann, John (1955). *Mathematical Foundations of Quantum Mechanics*. Princeton University Press. ISBN 0-691-02893-1.
- Hermann Weyl, 1950. *The Theory of Groups and Quantum Mechanics*, Dover Publications.
- D. Greenberger, K. Hentschel, F. Weinert, eds., 2009. *Compendium of quantum physics, Concepts, experiments, history and philosophy*, Springer-Verlag, Berlin, Heidelberg.

18.11 Further reading

- Bernstein, Jeremy (2009). *Quantum Leaps*. Cambridge, Massachusetts: Belknap Press of Harvard University Press. ISBN 978-0-674-03541-6.
- Bohm, David (1989). *Quantum Theory*. Dover Publications. ISBN 0-486-65969-0.
- Eisberg, Robert; Resnick, Robert (1985). *Quantum Physics of Atoms, Molecules, Solids, Nuclei, and Particles (2nd ed.)*. Wiley. ISBN 0-471-87373-X.
- Liboff, Richard L. (2002). *Introductory Quantum Mechanics*. Addison-Wesley. ISBN 0-8053-8714-5.
- Merzbacher, Eugen (1998). *Quantum Mechanics*. Wiley, John & Sons, Inc. ISBN 0-471-88702-1.
- Sakurai, J. J. (1994). *Modern Quantum Mechanics*. Addison Wesley. ISBN 0-201-53929-2.
- Shankar, R. (1994). *Principles of Quantum Mechanics*. Springer. ISBN 0-306-44790-8.
- Stone, A. Douglas (2013). *Einstein and the Quantum*. Princeton University Press. ISBN 978-0-691-13968-5.
- Martinus J. G. Veltman, 2003 *Facts and Mysteries in Elementary Particle Physics*.
- Shushi, Tomer (2014). *The Influence of Particle Interactions on the Existence of Quantum Particles Properties* (PDF). Haifa, Israel: Journal of Physical Science and Application.

18.12 External links

- 3D animations, applications and research for basic quantum effects (animations also available in commons.wikimedia.org (Université paris Sud))
- Quantum Cook Book by R. Shankar, Open Yale PHYS 201 material (4pp)
- The Modern Revolution in Physics - an online textbook.
- J. O'Connor and E. F. Robertson: A history of quantum mechanics.
- Introduction to Quantum Theory at Quantiki.
- Quantum Physics Made Relatively Simple: three video lectures by Hans Bethe
- H is for h-bar.
- Quantum Mechanics Books Collection: Collection of free books

Course material

- Quantum Physics Database - Fundamentals and Historical Background of Quantum Theory.

- Doron Cohen: Lecture notes in Quantum Mechanics (comprehensive, with advanced topics).
- MIT OpenCourseWare: Chemistry.
- MIT OpenCourseWare: Physics. See 8.04
- Stanford Continuing Education PHY 25: Quantum Mechanics by Leonard Susskind, see course description Fall 2007
- 5½ Examples in Quantum Mechanics
- Imperial College Quantum Mechanics Course.
- Spark Notes - Quantum Physics.
- Quantum Physics Online : interactive introduction to quantum mechanics (RS applets).
- Experiments to the foundations of quantum physics with single photons.
- AQME : Advancing Quantum Mechanics for Engineers — by T.Barzso, D.Vasileska and G.Klimeck online learning resource with simulation tools on nanohub
- Quantum Mechanics by Martin Plenio
- Quantum Mechanics by Richard Fitzpatrick
- Online course on *Quantum Transport*

FAQs

- Many-worlds or relative-state interpretation.
- Measurement in Quantum mechanics.

Media

- PHYS 201: Fundamentals of Physics II by Ramamurti Shankar, Open Yale Course
- Lectures on Quantum Mechanics by Leonard Susskind
- Everything you wanted to know about the quantum world — archive of articles from *New Scientist*.
- Quantum Physics Research from *Science Daily*
- Overbye, Dennis (December 27, 2005). "Quantum Trickery: Testing Einstein's Strangest Theory". *The New York Times*. Retrieved April 12, 2010.
- Audio: Astronomy Cast Quantum Mechanics — June 2009. Fraser Cain interviews Pamela L. Gay.

Philosophy

- Jenann Ismael. ""Quantum Mechanics"". *Stanford Encyclopedia of Philosophy*.
- Henry Krips. ""Measurement in Quantum Theory"". *Stanford Encyclopedia of Philosophy*.

Chapter 19

Quantum field theory

"Relativistic quantum field theory" redirects here. For other uses, see Relativity.

In theoretical physics, **quantum field theory** (**QFT**) is a theoretical framework for constructing quantum mechanical models of subatomic particles in particle physics and quasiparticles in condensed matter physics. A QFT treats particles as excited states of an underlying physical field, so these are called field quanta.

In quantum field theory, quantum mechanical interactions between particles are described by interaction terms between the corresponding underlying quantum fields. These interactions are conveniently visualized by Feynman diagrams, that also serve as a formal tool to evaluate various processes.

Historically the development began in the 1920s with the quantization of the electromagnetic field, the quantization being based on an analogy with the eigenmode expansion of a vibrating string with fixed endpoints. In Weinberg (2005), QFT is brought forward as an unavoidable consequence of the reconciliation of quantum mechanics with special relativity.

Max Born (1882–1970), one of the founders of quantum field theory.
He is also known for the Born rule that introduced the probabilistic interpretation in quantum mechanics. He received the 1954 Nobel Prize in Physics together with Walther Bothe.

19.1 History

Main article: History of quantum field theory

19.1.1 Early development

The first achievement of quantum field theory, namely quantum electrodynamics (QED), is "still the paradigmatic example of a successful quantum field theory" according to Weinberg (2005). Ordinary QM cannot give an account of photons, which constitute the prime case of relativistic 'particles'. Since photons have rest mass zero, and correspondingly travel in the vacuum at the speed c, a non-relativistic theory such as ordinary QM cannot give even an approximate description. Photons are implicit in the emission and absorption processes which have to be postulated, for instance, when one of an atom's electrons makes a transition between energy levels. The formalism of QFT is needed for an explicit description of photons. In fact most topics in the early development of quantum theory (the so-called old quantum theory, 1900–25) were related to the interaction of radiation and matter and thus should be treated by quantum field theoretical methods. However, quantum mechanics as formulated by Dirac, Heisenberg, and Schrödinger in 1926–27 started from atomic spectra and did not focus much on problems of radiation.

As soon as the conceptual framework of quantum mechanics was developed, a small group of theoreticians tried to extend quantum methods to electromagnetic fields. A good example is the famous paper by Born, Jordan & Heisenberg (1926). P. Jordan was especially acquainted with the literature on light quanta and made important contributions to QFT. The basic idea was that in QFT the electromagnetic field should be represented by matrices in the same way that position and momentum were represented in QM by matrices in matrix mechanics. The ideas of QM were extended to systems having an infinite number of degrees of freedom.

The inception of QFT is usually considered to be Dirac's famous 1927 paper on "The quantum theory of the emission and absorption of radiation".[1] Here Dirac coined the name "quantum electrodynamics" (QED) for the part of QFT that was developed first. Dirac supplied a systematic procedure for transferring the characteristic quantum phenomenon of discreteness of physical quantities from the quantum-mechanical treatment of particles to a corresponding treatment of fields. Employing the theory of the quantum harmonic oscillator, Dirac gave a theoretical description of how photons appear in the quantization of the electromagnetic radiation field. Later, Dirac's procedure became a model for the quantization of other fields as well. These first approaches to QFT were further developed during the following three years. P. Jordan introduced creation and annihilation operators for fields obeying Fermi–Dirac statistics. These differ from the corresponding operators for Bose–Einstein statistics in that the former satisfy *anti-commutation relations* while the latter satisfy commutation relations.

The methods of QFT could be applied to derive equations resulting from the quantum-mechanical (field-like) treatment of particles, e.g. the Dirac equation, the Klein–Gordon equation and the Maxwell equations. Schweber points out[2] that the idea and procedure of second quantization goes back to Jordan, in a number of papers from 1927,[3] while the expression itself was coined by Dirac. Some difficult problems concerning commutation relations, statistics, and Lorentz invariance were eventually solved. The first comprehensive account of a general theory of quantum fields, in particular, the method of canonical quantization, was presented by Heisenberg & Pauli in 1929. Whereas Jordan's second quantization procedure applied to the coefficients of the normal modes of the field, Heisenberg & Pauli started with the fields themselves and subjected them to the canonical procedure. Heisenberg and Pauli thus established the basic structure of QFT as presented in modern introductions to QFT. Fermi and Dirac, as well as Fock and Podolsky, presented different formulations which played a heuristic role in the following years.

Quantum electrodynamics rests on two pillars, see e.g., the short and lucid "Historical Introduction" of Scharf (2014). The first pillar is the quantization of the electromagnetic field, i.e., it is about photons as the quantized excitations or 'quanta' of the electromagnetic field. This procedure will be described in some more detail in the section on the particle interpretation. As Weinberg points out the "photon is the only particle that was known as a field before it was detected as a particle" so that it is natural that QED began with the analysis of the radiation field.[4] The second pillar of QED consists of the relativistic theory of the electron, centered on the Dirac equation.

19.1.2 The problem of infinities

The emergence of infinities

Pascual Jordan (1902–1980), doctoral student of Max Born, was a pioneer in quantum field theory, coauthoring a number of seminal papers with Born and Heisenberg.
Jordan algebras were introduced by him to formalize the notion of an algebra of observables in quantum mechanics. He was awarded the Max Planck medal 1954.

Quantum field theory started with a theoretical framework that was built in analogy to quantum mechanics. Although there was no unique and fully developed theory, quantum field theoretical tools could be applied to concrete processes. Examples are the scattering of radiation by free electrons, Compton scattering, the collision between relativistic electrons or the production of electron-positron pairs by photons. Calculations to the first order of approximation were quite successful, but most people working in the field thought that QFT still had to undergo a major change. On the one side, some calculations of effects for cosmic rays clearly differed from measurements. On the other side and, from a theoretical point of view more threatening, calculations of higher orders of the perturbation series led to infinite results. The self-energy of the electron as well as vacuum fluctuations of the electromagnetic field seemed to be infinite. The perturbation expansions did not converge to a finite sum and even most individual terms were divergent.

The various forms of infinities suggested that the divergences were more than failures of specific calculations. Many physicists tried to avoid the divergences by formal tricks (truncating the integrals at some value of momentum, or even ignoring infinite terms) but such rules were not reliable, violated the requirements of relativity and were not considered as satisfactory. Others came up with first ideas of coping with infinities by a redefinition of the parameters of the theory and using a measured finite value, for example of the charge of the electron, instead of the infinite 'bare' value. This process is called renormalization.

From the point of view of the philosophy of science, it is remarkable that these divergences did not give enough reason to discard the theory. The years from 1930 to the beginning of World War II were characterized by a variety of attitudes towards QFT. Some physicists tried to circumvent the infinities by more-or-less arbitrary prescriptions, others worked on transformations and improvements of the theoretical framework. Most of the theoreticians believed that QED would break down at high energies. There was also a considerable number of proposals in favor of alternative approaches. These proposals included changes in the basic concepts e.g. negative probabilities and interactions at a distance instead of a field theoretical approach, and a methodological change to phenomenological methods that focusses on relations between observable quantities without an analysis of the microphysical details of the interaction, the so-called S-matrix theory where the basic elements are amplitudes for various scattering processes.

Despite the feeling that QFT was imperfect and lacking rigor, its methods were extended to new areas of applications. In 1933 Fermi's theory of the beta decay started with conceptions describing the emission and absorption of photons, transferred them to beta radiation and analyzed the creation and annihilation of electrons and neutrinos described by the weak interaction. Further applications of QFT outside of quantum electrodynamics succeeded in nuclear physics with the strong interaction. In 1934 Pauli & Weisskopf showed that a new type of fields (scalar fields), described by the Klein–Gordon equation, could be quantized. This is another example of second quantization. This new theory for matter fields could be applied a decade later when new particles, pions, were detected.

The taming of infinities

Werner Heisenberg (1901–1976), doctoral student of Arnold Sommerfeld, was one of the founding fathers of quantum mechanics. In particular, he introduced the version of quantum mechanics known as matrix mechanics, but is now more known for the Heisenberg uncertainty relations. He was awarded the Nobel prize in physics 1932 together with Erwin Schrödinger and Paul Dirac.

After the end of World War II more reliable and effective methods for dealing with infinities in QFT were developed, namely coherent and systematic rules for performing relativistic field theoretical calculations, and a general renormalization theory. On three famous conferences, the Shelter Island Conference 1947, the Pocono Conference 1948, and the 1949 Oldstone Conference, developments in theoretical physics were confronted with relevant new experimental results. In the late forties, there were two different ways to address the problem of divergences. One of these was discovered by Richard Feynman, the other one (based on an operator formalism) by Julian Schwinger and independently by Sin-Itiro Tomonaga. In 1949 Freeman Dyson showed that the two approaches are in fact equivalent. Thus, Freeman Dyson, Feynman, Schwinger, and Tomonaga became the inventors of renormalization theory. The most spectacular experimental successes of renormalization theory were the calculations of the anomalous magnetic moment of electron and the Lamb shift in the spectrum of hydrogen. These successes were so outstanding because the theoretical results were in better agreement with high precision experiments than anything in physics before. Nevertheless, mathematical problems lingered on and prompted a search for rigorous formulations (to be discussed in the main article).

Richard Feynman (1918–1988)
His 1945 PhD thesis developed the path integral formulation of ordinary quantum mechanics. This was later generalized to field theory.

The basic idea of renormalization is to avoid divergences that appear in physical predictions by shifting them into a part of the theory where they do not influence empirical propositions. Dyson could show that a rescaling of charge and mass ('renormalization') is sufficient to remove all divergences in QED to all orders of perturbation theory. In general, a QFT is called renormalizable, if all infinities can be absorbed into a redefinition of a finite number of coupling constants and masses. A consequence is that the physical charge and mass of the electron must be measured and cannot be computed from first principles. Perturbation theory gives well-defined predictions only in renormalizable quantum field theories, and luckily QED, the first fully developed QFT, belonged to this class of renormalizable theories. There are various technical procedures to renormalize a theory. One way is to cut off the integrals in the calculations at a certain value Λ of the momentum which is large but finite. This cut-off procedure is successful if, after taking the limit $\Lambda \to \infty$, the resulting quantities are independent of Λ. Part II of Peskin & Schroeder (1995) gives an extensive description of renormalization.

Feynman's formulation of QED is of special interest from a philosophical point of view. His so-called space-time approach is visualized by the famous Feynman diagrams that look like depicting paths of particles. Feynman's method of calculating scattering amplitudes is based on the functional integral formulation of field theory.[5] A set of graphical rules can be derived so that the probability of a specific scattering process can be calculated by drawing a diagram of that process and then using the diagram to write down the mathematical expressions for calculating its amplitude. The diagrams provide an effective way to organize and visualize the various terms in the perturbation series, and they seem to display the flow of electrons and photons during the scattering process. External lines in the diagrams represent incoming and outgoing particles, internal lines are connected with virtual particles and vertices with interactions. Each of these graphical elements is associated with mathematical expressions that contribute to the amplitude of the respective process. The diagrams are part of Feynman's very efficient and elegant algorithm for computing the probability of scattering processes. The idea of particles traveling from one point to another was heuristically useful in constructing the theory. This heuristics, based on Huygen's principle, is useful for concrete calculations and actually give the correct particle propagators as derived more rigorously.[6] Nevertheless, an analysis of the theoretical justification of the space-time approach shows that its success does not imply that particle paths have to be taken seriously. General arguments against a particle interpretation of QFT clearly exclude that the diagrams represent paths of particles in the interaction area. Feynman himself was not particularly interested in ontological questions.

19.1.3 Gauge theory and the standard model

In the beginning of the 1950s, QED had become a reliable theory which no longer counted as preliminary. It took two decades from writing down the first equations until QFT could be applied to interesting physical problems in a systematic way. The new developments made it possible to apply QFT to new particles and new interactions. In the following decades QFT was extended to describe not only the electromagnetic force but also weak and strong interaction so that new Lagrangians had to be found which contain new classes of 'particles' or quantum fields. The research aimed at a more comprehensive theory of matter and in the end at a *unified theory of all interactions*.

New theoretical concepts had to be introduced, mainly connected with non-Abelian gauge theories (the effort of developing such theories started in 1954 with the work of Yang and Mills) and spontaneous symmetry breaking. Today there are trustworthy theories of the strong, weak, and electromagnetic interactions of elementary particles which have a similar structure as QED. A combined theory associated with the gauge group SU(3) × SU(2) × U(1) is considered as the *standard model of elementary particle physics* which was achieved by Sheldon Glashow, Steven Weinberg and Abdul Salam in 1968, and Frank Wilczek, David Gross and David Politzer in 1973.

According to the standard model there are, on the one side, six types of leptons (e.g. the electron and its neutrino) and six types of quarks, where the members of both groups are all fermions with spin 1/2. On the other side, there are spin 1 particles (thus bosons) that mediate the interaction between elementary particles and the fundamental forces, namely the photon for electromagnetic interaction, two W and one Z-boson for weak interaction, and the gluon for strong interaction. Altogether there is good agreement with experimental data, for example, the masses of W+ and W− bosons (detected in 1983) confirmed the theoretical prediction within one percent deviation.

19.1.4 Common trends in particle, condensed matter and statistical physics

Renormalization group theory

Main article: History of renormalization group theory

Parallel developments in the understanding of phase transitions in condensed matter physics led to the study of the renormalization group. This involved the work of Leo Kadanoff (1966) and Michael Fisher (1973), which led to the seminal reformulation of quantum field theory by Kenneth G. Wilson in 1975.

Conformal field theory

During the same period, Kadanoff (1969) introduced an operator algebra formalism for the two-dimensional Ising model, a widely studied mathematical model of ferromagnetism in statistical physics. This development suggested that quantum field theory describes its scaling limit. Later, there developed the idea that a finite number of generating operators could represent all the correlation functions of the Ising model. In the 1980s, the existence of a much stronger symmetry for the scaling limit of two-dimensional critical systems was suggested by Alexander Belavin, Alexander Polyakov and Alexander Zamolodchikov in 1984, which eventually led to the development of conformal field theory,[7] a special case of quantum field theory, which is presently employed successfully in different areas of particle physics and condensed matter physics.

19.1.5 Historiography

The first chapter in Weinberg (1995) is a very good short description of the earlier history of QFT. Detailed accounts of the historical development of QFT can be found, e.g., in Darrigol 1986, Schweber (1994) and Cao 1997a. Various historical and conceptual studies of the standard model are gathered in Hoddeson et al. 1997 and of renormalization theory in Brown 1993.

19.2 Definition

Quantum electrodynamics (QED) has one electron field and one photon field; quantum chromodynamics (QCD) has one field for each type of quark; and, in condensed matter, there is an atomic displacement field that gives rise to phonon particles.

19.2.1 Dynamics

See also: Relativistic dynamics

Ordinary quantum mechanical systems have a fixed number of particles, with each particle having a finite number of degrees of freedom. In contrast, the excited states of a quantum field can represent any number of particles. This makes quantum field theories especially useful for describing systems where the particle count/number may change over time, a crucial feature of relativistic dynamics.

19.2.2 States

QFT interaction terms are similar in spirit to those between charges with electric and magnetic fields in Maxwell's equations. However, unlike the classical fields of Maxwell's theory, fields in QFT generally exist in quantum superpositions of states and are subject to the laws of quantum mechanics.

Because the fields are continuous quantities over space, there exist excited states with arbitrarily large numbers of particles in them, providing QFT systems with effectively an infinite number of degrees of freedom. Infinite degrees of freedom can easily lead to divergences of calculated quantities (e.g., the quantities become infinite). Techniques such as renormalization of QFT parameters or discretization of spacetime, as in lattice QCD, are often used to avoid such infinities so as to yield physically plausible results.

19.2.3 Fields and radiation

The gravitational field and the electromagnetic field are the only two fundamental fields in nature that have infinite range and a corresponding classical low-energy limit, which greatly diminishes and hides their "particle-like" excitations. Albert Einstein in 1905, attributed "particle-like" and discrete exchanges of momenta and energy, characteristic of "field quanta", to the electromagnetic field. Originally, his principal motivation was to explain the thermodynamics of radiation. Although the photoelectric effect and Compton scattering strongly suggest the existence of the photon, it might alternatively be explained by a mere quantization of emission; more definitive evidence of the quantum nature of radiation is now taken up into modern quantum optics as in the antibunching effect.[8]

19.3 Varieties of approaches

There is currently no complete quantum theory of the remaining fundamental force, gravity. Many of the proposed theories to describe gravity as a QFT postulate the existence of a graviton particle that mediates the gravitational force. Presumably, the as yet unknown correct quantum field-theoretic treatment of the gravitational field will behave like Einstein's general theory of relativity in the low-energy limit. Quantum field theory of the fundamental forces itself has been postulated to be the low-energy effective field theory limit of a more fundamental theory such as superstring theory.

Most theories in standard particle physics are formulated as *relativistic quantum field theories*, such as QED, QCD, and the Standard Model. QED, the quantum field-theoretic description of the electromagnetic field, approximately reproduces Maxwell's theory of electrodynamics in the low-energy limit, with small non-linear corrections to the Maxwell equations required due to virtual electron–positron pairs.

In the perturbative approach to quantum field theory, the full field interaction terms are approximated as a perturbative expansion in the number of particles involved. Each term in the expansion can be thought of as forces between particles being mediated by other particles. In QED, the electromagnetic force between two electrons is caused by an exchange of photons. Similarly, intermediate vector bosons mediate the weak force and gluons mediate the strong force in QCD. The notion of a force-mediating particle comes from perturbation theory, and does not make sense in the context of non-perturbative approaches to QFT, such as with bound states.

19.4 Principles

19.4.1 Classical and quantum fields

Main article: Classical field theory

A classical field is a function defined over some region of space and time.[9] Two physical phenomena which are described by classical fields are Newtonian gravitation, described by Newtonian gravitational field $\mathbf{g}(\mathbf{x}, t)$, and classical electromagnetism, described by the electric and magnetic fields $\mathbf{E}(\mathbf{x}, t)$ and $\mathbf{B}(\mathbf{x}, t)$. Because such fields can in principle take on distinct values at each point in space, they are said to have infinite degrees of freedom.[9]

Classical field theory does not, however, account for the quantum-mechanical aspects of such physical phenomena. For instance, it is known from quantum mechanics that certain aspects of electromagnetism involve discrete particles—photons—rather than continuous fields. The business of *quantum* field theory is to write down a field that is, like a classical field, a function defined over space and time, but which also accommodates the observations of quantum mechanics. This is a *quantum field*.

It is not immediately clear *how* to write down such a quantum field, since quantum mechanics has a structure very unlike a field theory. In its most general formulation, quantum mechanics is a theory of abstract operators (observables) acting on an abstract state space (Hilbert space), where the observables represent physically observable quantities and the state space represents the possible states of the system under study.[10] For instance, the fundamental observables associated with the motion of a single quantum mechanical particle are the position and momentum operators $\hat{x}$ and $\hat{p}$. Field theory, in contrast, treats x as a way to index the field rather than as an operator.[11]

There are two common ways of developing a quantum field: the path integral formalism and canonical quantization.[12] The latter of these is pursued in this article.

Lagrangian formalism

Quantum field theory frequently makes use of the Lagrangian formalism from classical field theory. This formalism is analogous to the Lagrangian formalism used in classical mechanics to solve for the motion of a particle under the influence of a field. In classical field theory, one writes down a Lagrangian density, $\mathcal{L}$, involving a field, $\varphi(\mathbf{x},t)$, and possibly its first derivatives ($\partial\varphi/\partial t$ and $\nabla\varphi$), and then applies a field-theoretic form of the Euler–Lagrange equation. Writing coordinates $(t, \mathbf{x}) = (x^0, x^1, x^2, x^3) = x^\mu$, this form of the Euler–Lagrange equation is[9]

$$\frac{\partial}{\partial x^\mu}\left[\frac{\partial\mathcal{L}}{\partial(\partial\varphi/\partial x^\mu)}\right]-\frac{\partial\mathcal{L}}{\partial\varphi}=0,$$

where a sum over μ is performed according to the rules of Einstein notation.

By solving this equation, one arrives at the "equations of motion" of the field.[9] For example, if one begins with the Lagrangian density

$$\mathcal{L}(\varphi,\nabla\varphi)=-\rho(t,\mathbf{x})\,\varphi(t,\mathbf{x})-\frac{1}{8\pi G}|\nabla\varphi|^2,$$

and then applies the Euler–Lagrange equation, one obtains the equation of motion

$$4\pi G\rho(t,\mathbf{x})=\nabla^2\varphi.$$

This equation is Newton's law of universal gravitation, expressed in differential form in terms of the gravitational potential $\varphi(t, \mathbf{x})$ and the mass density $\rho(t, \mathbf{x})$. Despite the nomenclature, the "field" under study is the gravitational potential, φ, rather than the gravitational field, **g**. Similarly, when classical field theory is used to study electromagnetism, the "field" of interest is the electromagnetic four-potential (V/c, **A**), rather than the electric and magnetic fields **E** and **B**.

Quantum field theory uses this same Lagrangian procedure to determine the equations of motion for quantum fields. These equations of motion are then supplemented by commutation relations derived from the canonical quantization procedure described below, thereby incorporating quantum mechanical effects into the behavior of the field.

19.4.2 Single- and many-particle quantum mechanics

Main articles: Quantum mechanics and First quantization

In non-relativistic quantum mechanics, a particle (such as an electron or proton) is described by a complex wavefunction, $\psi(x, t)$, whose time-evolution is governed by the Schrödinger equation:

$$-\frac{\hbar^2}{2m}\frac{\partial^2}{\partial x^2}\psi(x,t)+V(x)\psi(x,t)=i\hbar\frac{\partial}{\partial t}\psi(x,t).$$

Here m is the particle's mass and $V(x)$ is the applied potential. Physical information about the behavior of the particle is extracted from the wavefunction by constructing expected values for various quantities; for example, the expected value of the particle's position is given by integrating $\psi^*(x)\, x\, \psi(x)$ over all space, and the expected value of the particle's momentum is found by integrating $-i\hbar\psi^*(x)d\psi/dx$. The quantity $\psi^*(x)\psi(x)$ is itself in the Copenhagen interpretation of quantum mechanics interpreted as a probability density function. This treatment of quantum mechanics, where a particle's wavefunction evolves against a classical background potential $V(x)$, is sometimes called *first quantization*.

This description of quantum mechanics can be extended to describe the behavior of multiple particles, so long as the number and the type of particles remain fixed. The particles are described by a wavefunction $\psi(x_1, x_2, \ldots, xN, t)$, which is governed by an extended version of the Schrödinger equation.

Often one is interested in the case where N particles are all of the same type (for example, the 18 electrons orbiting a neutral argon nucleus). As described in the article on identical particles, this implies that the state of the entire system must be either symmetric (bosons) or antisymmetric (fermions) when the coordinates of its constituent particles are exchanged. This is achieved by using a Slater determinant as the wavefunction of a fermionic system (and a Slater permanent for a bosonic system), which is equivalent to an element of the symmetric or antisymmetric subspace of a tensor product.

For example, the general quantum state of a system of N bosons is written as

$$|\phi_1\cdots\phi_N\rangle=\sqrt{\frac{\prod_j N_j!}{N!}}\sum_{p\in S_N}|\phi_{p(1)}\rangle\otimes\cdots\otimes|\phi_{p(N)}\rangle,$$

where $|\phi_i\rangle$ are the single-particle states, Nj is the number of particles occupying state j, and the sum is taken over all possible permutations p acting on N elements. In general, this is a sum of $N!$ (N factorial) distinct terms. $\sqrt{\frac{\prod_j N_j!}{N!}}$ is a normalizing factor.

There are several shortcomings to the above description of quantum mechanics, which are addressed by quantum field theory. First, it is unclear how to extend quantum mechanics to include the effects of special relativity.[13] Attempted replacements for the Schrödinger equation, such as the Klein–Gordon equation or the Dirac equation, have many unsatisfactory qualities; for instance, they possess energy eigenvalues that extend to $-\infty$, so that there seems to be no easy definition of a ground state. It turns out that such inconsistencies arise from relativistic wavefunctions not having a well-defined probabilistic interpretation in position space, as probability conservation is not a relativistically covariant concept. The second shortcoming, related to the first, is that in quantum mechanics there is no mechanism to describe particle creation and annihilation;[14] this is crucial for describing phenomena such as pair production, which result from the conversion between mass and energy according to the relativistic relation $E = mc^2$.

19.4.3 Second quantization

Main article: Second quantization

In this section, we will describe a method for constructing a quantum field theory called **second quantization**. This basically involves choosing a way to index the quantum mechanical degrees of freedom in the space of multiple identical-particle states. It is based on the Hamiltonian formulation of quantum mechanics.

Several other approaches exist, such as the Feynman path integral,[15] which uses a Lagrangian formulation. For an overview of some of these approaches, see the article on quantization.

Bosons

For simplicity, we will first discuss second quantization for bosons, which form perfectly symmetric quantum states. Let us denote the mutually orthogonal single-particle states which are possible in the system by $|\phi_1\rangle, |\phi_2\rangle, |\phi_3\rangle$, and so on. For example, the 3-particle state with one particle in state $|\phi_1\rangle$ and two in state $|\phi_2\rangle$ is

$$\frac{1}{\sqrt{3}}\left[|\phi_1\rangle|\phi_2\rangle|\phi_2\rangle + |\phi_2\rangle|\phi_1\rangle|\phi_2\rangle + |\phi_2\rangle|\phi_2\rangle|\phi_1\rangle\right].$$

The first step in second quantization is to express such quantum states in terms of **occupation numbers**, by listing the number of particles occupying each of the single-particle states $|\phi_1\rangle, |\phi_2\rangle$, etc. This is simply another way of labelling the states. For instance, the above 3-particle state is denoted as

$$|1, 2, 0, 0, 0, \dots\rangle.$$

An N-particle state belongs to a space of states describing systems of N particles. The next step is to combine the individual N-particle state spaces into an extended state space, known as Fock space, which can describe systems of any number of particles. This is composed of the state space of a system with no particles (the so-called vacuum state, written as $|0\rangle$), plus the state space of a 1-particle system, plus the state space of a 2-particle system, and so forth. States describing a definite number of particles are known as Fock states: a general element of Fock space will be a linear combination of Fock states. There is a one-to-one correspondence between the occupation number representation and valid boson states in the Fock space.

At this point, the quantum mechanical system has become a quantum field in the sense we described above. The field's elementary degrees of freedom are the occupation numbers, and each occupation number is indexed by a number j indicating which of the single-particle states $|\phi_1\rangle, |\phi_2\rangle, \dots, |\phi_j\rangle, \dots$ it refers to:

$$|N_1, N_2, N_3, \dots, N_j, \dots\rangle.$$

The properties of this quantum field can be explored by defining creation and annihilation operators, which add and subtract particles. They are analogous to ladder operators in the quantum harmonic oscillator problem, which added and subtracted energy quanta. However, these operators literally create and annihilate particles of a given quantum state. The bosonic annihilation operator a_2 and creation operator $a_2^\dagger$ are easily defined in the occupation number representation as having the following effects:

$$a_2|N_1, N_2, N_3, \dots\rangle = \sqrt{N_2}\mid N_1, (N_2 - 1), N_3, \dots\rangle,$$

$$a_2^\dagger|N_1, N_2, N_3, \dots\rangle = \sqrt{N_2 + 1}\mid N_1, (N_2 + 1), N_3, \dots\rangle.$$

It can be shown that these are operators in the usual quantum mechanical sense, i.e. linear operators acting on the Fock space. Furthermore, they are indeed Hermitian conjugates, which justifies the way we have written them. They can be shown to obey the commutation relation

$$[a_i, a_j] = 0 \quad , \quad \left[a_i^\dagger, a_j^\dagger\right] = 0 \quad , \quad \left[a_i, a_j^\dagger\right] = \delta_{ij},$$

where δ stands for the Kronecker delta. These are precisely the relations obeyed by the ladder operators for an infinite set of independent quantum harmonic oscillators, one for each single-particle state. Adding or removing bosons from each state is, therefore, analogous to exciting or de-exciting a quantum of energy in a harmonic oscillator.

Applying an annihilation operator a_k followed by its corresponding creation operator $a_k^\dagger$ returns the number N_k of particles in the kth single-particle eigenstate:

$$a_k^\dagger a_k|\ldots, N_k, \ldots\rangle = N_k|\ldots, N_k, \ldots\rangle.$$

The combination of operators $a_k^\dagger a_k$ is known as the number operator for the kth eigenstate.

The Hamiltonian operator of the quantum field (which, through the Schrödinger equation, determines its dynamics) can be written in terms of creation and annihilation operators. For instance, for a field of free (non-interacting) bosons, the total energy of the field is found by summing the energies of the bosons in each energy eigenstate. If the kth single-particle energy eigenstate has energy E_k and there are N_k bosons in this state, then the total energy of these bosons is $E_k N_k$. The energy in the *entire* field is then a sum over k :

$$E_{\text{tot}} = \sum_k E_k N_k$$

This can be turned into the Hamiltonian operator of the field by replacing N_k with the corresponding number operator, $a_k^\dagger a_k$. This yields

$$H = \sum_k E_k\, a_k^\dagger\, a_k.$$

Fermions

It turns out that a different definition of creation and annihilation must be used for describing fermions. According to the Pauli exclusion principle, fermions cannot share quantum states, so their occupation numbers *Ni* can only take on the value 0 or 1. The fermionic annihilation operators *c* and creation operators $c^\dagger$ are defined by their actions on a Fock state thus

$$c_j|N_1, N_2, \ldots, N_j = 0, \ldots\rangle = 0$$

$$c_j|N_1, N_2, \ldots, N_j = 1, \ldots\rangle = (-1)^{(N_1+\cdots+N_{j-1})}|N_1, N_2, \ldots, N_j = 0, \ldots\rangle$$

$$c_j^\dagger|N_1, N_2, \ldots, N_j = 0, \ldots\rangle = (-1)^{(N_1+\cdots+N_{j-1})}|N_1, N_2, \ldots, N_j = 1, \ldots\rangle$$

$$c_j^\dagger|N_1, N_2, \ldots, N_j = 1, \ldots\rangle = 0.$$

These obey an anticommutation relation:

$$\{c_i, c_j\} = 0 \quad , \quad \left\{c_i^\dagger, c_j^\dagger\right\} = 0 \quad , \quad \left\{c_i, c_j^\dagger\right\} = \delta_{ij}.$$

One may notice from this that applying a fermionic creation operator twice gives zero, so it is impossible for the particles to share single-particle states, in accordance with the exclusion principle.

Field operators

We have previously mentioned that there can be more than one way of indexing the degrees of freedom in a quantum field. Second quantization indexes the field by enumerating the single-particle quantum states. However, as we have discussed, it is more natural to think about a "field", such as the electromagnetic field, as a set of degrees of freedom indexed by position.

To this end, we can define *field operators* that create or destroy a particle at a particular point in space. In particle physics, these operators turn out to be more convenient to work with, because they make it easier to formulate theories that satisfy the demands of relativity.

Single-particle states are usually enumerated in terms of their momenta (as in the particle in a box problem.) We can construct field operators by applying the Fourier transform to the creation and annihilation operators for these states. For example, the bosonic field annihilation operator $\phi(\mathbf{r})$ is

$$\phi(\mathbf{r}) \stackrel{\text{def}}{=} \sum_j e^{i\mathbf{k}_j\cdot\mathbf{r}} a_j.$$

The bosonic field operators obey the commutation relation

$$[\phi(\mathbf{r}), \phi(\mathbf{r}')] = 0 \quad ,$$

$$\left[\phi^\dagger(\mathbf{r}), \phi^\dagger(\mathbf{r}')\right] = 0 \quad , \quad \left[\phi(\mathbf{r}), \phi^\dagger(\mathbf{r}')\right] = \delta^3(\mathbf{r}-\mathbf{r}')$$

where $\delta(x)$ stands for the Dirac delta function. As before, the fermionic relations are the same, with the commutators replaced by anticommutators.

The field operator is not the same thing as a single-particle wavefunction. The former is an operator acting on the Fock space, and the latter is a quantum-mechanical amplitude for finding a particle in some position. However, they are closely related and are indeed commonly denoted with the same symbol. If we have a Hamiltonian with a space representation, say

$$H = -\frac{\hbar^2}{2m}\sum_i \nabla_i^2 + \sum_{i<j} U(|\mathbf{r}_i - \mathbf{r}_j|)$$

where the indices i and j run over all particles, then the field theory Hamiltonian (in the non-relativistic limit and for negligible self-interactions) is

$$H = -\frac{\hbar^2}{2m}\int d^3r\,\phi^\dagger(\mathbf{r})\nabla^2\phi(\mathbf{r})+\frac{1}{2}\int d^3r\int d^3r'\,\phi^\dagger$$

$$(\mathbf{r})\phi^\dagger(\mathbf{r}')U(|\mathbf{r}-\mathbf{r}'|)\phi(\mathbf{r}')\phi(\mathbf{r}).$$

This looks remarkably like an expression for the expectation value of the energy, with ϕ playing the role of the wavefunction. This relationship between the field operators and wave functions makes it very easy to formulate field theories starting from space projected Hamiltonians.

19.4.4 Dynamics

Once the Hamiltonian operator is obtained as part of the canonical quantization process, the time dependence of the state is described with the Schrödinger equation, just as with other quantum theories. Alternatively, the Heisenberg picture can be used where the time dependence is in the operators rather than in the states.

19.4.5 Implications

Unification of fields and particles

The "second quantization" procedure that we have outlined in the previous section takes a set of single-particle quantum states as a starting point. Sometimes, it is impossible to define such single-particle states, and one must proceed directly to quantum field theory. For example, a quantum theory of the electromagnetic field *must* be a quantum field theory, because it is impossible (for various reasons) to define a wavefunction for a single photon.[16] In such situations, the quantum field theory can be constructed by examining the mechanical properties of the classical field and guessing the corresponding quantum theory. For free (non-interacting) quantum fields, the quantum field theories obtained in this way have the same properties as those obtained using second quantization, such as well-defined creation and annihilation operators obeying commutation or anticommutation relations.

Quantum field theory thus provides a unified framework for describing "field-like" objects (such as the electromagnetic field, whose excitations are photons) and "particle-like" objects (such as electrons, which are treated as excitations of an underlying electron field), so long as one can treat interactions as "perturbations" of free fields. There are still unsolved problems relating to the more general case of interacting fields that may or may not be adequately described by perturbation theory. For more on this topic, see Haag's theorem.

Physical meaning of particle indistinguishability

The second quantization procedure relies crucially on the particles being identical. We would not have been able to construct a quantum field theory from a distinguishable many-particle system, because there would have been no way of separating and indexing the degrees of freedom.

Many physicists prefer to take the converse interpretation, which is that *quantum field theory explains what identical particles are*. In ordinary quantum mechanics, there is not much theoretical motivation for using symmetric (bosonic) or antisymmetric (fermionic) states, and the need for such states is simply regarded as an empirical fact. From the point of view of quantum field theory, particles are identical if and only if they are excitations of the same underlying quantum field. Thus, the question "why are all electrons identical?" arises from mistakenly regarding individual electrons as fundamental objects, when in fact it is only the electron field that is fundamental.

Particle conservation and non-conservation

During second quantization, we started with a Hamiltonian and state space describing a fixed number of particles (N), and ended with a Hamiltonian and state space for an arbitrary number of particles. Of course, in many common situations N is an important and perfectly well-defined quantity, e.g. if we are describing a gas of atoms sealed in a box. From the point of view of quantum field theory, such situations are described by quantum states that are eigenstates of the number operator $\hat{N}$, which measures the total number of particles present. As with any quantum mechanical observable, $\hat{N}$ is conserved if it commutes with the Hamiltonian. In that case, the quantum state is trapped in the N-particle subspace of the total Fock space, and the situation could equally well be described by ordinary N-particle quantum mechanics. (Strictly speaking, this is only true in the noninteracting case or in the low energy density limit of renormalized quantum field theories)

For example, we can see that the free boson Hamiltonian described above conserves particle number. Whenever the Hamiltonian operates on a state, each particle destroyed by an annihilation operator a_k is immediately put back by the creation operator $a_k^\dagger$.

On the other hand, it is possible, and indeed common, to encounter quantum states that are *not* eigenstates of $\hat{N}$, which do not have well-defined particle numbers. Such states are difficult or impossible to handle using ordinary quantum mechanics, but they can be easily described in quantum field theory as quantum superpositions of states having different values of N. For example, suppose we have a bosonic field whose particles can be created or destroyed by interactions with a fermionic field. The Hamiltonian of the combined system would be given by the Hamiltonians of the free boson and free fermion fields, plus a "potential energy" term such as

$$H_I = \sum_{k,q} V_q(a_q + a_{-q}^\dagger)c_{k+q}^\dagger c_k,$$

where $a_k^\dagger$ and a_k denotes the bosonic creation and annihilation operators, $c_k^\dagger$ and c_k denotes the fermionic creation and annihilation operators, and V_q is a parameter that describes the strength of the interaction. This "interaction term" describes processes in which a fermion in state k either absorbs or emits a boson, thereby being kicked into a different eigenstate $k + q$. (In fact, this type of Hamiltonian is used to describe the interaction between conduction electrons and phonons in metals. The interaction between electrons and photons is treated in a similar way, but is a little more complicated because the role of spin must be taken into account.) One thing to notice here is that even if we start out with a fixed number of bosons, we will typically end up with a superposition of states with different numbers of bosons at later times. The number of fermions, however, is conserved in this case.

In condensed matter physics, states with ill-defined particle numbers are particularly important for describing the various superfluids. Many of the defining characteristics of a superfluid arise from the notion that its quantum state is a superposition of states with different particle numbers. In addition, the concept of a coherent state (used to model the laser and the BCS ground state) refers to a state with an ill-defined particle number but a well-defined phase.

19.4.6 Axiomatic approaches

The preceding description of quantum field theory follows the spirit in which most physicists approach the subject. However, it is not mathematically rigorous. Over the past several decades, there have been many attempts to put quantum field theory on a firm mathematical footing by formulating a set of axioms for it. These attempts fall into two broad classes.

The first class of axioms, first proposed during the 1950s, include the Wightman, Osterwalder–Schrader, and Haag–Kastler systems. They attempted to formalize the physicists' notion of an "operator-valued field" within the context of functional analysis and enjoyed limited success. It was possible to prove that any quantum field theory satisfying these axioms satisfied certain general theorems, such as the spin-statistics theorem and the CPT theorem. Unfortunately, it proved extraordinarily difficult to show that any realistic field theory, including the Standard Model, satisfied these axioms. Most of the theories that could be treated with these analytic axioms were physically trivial, being restricted to low-dimensions and lacking interesting dynamics. The construction of theories satisfying one of these sets of axioms falls in the field of constructive quantum field theory. Important work was done in this area in the 1970s by Segal, Glimm, Jaffe and others.

During the 1980s, the second set of axioms based on geometric ideas was proposed. This line of investigation, which restricts its attention to a particular class of quantum field theories known as topological quantum field theories, is associated most closely with Michael Atiyah and Graeme Segal, and was notably expanded upon by Edward Witten, Richard Borcherds, and Maxim Kontsevich. However, most of the physically relevant quantum field theories, such as the Standard Model, are not topological quantum field theories; the quantum field theory of the fractional quantum Hall effect is a notable exception. The main impact of axiomatic topological quantum field theory has been on mathematics, with important applications in representation theory, algebraic topology, and differential geometry.

Finding the proper axioms for quantum field theory is still an open and difficult problem in mathematics. One of the Millennium Prize Problems—proving the existence of a mass gap in Yang–Mills theory—is linked to this issue.

19.5 Associated phenomena

In the previous part of the article, we described the most general features of quantum field theories. Some of the quantum field theories studied in various fields of theoretical physics involve additional special ideas, such as renormalizability, gauge symmetry, and supersymmetry. These are described in the following sections.

19.5.1 Renormalization

Main article: Renormalization

Early in the history of quantum field theory, it was found that many seemingly innocuous calculations, such as the perturbative shift in the energy of an electron due to the presence of the electromagnetic field, give infinite results. The reason is that the perturbation theory for the shift in an energy involves a sum over all other energy levels, and there are infinitely many levels at short distances that each gives a finite contribution which results in a divergent series.

Many of these problems are related to failures in classical electrodynamics that were identified but unsolved in the

19th century, and they basically stem from the fact that many of the supposedly "intrinsic" properties of an electron are tied to the electromagnetic field that it carries around with it. The energy carried by a single electron—its self-energy—is not simply the bare value, but also includes the energy contained in its electromagnetic field, its attendant cloud of photons. The energy in a field of a spherical source diverges in both classical and quantum mechanics, but as discovered by Weisskopf with help from Furry, in quantum mechanics the divergence is much milder, going only as the logarithm of the radius of the sphere.

The solution to the problem, presciently suggested by Stueckelberg, independently by Bethe after the crucial experiment by Lamb, implemented at one loop by Schwinger, and systematically extended to all loops by Feynman and Dyson, with converging work by Tomonaga in isolated postwar Japan, comes from recognizing that all the infinities in the interactions of photons and electrons can be isolated into redefining a finite number of quantities in the equations by replacing them with the observed values: specifically the electron's mass and charge: this is called renormalization. The technique of renormalization recognizes that the problem is essentially purely mathematical, that extremely short distances are at fault. In order to define a theory on a continuum, first place a cutoff on the fields, by postulating that quanta cannot have energies above some extremely high value. This has the effect of replacing continuous space by a structure where very short wavelengths do not exist, as on a lattice. Lattices break rotational symmetry, and one of the crucial contributions made by Feynman, Pauli and Villars, and modernized by 't Hooft and Veltman, is a symmetry-preserving cutoff for perturbation theory (this process is called regularization). There is no known symmetrical cutoff outside of perturbation theory, so for rigorous or numerical work people often use an actual lattice.

On a lattice, every quantity is finite but depends on the spacing. When taking the limit of zero spacing, we make sure that the physically observable quantities like the observed electron mass stay fixed, which means that the constants in the Lagrangian defining the theory depend on the spacing. Hopefully, by allowing the constants to vary with the lattice spacing, all the results at long distances become insensitive to the lattice, defining a continuum limit.

The renormalization procedure only works for a certain class of quantum field theories, called *renormalizable quantum field theories*. A theory is *perturbatively renormalizable* when the constants in the Lagrangian only diverge at worst as logarithms of the lattice spacing for very short spacings. The continuum limit is then well defined in perturbation theory, and even if it is not fully well defined non-perturbatively, the problems only show up at distance scales that are exponentially small in the inverse coupling for weak couplings. The Standard Model of particle physics is perturbatively renormalizable, and so are its component theories (quantum electrodynamics/electroweak theory and quantum chromodynamics). Of the three components, quantum electrodynamics is believed to not have a continuum limit, while the asymptotically free SU(2) and SU(3) weak hypercharge and strong color interactions are nonperturbatively well defined.

The renormalization group describes how renormalizable theories emerge as the long distance low-energy effective field theory for any given high-energy theory. Because of this, renormalizable theories are insensitive to the precise nature of the underlying high-energy short-distance phenomena. This is a blessing because it allows physicists to formulate low energy theories without knowing the details of high energy phenomenon. It is also a curse, because once a renormalizable theory like the standard model is found to work, it gives very few clues to higher energy processes. The only way high energy processes can be seen in the standard model is when they allow otherwise forbidden events, or if they predict quantitative relations between the coupling constants.

19.5.2 Haag's theorem

See also: Haag's theorem

From a mathematically rigorous perspective, there exists no interaction picture in a Lorentz-covariant quantum field theory. This implies that the perturbative approach of Feynman diagrams in QFT is not strictly justified, despite producing vastly precise predictions validated by experiment. This is called Haag's theorem, but most particle physicists relying on QFT largely shrug it off.

19.5.3 Gauge freedom

A gauge theory is a theory that admits a symmetry with a local parameter. For example, in every quantum theory the global phase of the wave function is arbitrary and does not represent something physical. Consequently, the theory is invariant under a global change of phases (adding a constant to the phase of all wave functions, everywhere); this is a global symmetry. In quantum electrodynamics, the theory is also invariant under a *local* change of phase, that is – one may shift the phase of all wave functions so that the shift may be different at every point in space-time. This is a *local* symmetry. However, in order for a well-defined derivative operator to exist, one must introduce a new field, the gauge field, which also transforms in order for the local change of variables (the phase in our example) not to affect the derivative. In quantum electrodynamics, this gauge field is the electromagnetic field. The change of local gauge of variables is termed gauge transformation. It is worth noting that by Noether's theorem, for every such symmetry there exists an associated conserved current. The aforementioned symmetry of the wavefunction under global phase changes implies the conservation of electric charge.

In quantum field theory the excitations of fields represent particles. The particle associated with excitations of the gauge field is the gauge boson, which is the photon in the case of quantum electrodynamics.

The degrees of freedom in quantum field theory are local fluctuations of the fields. The existence of a gauge symmetry reduces the number of degrees of freedom, simply because some fluctuations of the fields can be transformed to zero by gauge transformations, so they are equivalent to having no fluctuations at all, and they, therefore, have no physical meaning. Such fluctuations are usually called "non-physical degrees of freedom" or *gauge artifacts*; usually, some of them have a negative norm, making them inadequate for a consistent theory. Therefore, if a classical field theory has a gauge symmetry, then its quantized version (i.e. the corresponding quantum field theory) will have this symmetry as well. In other words, a gauge symmetry cannot have a quantum anomaly. If a gauge symmetry is anomalous (i.e. not kept in the quantum theory) then the theory is non-consistent: for example, in quantum electrodynamics, had there been a gauge anomaly, this would require the appearance of photons with longitudinal polarization and polarization in the time direction, the latter having a negative norm, rendering the theory inconsistent; another possibility would be for these photons to appear only in intermediate processes but not in the final products of any interaction, making the theory non-unitary and again inconsistent (see optical theorem).

In general, the gauge transformations of a theory consist of several different transformations, which may not be commutative. These transformations are together described by a mathematical object known as a gauge group. Infinitesimal gauge transformations are the gauge group generators. Therefore, the number of gauge bosons is the group dimension (i.e. number of generators forming a basis).

All the fundamental interactions in nature are described by gauge theories. These are:

- Quantum chromodynamics, whose gauge group is **SU**(3). The gauge bosons are eight gluons.
- The electroweak theory, whose gauge group is **U**(1) × **SU**(2), (a direct product of **U**(1) and **SU**(2)).
- Gravity, whose classical theory is general relativity, admits the equivalence principle, which is a form of gauge symmetry. However, it is explicitly non-renormalizable.

19.5.4 Multivalued gauge transformations

The gauge transformations which leave the theory invariant involve, by definition, only single-valued gauge functions $\Lambda(x_i)$ which satisfy the Schwarz integrability criterion

$$\partial_{x_i x_j}\Lambda = \partial_{x_j x_i}\Lambda.$$

An interesting extension of gauge transformations arises if the gauge functions $\Lambda(x_i)$ are allowed to be multivalued functions which violate the integrability criterion. These are capable of changing the physical field strengths and are therefore not proper symmetry transformations. Nevertheless, the transformed field equations describe correctly the physical laws in the presence of the newly generated field strengths. See the textbook by H. Kleinert cited below for the applications to phenomena in physics.

19.5.5 Supersymmetry

Main article: Supersymmetry

Supersymmetry assumes that every fundamental fermion has a superpartner that is a boson and vice versa. It was introduced in order to solve the so-called Hierarchy Problem, that is, to explain why particles not protected by any symmetry (like the Higgs boson) do not receive radiative corrections to its mass driving it to the larger scales (GUT, Planck...). It was soon realized that supersymmetry has other interesting properties: its gauged version is an extension of general relativity (Supergravity), and it is a key ingredient for the consistency of string theory.

The way supersymmetry protects the hierarchies is the following: since for every particle there is a superpartner with the same mass, any loop in a radiative correction is cancelled by the loop corresponding to its superpartner, rendering the theory UV finite.

Since no superpartners have yet been observed, if supersymmetry exists it must be broken (through a so-called soft term, which breaks supersymmetry without ruining its helpful features). The simplest models of this breaking require that the energy of the superpartners not be too high; in these cases, supersymmetry is expected to be observed by experiments at the Large Hadron Collider. The Higgs particle has been detected at the LHC, and no such superparticles have been discovered.

19.6 See also

- Abraham–Lorentz force
- Basic concepts of quantum mechanics
- Common integrals in quantum field theory
- Einstein–Maxwell–Dirac equations
- Form factor (quantum field theory)
- Green–Kubo relations
- Green's function (many-body theory)
- Invariance mechanics
- List of quantum field theories
- Quantization of a field
- Quantum electrodynamics
- Quantum field theory in curved spacetime
- Quantum flavordynamics
- Quantum hydrodynamics
- Quantum triviality
- Relation between Schrödinger's equation and the path integral formulation of quantum mechanics
- Relationship between string theory and quantum field theory
- Schwinger–Dyson equation
- Static forces and virtual-particle exchange
- Symmetry in quantum mechanics
- Theoretical and experimental justification for the Schrödinger equation
- Ward–Takahashi identity
- Wheeler–Feynman absorber theory
- Wigner's classification
- Wigner's theorem

19.7 Notes

[1] Dirac 1927

[2] Schweber 1994, p. 28

[3] See references in Schweber (1994, pp. 695f)

[4] Weinberg 2005, p. 15

[5] Peskin & Schroeder (1995, Chapter4)

[6] Greiner & Reinhardt 1996

[7] Clément Hongler, "Conformal invariance of Ising model correlations", Ph.D. thesis, Université of Geneva, 2010, p. 9.

[8] Thorn et al. 2004

[9] Tong 2015, Chapter 1

[10] Srednicki 2007, p. 19

[11] Srednicki 2007, pp. 25–26

[12] Zee 2010, p. 61

[13] Tong 2015, Introduction

[14] Zee 2010, p. 3

[15] Pais 1994. Pais recounts how his astonishment at the rapidity with which Feynman could calculate using his method. Feynman's method is now part of the standard methods for physicists.

[16] Newton & Wigner 1949, pp. 400–406

19.8 References

Historical references

- Born, M.; Jordan, P.; Heisenberg, W. (1926). "Zur quantenmechanic II" [On Quantum mechanics II]. *Zeitschrift für Physik* (in German) (Springer Verlag) **35** (8). Bibcode:1926ZPhy...35..557B. doi:10.1007/BF01379806. ISSN 0044-3328. (subscription required (help)).
- Dirac, P. A. M. (1927). "The quantum theory of the emission and absorption of radiation". *Proc. R. Soc. Lond. A* (Royal Society Publishing) **114** (767): 243–265. Bibcode:1927RSPSA.114..243D. doi:10.1098/rspa.1927.0039. (subscription required (help)).

General reader level

- Pais, A. (1994) [1986]. *Inward Bound: Of Matter and Forces in the Physical World* (reprint ed.). Oxford, New York, Toronto: Oxford University Press. ISBN 978-0198519973.

- Schweber, S. S. (1994). *QED and the Men Who Made It: Dyson, Feynman, Schwinger, and Tomonaga*. Princeton University Press. ISBN 9780691033273.

Articles

- Newton, T. D.; Wigner, E.P. (1949). "Localized states for elementary systems". *Rev. Mod. Phys.* (APS) **21** (3). Bibcode:1949RvMP...21..400N. doi:10.1103/RevModPhys.21.400. ISSN 0034-6861.

- Thorn, J. J.; Neel, M. S.; Donato, W. V.; Bergreen, G. S.; Davies, R. E.; Beck, M.. (2004). "Observing the quantum behavior of light in an undergraduate laboratory" (PDF). *Am. J. Phys.* (American Association of Physics Teachers) **72** (1210): 243–265. Bibcode:2004AmJPh..72.1210T. doi:10.1119/1.1737397.

Introductory texts

- Greiner, W.; Reinhardt, J. (1996). *Field Quantization*. Springer Publishing. ISBN 3-540-59179-6.

- Peskin, M.; Schroeder, D. (1995). *An Introduction to Quantum Field Theory*. Westview Press. ISBN 0-201-50397-2.

- Scharf, Günter (2014) [1989]. *Finite Quantum Electrodynamics: The Causal Approach* (third ed.). Dover Publications. ISBN 978-0486492735.

- Srednicki, M. (2007). *Quantum Field Theory*. Cambridge University Press. ISBN 978-0521-8644-97.

- Tong, David (2015). "Lectures on Quantum Field Theory". Retrieved 2016-02-09.

- Zee, Anthony (2010). *Quantum Field Theory in a Nutshell* (2nd ed.). Princeton University Press. ISBN 978-0691140346.

Advanced texts

- Weinberg, S. (2005). *The Quantum Theory of Fields* **1**. Cambridge University Press. ISBN 978-0521670531.

19.9 Further reading

General readers

- Feynman, R.P. (2001) [1964]. *The Character of Physical Law*. MIT Press. ISBN 0-262-56003-8.

- Feynman, R.P. (2006) [1985]. *QED: The Strange Theory of Light and Matter*. Princeton University Press. ISBN 0-691-12575-9.

- Gribbin, J. (1998). *Q is for Quantum: Particle Physics from A to Z*. Weidenfeld & Nicolson. ISBN 0-297-81752-3.

- Schumm, Bruce A. (2004) *Deep Down Things*. Johns Hopkins Univ. Press. Chpt. 4.

Introductory texts

- McMahon, D. (2008). *Quantum Field Theory*. McGraw-Hill. ISBN 978-0-07-154382-8.

- Bogoliubov, N.; Shirkov, D. (1982). *Quantum Fields*. Benjamin-Cummings. ISBN 0-8053-0983-7.

- Frampton, P.H. (2000). *Gauge Field Theories. Frontiers in Physics (2nd ed.). Wiley.*

- Greiner, W; Müller, B. (2000). *Gauge Theory of Weak Interactions*. Springer. ISBN 3-540-67672-4.

- Itzykson, C.; Zuber, J.-B. (1980). *Quantum Field Theory*. McGraw-Hill. ISBN 0-07-032071-3.

- Kane, G.L. (1987). *Modern Elementary Particle Physics*. Perseus Books. ISBN 0-201-11749-5.

- Kleinert, H.; Schulte-Frohlinde, Verena (2001). *Critical Properties of φ^4-Theories*. World Scientific. ISBN 981-02-4658-7.

- Kleinert, H. (2008). *Multivalued Fields in Condensed Matter, Electrodynamics, and Gravitation* (PDF). World Scientific. ISBN 978-981-279-170-2.

- Loudon, R (1983). *The Quantum Theory of Light*. Oxford University Press. ISBN 0-19-851155-8.

- Mandl, F.; Shaw, G. (1993). *Quantum Field Theory*. John Wiley & Sons. ISBN 978-0-471-94186-6.

- Ryder, L.H. (1985). *Quantum Field Theory*. Cambridge University Press. ISBN 0-521-33859-X.

- Schwartz, M.D. (2014). *Quantum Field Theory and the Standard Model*. Cambridge University Press. ISBN 978-1107034730.

- Ynduráin, F.J. (1996). *Relativistic Quantum Mechanics and Introduction to Field Theory* (1st ed.). Springer. ISBN 978-3-540-60453-2.

Advanced texts

- Brown, Lowell S. (1994). *Quantum Field Theory*. Cambridge University Press. ISBN 978-0-521-46946-3.
- Bogoliubov, N.; Logunov, A.A.; Oksak, A.I.; Todorov, I.T. (1990). *General Principles of Quantum Field Theory*. Kluwer Academic Publishers. ISBN 978-0-7923-0540-8.

Articles

- 't Hooft, Gerard (2007). Butterfield, J.; Earman, John, eds. *Philosophy of Physics*. Part A. The Conceptual Basis of Quantum Field Theory: Elsevier. pp. 661–730 – via ScienceDirect. (subscription required (help)). On web at 't Hooft's university website
- Wilczek, frank (1999). "Quantum field theory". *Rev. Mod. Phys* **71** (S85–S95). arXiv:hep-th/9803075v2. Bibcode:1999RvMPS..71...85W. doi:10.1103/RevModPhys.71.S85.

19.10 External links

- Hazewinkel, Michiel, ed. (2001), "Quantum field theory", *Encyclopedia of Mathematics*, Springer, ISBN 978-1-55608-010-4
- *Stanford Encyclopedia of Philosophy*: "Quantum Field Theory", by Meinard Kuhlmann.
- Siegel, Warren, 2005. *Fields*. A free text, also available from arXiv:hep-th/9912205.
- Quantum Field Theory by P. J. Mulders

Chapter 20

Hamiltonian constraint

For a feature of the loop quantum gravity, see Hamiltonian constraint of LQG.

The **Hamiltonian constraint** arises from any theory that admits a Hamiltonian formulation and is reparametrisation-invariant. The Hamiltonian constraint of general relativity is an important non-trivial example.

In the context of general relativity, the Hamiltonian constraint technically refers to a linear combination of spatial and time diffeomorphism constraints reflecting the reparametrizability of the theory under both spatial as well as time coordinates. However, most of the time the term *Hamiltonian constraint* is reserved for the constraint that generates time diffeomorphisms.

20.1 Simplest example: the parametrized clock and pendulum system

20.1.1 Parametrization

In its usual presentation, classical mechanics appears to give time a special role as an independent variable. This is unnecessary, however. Mechanics can be formulated to treat the time variable on the same footing as the other variables in an extended phase space, by parameterizing the temporal variable(s) in terms of a common, albeit unspecified parameter variable. Phase space variables being on the same footing.

Say our system comprised a pendulum executing a simple harmonic motion and a clock. Whereas the system could be described classically by a position x=x(t), with x defined as a function of time, it is also possible to describe the same system as x(τ) and t(τ) where the relation between x and t is not directly specified. Instead, x and t are determined by the parameter τ , which is simply a parameter of the system, possibly having no objective meaning in its own right.

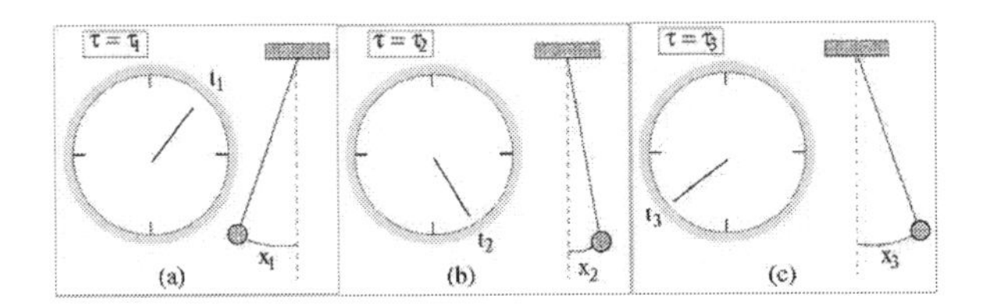

We introduce τ as an unphysical parameter labeling different possible correlations between the time reading t of the clock and the elongation x of the pendulum. τ is unphysical parameter and there are many different choices for it.

The system would be described by the position of a pendulum from the center, denoted x , and the reading on the clock, denoted t . We put these variables on the same footing by introducing a fictitious parameter τ ,

$x(\tau), \quad t(\tau)$

whose `evolution' with respect to τ takes us continuously through every possible correlation between the displacement and reading on the clock. Obviously the variable τ can be replaced by any monotonic function, $\tau' = f(\tau)$. This is what makes the system reparametrisation-invariant. Note that by this reparametrisation-invariance the theory cannot predict the value of $x(\tau)$ or $t(\tau)$ for a given value of τ but only the relationship between these quantities. Dynamics is then determined by this relationship.

20.1.2 Dynamics of this reparametrization-invariant system

The action for the parametrized Harmonic oscillator is then

$S = \int d\tau \Big[\frac{dx}{d\tau} p + \frac{dt}{d\tau} p_t - \lambda \Big(p_t + \frac{p^2}{2m} + \frac{1}{2} m\omega^2 x^2 \Big) \Big].$

where x and t are canonical coordinates and p and p_t are their conjugate momenta respectively and represent our extended phase space (we will show that we can recover the usual Newton's equations from this expression). Writing the action as

$S = \int d\tau \Big[\frac{dx}{d\tau} p + \frac{dt}{d\tau} p_t - \mathcal{H}(x, t; p, p_t) \Big]$

we identify the $\mathcal{H}$ as

$$\mathcal{H}(x,t,\lambda;p,p_t) = \lambda\Big(p_t + \frac{p^2}{2m} + \frac{1}{2}m\omega^2x^2\Big).$$

Hamilton's equations for λ are

$$\frac{\partial\mathcal{H}}{\partial\lambda} = 0$$

which gives a constraint,

$$C = p_t + \frac{p^2}{2m} + \frac{1}{2}m\omega^2x^2 = 0.$$

C is our Hamiltonian constraint! It could also be obtained from the Euler-Lagrange equation of motion, noting that the action depends on λ but not its τ derivative. Then the extended phase space variables x , t , p , and p_t are constrained to take values on this constraint-hypersurface of the extended phase space. We refer to λC as the 'smeared' Hamiltonian constraint where λ is an arbitrary number. The 'smeared' Hamiltonain constraint tells us how an extended phase space variable (or function thereof) evolves with respect to τ :

$$\frac{dx}{d\tau} = \{x,\lambda C\}, \qquad \frac{dp}{d\tau} = \{p,\lambda C\}$$

$$\frac{dt}{d\tau} = \{t,\lambda C\}, \quad \frac{dp_t}{d\tau} = \{p_t,\lambda C\}$$

(these are actually the other Hamilton's equations). These equations describe a flow or orbit in phase space. In general we have

$$\frac{dF(x,p,t,p_t)}{d\tau} = \{F(x,p,t,p_t),\lambda C\}$$

for any phase space function F . As the Hamiltonian constraint Poisson commutes with itself, it preserves itself and hence the constraint-hypersurface. The possible correlations between measurable quantities like $x(\tau)$ and $t(\tau)$ then correspond to 'orbits' generated by the constraint within the constraint surface, each particular orbit differentiated from each other by say also measuring the value of say $p(\tau)$ along with $x(\tau)$ and $t(\tau)$ at one τ -instant; after determining the particular orbit, for each measurement of $t(\tau)$ we can predict the value $x(\tau)$ will take.

20.1.3 Deparametrization

The other equations of Hamiltonian mechanics are

$$\frac{dx}{d\tau} = \frac{\partial\mathcal{H}}{\partial p}, \quad \frac{dp}{d\tau} = -\frac{\partial\mathcal{H}}{\partial x}; \qquad \frac{dt}{d\tau} = \frac{\partial\mathcal{H}}{\partial p_t}, \quad \frac{dp_t}{d\tau} = \frac{\partial\mathcal{H}}{\partial t}.$$

Upon substitution of our action these give,

$$\frac{dx}{d\tau} = \lambda\frac{p}{m}, \quad \frac{dp}{d\tau} = -\lambda m\omega^2x; \qquad \frac{dt}{d\tau} = \lambda, \quad \frac{dp_t}{d\tau} = 0,$$

These represent the fundamental equations governing our system.

In the case of the parametrized clock and pendulum system we can of course recover the usual equations of motion in which t is the independent variable:

Now dx/dt and dp/dt can be deduced by

$$\frac{dx}{dt} = \frac{dx}{d\tau}\Big/\frac{dt}{d\tau} = \frac{\lambda p/m}{\lambda} = \frac{p}{m}$$

$$\frac{dp}{dt} = \frac{dp}{d\tau}\Big/\frac{dt}{d\tau} = \frac{-\lambda m\omega^2x}{\lambda} = -m\omega^2x.$$

We recover the usual differential equation for the simple harmonic oscillator,

$$\frac{d^2x}{dt^2} = -\omega^2x.$$

We also have $dp_t/d\tau = dp_t/d\tau/dp_t/d\tau = 0$ or $p_t = Const.$

Our Hamiltonian constraint is then easily seen as the condition of constancy of energy! Deparametrization and the identification of a time variable with respect to which everything evolves is the opposite process of parametrization. It turns out in general that not all reparametrisation-invariant systems can be deparametrized. General relativity being a prime physical example (here the spacetime coordinates correspond to the unphysical τ and the Hamiltonian is a linear combination of constraints which generate spatial and time diffeomorphisms).

20.1.4 Reason why we could deparametrize here

The underlining reason why we could deparametrize (aside from the fact that we already know it was an artificial reparametrization in the first place) is the mathematical form of the constraint, namely,

$$C = p_t + C'(x,p) .$$

Substitute the Hamiltonian constraint into the original action we obtain

$$S = \int d\tau\Big[\frac{dx}{d\tau}p + \frac{dt}{d\tau}p_t - \lambda(p_t + C'(x,p))\Big]$$

$$= \int d\tau\Big[\frac{dx}{d\tau}p - \frac{dt}{d\tau}C'(x,p)\Big]$$

$$= \int dt\Big[\frac{dx}{dt}p - \frac{p^2}{2m} + \frac{1}{2}m\omega^2x^2\Big]$$

which is the standard action for the harmonic oscillator. General relativity is an example of a physical theory where the Hamiltonian constraint isn't of the above mathematical form in general, and so cannot be deparametrized in general.

20.2 Hamiltonian of classical general relativity

In the ADM formulation of general relativity one splits spacetime into spatial slices and time, the basic variables are taken to be the induced metric, $q_{ab}(x)$, on the spatial slice (the metric induced on the spatial slice by the spacetime metric), and its conjugate momentum variable related to the extrinsic curvature, $K^{ab}(x)$, (this tells us how the spatial slice curves with respect to spacetime and is a measure of how the induced metric evolves in time).[1] These are the metric canonical coordinates.

Dynamics such as time-evolutions of fields are controlled by the **Hamiltonian constraint**.

The identity of the Hamiltonian constraint is a major open question in quantum gravity, as is extracting of physical observables from any such specific constraint.

In 1986 Abhay Ashtekar introduced a new set of canonical variables, Ashtekar variables to represent an unusual way of rewriting the metric canonical variables on the three-dimensional spatial slices in terms of a SU(2) gauge field and its complementary variable.[2] The Hamiltonian was much simplified in this reformulation. This led to the loop representation of quantum general relativity[3] and in turn loop quantum gravity.

Within the loop quantum gravity representation Thiemann formulated a mathematically rigorous operator as a proposal as such a constraint.[4] Although this operator defines a complete and consistent quantum theory, doubts have been raised as to the physical reality of this theory due to inconsistencies with classical general relativity (the quantum constraint algebra closes, but it is not isomorphic to the classical constraint algebra of GR, which is seen as circumstantial evidence of inconsistencies definitely not a proof of inconsistencies), and so variants have been proposed.

20.2.1 Metric formulation

The idea was to quantize the canonical variables q_{ab} and $\pi^{ab} = \sqrt{q}(K^{ab} - q^{ab}K^c_c)$, making them into operators acting on wavefunctions on the space of 3-metrics, and then to quantize the Hamiltonian (and other constraints). However, this program soon became regarded as dauntingly difficult for various reasons, one being the non-polynomial nature of the Hamiltonian constraint:

$$H = \sqrt{det(q)}(K_{ab}K^{ab} - (K^a_a)^2 - {}^3R)$$

where ${}^3R$ is the scalar curvature of the three metric $q_{ab}(x)$. Being a non-polynomial expression in the canonical variables and their derivatives it is very difficult to promote to a quantum operator.

20.2.2 Expression using Ashtekar variables

The configuration variables of Ashtekar's variables behave like an $SU(2)$ gauge field or connection A^i_a . Its canonically conjugate momentum is $\tilde{E}^a_i$ is the densitized "electric" field or triad (densitized as $\tilde{E}^a_i = \sqrt{det(q)}E^a_i$). What do these variables have to do with gravity? The densitized triads can be used to reconstruct the spatial metric via

$$det(q)q^{ab} = \tilde{E}^a_i \tilde{E}^b_j \delta^{ij} .$$

The densitized triads are not unique, and in fact one can perform a local in space rotation with respect to the internal indices i . This is actually the origin of the $SU(2)$ gauge invariance. The connection can be use to reconstruct the extrinsic curvature. The relation is given by

$$A^i_a = \Gamma^i_a - iK^i_a$$

where Γ^i_a is related to the spin connection, $\Gamma_a{}^j{}_i$, by $\Gamma^i_a = \Gamma_{ajk}\epsilon^{jki}$ and $K^i_a = K_{ab}\tilde{E}^{ai}/\sqrt{det(q)}$.

In terms of Ashtekar variables the classical expression of the constraint is given by,

$$H = \frac{\epsilon_{ijk}F^k_{ab}\tilde{E}^a_i\tilde{E}^b_j}{\sqrt{det(q)}} .$$

where F^k_{ab} field strength tensor of the gauge field A^i_a . Due to the factor $1/\sqrt{det(q)}$ this is non-polynomial in the Ashtekar's variables. Since we impose the condition

$$H = 0 ,$$

we could consider the densitized Hamiltonian instead,

$$\tilde{H} = \sqrt{det(q)}H = \epsilon_{ijk}F^k_{ab}\tilde{E}^a_i\tilde{E}^b_j = 0 .$$

This Hamiltonian is now polynomial the Ashtekar's variables. This development raised new hopes for the canonical quantum gravity programme.[5] Although Ashtekar variables had the virtue of simplifying the Hamiltonian, it has the problem that the variables become complex. When one quantizes the theory it is a difficult task ensure that one recovers real general relativity as opposed to complex general relativity. Also there were also serious difficulties promoting the densitized Hamiltonian to a quantum operator.

A way of addressing the problem of reality conditions was noting that if we took the signature to be $(+,+,+,+)$, that is Euclidean instead of Lorentzian, then one can retain the simple form of the Hamiltonian for but for real variables. One can then define what is called a generalized Wick rotation to recover the Lorentzian theory.[6] Generalized as it is a Wick transformation in phase space and has nothing to do with analytical continuation of the time parameter t .

20.2.3 Expression for real formulation of Ashtekar variables

Thomas Thiemann addressed both the above problems.[4] He used the real connection

$$A^i_a = \Gamma^i_a + \beta K^i_a$$

In real Ashtekar variables the full Hamiltonian is

$H = -\zeta \frac{\epsilon_{ijk} F^k_{ab} \tilde{E}^a_i \tilde{E}^b_j}{\sqrt{det(q)}} + 2 \frac{\zeta\beta^2 - 1}{\beta^2} \frac{(\tilde{E}^a_i \tilde{E}^b_j - \tilde{E}^a_j \tilde{E}^b_i)}{\sqrt{det(q)}} (A^i_a - \Gamma^i_a)(A^j_b - \Gamma^j_b) = H_E + H'$.

where the constant β is the Barbero-Immirzi parameter.[7] The constant ζ is -1 for Lorentzian signature and +1 for Euclidean signature. The Γ^i_a have a complicated relationship with the desitized triads and causes serious problems upon quantization. Ashtekar variables can be seen as choosing $\beta = i$ to make the second more complicated term was made to vanish (the first term is denoted H_E because for the Euclidean theory this term remains for the real choice of $\beta = \pm 1$). Also we still have the problem of the $1/\sqrt{det(q)}$ factor.

Thiemann was able to make it work for real β . First he could simplify the troublesome $1/\sqrt{det(q)}$ by using the identity

$\{A^k_c, V\} = \frac{\epsilon_{abc} \epsilon^{ijk} \tilde{E}^a_i \tilde{E}^b_j}{\sqrt{det(q)}}$

where V is the volume,

$V = \int d^3x \sqrt{det(q)} = \frac{1}{6} \int d^3x \sqrt{|\tilde{E}^a_i \tilde{E}^b_j \tilde{E}^c_k \epsilon^{ijk} \epsilon_{abc}|}$.

The first term of the Hamiltonian constraint becomes

$H_E = \{A^k_c, V\} F^k_{ab} \tilde{\epsilon}^{abc}$

upon using Thiemann's identity. This Poisson bracket is replaced by a commutator upon quantization. It turns out that a similar trick can be used to teat the second term. Why are the Γ^i_a given by the densitized triads $\tilde{E}^a_i$? It actually come about from the Gauss Law

$D_a \tilde{E}^a_i = 0$.

We can solve this in much the same way as the Levi-Civita connection can be calculated from the equation $\nabla_c g_{ab} = 0$; by rotating the various indices and then adding and subtracting them. The result is complicated and non-linear. To circumvent the problems introduced by this complicated relationship Thiemann first defines the Gauss gauge invariant quantity

$K = \int d^3x K^i_a \tilde{E}^a_i$

where $K^i_a = K_{ab} \tilde{E}^{ai} / \sqrt{det(q)}$, and notes that

$K^i_a = \{A^i_a, K\}$.

We are then able to write

$A^i_a - \Gamma^i_a = \beta K^i_a = \beta \{A^i_a, K\}$

and as such find an expression in terms of the configuration variable A^i_a and K . We obtain for the second term of the Hamiltonian

$H' = \epsilon^{abc} \epsilon_{ijk} \{A^i_a, K\} \{A^j_b, K\} \{A^k_c, V\}$.

Why is it easier to quantize K ? This is because it can be rewritten in terms of quantities that we already know how to quantize. Specifically K can be rewritten as

$K = -\{V, \int d^3x H_E\}$

where we have used that the integrated densitized trace of the extrinsic curvature is the"time derivative of the volume".

20.2.4 Coupling to matter

20.3 References

[1] *Gravitation* by Charles W. Misner, Kip S. Thorne, John Archibald Wheeler, published by W. H. Freeman and company. New York.

[2] Ashtekar, A. (1986) *Phys. Rev. Lett. 57, 2244.*

[3] Rovelli, C. and Smolin, L. *Phys. Rev. Lett 61, 1155*

[4] *Anomaly-free formulation of non-perturbative, four-dimensional Lorentzian quantum gravity*, T. Thiemann, *Phys.Lett. B380 (1996) 257-264.*

[5] See the book *Lectures on Non-Perturbative Canonical Gravity* for more details on this and the subsequent development. First published in 1991. World Scientific Publishing Co. Pte. LtD.

[6] *Reality conditions inducing transforms for quantum gauge field theory and quantum gravity*, Thomas Thiemann, Class.Quant.Grav. 13 (1996) 1383-1404.

[7] *Real Ashtekar Variables for Lorentzian Signature Space-times*, J. Fernando, G. Barbero. Phys.Rev.D51:5507-5510, 1995

20.4 External links

- Overview by Carlo Rovelli
- Thiemann's paper in Physics Letters
- Good information on LQG

Chapter 21

Hamiltonian mechanics

Hamiltonian mechanics is a theory developed as a reformulation of classical mechanics and predicts the same outcomes as non-Hamiltonian classical mechanics. It uses a different mathematical formalism, providing a more abstract understanding of the theory. Historically, it was an important reformulation of classical mechanics, which later contributed to the formulation of statistical mechanics and quantum mechanics.

Hamiltonian mechanics was first formulated by William Rowan Hamilton in 1833, starting from Lagrangian mechanics, a previous reformulation of classical mechanics introduced by Joseph Louis Lagrange in 1788.

21.1 Overview

In Hamiltonian mechanics, a classical physical system is described by a set of canonical coordinates $\boldsymbol{r} = (\boldsymbol{q}, \boldsymbol{p})$, where each component of the coordinate q_i, p_i is indexed to the frame of reference of the system.

The time evolution of the system is uniquely defined by Hamilton's equations:[1]

where $\mathcal{H} = \mathcal{H}(\boldsymbol{q}, \boldsymbol{p}, t)$ is the Hamiltonian, which often corresponds to the total energy of the system.[2] For a closed system, it is the sum of the kinetic and potential energy in the system.

In Newtonian mechanics, the time evolution is obtained by computing the total force being exerted on each particle of the system, and from Newton's second law, the time-evolutions of both position and velocity are computed. In contrast, in Hamiltonian mechanics, the time evolution is obtained by computing the Hamiltonian of the system in the generalized coordinates and inserting it in the Hamiltonian equations. This approach is equivalent to the one used in Lagrangian mechanics. In fact, as is shown below, the Hamiltonian is the Legendre transform of the Lagrangian when holding q and t fixed and denoting p as the dual variable, and thus both approaches give the same equations for the same generalized momentum. The main motivation to use Hamiltonian mechanics instead of Lagrangian mechanics comes from the symplectic structure of Hamiltonian systems.

While Hamiltonian mechanics can be used to describe simple systems such as a bouncing ball, a pendulum or an oscillating spring in which energy changes from kinetic to potential and back again over time, its strength is shown in more complex dynamic systems, such as planetary orbits in celestial mechanics.[3] The more degrees of freedom the system has, the more complicated its time evolution is and, in most cases, it becomes chaotic.

21.1.1 Basic physical interpretation

A simple interpretation of the Hamilton mechanics comes from its application on a one-dimensional system consisting of one particle of mass *m*. The Hamiltonian represents the total energy of the system, which is the sum of kinetic and potential energy, traditionally denoted *T* and *V*, respectively. Here q is the space coordinate and p is the momentum, *mv*. Then

$$\mathcal{H} = T + V, \quad T = \frac{p^2}{2m}, \quad V = V(q).$$

Note that *T* is a function of *p* alone, while *V* is a function of *q* alone (i.e., *T* and *V* are scleronomic).

In this example, the time-derivative of the momentum p equals the *Newtonian force*, and so the first Hamilton equation means that the force equals the negative gradient of potential energy. The time-derivative of q is the velocity, and so the second Hamilton equation means that the particle's velocity equals the derivative of its kinetic energy with respect to its momentum.

21.1.2 Calculating a Hamiltonian from a Lagrangian

Given a Lagrangian in terms of the generalized coordinates q_i and generalized velocities $\dot{q}_i$ and time,

1. The momenta are calculated by differentiating the Lagrangian with respect to the (generalized) velocities: $p_i(q_i, \dot{q}_i, t) = \frac{\partial \mathcal{L}}{\partial \dot{q}_i}$.
2. The velocities $\dot{q}_i$ are expressed in terms of the momenta p_i by inverting the expressions in the previous step.
3. The Hamiltonian is calculated using the usual definition of $\mathcal{H}$ as the Legendre transformation of $\mathcal{L}$:

   $\mathcal{H} = \sum_i \dot{q}_i \frac{\partial \mathcal{L}}{\partial \dot{q}_i} - \mathcal{L} = \sum_i \dot{q}_i p_i - \mathcal{L}$.
   Then the velocities are substituted for through the above results.

21.2 Deriving Hamilton's equations

Hamilton's equations can be derived by looking at how the total differential of the Lagrangian depends on time, generalized positions q_i and generalized velocities $\dot{q}_i$: [4]

$$\mathrm{d}\mathcal{L} = \sum_i \left(\frac{\partial \mathcal{L}}{\partial q_i} \mathrm{d}q_i + \frac{\partial \mathcal{L}}{\partial \dot{q}_i} \mathrm{d}\dot{q}_i \right) + \frac{\partial \mathcal{L}}{\partial t} \mathrm{d}t \, .$$

The generalized momenta were defined as

$$p_i = \frac{\partial \mathcal{L}}{\partial \dot{q}_i} \, .$$

If this is substituted into the total differential of the Lagrangian, one gets

$$\mathrm{d}\mathcal{L} = \sum_i \left(\frac{\partial \mathcal{L}}{\partial q_i} \mathrm{d}q_i + p_i \mathrm{d}\dot{q}_i \right) + \frac{\partial \mathcal{L}}{\partial t} \mathrm{d}t \, .$$

This can be re-written as

$$\mathrm{d}\mathcal{L} = \sum_i \left(\frac{\partial \mathcal{L}}{\partial q_i} \mathrm{d}q_i + \mathrm{d}\left(p_i \dot{q}_i\right) - \dot{q}_i \mathrm{d}p_i \right) + \frac{\partial \mathcal{L}}{\partial t} \mathrm{d}t \, .$$

which after re-arranging leads to

$$\mathrm{d}\left(\sum_i p_i \dot{q}_i - \mathcal{L} \right) = \sum_i \left(-\frac{\partial \mathcal{L}}{\partial q_i} \mathrm{d}q_i + \dot{q}_i \mathrm{d}p_i \right) - \frac{\partial \mathcal{L}}{\partial t} \mathrm{d}t.$$

The term on the left-hand side is just the Hamiltonian that defined before, therefore

$$\mathrm{d}\mathcal{H} = \sum_i \left(-\frac{\partial \mathcal{L}}{\partial q_i} \mathrm{d}q_i + \dot{q}_i \mathrm{d}p_i \right) - \frac{\partial \mathcal{L}}{\partial t} \mathrm{d}t \, .$$

It is also possible to calculate the total differential of the Hamiltonian $\mathcal{H}$ with respect to time directly, similar to what was carried on with the Lagrangian $\mathcal{L}$ above, yielding:

$$\mathrm{d}\mathcal{H} = \sum_i \left(\frac{\partial \mathcal{H}}{\partial q_i} \mathrm{d}q_i + \frac{\partial \mathcal{H}}{\partial p_i} \mathrm{d}p_i \right) + \frac{\partial \mathcal{H}}{\partial t} \mathrm{d}t \, .$$

It follows from the previous two independent equations that their right-hand sides are equal with each other. The result is

$$\sum_i \left(-\frac{\partial \mathcal{L}}{\partial q_i} \mathrm{d}q_i + \dot{q}_i \mathrm{d}p_i \right) - \frac{\partial \mathcal{L}}{\partial t} \mathrm{d}t$$

$$= \sum_i \left(\frac{\partial \mathcal{H}}{\partial q_i} \mathrm{d}q_i + \frac{\partial \mathcal{H}}{\partial p_i} \mathrm{d}p_i \right) + \frac{\partial \mathcal{H}}{\partial t} \mathrm{d}t \, .$$

Since this calculation was done off-shell, one can associate corresponding terms from both sides of this equation to yield:

$$\frac{\partial \mathcal{H}}{\partial q_i} = -\frac{\partial \mathcal{L}}{\partial q_i} \, , \quad \frac{\partial \mathcal{H}}{\partial p_i} = \dot{q}_i \, , \quad \frac{\partial \mathcal{H}}{\partial t} = -\frac{\partial \mathcal{L}}{\partial t} \, .$$

On-shell, Lagrange's equations indicate that

$$\frac{\mathrm{d}}{\mathrm{d}t} \frac{\partial \mathcal{L}}{\partial \dot{q}_i} - \frac{\partial \mathcal{L}}{\partial q_i} = 0 \, .$$

A re-arrangement of this yields

$$\frac{\partial \mathcal{L}}{\partial q_i} = \dot{p}_i \, .$$

Thus Hamilton's equations hold on-shell:

$$\frac{\partial \mathcal{H}}{\partial q_j} = -\dot{p}_j \, , \quad \frac{\partial \mathcal{H}}{\partial p_j} = \dot{q}_j \, , \quad \frac{\partial \mathcal{H}}{\partial t} = -\frac{\partial \mathcal{L}}{\partial t} \, .$$

21.3 As a reformulation of Lagrangian mechanics

Starting with Lagrangian mechanics, the equations of motion are based on generalized coordinates

$$\{q_j \mid j = 1, \ldots, N\}$$

and matching generalized velocities

$$\{\dot{q}_j \mid j = 1, \ldots, N\}.$$

We write the Lagrangian as

$$\mathcal{L}(q_j, \dot{q}_j, t)$$

with the subscripted variables understood to represent all N variables of that type. Hamiltonian mechanics aims to replace the generalized velocity variables with generalized momentum variables, also known as *conjugate momenta*. By doing so, it is possible to handle certain systems, such as aspects of quantum mechanics, that would otherwise be even more complicated.

For each generalized velocity, there is one corresponding conjugate momentum, defined as:

$$p_j = \frac{\partial \mathcal{L}}{\partial \dot{q}_j}.$$

In Cartesian coordinates, the generalized momenta are precisely the physical linear momenta. In circular polar coordinates, the generalized momentum corresponding to the angular velocity is the physical angular momentum. For an arbitrary choice of generalized coordinates, it may not be possible to obtain an intuitive interpretation of the conjugate momenta.

One thing which is not too obvious in this coordinate dependent formulation is that different generalized coordinates are really nothing more than different coordinate patches on the same symplectic manifold (see *Mathematical formalism*, below).

The *Hamiltonian* is the Legendre transform of the Lagrangian:

$$\mathcal{H}\left(q_j, p_j, t\right) = \left(\sum_i \dot{q}_i p_i\right) - \mathcal{L}(q_j, \dot{q}_j, t).$$

If the transformation equations defining the generalized coordinates are independent of *t*, and the Lagrangian is a sum of products of functions (in the generalized coordinates) which are homogeneous of order 0, 1 or 2, then it can be shown that *H* is equal to the total energy *E* = *T* + *V*.

Each side in the definition of $\mathcal{H}$ produces a differential:

$$\mathrm{d}\mathcal{H} = \sum_i \left[\left(\frac{\partial \mathcal{H}}{\partial q_i}\right) \mathrm{d}q_i + \left(\frac{\partial \mathcal{H}}{\partial p_i}\right) \mathrm{d}p_i\right] + \left(\frac{\partial \mathcal{H}}{\partial t}\right) \mathrm{d}t$$

$$= \sum_i \left[\dot{q}_i \,\mathrm{d}p_i + p_i \,\mathrm{d}\dot{q}_i - \left(\frac{\partial \mathcal{L}}{\partial q_i}\right) \mathrm{d}q_i - \left(\frac{\partial \mathcal{L}}{\partial \dot{q}_i}\right) \mathrm{d}\dot{q}_i\right] - \left(\frac{\partial \mathcal{L}}{\partial t}\right) \mathrm{d}t.$$

Substituting the previous definition of the conjugate momenta into this equation and matching coefficients, we obtain the equations of motion of Hamiltonian mechanics, known as the canonical equations of Hamilton:

$$\frac{\partial \mathcal{H}}{\partial q_j} = -\dot{p}_j, \qquad \frac{\partial \mathcal{H}}{\partial p_j} = \dot{q}_j, \qquad \frac{\partial \mathcal{H}}{\partial t} = -\frac{\partial \mathcal{L}}{\partial t}.$$

Hamilton's equations consist of 2n first-order differential equations, while Lagrange's equations consist of n second-order equations. However, Hamilton's equations usually don't reduce the difficulty of finding explicit solutions. They still offer some advantages, since important theoretical results can be derived because coordinates and momenta are independent variables with nearly symmetric roles.

Hamilton's equations have another advantage over Lagrange's equations: if a system has a symmetry, such that a coordinate does not occur in the Hamiltonian, the corresponding momentum is conserved, and that coordinate can be ignored in the other equations of the set. Effectively, this reduces the problem from n coordinates to (n-1) coordinates. In the Lagrangian framework, of course the result that the corresponding momentum is conserved still follows immediately, but all the generalized velocities still occur in the Lagrangian - we still have to solve a system of equations in n coordinates.[2]

The Lagrangian and Hamiltonian approaches provide the groundwork for deeper results in the theory of classical mechanics, and for formulations of quantum mechanics.

21.4 Geometry of Hamiltonian systems

A Hamiltonian system may be understood as a fiber bundle *E* over time *R*, with the fibers *Et*, $t \in R$, being the position space. The Lagrangian is thus a function on the jet bundle *J* over *E*; taking the fiberwise Legendre transform of the Lagrangian produces a function on the dual bundle over time whose fiber at *t* is the cotangent space T^*Et, which comes equipped with a natural symplectic form, and this latter function is the Hamiltonian.

21.5 Generalization to quantum mechanics through Poisson bracket

Hamilton's equations above work well for classical mechanics, but not for quantum mechanics, since the differential equations discussed assume that one can specify the exact position and momentum of the particle simultaneously at any point in time. However, the equations can be further generalized to then be extended to apply to quantum mechanics as well as to classical mechanics, through the deformation of the Poisson algebra over p and q to the algebra of Moyal brackets.

Specifically, the more general form of the Hamilton's equation reads

$$\frac{\mathrm{d}f}{\mathrm{d}t} = \{f, \mathcal{H}\} + \frac{\partial f}{\partial t}$$

where f is some function of p and q, and H is the Hamiltonian. To find out the rules for evaluating a Poisson bracket without resorting to differential equations, see Lie algebra; a Poisson bracket is the name for the Lie bracket in a Poisson algebra. These Poisson brackets can then be extended to Moyal brackets comporting to an **inequivalent** Lie algebra, as proven by H. Groenewold, and thereby describe quantum mechanical diffusion in phase space (See the phase space formulation and the Wigner-Weyl transform). This more algebraic approach not only permits ultimately extending probability distributions in phase space to Wigner quasi-probability distributions, but, at the mere Poisson bracket classical setting, also provides more power in helping analyze the relevant conserved quantities in a system.

21.6 Mathematical formalism

Any smooth real-valued function H on a symplectic manifold can be used to define a Hamiltonian system. The function H is known as the **Hamiltonian** or the **energy function**. The symplectic manifold is then called the phase space. The Hamiltonian induces a special vector field on the symplectic manifold, known as the Hamiltonian vector field.

The Hamiltonian vector field (a special type of symplectic vector field) induces a Hamiltonian flow on the manifold. This is a one-parameter family of transformations of the manifold (the parameter of the curves is commonly called the **time**); in other words an isotopy of symplectomorphisms, starting with the identity. By Liouville's theorem, each symplectomorphism preserves the volume form on the phase space. The collection of symplectomorphisms induced by the Hamiltonian flow is commonly called the **Hamiltonian mechanics** of the Hamiltonian system.

The symplectic structure induces a Poisson bracket. The Poisson bracket gives the space of functions on the manifold the structure of a Lie algebra.

Given a function f

$$\frac{\mathrm{d}}{\mathrm{d}t} f = \frac{\partial}{\partial t} f + \{\, f, \mathcal{H} \,\}.$$

If we have a probability distribution, ρ, then (since the phase space velocity ($\dot{p}_i, \dot{q}_i$) has zero divergence, and probability is conserved) its convective derivative can be shown to be zero and so

$$\frac{\partial}{\partial t} \rho = -\{\, \rho, \mathcal{H} \,\}.$$

This is called Liouville's theorem. Every smooth function G over the symplectic manifold generates a one-parameter family of symplectomorphisms and if { G, H } = 0, then G is conserved and the symplectomorphisms are symmetry transformations.

A Hamiltonian may have multiple conserved quantities Gi. If the symplectic manifold has dimension $2n$ and there are n functionally independent conserved quantities Gi which are in involution (i.e., { Gi, Gj } = 0), then the Hamiltonian is Liouville integrable. The Liouville-Arnold theorem says that locally, any Liouville integrable Hamiltonian can be transformed via a symplectomorphism in a new Hamiltonian with the conserved quantities Gi as coordinates; the new coordinates are called *action-angle coordinates*. The transformed Hamiltonian depends only on the Gi, and hence the equations of motion have the simple form

$$\dot{G}_i = 0, \qquad \dot{\varphi}_i = F(G),$$

for some function F (Arnol'd et al., 1988). There is an entire field focusing on small deviations from integrable systems governed by the KAM theorem.

The integrability of Hamiltonian vector fields is an open question. In general, Hamiltonian systems are chaotic; concepts of measure, completeness, integrability and stability are poorly defined. At this time, the study of dynamical systems is primarily qualitative, and not a quantitative science.

21.7 Riemannian manifolds

An important special case consists of those Hamiltonians that are quadratic forms, that is, Hamiltonians that can be written as

$$\mathcal{H}(q,p) = \frac{1}{2}\langle p,p\rangle_q$$

where $\langle\cdot,\cdot\rangle_q$ is a smoothly varying inner product on the fibers T_q^*Q , the cotangent space to the point q in the configuration space, sometimes called a cometric. This Hamiltonian consists entirely of the kinetic term.

If one considers a Riemannian manifold or a pseudo-Riemannian manifold, the Riemannian metric induces a linear isomorphism between the tangent and cotangent bundles. (See Musical isomorphism). Using this isomorphism, one can define a cometric. (In coordinates, the matrix defining the cometric is the inverse of the matrix defining the metric.) The solutions to the Hamilton–Jacobi equations for this Hamiltonian are then the same as the geodesics on the manifold. In particular, the Hamiltonian flow in this case is the same thing as the geodesic flow. The existence of such solutions, and the completeness of the set of solutions, are discussed in detail in the article on geodesics. See also Geodesics as Hamiltonian flows.

21.8 Sub-Riemannian manifolds

When the cometric is degenerate, then it is not invertible. In this case, one does not have a Riemannian manifold, as one does not have a metric. However, the Hamiltonian still exists. In the case where the cometric is degenerate at every point q of the configuration space manifold Q, so that the rank of the cometric is less than the dimension of the manifold Q, one has a sub-Riemannian manifold.

The Hamiltonian in this case is known as a **sub-Riemannian Hamiltonian**. Every such Hamiltonian uniquely determines the cometric, and vice versa. This implies that every sub-Riemannian manifold is uniquely determined by its sub-Riemannian Hamiltonian, and that the converse is true: every sub-Riemannian manifold has a unique sub-Riemannian Hamiltonian. The existence of sub-Riemannian geodesics is given by the Chow–Rashevskii theorem.

The continuous, real-valued Heisenberg group provides a simple example of a sub-Riemannian manifold. For the Heisenberg group, the Hamiltonian is given by

$$\mathcal{H}(x,y,z,p_x,p_y,p_z) = \frac{1}{2}\left(p_x^2 + p_y^2\right).$$

p_z is not involved in the Hamiltonian.

21.9 Poisson algebras

Hamiltonian systems can be generalized in various ways. Instead of simply looking at the algebra of smooth functions over a symplectic manifold, Hamiltonian mechanics can be formulated on general commutative unital real Poisson algebras. A state is a continuous linear functional on the Poisson algebra (equipped with some suitable topology) such that for any element A of the algebra, A^2 maps to a nonnegative real number.

A further generalization is given by Nambu dynamics.

21.10 Charged particle in an electromagnetic field

A good illustration of Hamiltonian mechanics is given by the Hamiltonian of a charged particle in an electromagnetic field. In Cartesian coordinates (i.e. $q_i = x_i$), the Lagrangian of a non-relativistic classical particle in an electromagnetic field is (in SI Units):

$$\mathcal{L} = \sum_i \tfrac{1}{2}m\dot{x}_i^2 + \sum_i e\dot{x}_i A_i - e\phi,$$

where e is the electric charge of the particle (not necessarily the elementary charge), ϕ is the electric scalar potential, and the A_i are the components of the magnetic vector potential (these may be modified through a gauge transformation). This is called minimal coupling.

The generalized momenta are given by:

$$p_i = \frac{\partial\mathcal{L}}{\partial\dot{x}_i} = m\dot{x}_i + eA_i.$$

Rearranging, the velocities are expressed in terms of the momenta:

$$\dot{x}_i = \frac{p_i - eA_i}{m}.$$

If we substitute the definition of the momenta, and the definitions of the velocities in terms of the momenta, into the definition of the Hamiltonian given above, and then simplify and rearrange, we get:

$$\mathcal{H} = \left\{\sum_i \dot{x}_i p_i\right\} - \mathcal{L} = \sum_i \frac{(p_i - eA_i)^2}{2m} + e\phi.$$

This equation is used frequently in quantum mechanics.

21.11 Relativistic charged particle in an electromagnetic field

The Lagrangian for a relativistic charged particle is given by:

$$\mathcal{L}(t) = -mc^2\sqrt{1 - \frac{\dot{\vec{x}}(t)^2}{c^2}} - e\phi(\vec{x}(t), t) + e\dot{\vec{x}}(t)\cdot\vec{A}(\vec{x}(t), t) \,.$$

Thus the particle's canonical (total) momentum is

$$\vec{P}(t) = \frac{\partial\mathcal{L}(t)}{\partial\dot{\vec{x}}(t)} = \frac{m\dot{\vec{x}}(t)}{\sqrt{1 - \frac{\dot{\vec{x}}(t)^2}{c^2}}} + e\vec{A}(\vec{x}(t), t) \,,$$

that is, the sum of the kinetic momentum and the potential momentum.

Solving for the velocity, we get

$$\dot{\vec{x}}(t) = \frac{\vec{P}(t) - e\vec{A}(\vec{x}(t), t)}{\sqrt{m^2 + \frac{1}{c^2}\left(\vec{P}(t) - e\vec{A}(\vec{x}(t), t)\right)^2}} \,.$$

So the Hamiltonian is

$$\mathcal{H}(t) = \dot{\vec{x}}(t)\cdot\vec{P}(t) - \mathcal{L}(t) = c\sqrt{m^2c^2 + \left(\vec{P}(t) - e\vec{A}(\vec{x}(t), t)\right)^2} + e\phi(\vec{x}(t), t) \,.$$

From this we get the force equation (equivalent to the Euler–Lagrange equation)

$$\dot{\vec{P}} = -\frac{\partial\mathcal{H}}{\partial\vec{x}} = e(\vec{\nabla}\vec{A})\cdot\dot{\vec{x}} - e\vec{\nabla}\phi$$

from which one can derive

$$\frac{d}{dt}\left(\frac{m\dot{\vec{x}}}{\sqrt{1 - \frac{\dot{\vec{x}}^2}{c^2}}}\right) = e\vec{E} + e\dot{\vec{x}}\times\vec{B} \,.$$

An equivalent expression for the Hamiltonian as function of the relativistic (kinetic) momentum, $\vec{p} = \gamma m\dot{\vec{x}}(t)$, is

$$\mathcal{H}(t) = \dot{\vec{x}}(t)\cdot\vec{p}(t) + \frac{mc^2}{\gamma} + e\phi(\vec{x}(t), t) = \gamma mc^2 + e\phi(\vec{x}(t), t) = E + V \,.$$

This has the advantage that $\vec{p}$ can be measured experimentally whereas $\vec{P}$ cannot. Notice that the Hamiltonian (total energy) can be viewed as the sum of the relativistic energy (kinetic+rest), $E = \gamma mc^2$, plus the potential energy, $V = e\phi$.

21.12 See also

- Canonical transformation
- Classical field theory
- Hamiltonian field theory
- Covariant Hamiltonian field theory
- Classical mechanics
- Dynamical systems theory
- Hamilton–Jacobi equation
- Hamilton–Jacobi–Einstein equation
- Lagrangian mechanics
- Maxwell's equations
- Hamiltonian (quantum mechanics)
- Quantum Hamilton's equations
- Quantum field theory
- Hamiltonian optics
- De Donder–Weyl theory

21.13 References

21.13.1 Footnotes

[1] Hand, L. N.; Finch, J. D. (2008). *Analytical Mechanics*. Cambridge University Press. ISBN 978-0-521-57572-0.

[2] Goldstein, Herbert; Poole, Charles P., Jr.; Safko, John L. (2002), *Classical Mechanics* (3rd ed.), San Francisco, CA: Addison Wesley, pp. 347–349, ISBN 0-201-65702-3

[3] "16.3 The Hamiltonian", *MIT OpenCourseWare website 18.013A*, retrieved February 2007

21.13.2 Sources

- Arnol'd, V. I. (1989), *Mathematical Methods of Classical Mechanics*, Springer-Verlag, ISBN 0-387-96890-3
- Abraham, R.; Marsden, J.E. (1978), *Foundations of Mechanics*, London: Benjamin-Cummings, ISBN 0-8053-0102-X
- Arnol'd, V. I.; Kozlov, V. V.; Neĭshtadt, A. I. (1988), *Mathematical aspects of classical and celestial mechanics* **3**, Springer-Verlag
- Vinogradov, A. M.; Kupershmidt, B. A. (1981), *The structure of Hamiltonian mechanics* (DjVu), London Math. Soc. Lect. Notes Ser. **60**, London: Cambridge Univ. Press

21.14 External links

- Binney, James J., *Classical Mechanics (lecture notes)* (PDF), University of Oxford, retrieved 27 October 2010
- Tong, David, *Classical Dynamics (Cambridge lecture notes)*, University of Cambridge, retrieved 27 October 2010
- Hamilton, William Rowan, *On a General Method in Dynamics*, Trinity College Dublin

Chapter 22

Lie algebra

"Lie bracket" redirects here. For the operation on vector fields, see Lie bracket of vector fields.

In mathematics, a **Lie algebra** (/liː/, not /laɪ/) is a vector space together with a non-associative multiplication called "Lie bracket" $[x, y]$. When an algebraic product is defined on the space, the Lie bracket is the commutator $[x, y] = xy - yx$.

Lie algebras were introduced to study the concept of infinitesimal transformations. Hermann Weyl introduced the term "Lie algebra" (after Sophus Lie) in the 1930s. In older texts, the name "**infinitesimal group**" is used.

Lie algebras are closely related to Lie groups which are groups that are also smooth manifolds, with the property that the group operations of multiplication and inversion are smooth maps. Any Lie group gives rise to a Lie algebra. Conversely, to any finite-dimensional Lie algebra over real or complex numbers, there is a corresponding connected Lie group unique up to covering (Lie's third theorem). This correspondence between Lie groups and Lie algebras allows one to study Lie groups in terms of Lie algebras.

22.1 Definitions

A **Lie algebra** is a vector space $\mathfrak{g}$ over some field F together with a binary operation $[\cdot, \cdot] : \mathfrak{g} \times \mathfrak{g} \to \mathfrak{g}$ called the **Lie bracket** that satisfies the following axioms:

- Bilinearity,

$$[ax+by, z] = a[x, z]+b[y, z],$$

$$[z, ax+by] = a[z, x]+b[z, y]$$

  for all scalars a, b in F and all elements x, y, z in $\mathfrak{g}$.

- Alternativity,

$$[x, x] = 0$$

  for all x in $\mathfrak{g}$.

- The Jacobi identity,

$$[x, [y, z]] + [z, [x, y]] + [y, [z, x]] = 0$$

  for all x, y, z in $\mathfrak{g}$.

Using bilinearity to expand the Lie bracket $[x + y, x + y]$ and using alternativity shows that $[x, y] + [y, x] = 0$ for all elements x, y in $\mathfrak{g}$, showing that bilinearity and alternativity together imply

- Anticommutativity,

$$[x,y] = -[y,x],$$

  for all elements x, y in $\mathfrak{g}$. Anticommutativity only implies the alternating property if the field's characteristic is not 2.[1]

It is customary to express a Lie algebra in lower-case fraktur, like $\mathfrak{g}$. If a Lie algebra is associated with a Lie group, then the spelling of the Lie algebra is the same as that Lie group. For example, the Lie algebra of SU(n) is written as $\mathfrak{su}(n)$.

22.1.1 Generators and dimension

Elements of a Lie algebra $\mathfrak{g}$ are said to be **generators** of the Lie algebra if the smallest subalgebra of $\mathfrak{g}$ containing them is $\mathfrak{g}$ itself. The **dimension** of a Lie algebra is its dimension as a vector space over F. The cardinality of a minimal generating set of a Lie algebra is always less than or equal to its dimension.

22.1.2 Subalgebras, ideals and homomorphisms

The Lie bracket is not associative in general, meaning that $[[x, y], z]$ need not equal $[x, [y, z]]$. Nonetheless, much of the terminology that was developed in the theory of associative rings or associative algebras is commonly applied to Lie algebras. A subspace $\mathfrak{h} \subseteq \mathfrak{g}$ that is closed under the Lie bracket is called a **Lie subalgebra**. If a subspace $I \subseteq \mathfrak{g}$ satisfies a stronger condition that

$$[\mathfrak{g}, I] \subseteq I,$$

then I is called an **ideal** in the Lie algebra $\mathfrak{g}$.[2] A **homomorphism** between two Lie algebras (over the same base field) is a linear map that is compatible with the respective Lie brackets:

$$f : \mathfrak{g} \to \mathfrak{g}', \quad f([x, y]) = [f(x), f(y)],$$

for all elements x and y in $\mathfrak{g}$. As in the theory of associative rings, ideals are precisely the kernels of homomorphisms, given a Lie algebra $\mathfrak{g}$ and an ideal I in it, one constructs the **factor algebra** $\mathfrak{g}/I$, and the first isomorphism theorem holds for Lie algebras.

Let S be a subset of $\mathfrak{g}$. The set of elements x such that $[x, s] = 0$ for all s in S forms a subalgebra called the centralizer of S. The centralizer of $\mathfrak{g}$ itself is called the center of $\mathfrak{g}$. Similar to centralizers, if S is a subspace,[3] then the set of x such that $[x, s]$ is in S for all s in S forms a subalgebra called the normalizer of S.

22.1.3 Direct sum and semidirect product

Given two Lie algebras $\mathfrak{g}$ and $\mathfrak{g}'$, their direct sum is the Lie algebra consisting of the vector space $\mathfrak{g} \oplus \mathfrak{g}'$, of the pairs $(x, x'),\ x \in \mathfrak{g}, x' \in \mathfrak{g}'$, with the operation

$$[(x, x'), (y, y')] = ([x, y], [x', y']), \quad x, y \in \mathfrak{g},\ x', y' \in \mathfrak{g}'.$$

Let $\mathfrak{g}$ be a Lie algebra and $\mathfrak{i}$ its ideal. If the canonical map $\mathfrak{g} \to \mathfrak{g}/\mathfrak{i}$ splits (i.e., admits a section), then $\mathfrak{g}$ is said to be a semidirect product of $\mathfrak{i}$ and $\mathfrak{g}/\mathfrak{i}$, $\mathfrak{g} = \mathfrak{g}/\mathfrak{i} \ltimes \mathfrak{i}$.

Levi's theorem says that a finite-dimensional Lie algebra is a semidirect product of its radical and the complementary subalgebra (Levi subalgebra).

22.2 Properties

22.2.1 Admits an enveloping algebra

See also: Universal enveloping algebra

For any associative algebra A with multiplication $*$, one can construct a Lie algebra $L(A)$. As a vector space, $L(A)$ is the same as A. The Lie bracket of two elements of $L(A)$ is defined to be their commutator in A:

$$[a, b] = a * b - b * a.$$

The associativity of the multiplication $*$ in A implies the Jacobi identity of the commutator in $L(A)$. For example, the associative algebra of $n \times n$ matrices over a field F gives rise to the general linear Lie algebra $\mathfrak{gl}_n(F)$. The associative algebra A is called an **enveloping algebra** of the Lie algebra $L(A)$. Every Lie algebra can be embedded into one that arises from an associative algebra in this fashion; see universal enveloping algebra.

22.2.2 Representation

Given a vector space V, let $\mathfrak{gl}(V)$ denote the Lie algebra enveloped by the associative algebra of all linear endomorphisms of V. A representation of a Lie algebra $\mathfrak{g}$ on V is a Lie algebra homomorphism

$$\pi : \mathfrak{g} \to \mathfrak{gl}(V).$$

A representation is said to be faithful if its kernel is trivial. Every finite-dimensional Lie algebra has a faithful representation on a finite-dimensional vector space (Ado's theorem).[4]

For example,

$$\mathrm{ad} : \mathfrak{g} \to \mathfrak{gl}(\mathfrak{g})$$

given by $\mathrm{ad}(x)(y) = [x, y]$ is a representation of $\mathfrak{g}$ on the vector space $\mathfrak{g}$ called the adjoint representation. A derivation on the Lie algebra $\mathfrak{g}$ (in fact on any non-associative algebra) is a linear map $\delta : \mathfrak{g} \to \mathfrak{g}$ that obeys the Leibniz' law, that is,

$$\delta([x, y]) = [\delta(x), y] + [x, \delta(y)]$$

for all x and y in the algebra. For any x, $\mathrm{ad}(x)$ is a derivation; a consequence of the Jacobi identity. Thus, the image of ad lies in the subalgebra of $\mathfrak{gl}(\mathfrak{g})$ consisting of derivations on $\mathfrak{g}$. A derivation that happens to be in the image of ad is called an inner derivation. If $\mathfrak{g}$ is semisimple, every derivation on $\mathfrak{g}$ is inner.

22.3 Examples

22.3.1 Vector spaces

- Any vector space V endowed with the identically zero Lie bracket becomes a Lie algebra. Such Lie algebras are called abelian, cf. below. Any one-dimensional Lie algebra over a field is abelian, by the antisymmetry of the Lie bracket.
- The real vector space of all $n \times n$ skew-hermitian matrices is closed under the commutator and forms a real Lie algebra denoted $\mathfrak{u}(n)$. This is the Lie algebra of the unitary group $U(n)$.

22.3.2 Subspaces

- The subspace of the general linear Lie algebra $\mathfrak{gl}_n(F)$ consisting of matrices of trace zero is a subalgebra,[5] the special linear Lie algebra, denoted $\mathfrak{sl}_n(F)$.

22.3.3 Real matrix groups

- Any Lie group G defines an associated real Lie algebra $\mathfrak{g}$ =Lie(G). The definition in general is somewhat technical, but in the case of real matrix groups, it can be formulated via the exponential map, or the matrix exponent. The Lie algebra $\mathfrak{g}$ consists of those matrices X for which $\exp(tX) \in G$, $\forall$ real numbers t.

  The Lie bracket of $\mathfrak{g}$ is given by the commutator of matrices. As a concrete example, consider the special linear group SL(n,**R**), consisting of all $n \times n$ matrices with real entries and determinant 1. This is a matrix Lie group, and its Lie algebra consists of all $n \times n$ matrices with real entries and trace 0.

22.3.4 Three dimensions

- The three-dimensional Euclidean space $\mathbf{R}^3$ with the Lie bracket given by the cross product of vectors becomes a three-dimensional Lie algebra.
- The Heisenberg algebra H_3(R) is a three-dimensional Lie algebra generated by elements x, y and z with Lie brackets

$$[x,y]=z,\quad [x,z]=0,\quad [y,z]=0$$

It is explicitly realized as the space of 3×3 strictly upper-triangular matrices, with the Lie bracket given by the matrix commutator,

$$x=\begin{pmatrix}0&1&0\\0&0&0\\0&0&0\end{pmatrix},\quad y=\begin{pmatrix}0&0&0\\0&0&1\\0&0&0\end{pmatrix}$$

$$,\ \mathrm{z}\ =\ \begin{pmatrix}0&0&1\\0&0&0\\0&0&0\end{pmatrix}.$$

Any element of the Heisenberg group is thus representable as a product of group generators, i.e., matrix exponentials of these Lie algebra generators,

$$\begin{pmatrix}1&a&c\\0&1&b\\0&0&1\end{pmatrix}=e^{by}e^{cz}e^{ax}\ .$$

- The commutation relations between the x, y, and z components of the angular momentum operator in quantum mechanics are the same as those of $\mathfrak{su}(2)$ and $\mathfrak{so}(3)$,

$$[L_x,L_y]=i\hbar L_z$$

$$[L_y,L_z]=i\hbar L_x$$

$$[L_z,L_x]=i\hbar L_y$$

(The physicist convention for Lie algebras is used in the above equations, hence the factor of i.) The Lie algebra formed by these operators have, in fact, representations of all finite dimensions.

22.3.5 Infinite dimensions

- An important class of infinite-dimensional real Lie algebras arises in differential topology. The space of smooth vector fields on a differentiable manifold M forms a Lie algebra, where the Lie bracket is defined to be the commutator of vector fields. One way of expressing the Lie bracket is through the formalism of Lie derivatives, which identifies a vector field X with a first order partial differential operator LX acting on smooth functions by letting $LX(f)$ be the directional derivative of the function f in the direction of X. The Lie bracket $[X,Y]$ of two vector fields is the vector field defined through its action on functions by the formula:

$$L_{[X,Y]}f=L_X(L_Yf)-L_Y(L_Xf).$$

- A Kac–Moody algebra is an example of an infinite-dimensional Lie algebra.
- The Moyal algebra is an infinite-dimensional Lie algebra which contains all classical Lie algebras as subalgebras.

22.4 Structure theory and classification

Lie algebras can be classified to some extent. In particular, this has an application to the classification of Lie groups.

22.4.1 Abelian, nilpotent, and solvable

Analogously to abelian, nilpotent, and solvable groups, defined in terms of the derived subgroups, one can define abelian, nilpotent, and solvable Lie algebras.

A Lie algebra $\mathfrak{g}$ is **abelian** if the Lie bracket vanishes, i.e. $[x,y] = 0$, for all x and y in $\mathfrak{g}$. Abelian Lie algebras correspond to commutative (or abelian) connected Lie groups such as vector spaces K^n or tori T^n, and are all of the form $\mathfrak{k}^n$, meaning an n-dimensional vector space with the trivial Lie bracket.

A more general class of Lie algebras is defined by the vanishing of all commutators of given length. A Lie algebra $\mathfrak{g}$ is **nilpotent** if the lower central series

$$\mathfrak{g} > [\mathfrak{g}, \mathfrak{g}] > [[\mathfrak{g}, \mathfrak{g}], \mathfrak{g}] > [[[\mathfrak{g}, \mathfrak{g}], \mathfrak{g}], \mathfrak{g}] > \cdots$$

becomes zero eventually. By Engel's theorem, a Lie algebra is nilpotent if and only if for every u in $\mathfrak{g}$ the adjoint endomorphism

$$\mathrm{ad}(u) : \mathfrak{g} \to \mathfrak{g}, \quad \mathrm{ad}(u)v = [u, v]$$

is nilpotent.

More generally still, a Lie algebra $\mathfrak{g}$ is said to be **solvable** if the derived series:

$$\mathfrak{g} > [\mathfrak{g}, \mathfrak{g}] > [[\mathfrak{g}, \mathfrak{g}], [\mathfrak{g}, \mathfrak{g}]] > [[[\mathfrak{g}, \mathfrak{g}], [\mathfrak{g}, \mathfrak{g}]], [[\mathfrak{g}, \mathfrak{g}], [\mathfrak{g}, \mathfrak{g}]]] > \cdots$$

becomes zero eventually.

Every finite-dimensional Lie algebra has a unique maximal solvable ideal, called its radical. Under the Lie correspondence, nilpotent (respectively, solvable) connected Lie groups correspond to nilpotent (respectively, solvable) Lie algebras.

22.4.2 Simple and semisimple

A Lie algebra is "simple" if it has no non-trivial ideals and is not abelian. A Lie algebra $\mathfrak{g}$ is called **semisimple** if its radical is zero. Equivalently, $\mathfrak{g}$ is semisimple if it does not contain any non-zero abelian ideals. In particular, a simple Lie algebra is semisimple. Conversely, it can be proven that any semisimple Lie algebra is the direct sum of its minimal ideals, which are canonically determined simple Lie algebras.

The concept of semisimplicity for Lie algebras is closely related with the complete reducibility (semisimplicity) of their representations. When the ground field F has characteristic zero, any finite-dimensional representation of a semisimple Lie algebra is semisimple (i.e., direct sum of irreducible representations.) In general, a Lie algebra is called reductive if the adjoint representation is semisimple. Thus, a semisimple Lie algebra is reductive.

22.4.3 Cartan's criterion

Cartan's criterion gives conditions for a Lie algebra to be nilpotent, solvable, or semisimple. It is based on the notion of the Killing form, a symmetric bilinear form on $\mathfrak{g}$ defined by the formula

$$K(u, v) = \mathrm{tr}(\mathrm{ad}(u)\,\mathrm{ad}(v)),$$

where tr denotes the trace of a linear operator. A Lie algebra $\mathfrak{g}$ is semisimple if and only if the Killing form is nondegenerate. A Lie algebra $\mathfrak{g}$ is solvable if and only if $K(\mathfrak{g}, [\mathfrak{g}, \mathfrak{g}]) = 0$.

22.4.4 Classification

The Levi decomposition expresses an arbitrary Lie algebra as a semidirect sum of its solvable radical and a semisimple Lie algebra, almost in a canonical way. Furthermore, semisimple Lie algebras over an algebraically closed field have been completely classified through their root systems. However, the classification of solvable Lie algebras is a 'wild' problem, and cannot be accomplished in general.

22.5 Relation to Lie groups

See also: Lie group–Lie algebra correspondence

Although Lie algebras are often studied in their own right, historically they arose as a means to study Lie groups.

Lie's fundamental theorems describe a relation between Lie groups and Lie algebras. In particular, any Lie group gives rise to a canonically determined Lie algebra (concretely, *the tangent space at the identity*); and, conversely, for any finite-dimensional Lie algebra there is a corresponding connected Lie group (Lie's third theorem; see the Baker–Campbell–Hausdorff formula). This Lie group is not determined uniquely; however, any two connected Lie groups with the same Lie algebra are *locally isomorphic*, and in particular, have the same universal cover. For instance, the special orthogonal group SO(3) and the special unitary group SU(2) give rise to the same Lie algebra, which is isomorphic to $\mathbf{R}^3$ with the cross-product, while SU(2) is a simply-connected twofold cover of SO(3).

Given a Lie group, a Lie algebra can be associated to it either by endowing the tangent space to the identity with the differential of the adjoint map, or by considering the left-invariant vector fields as mentioned in the examples. In the case of real matrix groups, the Lie algebra $\mathfrak{g}$ consists of those matrices X for which $\exp(tX) \in G$ for all real numbers t, where exp is the exponential map.

Some examples of Lie algebras corresponding to Lie groups are the following:

- The Lie algebra $\mathfrak{gl}_n(\mathbb{C})$ for the group $\mathrm{GL}_n(\mathbb{C})$ is the algebra of complex $n{\times}n$ matrices
- The Lie algebra $\mathfrak{sl}_n(\mathbb{C})$ for the group $\mathrm{SL}_n(\mathbb{C})$ is the algebra of complex $n{\times}n$ matrices with trace 0
- The Lie algebras $\mathfrak{o}_n(\mathbb{R})$ for the group $\mathrm{O}_n(\mathbb{R})$ and $\mathfrak{so}_n(\mathbb{R})$ for $\mathrm{SO}_n(\mathbb{R})$ are both the algebra of real anti-symmetric $n{\times}n$ matrices (See Antisymmetric matrix: Infinitesimal rotations for a discussion)
- The Lie algebra $\mathfrak{u}_n(\mathbb{C})$ for the group $\mathrm{U}_n(\mathbb{C})$ is the algebra of skew-Hermitian complex $n{\times}n$ matrices while the Lie algebra $\mathfrak{su}_n(\mathbb{C})$ for $\mathrm{SU}_n(\mathbb{C})$ is the algebra of skew-Hermitian, traceless complex $n{\times}n$ matrices.

In the above examples, the Lie bracket $[X, Y]$ (for X and Y matrices in the Lie algebra) is defined as $[X, Y] = XY - YX$.

Given a set of generators T^a, the **structure constants** f^{abc} express the Lie brackets of pairs of generators as linear combinations of generators from the set, i.e., $[T^a, T^b] = f^{abc}\, T^c$. The structure constants determine the Lie brackets of elements of the Lie algebra, and consequently nearly completely determine the group structure of the Lie group. The structure of the Lie group near the identity element is displayed explicitly by the Baker–Campbell–Hausdorff formula, an expansion in Lie algebra elements X, Y and their Lie brackets, all nested together within a single exponent, $\exp(tX)\exp(tY) = \exp(tX+tY+½\, t^2[X,Y] + O(t^3))$.

The mapping from Lie groups to Lie algebras is functorial, which implies that homomorphisms of Lie groups lift to homomorphisms of Lie algebras, and various properties are satisfied by this lifting: it commutes with composition, it maps Lie subgroups, kernels, quotients and cokernels of Lie groups to subalgebras, kernels, quotients and cokernels of Lie algebras, respectively.

The functor **L** which takes each Lie group to its Lie algebra and each homomorphism to its differential is faithful and exact. It is however not an equivalence of categories: different Lie groups may have isomorphic Lie algebras (for example SO(3) and SU(2)), and there are (infinite dimensional) Lie algebras that are not associated to any Lie group.[6]

However, when the Lie algebra $\mathfrak{g}$ is finite-dimensional, one can associate to it a simply connected Lie group having $\mathfrak{g}$ as its Lie algebra. More precisely, the Lie algebra functor **L** has a left adjoint functor **Γ** from finite-dimensional (real) Lie algebras to Lie groups, factoring through the full subcategory of simply connected Lie groups.[7] In other words, there is a natural isomorphism of bifunctors

$$\mathrm{Hom}(\Gamma(\mathfrak{g}), H) \cong \mathrm{Hom}(\mathfrak{g}, \mathrm{L}(H)).$$

The adjunction $\mathfrak{g} \to \mathrm{L}(\Gamma(\mathfrak{g}))$ (corresponding to the identity on $\Gamma(\mathfrak{g})$) is an isomorphism, and the other adjunction $\Gamma(\mathrm{L}(H)) \to H$ is the projection homomorphism from the universal cover group of the identity component of H to H. It follows immediately that if G is simply connected, then the Lie algebra functor establishes a bijective correspondence between Lie group homomorphisms $G \to H$ and Lie algebra homomorphisms $\mathbf{L}(G) \to \mathbf{L}(H)$.

The universal cover group above can be constructed as the image of the Lie algebra under the exponential map. More generally, we have that the Lie algebra is homeomorphic to a neighborhood of the identity. But globally, if the Lie group is compact, the exponential will not be injective, and if the Lie group is not connected, simply connected or compact, the exponential map need not be surjective.

If the Lie algebra is infinite-dimensional, the issue is more subtle. In many instances, the exponential map is not even locally a homeomorphism (for example, in Diff($\mathbf{S}^1$), one may find diffeomorphisms arbitrarily close to the identity that are not in the image of exp). Furthermore, some infinite-dimensional Lie algebras are not the Lie algebra of any group.

The correspondence between Lie algebras and Lie groups is used in several ways, including in the classification of Lie groups and the related matter of the representation theory of Lie groups. Every representation of a Lie algebra lifts uniquely to a representation of the corresponding connected, simply connected Lie group, and conversely every representation of any Lie group induces a representation of the group's Lie algebra; the representations are in one to one correspondence. Therefore, knowing the representations of a Lie algebra settles the question of representations of the group.

As for classification, it can be shown that any connected Lie group with a given Lie algebra is isomorphic to the universal cover mod a discrete central subgroup. So classifying Lie groups becomes simply a matter of counting the discrete subgroups of the center, once the classification of Lie algebras is known (solved by Cartan et al. in the semisimple case).

22.6 Category theoretic definition

Using the language of category theory, a **Lie algebra** can be defined as an object A in **Vec***k*, the category of vector spaces over a field k of characteristic not 2, together with a morphism [.,.]: $A \otimes A \rightarrow A$, where $\otimes$ refers to the monoidal product of **Vec***k*, such that

- $[\cdot,\cdot] \circ (\mathrm{id} + \tau_{A,A}) = 0$
- $[\cdot,\cdot] \circ ([\cdot,\cdot] \otimes \mathrm{id}) \circ (\mathrm{id} + \sigma + \sigma^2) = 0$

where $\tau\ (a \otimes b) := b \otimes a$ and σ is the cyclic permutation braiding $(\mathrm{id} \otimes \tau A,A)\ ^\circ\ (\tau A,A \otimes \mathrm{id})$. In diagrammatic form:

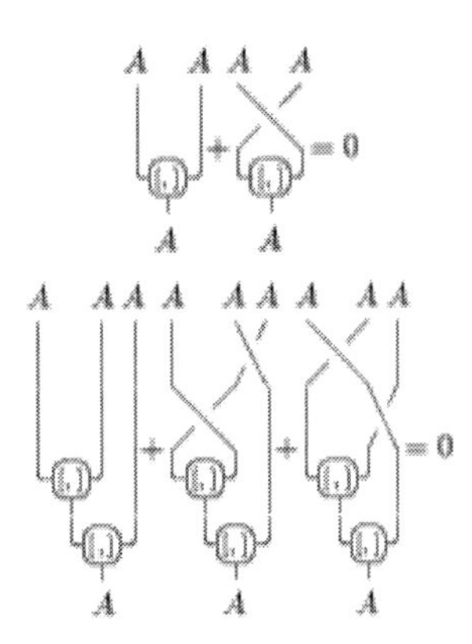

22.7 Lie ring

A **Lie ring** arises as a generalisation of Lie algebras, or through the study of the lower central series of groups. A **Lie ring** is defined as a nonassociative ring with multiplication that is anticommutative and satisfies the Jacobi identity. More specifically we can define a Lie ring L to be an abelian group with an operation $[\cdot,\cdot]$ that has the following properties:

- Bilinearity:

$$[x+y,z] = [x,z]+[y,z], \quad [z,x+y] = [z,x]+[z,y]$$

  for all $x, y, z \in L$.

- The *Jacobi identity*:

$$[x,[y,z]] + [y,[z,x]] + [z,[x,y]] = 0$$

  for all x, y, z in L.

- For all x in L:

$$[x,x] = 0$$

Lie rings need not be Lie groups under addition. Any Lie algebra is an example of a Lie ring. Any associative ring can be made into a Lie ring by defining a bracket operator $[x,y] = xy - yx$. Conversely to any Lie algebra there is a corresponding ring, called the universal enveloping algebra.

Lie rings are used in the study of finite p-groups through the Lazard correspondence. The lower central factors of a p-group are finite abelian p-groups, so modules over **Z**/p**Z**. The direct sum of the lower central factors is given the structure of a Lie ring by defining the bracket to be the commutator of two coset representatives. The Lie ring structure is enriched with another module homomorphism, then pth power map, making the associated Lie ring a so-called restricted Lie ring.

Lie rings are also useful in the definition of a p-adic analytic groups and their endomorphisms by studying Lie algebras over rings of integers such as the p-adic integers. The definition of finite groups of Lie type due to Chevalley involves restricting from a Lie algebra over the complex numbers to a Lie algebra over the integers, and the reducing modulo p to get a Lie algebra over a finite field.

22.7.1 Examples

- Any Lie algebra over a general ring instead of a field is an example of a Lie ring. Lie rings are *not* Lie groups under addition, despite the name.
- Any associative ring can be made into a Lie ring by defining a bracket operator $[x,y] = xy - yx$.

- For an example of a Lie ring arising from the study of groups, let G be a group with $(x, y) = x^{-1}y^{-1}xy$ the commutator operation, and let $G = G_0 \supseteq G_1 \supseteq G_2 \supseteq \cdots \supseteq G_n \supseteq \cdots$ be a central series in G — that is the commutator subgroup (G_i, G_j) is contained in G_{i+j} for any i, j . Then

$$L = \bigoplus G_i/G_{i+1}$$

is a Lie ring with addition supplied by the group operation (which will be commutative in each homogeneous part), and the bracket operation given by

$$[xG_i, yG_j] = (x, y)G_{i+j}$$

(x, y)

22.8 See also

22.9 Notes

[1] Humphreys p. 1

[2] Due to the anticommutativity of the commutator, the notions of a left and right ideal in a Lie algebra coincide.

[3] Jacobson 1962, pg. 28

[4] Jacobson 1962, Ch. VI

[5] Humphreys p.2

[6] Beltita 2005, pg. 75

[7] Adjoint property is discussed in more general context in Hofman & Morris (2007) (e.g., page 130) but is a straightforward consequence of, e.g., Bourbaki (1989) Theorem 1 of page 305 and Theorem 3 of page 310.

22.10 References

- Beltita, Daniel. *Smooth Homogeneous Structures in Operator Theory*, CRC Press, 2005. ISBN 978-1-4200-3480-6
- Boza, Luis; Fedriani, Eugenio M. & Núñez, Juan. *A new method for classifying complex filiform Lie algebras*, Applied Mathematics and Computation, 121 (2-3): 169–175, 2001
- Bourbaki, Nicolas. “Lie Groups and Lie Algebras - Chapters 1-3”, Springer, 1989, ISBN 3-540-64242-0
- Erdmann, Karin & Wildon, Mark. *Introduction to Lie Algebras*, 1st edition, Springer, 2006. ISBN 1-84628-040-0
- Hall, Brian C. *Lie Groups, Lie Algebras, and Representations: An Elementary Introduction*, Springer, 2003. ISBN 0-387-40122-9
- Hofman, Karl & Morris, Sidney. “The Lie Theory of Connected Pro-Lie Groups”, European Mathematical Society, 2007, ISBN 978-3-03719-032-6
- Humphreys, James E. *Introduction to Lie Algebras and Representation Theory*, Second printing, revised. Graduate Texts in Mathematics, 9. Springer-Verlag, New York, 1978. ISBN 0-387-90053-5
- Jacobson, Nathan, *Lie algebras*, Republication of the 1962 original. Dover Publications, Inc., New York, 1979. ISBN 0-486-63832-4
- Kac, Victor G. et al. *Course notes for MIT 18.745: Introduction to Lie Algebras*, math.mit.edu
- O'Connor, J.J. & Robertson, E.F. Biography of Sophus Lie, MacTutor History of Mathematics Archive, www-history.mcs.st-andrews.ac.uk
- O'Connor, J.J. & Robertson, E.F. Biography of Wilhelm Killing, MacTutor History of Mathematics Archive, www-history.mcs.st-andrews.ac.uk
- Serre, Jean-Pierre. “Lie Algebras and Lie Groups”, 2nd edition, Springer, 2006. ISBN 3-540-55008-9
- Steeb, W.-H. *Continuous Symmetries, Lie Algebras, Differential Equations and Computer Algebra*, second edition, World Scientific, 2007, ISBN 978-981-270-809-0
- Varadarajan, V.S. *Lie Groups, Lie Algebras, and Their Representations*, 1st edition, Springer, 2004. ISBN 0-387-90969-9.

22.11 External links

- Hazewinkel, Michiel, ed. (2001), “Lie algebra”, *Encyclopedia of Mathematics*, Springer, ISBN 978-1-55608-010-4
- McKenzie, Douglas, (2015), "An Elementary Introduction to Lie Algebras for Physicists"

Chapter 23

Lie group

In mathematics, a **Lie group** /ˈliː/ is a group that is also a differentiable manifold, with the property that the group operations are compatible with the smooth structure. Lie groups are named after Sophus Lie, who laid the foundations of the theory of continuous transformation groups. The term *groupes de Lie* first appeared in French in 1893 in the thesis of Lie's student Arthur Tresse, page 3.[1]

Lie groups represent the best-developed theory of continuous symmetry of mathematical objects and structures, which makes them indispensable tools for many parts of contemporary mathematics, as well as for modern theoretical physics. They provide a natural framework for analysing the continuous symmetries of differential equations (differential Galois theory), in much the same way as permutation groups are used in Galois theory for analysing the discrete symmetries of algebraic equations. An extension of Galois theory to the case of continuous symmetry groups was one of Lie's principal motivations.

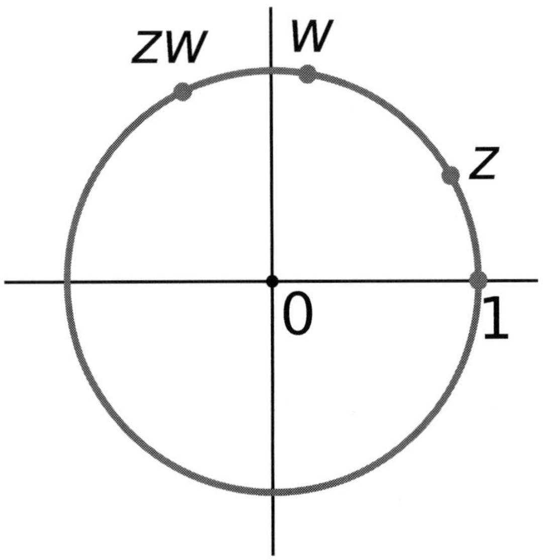

The circle of center 0 and radius 1 in the complex plane is a Lie group with complex multiplication.

23.1 Overview

Lie groups are smooth differentiable manifolds and as such can be studied using differential calculus, in contrast with the case of more general topological groups. One of the key ideas in the theory of Lie groups is to replace the *global* object, the group, with its *local* or linearized version, which Lie himself called its "infinitesimal group" and which has since become known as its Lie algebra.

Lie groups play an enormous role in modern geometry, on several different levels. Felix Klein argued in his Erlangen program that one can consider various "geometries" by specifying an appropriate transformation group that leaves certain geometric properties invariant. Thus Euclidean geometry corresponds to the choice of the group E(3) of distance-preserving transformations of the Euclidean space $\mathbf{R}^3$, conformal geometry corresponds to enlarging the group to the conformal group, whereas in projective geometry one is interested in the properties invariant under the projective group. This idea later led to the notion of a G-structure, where G is a Lie group of "local" symmetries of a manifold.

On a "global" level, whenever a Lie group acts on a geometric object, such as a Riemannian or a symplectic manifold, this action provides a measure of rigidity and yields a rich algebraic structure. The presence of continuous symmetries expressed via a Lie group action on a manifold places strong constraints on its geometry and facilitates analysis on the manifold. Linear actions of Lie groups are especially important, and are studied in representation theory.

In the 1940s–1950s, Ellis Kolchin, Armand Borel, and Claude Chevalley realised that many foundational results concerning Lie groups can be developed completely algebraically, giving rise to the theory of algebraic groups defined over an arbitrary field. This insight opened new possibilities in pure algebra, by providing a uniform construction for most finite simple groups, as well as in algebraic geometry. The theory of automorphic forms, an important branch of modern number theory, deals extensively with analogues of Lie groups over adele rings; p-adic Lie groups play an important role, via their connections with Galois representations in number theory.

23.2 Definitions and examples

A **real Lie group** is a group that is also a finite-dimensional real smooth manifold, in which the group operations of multiplication and inversion are smooth maps. Smoothness of the group multiplication

$$\mu : G \times G \to G \quad \mu(x, y) = xy$$

means that μ is a smooth mapping of the product manifold $G \times G$ into G. These two requirements can be combined to the single requirement that the mapping

$$(x, y) \mapsto x^{-1}y$$

be a smooth mapping of the product manifold into G.

23.2.1 First examples

- The 2×2 real invertible matrices form a group under multiplication, denoted by GL(2, **R**) or by GL*2*(**R**):

$$\mathrm{GL}(2, \mathbf{R}) = \left\{ A = \begin{pmatrix} a & b \\ c & d \end{pmatrix} : \det A = ad - bc \neq 0 \right\}.$$

  This is a four-dimensional noncompact real Lie group. This group is disconnected; it has two connected components corresponding to the positive and negative values of the determinant.

- The rotation matrices form a subgroup of GL(2, **R**), denoted by SO(2, **R**). It is a Lie group in its own right: specifically, a one-dimensional compact connected Lie group which is diffeomorphic to the circle. Using the rotation angle φ as a parameter, this group can be parametrized as follows:

$$\mathrm{SO}(2, \mathbf{R}) = \left\{ \begin{pmatrix} \cos\varphi & -\sin\varphi \\ \sin\varphi & \cos\varphi \end{pmatrix} : \varphi \in \mathbf{R}/2\pi\mathbf{Z} \right\}.$$

  Addition of the angles corresponds to multiplication of the elements of SO(2, **R**), and taking the opposite angle corresponds to inversion. Thus both multiplication and inversion are differentiable maps.

- The orthogonal group also forms an interesting example of a Lie group.

All of the previous examples of Lie groups fall within the class of classical groups.

23.2.2 Related concepts

A **complex Lie group** is defined in the same way using complex manifolds rather than real ones (example: SL(2, **C**)), and similarly, using an alternate metric completion of **Q**, one can define a ***p*-adic Lie group** over the *p*-adic numbers, a topological group in which each point has a *p*-adic neighborhood. Hilbert's fifth problem asked whether replacing differentiable manifolds with topological or analytic ones can yield new examples. The answer to this question turned out to be negative: in 1952, Gleason, Montgomery and Zippin showed that if *G* is a topological manifold with continuous group operations, then there exists exactly one analytic structure on *G* which turns it into a Lie group (see also Hilbert–Smith conjecture). If the underlying manifold is allowed to be infinite-dimensional (for example, a Hilbert manifold), then one arrives at the notion of an infinite-dimensional Lie group. It is possible to define analogues of many Lie groups over finite fields, and these give most of the examples of finite simple groups.

The language of category theory provides a concise definition for Lie groups: a Lie group is a group object in the category of smooth manifolds. This is important, because it allows generalization of the notion of a Lie group to Lie supergroups.

23.3 More examples of Lie groups

See also: Table of Lie groups and List of simple Lie groups

Lie groups occur in abundance throughout mathematics and physics. Matrix groups or algebraic groups are (roughly) groups of matrices (for example, orthogonal and symplectic groups), and these give most of the more common examples of Lie groups.

23.3.1 Examples with a specific number of dimensions

- The circle group $\mathbf{S}^1$ consisting of angles mod 2π under addition or, alternatively, the complex numbers with absolute value 1 under multiplication. This is a one-dimensional compact connected abelian Lie group.
- The 3-sphere $\mathbf{S}^3$ forms a Lie group by identification with the set of quaternions of unit norm, called versors. The only other spheres that admit the structure of a Lie group are the 0-sphere $\mathbf{S}^0$ (real numbers with absolute value 1) and the circle $\mathbf{S}^1$ (complex numbers with absolute value 1). For example, for even $n > 1$, $\mathbf{S}^n$ is not a Lie group because it does not admit a nonvanishing vector field and so *a fortiori* cannot be parallelizable as a differentiable manifold. Of the spheres only $\mathbf{S}^0$, $\mathbf{S}^1$, $\mathbf{S}^3$, and $\mathbf{S}^7$ are parallelizable. The last carries the structure of a Lie quasigroup (a nonassociative group), which can be identified with the set of unit octonions.
- The (3-dimensional) metaplectic group is a double cover of SL(2, **R**) playing an important role in the theory of modular forms. It is a connected Lie group that cannot be faithfully represented by matrices of finite size, i.e., a nonlinear group.
- The Heisenberg group is a connected nilpotent Lie group of dimension 3, playing a key role in quantum mechanics.
- The Lorentz group is a 6-dimensional Lie group of linear isometries of the Minkowski space.
- The Poincaré group is a 10-dimensional Lie group of affine isometries of the Minkowski space.
- The group U(1)×SU(2)×SU(3) is a Lie group of dimension 1+3+8=12 that is the gauge group of the Standard Model in particle physics. The dimensions of the factors correspond to the 1 photon + 3 vector bosons + 8 gluons of the standard model
- The exceptional Lie groups of types G_2, F_4, E_6, E_7, E_8 have dimensions 14, 52, 78, 133, and 248. Along with the A-B-C-D series of simple Lie groups, the exceptional groups complete the list of simple Lie groups. There is also a Lie group named $E_7½$ of dimension 190, but it is not a *simple* Lie group.

23.3.2 Examples with n dimensions

- Euclidean space $\mathbf{R}^n$ with ordinary vector addition as the group operation becomes an n-dimensional noncompact abelian Lie group.
- The Euclidean group E(n, **R**) is the Lie group of all Euclidean motions, i.e., isometric affine maps, of n-dimensional Euclidean space $\mathbf{R}^n$.
- The orthogonal group O(n, **R**), consisting of all $n \times n$ orthogonal matrices with real entries is an $n(n-1)/2$-dimensional Lie group. This group is disconnected, but it has a connected subgroup SO(n, **R**) of the same dimension consisting of orthogonal matrices of determinant 1, called the special orthogonal group (for n = 3, the rotation group SO(3)).
- The unitary group U(n) consisting of $n \times n$ unitary matrices (with complex entries) is a compact connected Lie group of dimension n^2. Unitary matrices of determinant 1 form a closed connected subgroup of dimension $n^2 - 1$ denoted SU(n), the special unitary group.
- Spin groups are double covers of the special orthogonal groups, used for studying fermions in quantum field theory (among other things).
- The group GL(n, **R**) of invertible matrices (under matrix multiplication) is a Lie group of dimension n^2, called the general linear group. It has a closed connected subgroup SL(n, **R**), the special linear group, consisting of matrices of determinant 1 which is also a Lie group.
- The symplectic group Sp($2n$, **R**) consists of all $2n \times 2n$ matrices preserving a *symplectic form* on $\mathbf{R}^{2n}$. It is a connected Lie group of dimension $2n^2 + n$.
- The group of invertible upper triangular n by n matrices is a solvable Lie group of dimension $n(n+1)/2$. (cf. Borel subgroup)
- The A-series, B-series, C-series and D-series, whose elements are denoted by An, Bn, Cn, and Dn, are infinite families of simple Lie groups.

23.3.3 Constructions

There are several standard ways to form new Lie groups from old ones:

- The product of two Lie groups is a Lie group.
- Any topologically closed subgroup of a Lie group is a Lie group. This is known as the Closed subgroup theorem or **Cartan's theorem**.
- The quotient of a Lie group by a closed normal subgroup is a Lie group.

- The universal cover of a connected Lie group is a Lie group. For example, the group **R** is the universal cover of the circle group $\mathbf{S}^1$. In fact any covering of a differentiable manifold is also a differentiable manifold, but by specifying *universal* cover, one guarantees a group structure (compatible with its other structures).

23.3.4 Related notions

Some examples of groups that are *not* Lie groups (except in the trivial sense that any group can be viewed as a 0-dimensional Lie group, with the discrete topology), are:

- Infinite-dimensional groups, such as the additive group of an infinite-dimensional real vector space. These are not Lie groups as they are not *finite-dimensional* manifolds.

- Some totally disconnected groups, such as the Galois group of an infinite extension of fields, or the additive group of the *p*-adic numbers. These are not Lie groups because their underlying spaces are not real manifolds. (Some of these groups are "*p*-adic Lie groups".) In general, only topological groups having similar local properties to $\mathbf{R}^n$ for some positive integer *n* can be Lie groups (of course they must also have a differentiable structure).

23.4 Basic concepts

23.4.1 The Lie algebra associated with a Lie group

Main article: Lie group–Lie algebra correspondence

To every Lie group we can associate a Lie algebra whose underlying vector space is the tangent space of the Lie group at the identity element and which completely captures the local structure of the group. Informally we can think of elements of the Lie algebra as elements of the group that are "infinitesimally close" to the identity, and the Lie bracket of the Lie algebra is related to the commutator of two such infinitesimal elements. Before giving the abstract definition we give a few examples:

- The Lie algebra of the vector space $\mathbf{R}^n$ is just $\mathbf{R}^n$ with the Lie bracket given by
  $[A, B] = 0$.
  (In general the Lie bracket of a connected Lie group is always 0 if and only if the Lie group is abelian.)

- The Lie algebra of the general linear group $GL(n, \mathbf{R})$ of invertible matrices is the vector space $M(n, \mathbf{R})$ of square matrices with the Lie bracket given by
  $[A, B] = AB - BA$.
  If G is a closed subgroup of $GL(n, \mathbf{R})$ then the Lie algebra of G can be thought of informally as the matrices m of $M(n, \mathbf{R})$ such that $1 + \varepsilon m$ is in G, where ε is an infinitesimal positive number with $\varepsilon^2 = 0$ (of course, no such real number ε exists). For example, the orthogonal group $O(n, \mathbf{R})$ consists of matrices A with $AA^T = 1$, so the Lie algebra consists of the matrices m with $(1 + \varepsilon m)(1 + \varepsilon m)^T = 1$, which is equivalent to $m + m^T = 0$ because $\varepsilon^2 = 0$.

- Formally, when working over the reals, as here, this is accomplished by considering the limit as $\varepsilon \to 0$; but the "infinitesimal" language generalizes directly to Lie groups over general rings.

The concrete definition given above is easy to work with, but has some minor problems: to use it we first need to represent a Lie group as a group of matrices, but not all Lie groups can be represented in this way, and it is not obvious that the Lie algebra is independent of the representation we use. To get around these problems we give the general definition of the Lie algebra of a Lie group (in 4 steps):

1. Vector fields on any smooth manifold M can be thought of as derivations X of the ring of smooth functions on the manifold, and therefore form a Lie algebra under the Lie bracket $[X, Y] = XY - YX$, because the Lie bracket of any two derivations is a derivation.

2. If G is any group acting smoothly on the manifold M, then it acts on the vector fields, and the vector space of vector fields fixed by the group is closed under the Lie bracket and therefore also forms a Lie algebra.

3. We apply this construction to the case when the manifold M is the underlying space of a Lie group G, with G acting on $G = M$ by left translations $Lg(h) = gh$. This shows that the space of left invariant vector fields (vector fields satisfying $Lg^*Xh = Xgh$ for every h in G, where Lg^* denotes the differential of Lg) on a Lie group is a Lie algebra under the Lie bracket of vector fields.

4. Any tangent vector at the identity of a Lie group can be extended to a left invariant vector field by left translating the tangent vector to other points of the manifold. Specifically, the left invariant extension of an element v of the tangent space at the identity is the vector field defined by $v\hat{}g = Lg^*v$. This identifies the tangent space TeG at the identity with the space of left invariant vector fields, and therefore makes the tangent

space at the identity into a Lie algebra, called the Lie algebra of G, usually denoted by a Fraktur $\mathfrak{g}$. Thus the Lie bracket on $\mathfrak{g}$ is given explicitly by $[v, w] = [v^\wedge, w^\wedge]e$.

This Lie algebra $\mathfrak{g}$ is finite-dimensional and it has the same dimension as the manifold G. The Lie algebra of G determines G up to "local isomorphism", where two Lie groups are called **locally isomorphic** if they look the same near the identity element. Problems about Lie groups are often solved by first solving the corresponding problem for the Lie algebras, and the result for groups then usually follows easily. For example, simple Lie groups are usually classified by first classifying the corresponding Lie algebras.

We could also define a Lie algebra structure on Te using right invariant vector fields instead of left invariant vector fields. This leads to the same Lie algebra, because the inverse map on G can be used to identify left invariant vector fields with right invariant vector fields, and acts as −1 on the tangent space Te.

The Lie algebra structure on Te can also be described as follows: the commutator operation

$$(x, y) \to xyx^{-1}y^{-1}$$

on $G \times G$ sends (e, e) to e, so its derivative yields a bilinear operation on TeG. This bilinear operation is actually the zero map, but the second derivative, under the proper identification of tangent spaces, yields an operation that satisfies the axioms of a Lie bracket, and it is equal to twice the one defined through left-invariant vector fields.

23.4.2 Homomorphisms and isomorphisms

If G and H are Lie groups, then a Lie group homomorphism $f : G \to H$ is a smooth group homomorphism. In the case of complex Lie groups, such a homomorphism is required to be a holomorphic map. However, these requirements are a bit stringent; every continuous homomorphism between real Lie groups turns out to be (real) analytic.

The composition of two Lie homomorphisms is again a homomorphism, and the class of all Lie groups, together with these morphisms, forms a category. Moreover, every Lie group homomorphism induces a homomorphism between the corresponding Lie algebras. Let $\phi\colon G \to H$ be a Lie group homomorphism and let ϕ_* be its derivative at the identity. If we identify the Lie algebras of G and H with their tangent spaces at the identity elements then ϕ_* is a map between the corresponding Lie algebras:

$$\phi_*\colon \mathfrak{g} \to \mathfrak{h}$$

One can show that ϕ_* is actually a Lie algebra homomorphism (meaning that it is a linear map which preserves the Lie bracket). In the language of category theory, we then have a covariant functor from the category of Lie groups to the category of Lie algebras which sends a Lie group to its Lie algebra and a Lie group homomorphism to its derivative at the identity.

Two Lie groups are called *isomorphic* if there exists a bijective homomorphism between them whose inverse is also a Lie group homomorphism. Equivalently, it is a diffeomorphism which is also a group homomorphism.

Ado's theorem says every finite-dimensional Lie algebra is isomorphic to a matrix Lie algebra. For every finite-dimensional matrix Lie algebra, there is a linear group (matrix Lie group) with this algebra as its Lie algebra. So every abstract Lie algebra is the Lie algebra of some (linear) Lie group.

The *global structure* of a Lie group is not determined by its Lie algebra; for example, if Z is any discrete subgroup of the center of G then G and G/Z have the same Lie algebra (see the table of Lie groups for examples). A *connected* Lie group is simple, semisimple, solvable, nilpotent, or abelian if and only if its Lie algebra has the corresponding property.

If we require that the Lie group be simply connected, then the global structure is determined by its Lie algebra: for every finite-dimensional Lie algebra $\mathfrak{g}$ over **F** there is a simply connected Lie group G with $\mathfrak{g}$ as Lie algebra, unique up to isomorphism. Moreover, every homomorphism between Lie algebras lifts to a unique homomorphism between the corresponding simply connected Lie groups.

23.4.3 The exponential map

Main article: Exponential map (Lie theory)

The exponential map from the Lie algebra $M(n, \mathbf{R})$ of the general linear group $GL(n, \mathbf{R})$ to $GL(n, \mathbf{R})$ is defined by the usual power series:

$$\exp(A) = 1 + A + \frac{A^2}{2!} + \frac{A^3}{3!} + \cdots$$

for matrices A. If G is any subgroup of $GL(n, \mathbf{R})$, then the exponential map takes the Lie algebra of G into G, so we have an exponential map for all matrix groups.

The definition above is easy to use, but it is not defined for Lie groups that are not matrix groups, and it is not clear that the exponential map of a Lie group does not depend on its representation as a matrix group. We can solve both problems using a more abstract definition of the exponential map that works for all Lie groups, as follows.

Every vector v in $\mathfrak{g}$ determines a linear map from **R** to $\mathfrak{g}$ taking 1 to v, which can be thought of as a Lie algebra homomorphism. Because **R** is the Lie algebra of the simply connected Lie group **R**, this induces a Lie group homomorphism $c : \mathbf{R} \to G$ so that

$$c(s+t) = c(s)c(t)$$

for all s and t. The operation on the right hand side is the group multiplication in G. The formal similarity of this formula with the one valid for the exponential function justifies the definition

$$\exp(v) = c(1).$$

This is called the **exponential map**, and it maps the Lie algebra $\mathfrak{g}$ into the Lie group G. It provides a diffeomorphism between a neighborhood of 0 in $\mathfrak{g}$ and a neighborhood of e in G. This exponential map is a generalization of the exponential function for real numbers (because **R** is the Lie algebra of the Lie group of positive real numbers with multiplication), for complex numbers (because **C** is the Lie algebra of the Lie group of non-zero complex numbers with multiplication) and for matrices (because M(n, **R**) with the regular commutator is the Lie algebra of the Lie group GL(n, **R**) of all invertible matrices).

Because the exponential map is surjective on some neighbourhood N of e, it is common to call elements of the Lie algebra **infinitesimal generators** of the group G. The subgroup of G generated by N is the identity component of G.

The exponential map and the Lie algebra determine the *local group structure* of every connected Lie group, because of the Baker–Campbell–Hausdorff formula: there exists a neighborhood U of the zero element of $\mathfrak{g}$, such that for u, v in U we have

$$\exp(u)\,\exp(v) = \exp\left(u + v + \tfrac{1}{2}[u,\,v\,] + \tfrac{1}{12}[\,[u,v],v] - \tfrac{1}{12}[\,[u,v],u] - \cdots\right),$$

where the omitted terms are known and involve Lie brackets of four or more elements. In case u and v commute, this formula reduces to the familiar exponential law exp(u) exp(v) = exp(u + v).

The exponential map relates Lie group homomorphisms. That is, if $\phi : G \to H$ is a Lie group homomorphism and $\phi_* : \mathfrak{g} \to \mathfrak{h}$ the induced map on the corresponding Lie algebras, then for all $x \in \mathfrak{g}$ we have

$$\phi(\exp(x)) = \exp(\phi_*(x)).$$

In other words, the following diagram commutes,[Note 1]

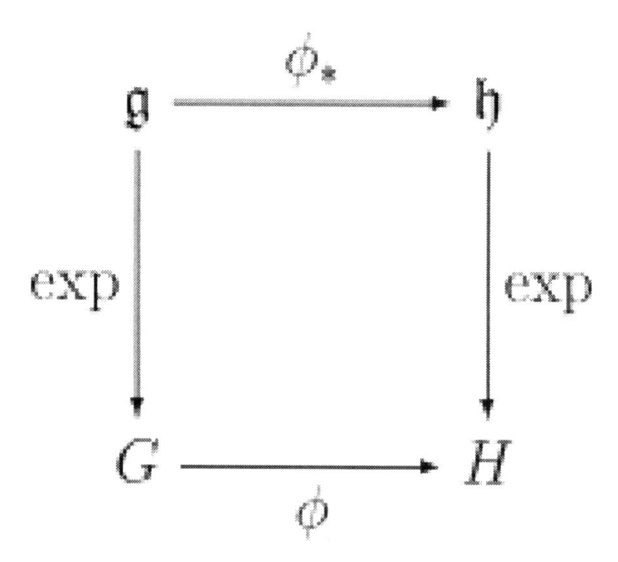

(In short, exp is a natural transformation from the functor Lie to the identity functor on the category of Lie groups.)

The exponential map from the Lie algebra to the Lie group is not always onto, even if the group is connected (though it does map onto the Lie group for connected groups that are either compact or nilpotent). For example, the exponential map of SL(2, **R**) is not surjective. Also, exponential map is not surjective nor injective for infinite-dimensional (see below) Lie groups modelled on C^∞ Fréchet space, even from arbitrary small neighborhood of 0 to corresponding neighborhood of 1.

See also: derivative of the exponential map and normal coordinates.

23.4.4 Lie subgroup

A **Lie subgroup** H of a Lie group G is a Lie group that is a subset of G and such that the inclusion map from H to G is an injective immersion and group homomorphism. According to Cartan's theorem, a closed subgroup of G admits a unique smooth structure which makes it an embedded Lie subgroup of G—i.e. a Lie subgroup such that the inclusion map is a smooth embedding.

Examples of non-closed subgroups are plentiful; for example take G to be a torus of dimension ≥ 2, and let H be a one-parameter subgroup of *irrational slope*, i.e. one that winds around in G. Then there is a Lie group homomorphism φ : **R** → G with H as its image. The closure of H will be a sub-torus in G.

The exponential map gives a one-to-one correspondence between the connected Lie subgroups of a connected Lie group G and the subalgebras of the Lie algebra of G.[2] Typically, the subgroup corresponding to a subalgebra is not a closed subgroup. There is no criterion solely based on the structure of g which determines which subalgebras correspond to closed subgroups.

23.5 Early history

According to the most authoritative source on the early history of Lie groups (Hawkins, p. 1), Sophus Lie himself considered the winter of 1873–1874 as the birth date of his theory of continuous groups. Hawkins, however, suggests that it was "Lie's prodigious research activity during the four-year period from the fall of 1869 to the fall of 1873" that led to the theory's creation (*ibid*). Some of Lie's early ideas were developed in close collaboration with Felix Klein. Lie met with Klein every day from October 1869 through 1872: in Berlin from the end of October 1869 to the end of February 1870, and in Paris, Göttingen and Erlangen in the subsequent two years (*ibid*, p. 2). Lie stated that all of the principal results were obtained by 1884. But during the 1870s all his papers (except the very first note) were published in Norwegian journals, which impeded recognition of the work throughout the rest of Europe (*ibid*, p. 76). In 1884 a young German mathematician, Friedrich Engel, came to work with Lie on a systematic treatise to expose his theory of continuous groups. From this effort resulted the three-volume *Theorie der Transformationsgruppen*, published in 1888, 1890, and 1893.

Lie's ideas did not stand in isolation from the rest of mathematics. In fact, his interest in the geometry of differential equations was first motivated by the work of Carl Gustav Jacobi, on the theory of partial differential equations of first order and on the equations of classical mechanics. Much of Jacobi's work was published posthumously in the 1860s, generating enormous interest in France and Germany (Hawkins, p. 43). Lie's *idée fixe* was to develop a theory of symmetries of differential equations that would accomplish for them what Évariste Galois had done for algebraic equations: namely, to classify them in terms of group theory. Lie and other mathematicians showed that the most important equations for special functions and orthogonal polynomials tend to arise from group theoretical symmetries. In Lie's early work, the idea was to construct a theory of *continuous groups*, to complement the theory of discrete groups that had developed in the theory of modular forms, in the hands of Felix Klein and Henri Poincaré. The initial application that Lie had in mind was to the theory of differential equations. On the model of Galois theory and polynomial equations, the driving conception was of a theory capable of unifying, by the study of symmetry, the whole area of ordinary differential equations. However, the hope that Lie Theory would unify the entire field of ordinary differential equations was not fulfilled. Symmetry methods for ODEs continue to be studied, but do not dominate the subject. There is a differential Galois theory, but it was developed by others, such as Picard and Vessiot, and it provides a theory of quadratures, the indefinite integrals required to express solutions.

Additional impetus to consider continuous groups came from ideas of Bernhard Riemann, on the foundations of geometry, and their further development in the hands of Klein. Thus three major themes in 19th century mathematics were combined by Lie in creating his new theory: the idea of symmetry, as exemplified by Galois through the algebraic notion of a group; geometric theory and the explicit solutions of differential equations of mechanics, worked out by Poisson and Jacobi; and the new understanding of geometry that emerged in the works of Plücker, Möbius, Grassmann and others, and culminated in Riemann's revolutionary vision of the subject.

Although today Sophus Lie is rightfully recognized as the creator of the theory of continuous groups, a major stride in the development of their structure theory, which was to have a profound influence on subsequent development of mathematics, was made by Wilhelm Killing, who in 1888 published the first paper in a series entitled *Die Zusammensetzung der stetigen endlichen Transformationsgruppen* (*The composition of continuous finite transformation groups*) (Hawkins, p. 100). The work of Killing, later refined and generalized by Élie Cartan, led to classification of semisimple Lie algebras, Cartan's theory of symmetric spaces, and Hermann Weyl's description of representations of compact and semisimple Lie groups using highest weights.

In 1900 David Hilbert challenged Lie theorists with his Fifth Problem presented at the International Congress of Mathematicians in Paris.

Weyl brought the early period of the development of the theory of Lie groups to fruition, for not only did he classify irreducible representations of semisimple Lie groups and connect the theory of groups with quantum mechanics, but he also put Lie's theory itself on firmer footing by clearly enunciating the distinction between Lie's *infinitesimal groups* (i.e., Lie algebras) and the Lie groups proper, and began investigations of topology of Lie groups.[3] The theory of Lie groups was systematically reworked in modern mathematical language in a monograph by Claude Chevalley.

23.6 The concept of a Lie group, and possibilities of classification

Lie groups may be thought of as smoothly varying families of symmetries. Examples of symmetries include rotation about an axis. What must be understood is the nature of 'small' transformations, for example, rotations through tiny angles, that link nearby transformations. The mathematical object capturing this structure is called a Lie algebra (Lie himself called them "infinitesimal groups"). It can be defined because Lie groups are manifolds, so have tangent spaces at each point.

The Lie algebra of any compact Lie group (very roughly: one for which the symmetries form a bounded set) can be decomposed as a direct sum of an abelian Lie algebra and some number of simple ones. The structure of an abelian Lie algebra is mathematically uninteresting (since the Lie bracket is identically zero); the interest is in the simple summands. Hence the question arises: what are the simple Lie algebras of compact groups? It turns out that they mostly fall into four infinite families, the "classical Lie algebras" A*n*, B*n*, C*n* and D*n*, which have simple descriptions in terms of symmetries of Euclidean space. But there are also just five "exceptional Lie algebras" that do not fall into any of these families. E_8 is the largest of these.

Lie groups are classified according to their algebraic properties (simple, semisimple, solvable, nilpotent, abelian), their connectedness (connected or simply connected) and their compactness.

- Compact Lie groups are all known: they are finite central quotients of a product of copies of the circle group $\mathbf{S}^1$ and simple compact Lie groups (which correspond to connected Dynkin diagrams).
- Any simply connected solvable Lie group is isomorphic to a closed subgroup of the group of invertible upper triangular matrices of some rank, and any finite-dimensional irreducible representation of such a group is 1-dimensional. Solvable groups are too messy to classify except in a few small dimensions.
- Any simply connected nilpotent Lie group is isomorphic to a closed subgroup of the group of invertible upper triangular matrices with 1's on the diagonal of some rank, and any finite-dimensional irreducible representation of such a group is 1-dimensional. Like solvable groups, nilpotent groups are too messy to classify except in a few small dimensions.
- Simple Lie groups are sometimes defined to be those that are simple as abstract groups, and sometimes defined to be connected Lie groups with a simple Lie algebra. For example, SL(2, **R**) is simple according to the second definition but not according to the first. They have all been classified (for either definition).
- Semisimple Lie groups are Lie groups whose Lie algebra is a product of simple Lie algebras.[4] They are central extensions of products of simple Lie groups.

The identity component of any Lie group is an open normal subgroup, and the quotient group is a discrete group. The universal cover of any connected Lie group is a simply connected Lie group, and conversely any connected Lie group is a quotient of a simply connected Lie group by a discrete normal subgroup of the center. Any Lie group G can be decomposed into discrete, simple, and abelian groups in a canonical way as follows. Write

G_{con} for the connected component of the identity

G_{sol} for the largest connected normal solvable subgroup

G_{nil} for the largest connected normal nilpotent subgroup

so that we have a sequence of normal subgroups

$$1 \subseteq G_{\mathrm{nil}} \subseteq G_{\mathrm{sol}} \subseteq G_{\mathrm{con}} \subseteq G.$$

Then

G/G_{con} is discrete

$G_{\mathrm{con}}/G_{\mathrm{sol}}$ is a central extension of a product of simple connected Lie groups.

$G_{\mathrm{sol}}/G_{\mathrm{nil}}$ is abelian. A connected abelian Lie group is isomorphic to a product of copies of **R** and the circle group S^1.

$G_{\mathrm{nil}}/1$ is nilpotent, and therefore its ascending central series has all quotients abelian.

This can be used to reduce some problems about Lie groups (such as finding their unitary representations) to the same problems for connected simple groups and nilpotent and solvable subgroups of smaller dimension.

- The diffeomorphism group of a Lie group acts transitively on the Lie group
- Every Lie group is parallelizable, and hence an orientable manifold (there is a bundle isomorphism between its tangent bundle and the product of itself with the tangent space at the identity)

23.7 Infinite-dimensional Lie groups

Lie groups are often defined to be finite-dimensional, but there are many groups that resemble Lie groups, except for being infinite-dimensional. The simplest way to define infinite-dimensional Lie groups is to model them on Banach spaces, and in this case much of the basic theory is similar to that of finite-dimensional Lie groups. However this is inadequate for many applications, because many natural examples of infinite-dimensional Lie groups are not Banach manifolds. Instead one needs to define Lie groups modeled on more general locally convex topological vector spaces. In this case the relation between the Lie algebra and the Lie group becomes rather subtle, and several results about finite-dimensional Lie groups no longer hold.

The literature is not entirely uniform in its terminology as to exactly which properties of infinite-dimensional groups qualify the group for the prefix *Lie* in *Lie group*. On the Lie algebra side of affairs, things are simpler since the qualifying criteria for the prefix *Lie* in *Lie algebra* are purely algebraic. For example, an infinite-dimensional Lie algebra may or may not have a corresponding Lie group. That is, there may be a group corresponding to the Lie algebra, but it might not be nice enough to be called a Lie group, or the connection between the group and the Lie algebra might not be nice enough (for example, failure of the exponential map to be onto a neighborhood of the identity). It is the "nice enough" that is not universally defined.

Some of the examples that have been studied include:

- The group of diffeomorphisms of a manifold. Quite a lot is known about the group of diffeomorphisms of the circle. Its Lie algebra is (more or less) the Witt algebra, which has a central extension called the Virasoro algebra, used in string theory and conformal field theory. Diffeomorphism groups of compact manifolds of larger dimension are regular Fréchet Lie groups; very little about their structure is known.

The diffeomorphism group of spacetime sometimes appears in attempts to quantize gravity.

- The group of smooth maps from a manifold to a finite-dimensional Lie group is an example of a gauge group (with operation of pointwise multiplication), and is used in quantum field theory and Donaldson theory. If the manifold is a circle these are called loop groups, and have central extensions whose Lie algebras are (more or less) Kac–Moody algebras.
- There are infinite-dimensional analogues of general linear groups, orthogonal groups, and so on. One important aspect is that these may have *simpler* topological properties: see for example Kuiper's theorem. In M-Theory theory, for example, a 10 dimensional SU(N) gauge theory becomes an 11 dimensional theory when N becomes infinite.
- A specific example is that $SU(\infty)$ is equal to the group of area preserving diffeomorphisms of a torus.

23.8 See also

- Lie subgroup
- E_8
- Adjoint representation of a Lie group
- Adjoint endomorphism
- Haar measure
- Homogeneous space
- List of Lie group topics
- List of simple Lie groups
- Moufang polygon
- Riemannian manifold
- Representations of Lie groups
- Table of Lie groups
- Lie algebra
- Symmetry in quantum mechanics
- Lie group action

23.9 Notes

23.9.1 Explanatory notes

[1] http://www.math.sunysb.edu/~{}vkiritch/MAT552/ProblemSet1.pdf

23.9.2 Citations

[1] Arthur Tresse (1893). "Sur les invariants différentiels des groupes continus de transformations". *Acta Mathematica* **18**: 1–88. doi:10.1007/bf02418270.

[2] Hall 2015 Theorem 5.20

[3] Borel (2001).

[4] Helgason, Sigurdur (1978). *Differential Geometry, Lie Groups, and Symmetric Spaces*. New York: Academic Press. p. 131. ISBN 0-12-338460-5.

23.10 References

- Adams, John Frank (1969), *Lectures on Lie Groups*, Chicago Lectures in Mathematics, Chicago: Univ. of Chicago Press, ISBN 0-226-00527-5, MR 0252560.
- Borel, Armand (2001), *Essays in the history of Lie groups and algebraic groups*, History of Mathematics **21**, Providence, R.I.: American Mathematical Society, ISBN 978-0-8218-0288-5, MR 1847105
- Bourbaki, Nicolas, *Elements of mathematics: Lie groups and Lie algebras*. Chapters 1–3 ISBN 3-540-64242-0, Chapters 4–6 ISBN 3-540-42650-7, Chapters 7–9 ISBN 3-540-43405-4
- Chevalley, Claude (1946), *Theory of Lie groups*, Princeton: Princeton University Press, ISBN 0-691-04990-4.
- P. M. Cohn (1957) *Lie Groups*, Cambridge Tracts in Mathematical Physics.
- J. L. Coolidge (1940) *A History of Geometrical Methods*, pp 304–17, Oxford University Press (Dover Publications 2003).
- Fulton, William; Harris, Joe (1991), *Representation theory. A first course*, Graduate Texts in Mathematics, Readings in Mathematics **129**, New York: Springer-Verlag, ISBN 978-0-387-97495-8, MR 1153249, ISBN 978-0-387-97527-6
- Robert Gilmore (2008) *Lie groups, physics, and geometry: an introduction for physicists, engineers and chemists*, Cambridge University Press ISBN 9780521884006 .
- Hall, Brian C. (2015), *Lie Groups, Lie Algebras, and Representations: An Elementary Introduction*, Graduate Texts in Mathematics **222** (2nd ed.), Springer, ISBN 0-387-40122-9.
- F. Reese Harvey (1990) *Spinors and calibrations*, Academic Press, ISBN 0-12-329650-1 .
- Hawkins, Thomas (2000), *Emergence of the theory of Lie groups*, Sources and Studies in the History of Mathematics and Physical Sciences, Berlin, New York: Springer-Verlag, ISBN 978-0-387-98963-1, MR 1771134 Borel's review
- Helgason, Sigurdur (2001), *Differential geometry, Lie groups, and symmetric spaces*, Graduate Studies in Mathematics **34**, Providence, R.I.: American Mathematical Society, ISBN 978-0-8218-2848-9, MR 1834454
- Knapp, Anthony W. (2002), *Lie Groups Beyond an Introduction*, Progress in Mathematics **140** (2nd ed.), Boston: Birkhäuser, ISBN 0-8176-4259-5.
- Nijenhuis, Albert (1959). "Review: *Lie groups*, by P. M. Cohn". *Bulletin of the American Mathematical Society* **65** (6): 338–341. doi:10.1090/s0002-9904-1959-10358-x.
- Rossmann, Wulf (2001), *Lie Groups: An Introduction Through Linear Groups*, Oxford Graduate Texts in Mathematics, Oxford University Press, ISBN 978-0-19-859683-7. The 2003 reprint corrects several typographical mistakes.
- Sattinger, David H.; Weaver, O. L. (1986). *Lie groups and algebras with applications to physics, geometry, and mechanics*. Springer-Verlag. ISBN 3-540-96240-9. MR 0835009.
- Serre, Jean-Pierre (1965), *Lie Algebras and Lie Groups: 1964 Lectures given at Harvard University*, Lecture notes in mathematics **1500**, Springer, ISBN 3-540-55008-9.
- Stillwell, John (2008). *Naive Lie Theory*. Springer. ISBN 978-0387782140.
- Heldermann Verlag Journal of Lie Theory
- Warner, Frank W. (1983), *Foundations of differentiable manifolds and Lie groups*, Graduate Texts in Mathematics **94**, New York Berlin Heidelberg: Springer-Verlag, ISBN 978-0-387-90894-6, MR 0722297
- Steeb, Willi-Hans (2007), *Continuous Symmetries, Lie algebras, Differential Equations and Computer Algebra: second edition*, World Scientific Publishing, ISBN 981-270-809-X, MR 2382250.
- Lie Groups. Representation Theory and Symmetric Spaces Wolfgang Ziller, Vorlesung 2010

Chapter 24

Lie derivative

In mathematics, the **Lie derivative** /ˈliː/, named after Sophus Lie by Władysław Ślebodziński,[1][2] evaluates the change of a tensor field (including scalar function, vector field and one-form), along the flow of another vector field. This change is coordinate invariant and therefore the Lie derivative is defined on any differentiable manifold.

Functions, tensor fields and forms can be differentiated with respect to a vector field. Since a vector field is a derivation of zero degree on the algebra of smooth functions, the Lie derivative of a function f along a vector field X is the evaluation $X(f)$, i.e., is simply the application of the vector field. The process of Lie differentiation extends to a derivation of zero degree on the algebra of tensor fields over a manifold *M*. It also commutes with contraction and the exterior derivative on differential forms. This uniquely determines the Lie derivative and it follows that for vector fields the Lie derivative is the commutator

$$\mathcal{L}_X(Y) = [X, Y]$$

It also shows that the Lie derivatives on *M* are an infinite-dimensional Lie algebra representation of the Lie algebra of vector fields with the Lie bracket defined by the commutator,

$$\mathcal{L}_{[X,Y]} = [\mathcal{L}_X, \mathcal{L}_Y].$$

Considering vector fields as infinitesimal generators of flows (active diffeomorphisms) on *M*, the Lie derivatives are the infinitesimal representation of the representation of the diffeomorphism group on tensor fields, analogous to Lie algebra representations as infinitesimal representations associated to group representation in Lie group theory.

Generalisations exist for spinor fields, fibre bundles with connection and vector-valued differential forms.

24.1 Definition

The Lie derivative may be defined in several equivalent ways. In this section, to keep things simple, we begin by defining the Lie derivative acting on scalar functions and vector fields. The Lie derivative can also be defined to act on general tensors, as developed later in the article.

24.1.1 The Lie derivative of a function

Note: the Einstein summation convention of summing on repeated indices is used below.

There are several equivalent definitions of a Lie derivative of a function.

- The Lie derivative can be defined in terms of the definition of vector fields as first order differential operators. Given a function $f : M \to \mathbf{R}$ and a vector field X defined on *M*, the Lie derivative $\mathcal{L}_X f$ of a function f along a vector field X is simply the application of the vector field. It can be interpreted as the directional derivative of f along X. Hence at a point $p \in M$ we have

  $$(\mathcal{L}_X f)(p) \triangleq X_p(f) \triangleq (Xf)(p).$$

  By the definition of the differential of a function on *M* the definition can also be written as

  $$(\mathcal{L}_X f)(p) \triangleq \mathrm{d}\, f_p\,(X_p).$$

  Choosing local coordinates x^a, and writing : $X = X^a \partial_a$, where the $\partial_a = \frac{\partial}{\partial x^a}$ are local basis vectors for the tangent bundle TM, we have locally

  $$(\mathcal{L}_X f)(p) = X^a(p)(\partial_a f)(p).$$

Likewise $\mathrm{d}f : M \to T^*M$ is the 1-form locally given by $\mathrm{d}f \triangleq \partial_a f \, \mathrm{d}x^a$. which implies

$$(\mathcal{L}_X f)(p) = \mathrm{d}f_p(X_p) = X^a(p)(\partial_b f)(p)\,dx^b(\partial_a) = X^a(p)(\partial_a f)(p)$$

recovering the original definition.

- Alternatively, the Lie derivative can be defined as

$$(\mathcal{L}_X f)(p) \triangleq \frac{\mathrm{d}}{\mathrm{d}t} f(\gamma(t))\Big|_{t=0}$$

where $\gamma(t)$ is any curve on *M* with $\gamma(0) = p$ and $\gamma'(0) = X_p$. One such curve is the flow of X through p: the solution of the first-order ordinary differential equation $\frac{d}{dt}\gamma(t) = X_{\gamma(t)}$, which exists by the Picard–Lindelöf theorem (see also the Frobenius theorem).

24.1.2 The Lie derivative of a vector field

The Lie derivative can be defined for vector fields by first defining the Lie bracket $[X, Y]$ of a pair of vector fields *X* and *Y*. There are several approaches to defining the Lie bracket, all of which are equivalent. Regardless of the chosen definition, one then defines the Lie derivative of the vector field *Y* to be equal to the Lie bracket of *X* and *Y*, that is,

$\mathcal{L}_X Y = [X, Y]$.

Other equivalent definitions are (here, Φ_t^X is the flow transformation and d the tangent map derivative operator):[3]

$$(\mathcal{L}_X Y)_x := \lim_{t\to 0} \frac{(\mathrm{d}\Phi_{-t}^X) Y_{\Phi_t^X(x)} - Y_x}{t} = \frac{\mathrm{d}}{\mathrm{d}t}\Big|_{t=0} (\mathrm{d}\Phi_{-t}^X) Y_{\Phi_t^X(x)}$$

$$\mathcal{L}_X Y := \frac{1}{2}\frac{\mathrm{d}^2}{\mathrm{d}t^2}\Big|_{t=0} \Phi_{-t}^Y \circ \Phi_{-t}^X \circ \Phi_t^Y \circ \Phi_t^X = \frac{\mathrm{d}}{\mathrm{d}t}\Big|_{t=0} \Phi_{-\sqrt{t}}^Y \circ \Phi_{-\sqrt{t}}^X \circ \Phi_{\sqrt{t}}^Y \circ \Phi_{\sqrt{t}}^X.$$

24.2 The Lie derivative of differential forms

The Lie derivative can also be defined on differential forms. In this context, it is closely related to the exterior derivative. Both the Lie derivative and the exterior derivative attempt to capture the idea of a derivative in different ways. These differences can be bridged by introducing the idea of an **antiderivation** or equivalently an interior product, after which the relationships fall out as a set of identities.

Let *M* be a manifold and *X* a vector field on *M*. Let $\omega \in \Lambda^{k+1}(M)$ be a (*k* + 1)-form. The **interior product** of *X* and ω is the *k*-form $i_X\omega$ defined as

$$(i_X\omega)(X_1, \ldots, X_k) = \omega(X, X_1, \ldots, X_k)$$

The differential form $i_X\omega$ is also called the **contraction** of ω with *X*. Note that

$$i_X : \Lambda^{k+1}(M) \to \Lambda^k(M)$$

and that i_X is a $\wedge$ -antiderivation. That is, i_X is **R**-linear, and

$$i_X(\omega \wedge \eta) = (i_X\omega) \wedge \eta + (-1)^k \omega \wedge (i_X\eta)$$

for $\omega \in \Lambda^k(M)$ and η another differential form. Also, for a function $f \in \Lambda^0(M)$, that is a real or complex-valued function on *M*, one has

$$i_{fX}\omega = f\, i_X\omega$$

where *fX* denotes the product of *f* and *X*. The relationship between exterior derivatives and Lie derivatives can then be summarized as follows. As discussed in a previous section, the Lie derivative of an ordinary function *f* is just the contraction of the exterior derivative with the vector field *X*:

$$\mathcal{L}_X f = i_X\, df$$

For a general differential form, the Lie derivative is likewise a contraction, taking into account the variation in *X*:

$$\mathcal{L}_X\omega = i_X d\omega + d(i_X\omega).$$

This identity is known variously as "Cartan's formula" or "Cartan's magic formula," and shows in particular that:

$$d\mathcal{L}_X\omega = \mathcal{L}_X(d\omega).$$

The derivative of products is distributed:

$$\mathcal{L}_{fX}\omega = f\mathcal{L}_X\omega + df \wedge i_X\omega$$

24.3 Properties

The Lie derivative has a number of properties. Let $\mathcal{F}(M)$ be the algebra of functions defined on the manifold *M*. Then

$$\mathcal{L}_X : \mathcal{F}(M) \to \mathcal{F}(M)$$

is a derivation on the algebra $\mathcal{F}(M)$. That is, $\mathcal{L}_X$ is **R**-linear and

$$\mathcal{L}_X(fg) = (\mathcal{L}_X f)g + f\mathcal{L}_X g.$$

Similarly, it is a derivation on $\mathcal{F}(M)\times\mathcal{X}(M)$ where $\mathcal{X}(M)$ is the set of vector fields on *M*:

$$\mathcal{L}_X(fY) = (\mathcal{L}_X f)Y + f\mathcal{L}_X Y$$

which may also be written in the equivalent notation

$$\mathcal{L}_X(f \otimes Y) = (\mathcal{L}_X f) \otimes Y + f \otimes \mathcal{L}_X Y$$

where the tensor product symbol $\otimes$ is used to emphasize the fact that the product of a function times a vector field is being taken over the entire manifold.

Additional properties are consistent with that of the Lie bracket. Thus, for example, considered as a derivation on a vector field,

$$\mathcal{L}_X[Y, Z] = [\mathcal{L}_X Y, Z] + [Y, \mathcal{L}_X Z]$$

one finds the above to be just the Jacobi identity. Thus, one has the important result that the space of vector fields over *M*, equipped with the Lie bracket, forms a Lie algebra.

The Lie derivative also has important properties when acting on differential forms. Let α and β be two differential forms on *M*, and let *X* and *Y* be two vector fields. Then

- $\mathcal{L}_X(\alpha \wedge \beta) = (\mathcal{L}_X \alpha) \wedge \beta + \alpha \wedge (\mathcal{L}_X \beta)$
- $[\mathcal{L}_X, \mathcal{L}_Y]\alpha := \mathcal{L}_X\mathcal{L}_Y\alpha - \mathcal{L}_Y\mathcal{L}_X\alpha = \mathcal{L}_{[X,Y]}\alpha$
- $[\mathcal{L}_X, i_Y]\alpha = [i_X, \mathcal{L}_Y]\alpha = i_{[X,Y]}\alpha$, where *i* denotes interior product defined above and it's clear whether $[\cdot,\cdot]$ denotes the commutator or the Lie bracket of vector fields.

24.4 Lie derivative of tensor fields

More generally, if we have a differentiable tensor field *T* of rank (q, r) and a differentiable vector field *Y* (i.e. a differentiable section of the tangent bundle *TM*), then we can define the Lie derivative of *T* along *Y*. Let, for some open interval I around 0, $\varphi: M \times \mathbf{I} \to M$ be the one-parameter semigroup of local diffeomorphisms of *M* induced by the vector flow of *Y* and denote $\varphi t(p) := \varphi(p, t)$. For each sufficiently small *t*, φt is a diffeomorphism from a neighborhood in *M* to another neighborhood in *M*, and φ_0 is the identity diffeomorphism. The Lie derivative of *T* is defined at a point *p* by

$$(\mathcal{L}_Y T)_p = \left.\frac{d}{dt}\right|_{t=0} \left((\varphi_{-t})_* T_{\varphi_t(p)}\right) = \left.\frac{d}{dt}\right|_{t=0} ((\varphi_t)^* T)_p .$$

where $(\varphi_t)_*$ is the pushforward along the diffeomorphism and $(\varphi_t)^*$ is the pullback along the diffeomorphism. Intuitively, if you have a tensor field *T* and a vector field *Y*, then $\mathcal{L}_Y T$ is the infinitesimal change you would see when you flow *T* using the vector field $-Y$, which is the same thing as the infinitesimal change you would see in T if you yourself flowed along the vector field *Y*.

We now give an algebraic definition. The algebraic definition for the Lie derivative of a tensor field follows from the following four axioms:

Axiom 1. The Lie derivative of a function is the directional derivative of the function. So if *f* is a real valued function on *M*, then

$$\mathcal{L}_Y f = Y(f)$$

Axiom 2. The Lie derivative obeys the Leibniz rule. For any tensor fields *S* and *T*, we have

$$\mathcal{L}_Y(S \otimes T) = (\mathcal{L}_Y S) \otimes T + S \otimes (\mathcal{L}_Y T).$$

Axiom 3. The Lie derivative obeys the Leibniz rule with respect to contraction

$$\mathcal{L}_X(T(Y_1, \ldots, Y_n)) = (\mathcal{L}_X T)(Y_1, \ldots, Y_n) + T((\mathcal{L}_X Y_1), \ldots, Y_n) + \cdots + T(Y_1, \ldots, (\mathcal{L}_X Y_n))$$

Axiom 4. The Lie derivative commutes with exterior derivative on functions

$$[\mathcal{L}_X, d] = 0$$

Taking the Lie derivative of the relation $df(Y) = Y(f)$ then easily shows that the Lie derivative of a vector field is the Lie bracket. So if *X* is a vector field, one has

$$\mathcal{L}_Y X = [Y, X].$$

The Lie derivative of a differential form is the anticommutator of the interior product with the exterior derivative. So if α is a differential form,

$$\mathcal{L}_Y \alpha = i_Y d\alpha + d i_Y \alpha.$$

This follows easily by checking that the expression commutes with exterior derivative, is a derivation (being an anticommutator of graded derivations) and does the right thing on functions.

Explicitly, let *T* be a tensor field of type (*p*,*q*). Consider *T* to be a differentiable multilinear map of smooth sections $\alpha^1, \alpha^2, \ldots, \alpha^q$ of the cotangent bundle *T*M* and of sections $X_1, X_2, \ldots X_p$ of the tangent bundle *TM*, written $T(\alpha^1, \alpha^2, \ldots, X_1, X_2, \ldots)$ into **R**. Define the Lie derivative of *T* along *Y* by the formula

$$(\mathcal{L}_Y T)(\alpha_1, \alpha_2, \ldots, X_1, X_2, \ldots) = Y(T(\alpha_1, \alpha_2, \ldots, X_1, X_2, \ldots))$$

$$-T(\mathcal{L}_Y \alpha_1, \alpha_2, \ldots, X_1, X_2, \ldots) - T(\alpha_1, \mathcal{L}_Y \alpha_2, \ldots, X_1, X_2, \ldots) - \ldots$$

$$-T(\alpha_1, \alpha_2, \ldots, \mathcal{L}_Y X_1, X_2, \ldots) - T(\alpha_1, \alpha_2, \ldots, X_1, \mathcal{L}_Y X_2, \ldots) - \ldots$$

The analytic and algebraic definitions can be proven to be equivalent using the properties of the pushforward and the Leibniz rule for differentiation. Note also that the Lie derivative commutes with the contraction.

24.5 Coordinate expressions

In local coordinate notation, for a type (r,s) tensor field T, the Lie derivative along X is

$$(\mathcal{L}_X T)^{a_1 \ldots a_r}{}_{b_1 \ldots b_s} = X^c(\partial_c T^{a_1 \ldots a_r}{}_{b_1 \ldots b_s})$$

$$-(\partial_c X^{a_1}) T^{c a_2 \ldots a_r}{}_{b_1 \ldots b_s} - \ldots - (\partial_c X^{a_r}) T^{a_1 \ldots a_{r-1} c}{}_{b_1 \ldots b_s}$$

$$+(\partial_{b_1} X^c) T^{a_1 \ldots a_r}{}_{c b_2 \ldots b_s} + \ldots + (\partial_{b_s} X^c) T^{a_1 \ldots a_r}{}_{b_1 \ldots b_{s-1} c}$$

here, the notation $\partial_a = \frac{\partial}{\partial x^a}$ means taking the partial derivative with respect to the coordinate x^a. Alternatively, if we are using a torsion-free connection (e.g. the Levi Civita connection), then the partial derivative ∂_a can be replaced with the covariant derivative ∇_a. The Lie derivative of a tensor is another tensor of the same type, i.e. even though the individual terms in the expression depend on the choice of coordinate system, the expression as a whole results in a tensor

$$(\mathcal{L}_X T)^{a_1 \ldots a_r}{}_{b_1 \ldots b_s} \partial_{a_1} \otimes \cdots \otimes \partial_{a_r} \otimes dx^{b_1} \otimes \cdots \otimes dx^{b_s}$$

which is independent of any coordinate system.

The definition can be extended further to tensor densities of weight *w* for any real *w*. If *T* is such a tensor density, then its Lie derivative is a tensor density of the same type and weight.

$$(\mathcal{L}_X T)^{a_1 \ldots a_r}{}_{b_1 \ldots b_s} = X^c(\partial_c T^{a_1 \ldots a_r}{}_{b_1 \ldots b_s}) - (\partial_c X^{a_1}) T^{c a_2 \ldots a_r}{}_{b_1 \ldots b_s} - \ldots - (\partial_c X^{a_r}) T^{a_1 \ldots a_{r-1} c}{}_{b_1 \ldots b_s} +$$

$$+(\partial_{b_1} X_c) T_{a_1 \ldots a_r c b_2 \ldots b_s} + \ldots + (\partial_{b_s} X_c)$$

$$T^{a_1 \ldots a_r}{}_{b_1 \ldots b_{s-1} c} + w(\partial_c X^c) T^{a_1 \ldots a_r}{}_{b_1 \ldots b_s}$$

Notice the new term at the end of the expression.

24.5.1 Examples

For clarity we now show the following examples in local coordinate notation.

For a scalar field $\phi(x^c) \in \mathcal{F}(M)$ we have:

$$(\mathcal{L}_X \phi) = X^a \partial_a \phi$$

For a covector field, i.e., a differential form, $A = A_a(x^b) dx^a$ we have:

$$(\mathcal{L}_X A)_a = X^b \partial_b A_a + A_b \partial_a (X^b)$$

For a covariant symmetric tensor field $g = g_{ab}(x^c) dx^a \otimes dx^b$ we have:

$$(\mathcal{L}_X g)_{ab} = X^c \partial_c g_{ab} + g_{cb} \partial_a X^c + g_{ca} \partial_b X^c$$

24.6 Generalizations

Various generalizations of the Lie derivative play an important role in differential geometry.

24.6.1 The Lie derivative of a spinor field

A definition for Lie derivatives of spinors along generic spacetime vector fields, not necessarily Killing ones, on a general (pseudo) Riemannian manifold was already proposed in 1972 by Yvette Kosmann.[4] Later, it was provided a geometric framework which justifies her *ad hoc* prescription within the general framework of Lie derivatives on fiber bundles[5] in the explicit context of gauge natural bundles which turn out to be the most appropriate arena for (gauge-covariant) field theories.[6]

In a given spin manifold, that is in a Riemannian manifold (M, g) admitting a spin structure, the Lie derivative of a spinor field ψ can be defined by first defining it with respect to infinitesimal isometries (Killing vector fields) via the André Lichnerowicz's local expression given in 1963:[7]

$$\mathcal{L}_X \psi := X^a \nabla_a \psi - \frac{1}{4} \nabla_a X_b \gamma^a \gamma^b \psi \,,$$

where $\nabla_a X_b = \nabla_{[a} X_{b]}$, as $X = X^a \partial_a$ is assumed to be a Killing vector field, and γ^a are Dirac matrices.

It is then possible to extend Lichnerowicz's definition to all vector fields (generic infinitesimal transformations) by retaining Lichnerowicz's local expression for a *generic* vector field X , but explicitly taking the antisymmetric part of $\nabla_a X_b$ only.[4] More explicitly, Kosmann's local expression given in 1972 is:[4]

$$\mathcal{L}_X \psi := X^a \nabla_a \psi - \frac{1}{8} \nabla_{[a} X_{b]} [\gamma^a, \gamma^b] \psi = \nabla_X \psi - \frac{1}{4} (dX^\flat) \cdot \psi \,,$$

where $[\gamma^a, \gamma^b] = \gamma^a \gamma^b - \gamma^b \gamma^a$ is the commutator, d is exterior derivative, $X^\flat = g(X, -)$ is the dual 1 form corresponding to X under the metric (i.e. with lowered indices) and $\cdot$ is Clifford multiplication. It is worth noting that the spinor Lie derivative is independent of the metric, and hence the connection. This is not obvious from the right-hand side of Kosmann's local expression, as the right-hand side seems to depend on the metric through the spin connection (covariant derivative), the dualisation of vector fields (lowering of the indices) and the Clifford multiplication on the spinor bundle. Such is not the case: the quantities on the right-hand side of Kosmann's local expression combine so as to make all metric and connection dependent terms cancel.

To gain a better understanding of the long-debated concept of Lie derivative of spinor fields one may refer to the original article,[8][9] where the definition of a Lie derivative of spinor fields is placed in the more general framework of the theory of Lie derivatives of sections of fiber bundles and the direct approach by Y. Kosmann to the spinor case is generalized to gauge natural bundles in the form of a new geometric concept called the Kosmann lift.

24.6.2 Covariant Lie derivative

If we have a principal bundle over the manifold M with G as the structure group, and we pick X to be a covariant vector field as section of the tangent space of the principal bundle (i.e. it has horizontal and vertical components), then the covariant Lie derivative is just the Lie derivative with respect to X over the principal bundle.

Now, if we're given a vector field *Y* over *M* (but not the principal bundle) but we also have a connection over the principal bundle, we can define a vector field X over the principal bundle such that its horizontal component matches *Y* and its vertical component agrees with the connection. This is the covariant Lie derivative.

See connection form for more details.

24.6.3 Nijenhuis–Lie derivative

Another generalization, due to Albert Nijenhuis, allows one to define the Lie derivative of a differential form along any section of the bundle $\Omega^k(M, TM)$ of differential forms with values in the tangent bundle. If $K \in \Omega^k(M, TM)$ and α is a differential *p*-form, then it is possible to define the interior product $i_K\alpha$ of *K* and α. The Nijenhuis–Lie derivative is then the anticommutator of the interior product and the exterior derivative:

$$\mathcal{L}_K \alpha = [d, i_K] \alpha = d i_K \alpha - (-1)^{k-1} i_K \, d\alpha.$$

24.7 History

In 1931, Władysław Ślebodziński introduced a new differential operator, later called by David van Dantzig that of Lie derivation, which can be applied to scalars, vectors, tensors and affine connections and which proved to be a powerful instrument in the study of groups of automorphisms.

The Lie derivatives of general geometric objects (i.e., sections of natural fiber bundles) were studied by A. Nijenhuis, Y. Tashiro and K. Yano.

For a quite long time, physicists had been using Lie derivatives, without reference to the work of mathematicians. In 1940, Léon Rosenfeld[10]—and before him Wolfgang Pauli[11]—introduced what he called a 'local variation' $\delta^* A$ of a geometric object A induced by an infinitesimal transformation of coordinates generated by a vector field X . One can easily prove that his $\delta^* A$ is $-\mathcal{L}_X(A)$.

24.8 See also

- Covariant derivative
- Connection (mathematics)
- Frölicher–Nijenhuis bracket
- Geodesic
- Killing field
- Derivative of the exponential map

24.9 Notes

[1] Trautman, A. (2008). "Remarks on the history of the notion of Lie differentiation". In Krupková, O.; Saunders, D. J. *Variations, Geometry and Physics: In honour of Demeter Krupka's sixty-fifth birthday*. New York: Nova Science. pp. 297–302. ISBN 978-1-60456-920-9.

[2] Ślebodziński, W. (1931). "Sur les équations de Hamilton". *Bull. Acad. Roy. d. Belg.* **17** (5): 864–870.

[3] Kolář, I.; Michor, P.; Slovák, J. (1993). *Natural Operations in Differential Geometry*. p. 21. ISBN 3-540-56235-4.

[4] Kosmann, Y. (1972). "Dérivées de Lie des spineurs". *Ann. Mat. Pura Appl.* **91** (4): 317–395. doi:10.1007/BF02428822.

[5] Trautman, A. (1972). "Invariance of Lagrangian Systems". In O'Raifeartaigh, L. *General Relativity: Papers in honour of J. L. Synge*. Oxford: Clarenden Press. p. 85. ISBN 0-19-851126-4.

[6] Fatibene, L.; Francaviglia, M. (2003). *Natural and Gauge Natural Formalism for Classical Field Theories*. Dordrecht: Kluwer Academic.

[7] Lichnerowicz, A. (1963). "Spineurs harmoniques". *C. R. Acad. Sci. Paris* **257**: 7–9.

[8] Fatibene, L.; Ferraris, M.; Francaviglia, M.; Godina, M. (1996). "A geometric definition of Lie derivative for Spinor Fields". In Janyska, J.; Kolář, I.; Slovák, J. *Proceedings of the 6th International Conference on Differential Geometry and Applications, August 28th–September 1st 1995 (Brno, Czech Republic)*. Brno: Masaryk University. pp. 549–558. ISBN 80-210-1369-9.

[9] Godina, M.; Matteucci, P. (2003). "Reductive G-structures and Lie derivatives". *Journal of Geometry and Physics* **47**: 66–86. doi:10.1016/S0393-0440(02)00174-2.

[10] Rosenfeld, L. (1940). "Sur le tenseur d'impulsion-énergie". *Mémoires Acad. Roy. d. Belg.* **18** (6): 1–30.

[11] Pauli, W. (1981) [1921]. *Theory of Relativity* (First ed.). New York: Dover. ISBN 978-0-486-64152-2. *See section 23*

24.10 References

- Abraham, Ralph; Marsden, Jerrold E. (1978). *Foundations of Mechanics*. London: Benjamin-Cummings. ISBN 0-8053-0102-X. *See section 2.2.*
- Bleecker, David (1981). *Gauge Theory and Variational Principles*. Addison-Wesley. ISBN 0-201-10096-7. *See Chapter 0.*
- Jost, Jürgen (2002). *Riemannian Geometry and Geometric Analysis*. Berlin: Springer. ISBN 3-540-42627-2. *See section 1.6.*
- Kolář, I.; Michor, P.; Slovák, J. (1993). *Natural operations in differential geometry*. Springer-Verlag. Extensive discussion of Lie brackets, and the general theory of Lie derivatives.
- Lang, S. (1995). *Differential and Riemannian manifolds*. Springer-Verlag. ISBN 978-0-387-94338-1. For generalizations to infinite dimensions.
- Lang, S. (1999). *Fundamentals of Differential Geometry*. Springer-Verlag. ISBN 978-0-387-98593-0. For generalizations to infinite dimensions.
- Yano, K. (1957). *The Theory of Lie Derivatives and its Applications*. North-Holland. ISBN 978-0-7204-2104-0. Classical approach using coordinates.

24.11 External links

- Hazewinkel, Michiel, ed. (2001), "Lie derivative", *Encyclopedia of Mathematics*, Springer, ISBN 978-1-55608-010-4

Chapter 25

Gauge theory

For a more accessible and less technical introduction to this topic, see Introduction to gauge theory.

In physics, a **gauge theory** is a type of field theory in which the Lagrangian is invariant under a continuous group of local transformations.

The term *gauge* refers to redundant degrees of freedom in the Lagrangian. The transformations between possible gauges, called *gauge transformations*, form a Lie group—referred to as the *symmetry group* or the *gauge group* of the theory. Associated with any Lie group is the Lie algebra of group generators. For each group generator there necessarily arises a corresponding field (usually a vector field) called the *gauge field*. Gauge fields are included in the Lagrangian to ensure its invariance under the local group transformations (called *gauge invariance*). When such a theory is quantized, the quanta of the gauge fields are called *gauge bosons*. If the symmetry group is non-commutative, the gauge theory is referred to as *non-abelian*, the usual example being the Yang–Mills theory.

Many powerful theories in physics are described by Lagrangians that are invariant under some symmetry transformation groups. When they are invariant under a transformation identically performed at *every* point in the spacetime in which the physical processes occur, they are said to have a global symmetry. Local symmetry, the cornerstone of gauge theories, is a stricter constraint. In fact, a global symmetry is just a local symmetry whose group's parameters are fixed in spacetime.

Gauge theories are important as the successful field theories explaining the dynamics of elementary particles. Quantum electrodynamics is an abelian gauge theory with the symmetry group U(1) and has one gauge field, the electromagnetic four-potential, with the photon being the gauge boson. The Standard Model is a non-abelian gauge theory with the symmetry group U(1)×SU(2)×SU(3) and has a total of twelve gauge bosons: the photon, three weak bosons and eight gluons.

Gauge theories are also important in explaining gravitation in the theory of general relativity. Its case is somewhat unique in that the gauge field is a tensor, the Lanczos tensor. Theories of quantum gravity, beginning with gauge gravitation theory, also postulate the existence of a gauge boson known as the graviton. Gauge symmetries can be viewed as analogues of the principle of general covariance of general relativity in which the coordinate system can be chosen freely under arbitrary diffeomorphisms of spacetime. Both gauge invariance and diffeomorphism invariance reflect a redundancy in the description of the system. An alternative theory of gravitation, gauge theory gravity, replaces the principle of general covariance with a true gauge principle with new gauge fields.

Historically, these ideas were first stated in the context of classical electromagnetism and later in general relativity. However, the modern importance of gauge symmetries appeared first in the relativistic quantum mechanics of electrons – quantum electrodynamics, elaborated on below. Today, gauge theories are useful in condensed matter, nuclear and high energy physics among other subfields.

25.1 History

The earliest field theory having a gauge symmetry was Maxwell's formulation, in 1864–65, of electrodynamics ("A Dynamical Theory of the Electromagnetic Field"). The importance of this symmetry remained unnoticed in the earliest formulations. Similarly unnoticed, Hilbert had derived the Einstein field equations by postulating the invariance of the action under a general coordinate transformation. Later Hermann Weyl, in an attempt to unify general relativity and electromagnetism, conjectured that *Eichinvarianz* or invariance under the change of scale (or "gauge") might also be a local symmetry of general relativity. After the development of quantum mechanics, Weyl, Vladimir Fock and Fritz London modified gauge by replacing the scale factor with a complex quantity and turned the scale transformation into a change of phase, which is a U(1) gauge symmetry. This explained the electromagnetic field effect on the wave function of a charged quantum mechanical particle. This was the first widely recognised gauge theory, popularised by Pauli in the 1940s.[1]

In 1954, attempting to resolve some of the great confusion in elementary particle physics, Chen Ning Yang and Robert Mills introduced **non-abelian gauge theories** as models to understand the strong interaction holding together nucleons in atomic nuclei. (Ronald Shaw, working under Abdus Salam, independently introduced the same notion in his doctoral thesis.) Generalizing the gauge invariance of electromagnetism, they attempted to construct a theory based on the action of the (non-abelian) SU(2) symmetry group on the isospin doublet of protons and neutrons. This is similar to the action of the U(1) group on the spinor fields of quantum electrodynamics. In particle physics the emphasis was on using quantized gauge theories.

This idea later found application in the quantum field theory of the weak force, and its unification with electromagnetism in the electroweak theory. Gauge theories became even more attractive when it was realized that non-abelian gauge theories reproduced a feature called asymptotic freedom. Asymptotic freedom was believed to be an important characteristic of strong interactions. This motivated searching for a strong force gauge theory. This theory, now known as quantum chromodynamics, is a gauge theory with the action of the SU(3) group on the color triplet of quarks. The Standard Model unifies the description of electromagnetism, weak interactions and strong interactions in the language of gauge theory.

In the 1970s, Sir Michael Atiyah began studying the mathematics of solutions to the classical Yang–Mills equations. In 1983, Atiyah's student Simon Donaldson built on this work to show that the differentiable classification of smooth 4-manifolds is very different from their classification up to homeomorphism. Michael Freedman used Donaldson's work to exhibit exotic $\mathbf{R}^4$s, that is, exotic differentiable structures on Euclidean 4-dimensional space. This led to an increasing interest in gauge theory for its own sake, independent of its successes in fundamental physics. In 1994, Edward Witten and Nathan Seiberg invented gauge-theoretic techniques based on supersymmetry that enabled the calculation of certain topological invariants (the Seiberg–Witten invariants). These contributions to mathematics from gauge theory have led to a renewed interest in this area.

The importance of gauge theories in physics is exemplified in the tremendous success of the mathematical formalism in providing a unified framework to describe the quantum field theories of electromagnetism, the weak force and the strong force. This theory, known as the Standard Model, accurately describes experimental predictions regarding three of the four fundamental forces of nature, and is a gauge theory with the gauge group SU(3) × SU(2) × U(1). Modern theories like string theory, as well as general relativity, are, in one way or another, gauge theories.

See Pickering[2] for more about the history of gauge and quantum field theories.

25.2 Description

25.2.1 Global and local symmetries

Global symmetry

In physics, the mathematical description of any physical situation usually contains excess degrees of freedom; the same physical situation is equally well described by many equivalent mathematical configurations. For instance, in Newtonian dynamics, if two configurations are related by a Galilean transformation (an inertial change of reference frame) they represent the same physical situation. These transformations form a group of "symmetries" of the theory, and a physical situation corresponds not to an individual mathematical configuration but to a class of configurations related to one another by this symmetry group.

This idea can be generalized to include local as well as global symmetries, analogous to much more abstract "changes of coordinates" in a situation where there is no preferred "inertial" coordinate system that covers the entire physical system. A gauge theory is a mathematical model that has symmetries of this kind, together with a set of techniques for making physical predictions consistent with the symmetries of the model.

Example of global symmetry

When a quantity occurring in the mathematical configuration is not just a number but has some geometrical significance, such as a velocity or an axis of rotation, its representation as numbers arranged in a vector or matrix is also changed by a coordinate transformation. For instance, if one description of a pattern of fluid flow states that the fluid velocity in the neighborhood of (x=1, y=0) is 1 m/s in the positive x direction, then a description of the same situation in which the coordinate system has been rotated clockwise by 90 degrees states that the fluid velocity in the neighborhood of (x=0, y=1) is 1 m/s in the positive y direction. The coordinate transformation has affected both the coordinate system used to identify the *location* of the measurement and the basis in which its *value* is expressed. As long as this transformation is performed globally (affecting the coordinate basis in the same way at every point), the effect on values that represent the *rate of change* of some quantity along some path in space and time as it passes through point *P* is the same as the effect on values that are truly local to *P*.

Local symmetry

Use of fiber bundles to describe local symmetries In order to adequately describe physical situations in more complex theories, it is often necessary to introduce a "coordinate basis" for some of the objects of the theory that do not have this simple relationship to the coordinates used to label points in space and time. (In mathematical terms, the theory involves a fiber bundle in which the fiber at each point of the base space consists of possible coordinate bases for use when describing the values of objects at that point.) In order to spell out a mathematical configuration, one must choose a particular coordinate basis at each point (a *local section* of the fiber bundle) and express the values of the objects of the theory (usually "fields" in the physicist's sense) using this basis. Two such mathematical configurations are equivalent (describe the same physical situation) if they are related by a transformation of this abstract coordinate basis (a change of local section, or *gauge transformation*).

In most gauge theories, the set of possible transformations of the abstract gauge basis at an individual point in space and time is a finite-dimensional Lie group. The simplest such group is U(1), which appears in the modern formulation of quantum electrodynamics (QED) via its use of complex numbers. QED is generally regarded as the first, and simplest, physical gauge theory. The set of possible gauge transformations of the entire configuration of a given gauge theory also forms a group, the *gauge group* of the theory. An element of the gauge group can be parameterized by a smoothly varying function from the points of spacetime to the (finite-dimensional) Lie group, such that the value of the function and its derivatives at each point represents the action of the gauge transformation on the fiber over that point.

A gauge transformation with constant parameter at every point in space and time is analogous to a rigid rotation of the geometric coordinate system; it represents a global symmetry of the gauge representation. As in the case of a rigid rotation, this gauge transformation affects expressions that represent the rate of change along a path of some gauge-dependent quantity in the same way as those that represent a truly local quantity. A gauge transformation whose parameter is *not* a constant function is referred to as a local symmetry; its effect on expressions that involve a derivative is qualitatively different from that on expressions that don't. (This is analogous to a non-inertial change of reference frame, which can produce a Coriolis effect.)

25.2.2 Gauge fields

The "gauge covariant" version of a gauge theory accounts for this effect by introducing a gauge field (in mathematical language, an Ehresmann connection) and formulating all rates of change in terms of the covariant derivative with respect to this connection. The gauge field becomes an essential part of the description of a mathematical configuration. A configuration in which the gauge field can be eliminated by a gauge transformation has the property that its field strength (in mathematical language, its curvature) is zero everywhere; a gauge theory is *not* limited to these configurations. In other words, the distinguishing characteristic of a gauge theory is that the gauge field does not merely compensate for a poor choice of coordinate system; there is generally no gauge transformation that makes the gauge field vanish.

When analyzing the dynamics of a gauge theory, the gauge field must be treated as a dynamical variable, similarly to other objects in the description of a physical situation. In addition to its interaction with other objects via the covariant derivative, the gauge field typically contributes energy in the form of a "self-energy" term. One can obtain the equations for the gauge theory by:

- starting from a naïve ansatz without the gauge field (in which the derivatives appear in a "bare" form);
- listing those global symmetries of the theory that can be characterized by a continuous parameter (generally an abstract equivalent of a rotation angle);
- computing the correction terms that result from allowing the symmetry parameter to vary from place to place; and
- reinterpreting these correction terms as couplings to one or more gauge fields, and giving these fields appropriate self-energy terms and dynamical behavior.

This is the sense in which a gauge theory "extends" a global symmetry to a local symmetry, and closely resembles the historical development of the gauge theory of gravity known as general relativity.

25.2.3 Physical experiments

Gauge theories are used to model the results of physical experiments, essentially by:

- limiting the universe of possible configurations to those consistent with the information used to set up the experiment, and then

- computing the probability distribution of the possible outcomes that the experiment is designed to measure.

The mathematical descriptions of the "setup information" and the "possible measurement outcomes" (loosely speaking, the "boundary conditions" of the experiment) are generally not expressible without reference to a particular coordinate system, including a choice of gauge. (If nothing else, one assumes that the experiment has been adequately isolated from "external" influence, which is itself a gauge-dependent statement.) Mishandling gauge dependence in boundary conditions is a frequent source of anomalies in gauge theory calculations, and gauge theories can be broadly classified by their approaches to anomaly avoidance.

25.2.4 Continuum theories

The two gauge theories mentioned above (continuum electrodynamics and general relativity) are examples of continuum field theories. The techniques of calculation in a continuum theory implicitly assume that:

- given a completely fixed choice of gauge, the boundary conditions of an individual configuration can in principle be completely described;
- given a completely fixed gauge and a complete set of boundary conditions, the principle of least action determines a unique mathematical configuration (and therefore a unique physical situation) consistent with these bounds;
- the likelihood of possible measurement outcomes can be determined by:
    - establishing a probability distribution over all physical situations determined by boundary conditions that are consistent with the setup information,
    - establishing a probability distribution of measurement outcomes for each possible physical situation, and
    - convolving these two probability distributions to get a distribution of possible measurement outcomes consistent with the setup information; and
- fixing the gauge introduces no anomalies in the calculation, due either to gauge dependence in describing partial information about boundary conditions or to incompleteness of the theory.

These assumptions are close enough to be valid across a wide range of energy scales and experimental conditions, to allow these theories to make accurate predictions about almost all of the phenomena encountered in daily life, from light, heat, and electricity to eclipses and spaceflight. They fail only at the smallest and largest scales (due to omissions in the theories themselves) and when the mathematical techniques themselves break down (most notably in the case of turbulence and other chaotic phenomena).

25.2.5 Quantum field theories

Other than these classical continuum field theories, the most widely known gauge theories are quantum field theories, including quantum electrodynamics and the Standard Model of elementary particle physics. The starting point of a quantum field theory is much like that of its continuum analog: a gauge-covariant action integral that characterizes "allowable" physical situations according to the principle of least action. However, continuum and quantum theories differ significantly in how they handle the excess degrees of freedom represented by gauge transformations. Continuum theories, and most pedagogical treatments of the simplest quantum field theories, use a gauge fixing prescription to reduce the orbit of mathematical configurations that represent a given physical situation to a smaller orbit related by a smaller gauge group (the global symmetry group, or perhaps even the trivial group).

More sophisticated quantum field theories, in particular those that involve a non-abelian gauge group, break the gauge symmetry within the techniques of perturbation theory by introducing additional fields (the Faddeev–Popov ghosts) and counterterms motivated by anomaly cancellation, in an approach known as BRST quantization. While these concerns are in one sense highly technical, they are also closely related to the nature of measurement, the limits on knowledge of a physical situation, and the interactions between incompletely specified experimental conditions and incompletely understood physical theory. The mathematical techniques that have been developed in order to make gauge theories tractable have found many other applications, from solid-state physics and crystallography to low-dimensional topology.

25.3 Classical gauge theory

25.3.1 Classical electromagnetism

Historically, the first example of gauge symmetry discovered was classical electromagnetism. In electrostatics, one can either discuss the electric field, **E**, or its corresponding electric potential, V. Knowledge of one makes it possible to find the other, except that potentials differing by a constant, $V \rightarrow V + C$, correspond to the same electric field. This is because the electric field relates to *changes* in the potential from one point in space to another, and the constant C would cancel out when subtracting to find the change in potential. In terms of vector calculus, the electric field is the gradient of the potential, $\mathbf{E} = -\nabla V$. Generalizing from static electricity to electromagnetism, we have a second potential, the vector potential $\mathbf{A}$, with

$$\mathbf{E} = -\nabla V - \frac{\partial \mathbf{A}}{\partial t}$$
$$\mathbf{B} = \nabla \times \mathbf{A}$$

The general gauge transformations now become not just $V \to V + C$ but

$$\mathbf{A} \to \mathbf{A} + \nabla f$$
$$V \to V - \frac{\partial f}{\partial t}$$

where f is any function that depends on position and time. The fields remain the same under the gauge transformation, and therefore Maxwell's equations are still satisfied. That is, Maxwell's equations have a gauge symmetry.

25.3.2 An example: Scalar O(*n*) gauge theory

The remainder of this section requires some familiarity with classical or quantum field theory, and the use of Lagrangians.

Definitions in this section: gauge group, gauge field, interaction Lagrangian, gauge boson.

The following illustrates how local gauge invariance can be "motivated" heuristically starting from global symmetry properties, and how it leads to an interaction between originally non-interacting fields.

Consider a set of n non-interacting real scalar fields, with equal masses m. This system is described by an action that is the sum of the (usual) action for each scalar field φ_i

$$\mathcal{S} = \int \mathrm{d}^4x \sum_{i=1}^{n} \left[\frac{1}{2}\partial_\mu\varphi_i\partial^\mu\varphi_i - \frac{1}{2}m^2\varphi_i^2\right]$$

The Lagrangian (density) can be compactly written as

$$\mathcal{L} = \frac{1}{2}(\partial_\mu\Phi)^T\partial^\mu\Phi - \frac{1}{2}m^2\Phi^T\Phi$$

by introducing a vector of fields

$$\Phi = (\varphi_1, \varphi_2, \ldots, \varphi_n)^T$$

The term ∂_μ is Einstein notation for the partial derivative of Φ in each of the four dimensions.

It is now transparent that the Lagrangian is invariant under the transformation

$$\Phi \mapsto \Phi' = G\Phi$$

whenever G is a *constant* matrix belonging to the n-by-n orthogonal group O(n). This is seen to preserve the Lagrangian, since the derivative of Φ transforms identically to Φ and both quantities appear inside dot products in the Lagrangian (orthogonal transformations preserve the dot product).

$$(\partial_\mu\Phi) \mapsto (\partial_\mu\Phi)' = G\partial_\mu\Phi$$

This characterizes the *global* symmetry of this particular Lagrangian, and the symmetry group is often called the **gauge group**; the mathematical term is **structure group**, especially in the theory of G-structures. Incidentally, Noether's theorem implies that invariance under this group of transformations leads to the conservation of the *currents*

$$J^a_\mu = i\partial_\mu\Phi^T T^a\Phi$$

where the T^a matrices are generators of the SO(n) group. There is one conserved current for every generator.

Now, demanding that this Lagrangian should have *local* O(n)-invariance requires that the G matrices (which were earlier constant) should be allowed to become functions of the space-time coordinates x.

In this case, the G matrices do not "pass through" the derivatives, when $G = G(x)$,

$$\partial_\mu(G\Phi) \neq G(\partial_\mu\Phi)$$

The failure of the derivative to commute with "G" introduces an additional term (in keeping with the product rule), which spoils the invariance of the Lagrangian. In order to rectify this we define a new derivative operator such that the derivative of Φ again transforms identically with Φ

$$(D_\mu\Phi)' = GD_\mu\Phi$$

This new "derivative" is called a (gauge) covariant derivative and takes the form

$$D_\mu = \partial_\mu + igA_\mu$$

Where g is called the coupling constant; a quantity defining the strength of an interaction. After a simple calculation we can see that the **gauge field** $A(x)$ must transform as follows

$$A'_\mu = GA_\mu G^{-1} + \frac{i}{g}(\partial_\mu G)G^{-1}$$

The gauge field is an element of the Lie algebra, and can therefore be expanded as

$$A_\mu = \sum_a A_\mu^a T^a$$

There are therefore as many gauge fields as there are generators of the Lie algebra.

Finally, we now have a *locally gauge invariant* Lagrangian

$$\mathcal{L}_{\text{loc}} = \frac{1}{2}(D_\mu \Phi)^T D^\mu \Phi - \frac{1}{2}m^2 \Phi^T \Phi$$

Pauli uses the term *gauge transformation of the first type* to mean the transformation of Φ , while the compensating transformation in A is called a *gauge transformation of the second type*.

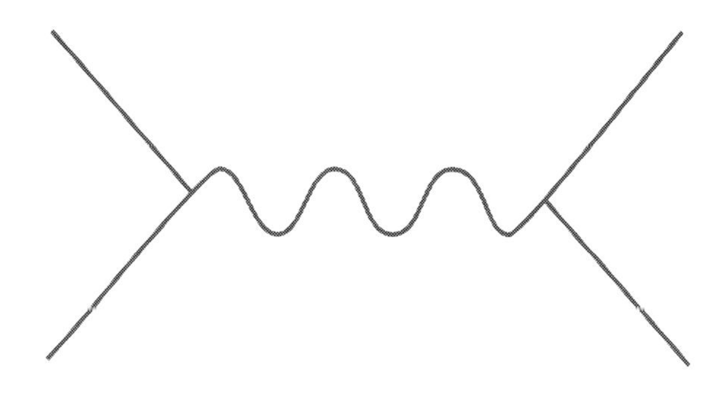

Feynman diagram of scalar bosons interacting via a gauge boson

The difference between this Lagrangian and the original *globally gauge-invariant* Lagrangian is seen to be the **interaction Lagrangian**

$$\mathcal{L}_{\text{int}} = i\frac{g}{2}\Phi^T A_\mu^T \partial^\mu \Phi + i\frac{g}{2}(\partial_\mu \Phi)^T A^\mu \Phi - \frac{g^2}{2}(A_\mu \Phi)^T A^\mu \Phi$$

This term introduces interactions between the n scalar fields just as a consequence of the demand for local gauge invariance. However, to make this interaction physical and not completely arbitrary, the mediator $A(x)$ needs to propagate in space. That is dealt with in the next section by adding yet another term, $\mathcal{L}_{\text{gf}}$, to the Lagrangian. In the quantized version of the obtained classical field theory, the quanta of the gauge field $A(x)$ are called gauge bosons. The interpretation of the interaction Lagrangian in quantum field theory is of scalar bosons interacting by the exchange of these gauge bosons.

25.3.3 The Yang–Mills Lagrangian for the gauge field

Main article: Yang–Mills theory

The picture of a classical gauge theory developed in the previous section is almost complete, except for the fact that to define the covariant derivatives D, one needs to know the value of the gauge field $A(x)$ at all space-time points. Instead of manually specifying the values of this field, it can be given as the solution to a field equation. Further requiring that the Lagrangian that generates this field equation is locally gauge invariant as well, one possible form for the gauge field Lagrangian is (conventionally) written as

$$\mathcal{L}_{\text{gf}} = -\frac{1}{2}\operatorname{Tr}(F^{\mu\nu} F_{\mu\nu})$$

with

$$F_{\mu\nu} = \frac{1}{ig}[D_\mu, D_\nu]$$

and the trace being taken over the vector space of the fields. This is called the **Yang–Mills action**. Other gauge invariant actions also exist (e.g., nonlinear electrodynamics, Born–Infeld action, Chern–Simons model, theta term, etc.).

Note that in this Lagrangian term there is no field whose transformation counterweighs the one of A . Invariance of this term under gauge transformations is a particular case of *a priori* classical (geometrical) symmetry. This symmetry must be restricted in order to perform quantization, the procedure being denominated gauge fixing, but even after restriction, gauge transformations may be possible.[3]

The complete Lagrangian for the gauge theory is now

$$\mathcal{L} = \mathcal{L}_{\text{loc}} + \mathcal{L}_{\text{gf}} = \mathcal{L}_{\text{global}} + \mathcal{L}_{\text{int}} + \mathcal{L}_{\text{gf}}$$

25.3.4 An example: Electrodynamics

As a simple application of the formalism developed in the previous sections, consider the case of electrodynamics, with only the electron field. The bare-bones action that generates the electron field's Dirac equation is

$$\mathcal{S} = \int \bar{\psi}(i\hbar c\,\gamma^\mu \partial_\mu - mc^2)\psi \,\mathrm{d}^4x$$

The global symmetry for this system is

$$\psi \mapsto e^{i\theta}\psi$$

The gauge group here is U(1), just rotations of the phase angle of the field, with the particular rotation determined by the constant θ.

"Localising" this symmetry implies the replacement of θ by θ(x). An appropriate covariant derivative is then

$$D_\mu = \partial_\mu - i\frac{e}{\hbar}A_\mu$$

Identifying the "charge" e (not to be confused with the mathematical constant e in the symmetry description) with the usual electric charge (this is the origin of the usage of the term in gauge theories), and the gauge field $A(x)$ with the four-vector potential of electromagnetic field results in an interaction Lagrangian

$$\mathcal{L}_{\text{int}} = \frac{e}{\hbar}\bar{\psi}(x)\gamma^\mu\psi(x)A_\mu(x) = J^\mu(x)A_\mu(x)$$

where $J^\mu(x)$ is the usual four vector electric current density. The gauge principle is therefore seen to naturally introduce the so-called minimal coupling of the electromagnetic field to the electron field.

Adding a Lagrangian for the gauge field $A_\mu(x)$ in terms of the field strength tensor exactly as in electrodynamics, one obtains the Lagrangian used as the starting point in quantum electrodynamics.

$$\mathcal{L}_{\text{QED}} = \bar{\psi}(i\hbar c\,\gamma^\mu D_\mu - mc^2)\psi - \frac{1}{4\mu_0}F_{\mu\nu}F^{\mu\nu}$$

See also: Dirac equation, Maxwell's equations, Quantum electrodynamics

25.4 Mathematical formalism

Gauge theories are usually discussed in the language of differential geometry. Mathematically, a *gauge* is just a choice of a (local) section of some principal bundle. A **gauge transformation** is just a transformation between two such sections.

Although gauge theory is dominated by the study of connections (primarily because it's mainly studied by high-energy physicists), the idea of a connection is not central to gauge theory in general. In fact, a result in general gauge theory shows that affine representations (i.e., affine modules) of the gauge transformations can be classified as sections of a jet bundle satisfying certain properties. There are representations that transform covariantly pointwise (called by physicists gauge transformations of the first kind), representations that transform as a connection form (called by physicists gauge transformations of the second kind, an affine representation)—and other more general representations, such as the B field in BF theory. There are more general nonlinear representations (realizations), but these are extremely complicated. Still, nonlinear sigma models transform nonlinearly, so there are applications.

If there is a principal bundle P whose base space is space or spacetime and structure group is a Lie group, then the sections of P form a principal homogeneous space of the group of gauge transformations.

Connections (gauge connection) define this principal bundle, yielding a covariant derivative ∇ in each associated vector bundle. If a local frame is chosen (a local basis of sections), then this covariant derivative is represented by the connection form A, a Lie algebra-valued 1-form, which is called the **gauge potential** in physics. This is evidently not an intrinsic but a frame-dependent quantity. The curvature form F, a Lie algebra-valued 2-form that is an intrinsic quantity, is constructed from a connection form by

$$\mathbf{F} = \mathrm{d}\mathbf{A} + \mathbf{A} \wedge \mathbf{A}$$

where d stands for the exterior derivative and $\wedge$ stands for the wedge product. (**A** is an element of the vector space spanned by the generators T^a , and so the components of **A** do not commute with one another. Hence the wedge product **A** $\wedge$ **A** does not vanish.)

Infinitesimal gauge transformations form a Lie algebra, which is characterized by a smooth Lie-algebra-valued scalar, ε. Under such an infinitesimal gauge transformation,

$$\delta_\varepsilon \mathbf{A} = [\varepsilon, \mathbf{A}] - \mathrm{d}\varepsilon$$

where $[\cdot,\cdot]$ is the Lie bracket.

One nice thing is that if $\delta_\varepsilon X = \varepsilon X$, then $\delta_\varepsilon DX = \varepsilon DX$ where D is the covariant derivative

$$DX \stackrel{\text{def}}{=} \mathrm{d}X + \mathbf{A}X$$

Also, $\delta_\varepsilon \mathbf{F} = \varepsilon \mathbf{F}$, which means **F** transforms covariantly.

Not all gauge transformations can be generated by infinitesimal gauge transformations in general. An example is when the base manifold is a compact manifold without boundary such that the homotopy class of mappings from that manifold to the Lie group is nontrivial. See instanton for an example.

The *Yang–Mills action* is now given by

$$\frac{1}{4g^2}\int \mathrm{Tr}[*F \wedge F]$$

where * stands for the Hodge dual and the integral is defined as in differential geometry.

A quantity which is **gauge-invariant** (i.e., invariant under gauge transformations) is the Wilson loop, which is defined over any closed path, γ, as follows:

$$\chi^{(\rho)}\left(\mathcal{P}\left\{e^{\int_\gamma A}\right\}\right)$$

where χ is the character of a complex representation ρ and $\mathcal{P}$ represents the path-ordered operator.

25.5 Quantization of gauge theories

Main article: Quantum gauge theory

Gauge theories may be quantized by specialization of methods which are applicable to any quantum field theory. However, because of the subtleties imposed by the gauge constraints (see section on Mathematical formalism, above) there are many technical problems to be solved which do not arise in other field theories. At the same time, the richer structure of gauge theories allows simplification of some computations: for example Ward identities connect different renormalization constants.

25.5.1 Methods and aims

The first gauge theory quantized was quantum electrodynamics (QED). The first methods developed for this involved gauge fixing and then applying canonical quantization. The Gupta–Bleuler method was also developed to handle this problem. Non-abelian gauge theories are now handled by a variety of means. Methods for quantization are covered in the article on quantization.

The main point to quantization is to be able to compute quantum amplitudes for various processes allowed by the theory. Technically, they reduce to the computations of certain correlation functions in the vacuum state. This involves a renormalization of the theory.

When the running coupling of the theory is small enough, then all required quantities may be computed in perturbation theory. Quantization schemes intended to simplify such computations (such as canonical quantization) may be called **perturbative quantization schemes**. At present some of these methods lead to the most precise experimental tests of gauge theories.

However, in most gauge theories, there are many interesting questions which are non-perturbative. Quantization schemes suited to these problems (such as lattice gauge theory) may be called **non-perturbative quantization schemes**. Precise computations in such schemes often require supercomputing, and are therefore less well-developed currently than other schemes.

25.5.2 Anomalies

Some of the symmetries of the classical theory are then seen not to hold in the quantum theory; a phenomenon called an **anomaly**. Among the most well known are:

- The scale anomaly, which gives rise to a *running coupling constant*. In QED this gives rise to the phenomenon of the Landau pole. In Quantum Chromodynamics (QCD) this leads to asymptotic freedom.
- The chiral anomaly in either chiral or vector field theories with fermions. This has close connection with topology through the notion of instantons. In QCD this anomaly causes the decay of a pion to two photons.
- The gauge anomaly, which must cancel in any consistent physical theory. In the electroweak theory this cancellation requires an equal number of quarks and leptons.

25.6 Pure gauge

A pure gauge is the set of field configurations obtained by a gauge transformation on the null-field configuration, i.e., a gauge-transform of zero. So it is a particular "gauge orbit" in the field configuration's space.

Thus, in the abelian case, where $A_\mu(x) \to A'_\mu(x) = A_\mu(x) + \partial_\mu f(x)$, the pure gauge is just the set of field configurations $A'_\mu(x) = \partial_\mu f(x)$ for all $f(x)$.

25.7 See also

25.8 References

[1] Wolfgang Pauli (1941). "Relativistic Field Theories of Elementary Particles," *Rev. Mod. Phys.* **13**: 203–32.

[2] Pickering, A. (1984). *Constructing Quarks*. University of Chicago Press. ISBN 0-226-66799-5.

[3] J. J. Sakurai, *Advanced Quantum Mechanics*, Addison-Wesley, 1967, sect. 1–4.

25.9 Bibliography

General readers

- Schumm, Bruce (2004) *Deep Down Things*. Johns Hopkins University Press. Esp. chpt. 8. A serious attempt by a physicist to explain gauge theory and the Standard Model with little formal mathematics.

Texts

- Bromley, D.A. (2000). *Gauge Theory of Weak Interactions*. Springer. ISBN 3-540-67672-4.
- Cheng, T.-P.; Li, L.-F. (1983). *Gauge Theory of Elementary Particle Physics*. Oxford University Press. ISBN 0-19-851961-3.
- Frampton, P. (2008). *Gauge Field Theories* (3rd ed.). Wiley-VCH.
- Kane, G.L. (1987). *Modern Elementary Particle Physics*. Perseus Books. ISBN 0-201-11749-5.

Articles

- Becchi, C. (1997). “Introduction to Gauge Theories”. arXiv:hep-ph/9705211.
- Gross, D. (1992). “Gauge theory – Past, Present and Future”. Retrieved 2009-04-23.
- Jackson, J.D. (2002). “From Lorenz to Coulomb and other explicit gauge transformations”. *Am. J. Phys.* **70** (9): 917–928. arXiv:physics/0204034. Bibcode:2002AmJPh..70..917J. doi:10.1119/1.1491265.
- Svetlichny, George (1999). “Preparation for Gauge Theory”. arXiv:math-ph/9902027.

25.10 External links

- Hazewinkel, Michiel, ed. (2001), “Gauge transformation”, *Encyclopedia of Mathematics*, Springer, ISBN 978-1-55608-010-4
- Yang–Mills equations on DispersiveWiki
- Gauge theories on Scholarpedia

Chapter 26

Holonomy

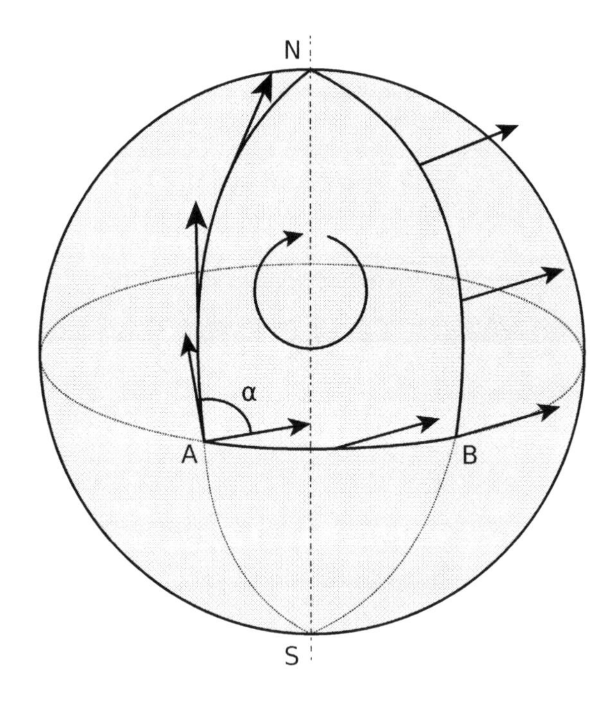

Parallel transport on a sphere depends on the path. Transporting from A → N → B → A *yields a vector different from the initial vector. This failure to return to the initial vector is measured by the holonomy of the connection.*

In differential geometry, the **holonomy** of a connection on a smooth manifold is a general geometrical consequence of the curvature of the connection measuring the extent to which parallel transport around closed loops fails to preserve the geometrical data being transported. For flat connections, the associated holonomy is a type of monodromy, and is an inherently global notion. For curved connections, holonomy has nontrivial local and global features.

Any kind of connection on a manifold gives rise, through its parallel transport maps, to some notion of holonomy. The most common forms of holonomy are for connections possessing some kind of symmetry. Important examples include: holonomy of the Levi-Civita connection in Riemannian geometry (called **Riemannian holonomy**), holonomy of connections in vector bundles, holonomy of Cartan connections, and holonomy of connections in principal bundles. In each of these cases, the holonomy of the connection can be identified with a Lie group, the **holonomy group**. The holonomy of a connection is closely related to the curvature of the connection, via the *Ambrose–Singer theorem*.

The study of Riemannian holonomy has led to a number of important developments. The holonomy was introduced by Cartan (1926) in order to study and classify symmetric spaces. It was not until much later that holonomy groups would be used to study Riemannian geometry in a more general setting. In 1952 Georges de Rham proved the *de Rham decomposition theorem*, a principle for splitting a Riemannian manifold into a Cartesian product of Riemannian manifolds by splitting the tangent bundle into irreducible spaces under the action of the local holonomy groups. Later, in 1953, M. Berger classified the possible irreducible holonomies. The decomposition and classification of Riemannian holonomy has applications to physics and to string theory.

26.1 Definitions

26.1.1 Holonomy of a connection in a vector bundle

Let E be a rank k vector bundle over a smooth manifold M and let ∇ be a connection on E. Given a piecewise smooth loop $\gamma : [0,1] \to M$ based at x in M, the connection defines a parallel transport map P_γ: $Ex \to Ex$. This map is both linear and invertible and so defines an element of the General linear group GL(Ex). The **holonomy group** of ∇ based at x is defined as

$$\mathrm{Hol}_x(\nabla) = \{P_\gamma \in \mathrm{GL}(E_x) \mid \gamma \text{ at based loop a is } x\}.$$

The **restricted holonomy group** based at x is the subgroup $\mathrm{Hol}^0x(\nabla)$ coming from contractible loops γ.

If M is connected then the holonomy group depends on the basepoint x only up to conjugation in GL(k, **R**). Explicitly, if γ is a path from x to y in M then

$$\mathrm{Hol}_y(\nabla) = P_\gamma \mathrm{Hol}_x(\nabla) P_\gamma^{-1}.$$

Choosing different identifications of Ex with $\mathbf{R}^k$ also gives conjugate subgroups. Sometimes, particularly in general or informal discussions (such as below), one may drop reference to the basepoint, with the understanding that the definition is good up to conjugation.

Some important properties of the holonomy group include:

- $\mathrm{Hol}^0(\nabla)$ is a connected, Lie subgroup of GL(k, **R**).
- $\mathrm{Hol}^0(\nabla)$ is the identity component of $\mathrm{Hol}(\nabla)$.
- There is a natural, surjective group homomorphism $\pi_1(M) \to \mathrm{Hol}(\nabla)/\mathrm{Hol}^0(\nabla)$, where $\pi_1(M)$ is the fundamental group of M, which sends the homotopy class $[\gamma]$ to the coset $P_\gamma \cdot \mathrm{Hol}^0(\nabla)$.
- If M is simply connected then $\mathrm{Hol}(\nabla) = \mathrm{Hol}^0(\nabla)$.
- ∇ is flat (i.e. has vanishing curvature) if and only if $\mathrm{Hol}^0(\nabla)$ is trivial.

26.1.2 Holonomy of a connection in a principal bundle

The definition for holonomy of connections on principal bundles proceeds in parallel fashion. Let G be a Lie group and P a principal G-bundle over a smooth manifold M which is paracompact. Let ω be a connection on P. Given a piecewise smooth loop $\gamma : [0,1] \to M$ based at x in M and a point p in the fiber over x, the connection defines a unique *horizontal lift* $\tilde{\gamma}\colon [0,1] \to P$ such that $\tilde{\gamma}(0) = p$. The end point of the horizontal lift, $\tilde{\gamma}(1)$, will not generally be p but rather some other point $p{\cdot}g$ in the fiber over x. Define an equivalence relation ~ on P by saying that $p \sim q$ if they can be joined by a piecewise smooth horizontal path in P.

The **holonomy group** of ω based at p is then defined as

$$\mathrm{Hol}_p(\omega) = \{g \in G \mid p \sim p \cdot g\}.$$

The **restricted holonomy group** based at p is the subgroup $\mathrm{Hol}^0p(\omega)$ coming from horizontal lifts of contractible loops γ.

If M and P are connected then the holonomy group depends on the basepoint p only up to conjugation in G. Explicitly, if q is any other chosen basepoint for the holonomy, then there exists a unique $g \in G$ such that $q \sim p\,g$. With this value of g,

$$\mathrm{Hol}_q(\omega) = g^{-1}\mathrm{Hol}_p(\omega)g.$$

In particular,

$$\mathrm{Hol}_{p\cdot g}(\omega) = g^{-1}\mathrm{Hol}_p(\omega)g,$$

Moreover, if $p \sim q$ then $\mathrm{Hol}p(\omega) = \mathrm{Hol}q(\omega)$. As above, sometimes one drops reference to the basepoint of the holonomy group, with the understanding that the definition is good up to conjugation.

Some important properties of the holonomy and restricted holonomy groups include:

- $\mathrm{Hol}^0p(\omega)$ is a connected Lie subgroup of G.
- $\mathrm{Hol}^0p(\omega)$ is the identity component of $\mathrm{Hol}p(\omega)$.
- There is a natural, surjective group homomorphism $\pi_1(M) \to \mathrm{Hol}p(\omega)/\mathrm{Hol}^0p(\omega)$.
- If M is simply connected then $\mathrm{Hol}_p(\omega) = \mathrm{Hol}^0p(\omega)$.
- ω is flat (i.e. has vanishing curvature) if and only if $\mathrm{Hol}^0{}_p(\omega)$ is trivial.

26.1.3 Holonomy bundles

Let M be a connected paracompact smooth manifold and P a principal G-bundle with connection ω, as above. Let $p \in P$ be an arbitrary point of the principal bundle. Let $H(p)$ be the set of points in P which can be joined to p by a horizontal curve. Then it can be shown that $H(p)$, with the evident projection map, is a principal bundle over M with structure group $\mathrm{Hol}_p(\omega)$. This principal bundle is called the **holonomy bundle** (through p) of the connection. The connection ω restricts to a connection on $H(p)$, since its parallel transport maps preserve $H(p)$. Thus $H(p)$ is a reduced bundle for the connection. Furthermore, since no subbundle of $H(p)$ is preserved by parallel transport, it is the minimal such reduction.[1]

As with the holonomy groups, the holonomy bundle also transforms equivariantly within the ambient principal bundle P. In detail, if $q \in P$ is another chosen basepoint for the holonomy, then there exists a unique $g \in G$ such that $q \sim p\,g$ (since, by assumption, M is path-connected). Hence $H(q) = H(p)\,g$. As a consequence, the induced connections on holonomy bundles corresponding to different choices of basepoint are compatible with one another: their parallel transport maps will differ by precisely the same element g.

26.1.4 Monodromy

The holonomy bundle $H(p)$ is a principal bundle for $\mathrm{Hol}_p(\omega)$, and so also admits an action of the restricted holonomy group $\mathrm{Hol}^0{}_p(\omega)$ (which is a normal subgroup of the full holonomy group). The discrete group $\mathrm{Hol}_p(\omega)/\mathrm{Hol}^0{}_p(\omega)$ is called the monodromy group of the connection; it acts on the quotient bundle $H(p)/\mathrm{Hol}^0{}_p(\omega)$. There is a surjective homomorphism $\varphi : \pi_1(M) \to \mathrm{Hol}_p(\omega)/\mathrm{Hol}^0{}_p(\omega)$, so that $\varphi(\pi_1(M))$ acts on $H(p)/\mathrm{Hol}^0{}_p(\omega)$. This action of the fundamental group is a **monodromy representation** of the fundamental group.[2]

26.1.5 Local and infinitesimal holonomy

If $\pi: P \to M$ is a principal bundle, and ω is a connection in P, then the holonomy of ω can be restricted to the fibre over an open subset of M. Indeed, if U is a connected open subset of M, then ω restricts to give a connection in the bundle $\pi^{-1}U$ over U. The holonomy (resp. restricted holonomy) of this bundle will be denoted by $Holp(\omega, U)$ (resp. $Holp^0(\omega, U)$) for each p with $\pi(p) \in U$.

If $U \subset V$ are two open sets containing $\pi(p)$, then there is an evident inclusion

$$\mathrm{Hol}_p^0(\omega, U) \subset \mathrm{Hol}_p^0(\omega, V).$$

The **local holonomy group** at a point p is defined by

$$\mathrm{Hol}^*(\omega) = \cap_{k=1}^{\infty} \mathrm{Hol}^0(\omega, U_k)$$

for any family of nested connected open sets Uk with $\cap_k U_k = \pi(p)$.

The local holonomy group has the following properties:

1. It is a connected Lie subgroup of the restricted holonomy group $\mathrm{Hol}p^0(\omega)$.
2. Every point p has a neighborhood V such that $\mathrm{Hol}p^*(\omega)=\mathrm{Hol}p^0(\omega, V)$. In particular, the local holonomy group depends only on the point p, and not the choice of sequence Uk used to define it.
3. The local holonomy is equivariant with respect to translation by elements of the structure group G of P; i.e., $\mathrm{Hol}^*pg(\omega) = \mathrm{Ad}(g^{-1})\mathrm{Hol}^*p(\omega)$ for all $g \in G$. (Note that, by property 1., the local holonomy group is a connected Lie subgroup of G, so the adjoint is well-defined.)

The local holonomy group is not well-behaved as a global object. In particular, its dimension may fail to be constant. However, the following theorem holds:

- If the dimension of the local holonomy group is constant, then the local and restricted holonomy agree: $\mathrm{Hol}^*p(\omega) = \mathrm{Hol}p^0(\omega)$.

26.2 Ambrose–Singer theorem

The Ambrose–Singer theorem relates the holonomy of a connection in a principal bundle with the curvature form of the connection. To make this theorem plausible, consider the familiar case of an affine connection (or a connection in the tangent bundle — the Levi-Civita connection, for example). The curvature arises when one travels around an infinitesimal parallelogram.

In detail, if σ: $[0, 1] \times [0, 1] \to M$ is a surface in M parametrized by a pair of variables x and y, then a vector V may be transported around the boundary of σ: first along $(x, 0)$, then along $(1, y)$, followed by $(x, 1)$ going in the negative direction, and then $(0, y)$ back to the point of origin. This is a special case of a holonomy loop: the vector V is acted upon by the holonomy group element corresponding to the lift of the boundary of σ. The curvature enters explicitly when the parallelogram is shrunk to zero, by traversing the boundary of smaller parallelograms over $[0, x] \times [0, y]$. This corresponds to taking a derivative of the parallel transport maps at $x = y = 0$:

$$\frac{D}{dx}\frac{D}{dy}V - \frac{D}{dy}\frac{D}{dx}V = R\left(\frac{\partial \sigma}{\partial x}, \frac{\partial \sigma}{\partial y}\right)V$$

where R is the curvature tensor.[3] So, roughly speaking, the curvature gives the infinitesimal holonomy over a closed loop (the infinitesimal parallelogram). More formally, the curvature is the differential of the holonomy action at the identity of the holonomy group. In other words, $R(X, Y)$ is an element of the Lie algebra of $\mathrm{Hol}p(\omega)$.

In general, consider the holonomy of a connection in a principal bundle $P \to M$ over P with structure group G. Denoting the Lie algebra of G by $\mathbf{g}$, the curvature form of the connection is a $\mathbf{g}$-valued 2-form Ω on P. The Ambrose–Singer theorem states:[4]

- The Lie algebra of $\mathrm{Hol}p(\omega)$ is spanned by all the elements of $\mathbf{g}$ of the form $\Omega_q(X,Y)$ as q ranges over all points which can be joined to p by a horizontal curve ($q \sim p$), and X and Y are horizontal tangent vectors at q. Alternatively, the theorem can be restated in terms of the holonomy bundle:[5]
- The Lie algebra of $\mathrm{Hol}_p(\omega)$ is the subspace of $\mathbf{g}$ spanned by elements of the form $\Omega q(X, Y)$ where $q \in H(p)$ and X and Y are horizontal vectors at q.

26.3 Riemannian holonomy

The holonomy of a Riemannian manifold (M, g) is just the holonomy group of the Levi-Civita connection on the tangent bundle to M. A 'generic' n-dimensional Riemannian manifold has an $O(n)$ holonomy, or $SO(n)$ if it is orientable. Manifolds whose holonomy groups are proper subgroups of $O(n)$ or $SO(n)$ have special properties.

One of the earliest fundamental results on Riemannian holonomy is the theorem of Borel & Lichnerowicz (1952), which asserts that the holonomy group is a closed Lie subgroup of $O(n)$. In particular, it is compact.

26.3.1 Reducible holonomy and the de Rham decomposition

Let $x \in M$ be an arbitrary point. Then the holonomy group Hol(M) acts on the tangent space T_xM. This action may either be irreducible as a group representation, or reducible in the sense that there is a splitting of T_xM into orthogonal subspaces $T_xM = T'_xM \oplus T''_xM$, each of which is invariant under the action of Hol(M). In the latter case, M is said to be **reducible**.

Suppose that M is a reducible manifold. Allowing the point x to vary, the bundles $T'M$ and $T''M$ formed by the reduction of the tangent space at each point are smooth distributions which are integrable in the sense of Frobenius. The integral manifolds of these distributions are totally geodesic submanifolds. So M is locally a Cartesian product $M' \times M''$. The (local) de Rham isomorphism follows by continuing this process until a complete reduction of the tangent space is achieved:[6]

- Let M be a simply connected Riemannian manifold,[7] and $TM = T^{(0)}M \oplus T^{(1)}M \oplus \ldots \oplus T^{(k)}M$ be the complete reduction of the tangent bundle under the action of the holonomy group. Suppose that $T^{(0)}M$ consists of vectors invariant under the holonomy group (i.e., such that the holonomy representation is trivial). Then locally M is isometric to a product

  $$V_0 \times V_1 \times \cdots \times V_k,$$

  where V_0 is an open set in a Euclidean space, and each Vi is an integral manifold for $T^{(i)}M$. Furthermore, Hol(M) splits as a direct product of the holonomy groups of each Mi.

If, moreover, M is assumed to be geodesically complete, then the theorem holds globally, and each Mi is a geodesically complete manifold.[8]

26.3.2 The Berger classification

In 1955, M. Berger gave a complete classification of possible holonomy groups for simply connected, Riemannian manifolds which are irreducible (not locally a product space) and nonsymmetric (not locally a Riemannian symmetric space). **Berger's list** is as follows:

Manifolds with holonomy $Sp(n)\cdot Sp(1)$ were simultaneously studied in 1965 by Edmond Bonan and Vivian Yoh Kraines and they constructed the parallel 4-form.

Manifolds with holonomy G_2 or Spin(7) were firstly introduced by Edmond Bonan in 1966, who constructed all the parallel forms and showed that those manifolds were Ricci-flat.

(Berger's original list also included the possibility of Spin(9) as a subgroup of SO(16). Riemannian manifolds with such holonomy were later shown independently by D. Alekseevski and Brown-Gray to be necessarily locally symmetric, i.e., locally isometric to the Cayley plane F_4/Spin(9) or locally flat. See below.) It is now known that all of these possibilities occur as holonomy groups of Riemannian manifolds. The last two exceptional cases were the most difficult to find. See G_2 manifold and Spin(7) manifold.

Note that $Sp(n) \subset SU(2n) \subset U(2n) \subset SO(4n)$, so every hyperkähler manifold is a Calabi–Yau manifold, every Calabi–Yau manifold is a Kähler manifold, and every Kähler manifold is orientable.

The strange list above was explained by Simons's proof of Berger's theorem. A simple and geometric proof of Berger's theorem was given by Carlos E. Olmos in 2005. One first shows that if a Riemannian manifold is *not* a locally symmetric space and the reduced holonomy acts irreducibly on the tangent space, then it acts transitively on the unit sphere. The Lie groups acting transitively on spheres are known: they consist of the list above, together with 2 extra cases: the group Spin(9) acting on $\mathbf{R}^{16}$, and the group $T \cdot Sp(m)$ acting on $\mathbf{R}^{4m}$. Finally one checks that the first of these two extra cases only occurs as a holonomy group for locally symmetric spaces (that are locally isomorphic to the Cayley projective plane), and the second does not occur at all as a holonomy group.

Berger's original classification also included non-positive-definite pseudo-Riemannian metric non-locally symmetric holonomy. That list consisted of $SO(p,q)$ of signature (p, q), $U(p, q)$ and $SU(p, q)$ of signature $(2p, 2q)$, $Sp(p, q)$ and $Sp(p, q)\cdot Sp(1)$ of signature $(4p, 4q)$, $SO(n, \mathbf{C})$ of signature (n, n), $SO(n, \mathbf{H})$ of signature $(2n, 2n)$, split G_2 of signature (4, 3), $G_2(\mathbf{C})$ of signature (7, 7), Spin(4, 3) of signature (4, 4), Spin(7, $\mathbf{C}$) of signature (7,7), Spin(5,4) of signature (8,8) and, lastly, Spin(9, $\mathbf{C}$) of signature (16,16). The split and complexified Spin(9) are necessarily locally symmetric as above and should not have been on the list. The complexified holonomies SO(n, **C**), G_2(**C**), and Spin(7,**C**) may be realized from complexifying real analytic Riemannian manifolds. The last case, manifolds with holonomy contained in SO(n, **H**), were shown to be locally flat by R. McLean.

Riemannian symmetric spaces, which are locally isometric to homogeneous spaces G/H have local holonomy isomorphic to H. These too have been completely classified.

Finally, Berger's paper lists possible holonomy groups of manifolds with only a torsion-free affine connection; this is discussed below.

26.3.3 Special holonomy and spinors

Manifolds with special holonomy are characterized by the presence of parallel spinors, meaning spinor fields with vanishing covariant derivative.[9] In particular, the following facts hold:

- $\mathrm{Hol}(\omega) \subset U(\mathrm{n})$ if and only if M admits a covariantly constant (or *parallel*) projective pure spinor field.
- If M is a spin manifold, then $\mathrm{Hol}(\omega) \subset SU(\mathrm{n})$ if and only if M admits at least two linearly independent parallel pure spinor fields. In fact, a parallel pure spinor field determines a canonical reduction of the structure group to $SU(n)$.
- If M is a seven-dimensional spin manifold, then M carries a non-trivial parallel spinor field if and only if the holonomy is contained in G_2.
- If M is an eight-dimensional spin manifold, then M carries a non-trivial parallel spinor field if and only if the holonomy is contained in Spin(7).

The unitary and special unitary holonomies are often studied in connection with twistor theory,[10] as well as in the study of almost complex structures.[9]

Applications to string theory

Riemannian manifolds with special holonomy play an important role in string theory compactifications. [11] This is because special holonomy manifolds admit covariantly constant (parallel) spinors and thus preserve some fraction of the original supersymmetry. Most important are compactifications on Calabi–Yau manifolds with SU(2) or SU(3) holonomy. Also important are compactifications on G_2 manifolds.

26.4 Affine holonomy

Affine holonomy groups are the groups arising as holonomies of torsion-free affine connections; those which are not Riemannian or pseudo-Riemannian holonomy groups are also known as non-metric holonomy groups. The deRham decomposition theorem does not apply to affine holonomy groups, so a complete classification is out of reach. However, it is still natural to classify irreducible affine holonomies.

On the way to his classification of Riemannian holonomy groups, Berger developed two criteria that must be satisfied by the Lie algebra of the holonomy group of a torsion-free affine connection which is not locally symmetric: one of them, known as *Berger's first criterion*, is a consequence of the Ambrose–Singer theorem, that the curvature generates the holonomy algebra; the other, known as *Berger's second criterion*, comes from the requirement that the connection should not be locally symmetric. Berger presented a list of groups acting irreducibly and satisfying these two criteria; this can be interpreted as a list of possibilities for irreducible affine holonomies.

Berger's list was later shown to be incomplete: further examples were found by R. Bryant (1991) and by Q. Chi, S. Merkulov, and L. Schwachhöfer (1996). These are sometimes known as *exotic holonomies*. The search for examples ultimately led to a complete classification of irreducible affine holonomies by Merkulov and Schwachhöfer (1999), with Bryant (2000) showing that every group on their list occurs as an affine holonomy group.

The Merkulov–Schwachhöfer classification has been clarified considerably by a connection between the groups on the list and certain symmetric spaces, namely the hermitian symmetric spaces and the quaternion-Kähler symmetric spaces. The relationship is particularly clear in the case of complex affine holonomies, as demonstrated by Schwachhöfer (2001).

Let V be a finite-dimensional complex vector space, let $H \subset \mathrm{Aut}(V)$ be an irreducible semisimple complex connected Lie subgroup and let $K \subset H$ be a maximal compact subgroup.

1. If there is an irreducible hermitian symmetric space of the form $G/(\mathrm{U}(1) \cdot K)$, then both H and $\mathbf{C}^* \cdot H$ are non-symmetric irreducible affine holonomy groups, where V the tangent representation of K.
2. If there is an irreducible quaternion-Kähler symmetric space of the form $G/(\mathrm{Sp}(1) \cdot K)$, then H is a non-symmetric irreducible affine holonomy groups, as is $\mathbf{C}^* \cdot H$ if $\dim V = 4$. Here the complexified tangent representation of $\mathrm{Sp}(1) \cdot K$ is $\mathbf{C}^2 \otimes V$, and H preserves a complex symplectic form on V.

These two families yield all non-symmetric irreducible complex affine holonomy groups apart from the following:

$$\begin{aligned} &\mathrm{Sp}(2,\mathbf{C})\cdot\mathrm{Sp}(2n,\mathbf{C}) && \subset \mathrm{Aut}(\mathbf{C}^2\otimes\mathbf{C}^{2n})\\ &G_2(\mathbf{C}) && \subset \mathrm{Aut}(\mathbf{C}^7)\\ &\mathrm{Spin}(7,\mathbf{C}) && \subset \mathrm{Aut}(\mathbf{C}^8). \end{aligned}$$

Using the classification of hermitian symmetric spaces, the first family gives the following complex affine holonomy groups:

$$\begin{aligned} &Z_{\mathbf{C}}\cdot\mathrm{SL}(m,\mathbf{C})\cdot\mathrm{SL}(n,\mathbf{C}) && \subset \mathrm{Aut}(\mathbf{C}^m\otimes\mathbf{C}^n)\\ &Z_{\mathbf{C}}\cdot\mathrm{SL}(n,\mathbf{C}) && \subset \mathrm{Aut}(\Lambda^2\mathbf{C}^n)\\ &Z_{\mathbf{C}}\cdot\mathrm{SL}(n,\mathbf{C}) && \subset \mathrm{Aut}(S^2\mathbf{C}^n)\\ &Z_{\mathbf{C}}\cdot\mathrm{SO}(n,\mathbf{C}) && \subset \mathrm{Aut}(\mathbf{C}^n)\\ &Z_{\mathbf{C}}\cdot\mathrm{Spin}(10,\mathbf{C}) && \subset \mathrm{Aut}(\Delta_{10}^{+})\cong\mathrm{Aut}(\mathbf{C}^{16})\\ &Z_{\mathbf{C}}\cdot E_6(\mathbf{C}) && \subset \mathrm{Aut}(\mathbf{C}^{27}) \end{aligned}$$

where **ZC** is either trivial, or the group **C***.

Using the classification of quaternion-Kähler symmetric spaces, the second family gives the following complex symplectic holonomy groups:

$$\begin{aligned} &\mathrm{Sp}(2,\mathbf{C})\cdot\mathrm{SO}(n,\mathbf{C}) && \subset \mathrm{Aut}(\mathbf{C}^2\otimes\mathbf{C}^n)\\ &(Z_{\mathbf{C}}\,\cdot)\,\mathrm{Sp}(2n,\mathbf{C}) && \subset \mathrm{Aut}(\mathbf{C}^{2n})\\ &Z_{\mathbf{C}}\cdot\mathrm{SL}(2,\mathbf{C}) && \subset \mathrm{Aut}(S^3\mathbf{C}^2)\\ &\mathrm{Sp}(6,\mathbf{C}) && \subset \mathrm{Aut}(\Lambda_0^3\mathbf{C}^6)\cong\mathrm{Aut}(\mathbf{C}^{14})\\ &\mathrm{SL}(6,\mathbf{C}) && \subset \mathrm{Aut}(\Lambda^3\mathbf{C}^6)\\ &\mathrm{Spin}(12,\mathbf{C}) && \subset \mathrm{Aut}(\Delta_{12}^{+})\cong\mathrm{Aut}(\mathbf{C}^{32})\\ &E_7(\mathbf{C}) && \subset \mathrm{Aut}(\mathbf{C}^{56}) \end{aligned}$$

(In the second row, **ZC** must be trivial unless $n = 2$.)

From these lists, an analogue of Simon's result that Riemannian holonomy groups act transitively on spheres may be observed: the complex holonomy representations are all prehomogeneous vector spaces. A conceptual proof of this fact is not known.

The classification of irreducible real affine holonomies can be obtained from a careful analysis, using the lists above and the fact that real affine holonomies complexify to complex ones.

26.5 Etymology

There is a similar word, "holomorphic", that was introduced by two of Cauchy's students, Briot (1817–1882) and Bouquet (1819–1895), and derives from the Greek ὅλος (*holos*) meaning "entire", and μορφή (*morphē*) meaning "form" or "appearance".[12] The etymology of "holonomy" shares the first part with "holomorphic" (*holos*). About the second part:

> "It is remarkably hard to find the etymology of holonomic (or holonomy) on the web. I found the following (thanks to John Conway of Princeton):
>
> *'I believe it was first used by Poinsot in his analysis of the motion of a rigid body. In this theory, a system is called "holonomic" if, in a certain sense, one can recover global information from local information, so the meaning "entire-law" is quite appropriate. The rolling of a ball on a table is non-holonomic, because one rolling along different paths to the same point can put it into different orientations. However, it is perhaps a bit too simplistic to say that "holonomy" means "entire-law". The "nom" root has many intertwined meanings in Greek, and perhaps more often refers to "counting". It comes from the same Indo-European root as our word "number." ' "*
>
> — S.Golwala, [13]

See νόμος (*nomos*) and -nomy.

26.6 Notes

[1] Kobayashi & Nomizu 1963, §II.7

[2] Sharpe 1997, §3.7

[3] Spivak 1999, p. 241

[4] Sternberg 1964, Theorem VII.1.2

[5] Kobayashi & Nomizu 1963, Volume I, §II.8

[6] Kobayashi Nomizu, §IV.5

[7] This theorem generalizes to non-simply connected manifolds, but the statement is more complicated.

[8] Kobayashi & Nomizu §IV.6

[9] Lawson & Michelsohn 1989, §IV.9–10

[10] Baum 1991

[11] Gubser, S., "Special holonomy in string theory and M-theory", in Gubser S.; et al., *Special holonomy in string theory and M-theory +Strings, branes and extra dimensions, TASI 2001. Lectures presented at the 2001 TASI school, Boulder, Colorado, USA, 4–29 June 2001.* (PDF), River Edge, NJ: World Scientific, 2004, pp. 197–233, ISBN 981-238-788-9.

[12] Markushevich, A.I. 2005

[13] Golwala 2007, pp. 65–66

26.7 References

- Agricola, Ilka (2006), "The Srni lectures on non-integrable geometries with torsion", *Arch. Math.* **42**: 5–84
- Ambrose, W.; Singer, I. M. (1953), "A theorem on holonomy", *Trans. Amer. Math. Soc.* (American Mathematical Society) **75** (3): 428–443, doi:10.2307/1990721, JSTOR 1990721
- Baum, H.; Friedrich, Th.; Grunewald, R.; Kath, I. (1991), *Twistors and Killing spinors on Riemannian manifolds*, B.G. Teubner
- Berger, M. (1953), "Sur les groupes d'holonomie homogènes des variétés a connexion affines et des variétés riemanniennes", *Bull. Soc. Math. France* **83**: 279–330, MR 0079806
- Besse, Arthur L. (1987), *Einstein manifolds*, Ergebnisse der Mathematik und ihrer Grenzgebiete (3) [Results in Mathematics and Related Areas (3)], vol. 10, Berlin, New York: Springer-Verlag, pp. xii+510, ISBN 978-3-540-15279-8
- Bonan, Edmond (1965), "Structure presque quaternale sur une variété différentiable", *C. R. Acad. Sci. Paris* **261**: 5445–5448.
- Bonan, Edmond (1966), "Sur les variétés riemanniennes à groupe d'holonomie G2 ou Spin(7)", *C. R. Acad. Sci. Paris* **320**: 127–129].
- Borel, Armand; Lichnerowicz, André (1952), "Groupes d'holonomie des variétés riemanniennes", *Les Comptes rendus de l'Académie des sciences* **234**: 1835–1837, MR 0048133
- Bryant, Robert L. (1987), "Metrics with exceptional holonomy", *Annals of Mathematics* (Annals of Mathematics) **126** (3): 525–576, doi:10.2307/1971360, JSTOR 1971360.
- Bryant, Robert L. (1991), "Two exotic holonomies in dimension four, path geometries, and twistor theory", *Amer. Math. Soc. Proc. Symp. Pure Math.* **53**: 33–88, doi:10.1090/pspum/053/1141197
- Bryant, Robert L. (2000), "Recent Advances in the Theory of Holonomy", *Asterisque*, Séminaire Bourbaki 1998–1999 **266**: 351–374 arXiv:math.DG/9910059.
- Cartan, Élie (1926), "Sur une classe remarquable d'espaces de Riemann", *Bulletin de la Société Mathématique de France* **54**: 214–264, ISSN 0037-9484, MR 1504900
- Cartan, Élie (1927), "Sur une classe remarquable d'espaces de Riemann", *Bulletin de la Société Mathématique de France* **55**: 114–134, ISSN 0037-9484
- Chi, Quo-Shin; Merkulov, Sergey A.; Schwachhöfer, Lorenz J. (1996), "On the Incompleteness of Berger's List of Holonomy Representations", *Invent. Math.* **126**: 391–411, arXiv:dg-da/9508014, doi:10.1007/s002220050104
- Golwala, S. (2007), *Lecture Notes on Classical Mechanics for Physics 106ab* (PDF)
- Joyce, D. (2000), *Compact Manifolds with Special Holonomy*, Oxford University Press, ISBN 0-19-850601-5
- Kobayashi, S.; Nomizu, K. (1963), *Foundations of Differential Geometry, Vol. 1 & 2* (New ed.), Wiley-Interscience (published 1996), ISBN 0-471-15733-3
- Kraines, Vivian Yoh (1965), "Topology of quaternionic manifolds", *Bull. Amer. Math. Soc*, 71,3, 1: 526–527.
- Lawson, H. B.; Michelsohn, M-L. (1989), *Spin Geometry*, Princeton University Press, ISBN 0-691-08542-0
- Markushevich, A.I.; Silverman, Richard A. (ed.) (2005) [1977], *Theory of functions of a Complex Variable* (2nd ed.), New York: American Mathematical Society, p. 112, ISBN 0-8218-3780-X Cite uses deprecated parameter |coauthors= (help)
- Merkulov, Sergei A.; Schwachhöfer, Lorenz J. (1999), "Classification of irreducible holonomies of torsion-free affine connections", *Ann. of Math.* (Annals of Mathematics) **150** (1): 77–149, doi:10.2307/121098, JSTOR 121098 arXiv:math.DG/9907206; "Addendum", *Ann. of Math.* **150** (3), 1999: 1177–1179, doi:10.2307/121067, JSTOR 121067. arXiv:math.DG/9911266.
- Olmos, C. (2005), "A geometric proof of the Berger Holonomy Theorem", *Annals of Mathematics* **161** (1): 579–588, doi:10.4007/annals.2005.161.579
- Sharpe, R.W. (1997), *Differential Geometry: Cartan's Generalization of Klein's Erlangen Program*, Springer-Verlag, New York, ISBN 0-387-94732-9
- Schwachhöfer, Lorenz J. (2001), "Connections with irreducible holonomy representations", *Advances in Mathematics* **160** (1): 1–80, doi:10.1006/aima.2000.1973

- Simons, J. (1962), "On the transitivity of holonomy systems", *Annals of Mathematics* (Annals of Mathematics) **76** (2): 213–234, doi:10.2307/1970273, JSTOR 1970273

- Spivak, M. (1999), *A comprehensive introduction to differential geometry, Volume II*, Houston, Texas: Publish or Perish, ISBN 0-914098-71-3

- Sternberg, S. (1964), *Lectures on differential geometry*, New York: Chelsea, ISBN 0-8284-0316-3

26.8 Further reading

- Literature about manifolds of special holonomy, a bibliography by Frederik Witt.

Chapter 27

Wheeler–DeWitt equation

The **Wheeler–DeWitt equation**[1] is an attempt to combine mathematically the ideas of quantum mechanics and general relativity, a step toward a theory of quantum gravity. In this approach, time plays no role in the equation, leading to the problem of time.[2] More specifically, the equation describes the quantum version of the Hamiltonian constraint using metric variables. Its commutation relations with the diffeomorphism constraints generate the Bergmann-Komar "group" (which *is* the diffeomorphism group on-shell, but differs off-shell).

Because of its connections with the low-energy effective field theory, it inherits all the problems of the naively quantized GR, and thus it cannot be used at multi-loop level, etc., at least not according to the current knowledge.

The equation has not played a role in string theory thus far, since all properly defined and understood descriptions of string/M-theory deal with some fixed asymptotic conditions on the background. Thus, at infinity, the "right" choice of the time coordinate "t" is determined in every description, so there is a preferred definition of the Hamiltonian (with nonzero eigenvalues) to evolve states of the system forward in time. This avoids all the issues of the Wheeler-de Witt equation to dynamically generate a time dimension.

But at the end, there could exist a Wheeler-de Witt style manner to describe the bulk dynamics of quantum theory of gravity. Some experts believe that this equation still holds the potential for understanding quantum gravity; however, decades after the equation was first written down, it has not brought physicists as clear results about quantum gravity as some of the results building on completely different approaches, such as string theory.

27.1 Motivation and background

In canonical gravity, spacetime is foliated into spacelike submanifolds. The three-metric (i.e., metric on the hypersurface) is γ_{ij} and given by

$$g_{\mu\nu}\, \mathrm{d}x^{\mu}\, \mathrm{d}x^{\nu} = (-\,N^2+\beta_k\beta^k)\, \mathrm{d}t^2+2\beta_k\, \mathrm{d}x^k\, \mathrm{d}t+\gamma_{ij}\, \mathrm{d}x^i\, \mathrm{d}x^j.$$

In that equation the Roman indices run over the values 1, 2, 3 and the Greek indices run over the values 1, 2, 3, 4. The three-metric γ_{ij} is the field, and we denote its conjugate momenta as π^{kl}. The Hamiltonian is a constraint (characteristic of most relativistic systems)

$$\mathcal{H} = \frac{1}{2\sqrt{\gamma}} G_{ijkl}\pi^{ij}\pi^{kl} - \sqrt{\gamma}\,^{(3)}R = 0$$

where $\gamma = \det(\gamma_{ij})$ and $G_{ijkl} = (\gamma_{ik}\gamma_{jl}+\gamma_{il}\gamma_{jk}-\gamma_{ij}\gamma_{kl})$ is the Wheeler-DeWitt metric.

Quantization "puts hats" on the momenta and field variables; that is, the functions of numbers in the classical case become operators that modify the state function in the quantum case. Thus we obtain the operator

$$\widehat{\mathcal{H}} = \frac{1}{2\sqrt{\gamma}} \widehat{G}_{ijkl}\widehat{\pi}^{ij}\widehat{\pi}^{kl} - \sqrt{\gamma}\,^{(3)}\widehat{R}.$$

Working in "position space", these operators are

$$\hat{\gamma}_{ij}(t,x^k) \to \gamma_{ij}(t,x^k)$$
$$\hat{\pi}^{ij}(t,x^k) \to -i\frac{\delta}{\delta\gamma_{ij}(t,x^k)}.$$

One can apply the operator to a general wave functional of the metric $\widehat{\mathcal{H}}\Psi[\gamma] = 0$ where:

$$\Psi[\gamma] = a+\int \psi(x)\gamma(x)dx^3+\int\int \psi(x,y)\gamma(x)\gamma(y)dx^3dy^3+...$$

Which would give a set of constraints amongst the coefficients $\psi(x,y,...)$. Which means the amplitudes for N gravitons at certain positions is related to the amplitudes for a different number of gravitons at different positions. Or one could use the two field formalism treating $\omega(g)$ as an independent field so the wave function is $\Psi[\gamma,\omega]$

27.2 Derivation from path integral

The Wheeler–DeWitt equation can be derived from a path integral using the gravitational action in the Euclidean quantum gravity paradigm:[3]

$$Z = \int_C \mathrm{e}^{-I[g_{\mu\nu},\phi]} \mathcal{D}\mathbf{g}\, \mathcal{D}\phi$$

where one integrates over a class of *Riemannian* four-metrics and matter fields matching certain boundary conditions. Because the concept of a universal time coordinate seems unphysical, and at odds with the principles of general relativity, the action is evaluated around a 3-metric which we take as the boundary of the classes of four-metrics and on which a certain configuration of matter fields exists. This latter might for example be the current configuration of matter in our universe as we observe it today. Evaluating the action so that it only depends on the 3-metric and the matter fields is sufficient to remove the need for a time coordinate as it effectively fixes a point in the evolution of the universe.

We obtain the Hamiltonian constraint from

$$\frac{\delta I_{EH}}{\delta N} = 0$$

where I_{EH} is the Einstein-Hilbert action, and N is the lapse function (i.e., the Lagrange multiplier for the Hamiltonian constraint). This is purely classical so far. We can recover the Wheeler–DeWitt equation from

$$\frac{\delta Z}{\delta N} = 0 = \int \left. \frac{\delta I[g_{\mu\nu},\phi]}{\delta N} \right|_{\Sigma} \exp\left(-I[g_{\mu\nu},\phi]\right) \mathcal{D}\mathbf{g}\, \mathcal{D}\phi$$

where Σ is the three-dimensional boundary. Observe that this expression vanishes, implying that the functional derivative also vanishes, giving us the Wheeler–DeWitt equation. A similar statement may be made for the diffeomorphism constraint (take functional derivative with respect to the shift functions instead).

27.3 Mathematical formalism

The Wheeler–DeWitt equation[1] is a functional differential equation. It is ill defined in the general case, but very important in theoretical physics, especially in quantum gravity. It is a functional differential equation on the space of three dimensional spatial metrics. The Wheeler–DeWitt equation has the form of an operator acting on a wave functional, the functional reduces to a function in cosmology. Contrary to the general case, the Wheeler–DeWitt equation is well defined in mini-superspaces like the configuration space of cosmological theories. An example of such a wave function is the Hartle–Hawking state. Bryce DeWitt first published this equation in 1967 under the name "Einstein–Schrödinger equation"; it was later renamed the "Wheeler–DeWitt equation".[4]

Simply speaking, the Wheeler–DeWitt equation says

where $\hat{H}(x)$ is the Hamiltonian constraint in quantized general relativity and $|\psi\rangle$ stands for the wave function of the universe. Unlike ordinary quantum field theory or quantum mechanics, the Hamiltonian is a first class constraint on physical states. We also have an independent constraint for each point in space.

Although the symbols $\hat{H}$ and $|\psi\rangle$ may appear familiar, their interpretation in the Wheeler–DeWitt equation is substantially different from non-relativistic quantum mechanics. $|\psi\rangle$ is no longer a spatial wave function in the traditional sense of a complex-valued function that is defined on a 3-dimensional space-like surface and normalized to unity. Instead it is a functional of field configurations on all of spacetime. This wave function contains all of the information about the geometry and matter content of the universe. $\hat{H}$ is still an operator that acts on the Hilbert space of wave functions, but it is not the same Hilbert space as in the nonrelativistic case, and the Hamiltonian no longer determines evolution of the system, so the Schrödinger equation $\hat{H}|\psi\rangle = i\hbar\partial/\partial t|\psi\rangle$ no longer applies. This property is known as timelessness. The reemergence of time requires the tools of decoherence and clock operators (or the use of a scalar field).

We also need to augment the Hamiltonian constraint with momentum constraints

$$\vec{\mathcal{P}}(x)\,|\psi\rangle = 0$$

associated with spatial diffeomorphism invariance.

In minisuperspace approximations, we only have one Hamiltonian constraint (instead of infinitely many of them).

In fact, the principle of general covariance in general relativity implies that global evolution per se does not exist; the time t is just a label we assign to one of the coordinate axes. Thus, what we think about as time evolution of any physical system is just a gauge transformation, similar to that of QED induced by U(1) local gauge transformation $\psi \to e^{i\theta(\vec{r})}\psi$ where $\theta(\vec{r})$ plays the role of local time. The role of a Hamiltonian is simply to restrict the space of the "kinematic" states of the Universe to that of "physical" states - the ones that follow gauge orbits. For this reason we call it a "Hamiltonian constraint." Upon quantization, physical states become wave functions that lie in the kernel of the Hamiltonian operator.

In general, the Hamiltonian vanishes for a theory with general covariance or time-scaling invariance.

27.4 See also

- ADM formalism
- Diffeomorphism constraint
- Euclidean quantum gravity
- Regge Calculus
- Canonical quantum gravity
- Peres metric
- Loop quantum gravity

27.5 References

[1] DeWitt, B. S. (1967). "Quantum Theory of Gravity. I. The Canonical Theory". *Phys. Rev.* **160** (5): 1113–1148. Bibcode:1967PhRv..160.1113D. doi:10.1103/PhysRev.160.1113.

[2] https://medium.com/the-physics-arxiv-blog/d5d3dc850933

[3] See J. B. Hartle and S. W. Hawking, "Wave function of the Universe." *Phys. Rev. D* **28** (1983) 2960–2975, eprint.

[4] Go to Arxiv.org to read "Notes for a Brief History of Quantum Gravity" by Carlo Rovelli

- Herbert W. Hamber and Ruth M. Williams (2011). "Discrete Wheeler-DeWitt Equation". *Physical Review D* **84**: 104033. arXiv:1109.2530. Bibcode:2011PhRvD..84j4033H. doi:10.1103/PhysRevD.84.104033. Available at .
- Herbert W. Hamber, Reiko Toriumi and Ruth M. Williams (2012). "Wheeler-DeWitt Equation in 2+1 Dimensions". *Physical Review D* **86**: 084010. arXiv:1207.3759. Bibcode:2012PhRvD..86h4010H. doi:10.1103/PhysRevD.86.084010. Available at .

Chapter 28

Graph (discrete mathematics)

This article is about sets of vertices connected by edges. For graphs of mathematical functions, see Graph of a function. For other uses, see Graph (disambiguation).

In mathematics, and more specifically in graph theory, a

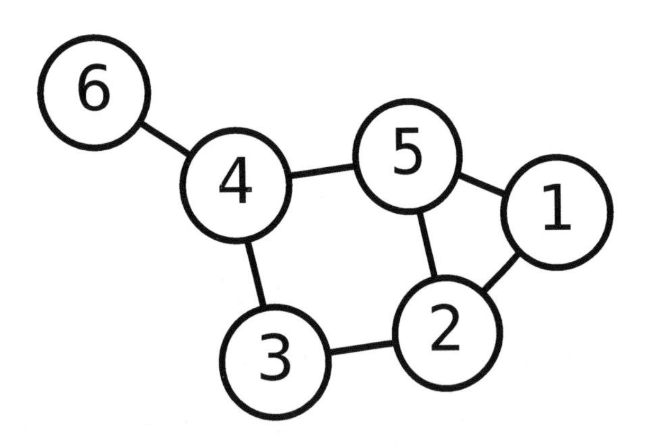

A drawing of a labeled graph on 6 vertices and 7 edges.

graph is a representation of a set of objects where some pairs of objects are connected by links. The interconnected objects are represented by mathematical abstractions called *vertices* (also called *nodes* or *points*), and the links that connect some pairs of vertices are called *edges* (also called *arcs* or *lines*).[1] Typically, a graph is depicted in diagrammatic form as a set of dots for the vertices, joined by lines or curves for the edges. Graphs are one of the objects of study in discrete mathematics.

The edges may be directed or undirected. For example, if the vertices represent people at a party, and there is an edge between two people if they shake hands, then this is an undirected graph, because if person A shook hands with person B, then person B also shook hands with person A. In contrast, if there is an edge from person A to person B when person A knows of person B, then this graph is directed, because knowledge of someone is not necessarily a symmetric relation (that is, one person knowing another person does not necessarily imply the reverse; for example, many fans may know of a celebrity, but the celebrity is unlikely to know of all their fans). The former type of graph is called an *undirected graph* and the edges are called *undirected edges* while the latter type of graph is called a *directed graph* and the edges are called *directed edges*.

Graphs are the basic subject studied by graph theory. The word "graph" was first used in this sense by J. J. Sylvester in 1878.[2][3]

28.1 Definitions

Definitions in graph theory vary. The following are some of the more basic ways of defining graphs and related mathematical structures.

28.1.1 Graph

In the most common sense of the term,[4] a *graph* is an ordered pair $G = (V, E)$ comprising a set V of *vertices*, *nodes* or *points* together with a set E of *edges*, *arcs* or *lines*, which are 2-element subsets of V (i.e., an edge is related with two vertices, and the relation is represented as an unordered pair of the vertices with respect to the particular edge). To avoid ambiguity, this type of graph may be described precisely as *undirected* and *simple*.

Other senses of *graph* stem from different conceptions of the edge set. In one more generalized notion,[5] E is a set together with a relation of *incidence* that associates with each edge two vertices. In another generalized notion, E is a multiset of unordered pairs of (not necessarily distinct) vertices. Many authors call this type of object a multigraph or pseudograph.

All of these variants and others are described more fully below.

The vertices belonging to an edge are called the *ends* or *end vertices* of the edge. A vertex may exist in a graph and not belong to an edge.

V and E are usually taken to be finite, and many of the well-known results are not true (or are rather different) for *infinite graphs* because many of the arguments fail in the infinite case. Moreover, V is often assumed to be non-empty, but E is allowed to be the empty set. The *order* of a graph is $|V|$, its number of vertices. The *size* of a graph is $|E|$, its number of edges. The *degree* or *valency* of a vertex is the number of edges that connect to it, where an edge that connects to the vertex at both ends (a loop) is counted twice.

For an edge $\{x, y\}$, graph theorists usually use the somewhat shorter notation xy.

28.1.2 Adjacency relation

The edges E of an undirected graph G induce a symmetric binary relation ~ on V that is called the *adjacency relation* of G. Specifically, for each edge $\{x, y\}$, the vertices x and y are said to be *adjacent* to one another, which is denoted $x \sim y$.

28.2 Types of graphs

28.2.1 Distinction in terms of the main definition

As stated above, in different contexts it may be useful to refine the term *graph* with different degrees of generality. Whenever it is necessary to draw a strict distinction, the following terms are used. Most commonly, in modern texts in graph theory, unless stated otherwise, *graph* means "undirected simple finite graph" (see the definitions below).

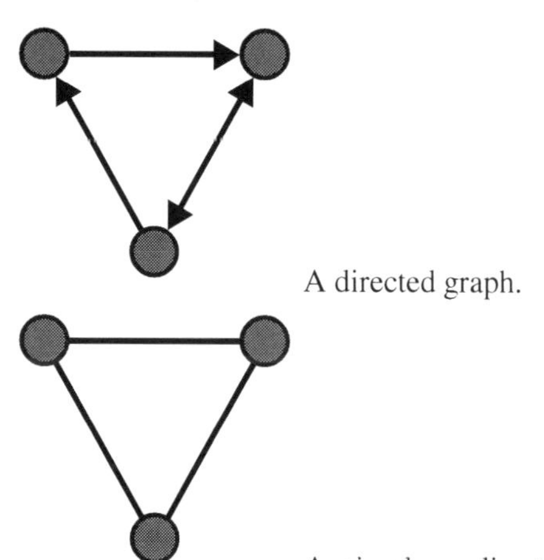

A directed graph.

A simple undirected graph with three vertices and three edges. Each vertex has degree two, so this is also a regular graph.

Undirected graph

An *undirected graph* is a graph in which edges have no orientation. The edge (x, y) is identical to the edge (y, x), i.e., they are not ordered pairs, but sets $\{x, y\}$ (or 2-multisets) of vertices. The maximum number of edges in an undirected graph without a loop is $n(n - 1)/2$.

Directed graph

Main article: Directed graph

A *directed graph* or *digraph* is a graph in which edges have orientations. It is written as an ordered pair $G = (V, A)$ (sometimes $G = (V, E)$) with

- V a set whose elements are called *vertices*, *nodes*, or *points*;
- A a set of ordered pairs of vertices, called *arrows*, *directed edges* (sometimes simply *edges* with the corresponding set named E instead of A), *directed arcs*, or *directed lines*.

An arrow (x, y) is considered to be directed *from x to y*; y is called the *head* and x is called the *tail* of the arrow; y is said to be a *direct successor* of x and x is said to be a *direct predecessor* of y. If a path leads from x to y, then y is said to be a *successor* of x and *reachable* from x, and x is said to be a *predecessor* of y. The arrow (y, x) is called the *inverted arrow* of (x, y).

A directed graph G is called *symmetric* if, for every arrow in G, the corresponding inverted arrow also belongs to G. A symmetric loopless directed graph $G = (V, A)$ is equivalent to a simple undirected graph $G' = (V, E)$, where the pairs of inverse arrows in A correspond one-to-one with the edges in E; thus the number of edges in G' is $| E | = | A |/2$, that is half the number of arrows in G.

Oriented graph

An *oriented graph* is a directed graph in which at most one of (x, y) and (y, x) may be arrows of the graph. That is, it is a directed graph that can be formed as an orientation of an undirected graph. However, some authors use "oriented graph" to mean the same as "directed graph".

Mixed graph

Main article: Mixed graph

A *mixed graph* is a graph in which some edges may be directed and some may be undirected. It is written as an ordered triple $G = (V, E, A)$ with V, E, and A defined as above. Directed and undirected graphs are special cases.

Multigraph

Main article: Multigraph

Multiple edges are two or more edges that connect the same two vertices. A *loop* is an edge (directed or undirected) that connects a vertex to itself; it may be permitted or not, according to the application. In this context, an edge with two different ends is called a *link*.

A *multigraph*, as opposed to a simple graph, is an undirected graph in which multiple edges (and sometimes loops) are allowed.

Where graphs are defined so as to *disallow* both multiple edges and loops, a multigraph is often defined to mean a graph which can have both multiple edges and loops,[6] although many use the term *pseudograph* for this meaning.[7] Where graphs are defined so as to *allow* both multiple edges and loops, a multigraph is often defined to mean a graph without loops.[8]

Simple graph

A *simple graph*, as opposed to a multigraph, is an undirected graph in which both multiple edges and loops are disallowed. In a simple graph the edges form a *set* (rather than a multiset) and each edge is an unordered pair of *distinct* vertices. In a simple graph with n vertices, the degree of every vertex is at most $n - 1$.

Quiver

Main article: Quiver (mathematics)

A *quiver* or *multidigraph* is a directed multigraph. A quiver may also have directed loops in it.

Weighted graph

A *weighted graph* is a graph in which a number (the weight) is assigned to each edge.[9] Such weights might represent for example costs, lengths or capacities, depending on the problem at hand. Some authors call such a graph a *network*.[10] Weighted correlation networks can be defined by soft-thresholding the pairwise correlations among variables (e.g. gene measurements). Such graphs arise in many contexts, for example in shortest path problems such as the traveling salesman problem.

Half-edges, loose edges

In certain situations it can be helpful to allow edges with only one end, called *half-edges*, or no ends, called *loose edges*; see the articles Signed graphs and Biased graphs.

28.2.2 Important classes of graph

Regular graph

Main article: Regular graph

A *regular graph* is a graph in which each vertex has the same number of neighbours, i.e., every vertex has the same degree. A regular graph with vertices of degree k is called a k-regular graph or regular graph of degree k.

Complete graph

Main article: Complete graph

A *complete graph* is a graph in which each pair of vertices is joined by an edge. A complete graph contains all possible edges.

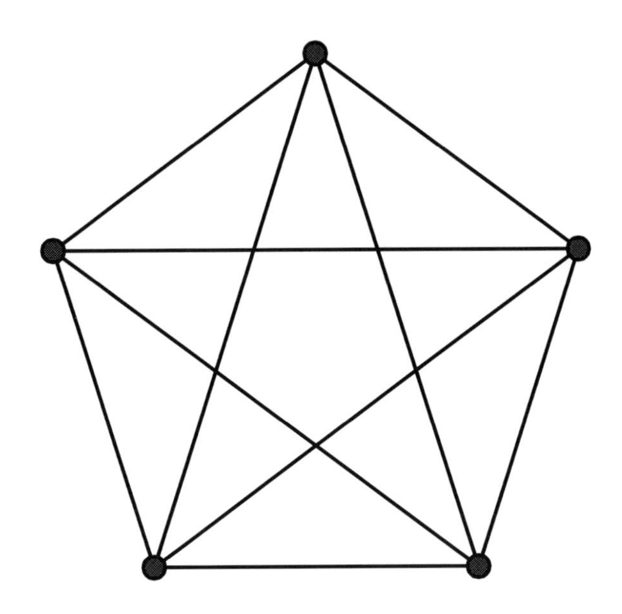

A complete graph with 5 vertices. Each vertex has an edge to every other vertex.

Finite graph

A *finite graph* is a graph in which the vertex set and the edge set are finite sets. Otherwise, it is called an *infinite graph.*

Most commonly in graph theory it is implied that the graphs discussed are finite. If the graphs are infinite, that is usually specifically stated.

Connected graph

Main article: Connectivity (graph theory)

In an undirected graph, an unordered pair of vertices $\{x, y\}$ is called *connected* if a path leads from x to y. Otherwise, the unordered pair is called *disconnected.*

A *connected graph* is an undirected graph in which every unordered pair of vertices in the graph is connected. Otherwise, it is called a *disconnected graph.*

In a directed graph, an ordered pair of vertices (x, y) is called *strongly connected* if a directed path leads from x to y. Otherwise, the ordered pair is called *weakly connected* if an undirected path leads from x to y after replacing all of its directed edges with undirected edges. Otherwise, the ordered pair is called *disconnected.*

A *strongly connected graph* is a directed graph in which every ordered pair of vertices in the graph is strongly connected. Otherwise, it is called a *weakly connected graph* if every ordered pair of vertices in the graph is weakly connected. Otherwise it is called a *disconnected graph.*

A *k-vertex-connected graph* or *k-edge-connected graph* is a graph in which no set of $k - 1$ vertices (respectively, edges) exists that, when removed, disconnects the graph. A *k*-vertex-connected graph is often called simply a *k-connected graph.*

Bipartite graph

Main article: Bipartite graph

A *bipartite graph* is a graph in which the vertex set can be partitioned into two sets, W and X, so that no two vertices in W share a common edge and no two vertices in X share a common edge. Alternatively, it is a graph with a chromatic number of 2.

In a complete bipartite graph, the vertex set is the union of two disjoint sets, W and X, so that every vertex in W is adjacent to every vertex in X but there are no edges within W or X.

Path graph

Main article: Path graph

A *path graph* or *linear graph* of order $n \geq 2$ is a graph in which the vertices can be listed in an order $v_1, v_2, \ldots, vn$ such that the edges are the $\{ vi, vi_{+1} \}$ where $i = 1, 2, \ldots, n - 1$. Path graphs can be characterized as connected graphs in which the degree of all but two vertices is 2 and the degree of the two remainding vertices is 1. If a path graph occurs as a subgraph of another graph, it is a path in that graph.

Planar graph

Main article: Planar graph

A *planar graph* is a graph whose vertices and edges can be drawn in a plane such that no two of the edges intersect.

Cycle graph

Main article: Cycle graph

A *cycle graph* or *circular graph* of order $n \geq 3$ is a graph in which the vertices can be listed in an order $v_1, v_2, \ldots, vn$ such that the edges are the $\{ vi, vi_{+1} \}$ where $i = 1, 2, \ldots, n - 1$, plus the edge $\{ vn, v_1 \}$. Cycle graphs can be characterized as connected graphs in which the degree of all vertices is 2. If a cycle graph occurs as a subgraph of another graph, it is a cycle or circuit in that graph.

Tree

Main article: Tree (graph theory)

A *tree* is a connected graph with no cycles.

A *forest* is a graph with no cycles, i.e. the disjoint union of one or more trees.

Advanced classes

More advanced kinds of graphs are:

- Petersen graph and its generalizations;
- perfect graphs;
- cographs;
- chordal graphs;

- other graphs with large automorphism groups: vertex-transitive, arc-transitive, and distance-transitive graphs;
- strongly regular graphs and their generalizations distance-regular graphs.

28.3 Properties of graphs

See also: Glossary of graph theory and Graph property

Two edges of a graph are called *adjacent* if they share a common vertex. Two arrows of a directed graph are called *consecutive* if the head of the first one is the tail of the second one. Similarly, two vertices are called *adjacent* if they share a common edge (*consecutive* if the first one is the tail and the second one is the head of an arrow), in which case the common edge is said to *join* the two vertices. An edge and a vertex on that edge are called *incident*.

The graph with only one vertex and no edges is called the *trivial graph*. A graph with only vertices and no edges is known as an *edgeless graph*. The graph with no vertices and no edges is sometimes called the *null graph* or *empty graph*, but the terminology is not consistent and not all mathematicians allow this object.

Normally, the vertices of a graph, by their nature as elements of a set, are distinguishable. This kind of graph may be called *vertex-labeled*. However, for many questions it is better to treat vertices as indistinguishable. (Of course, the vertices may be still distinguishable by the properties of the graph itself, e.g., by the numbers of incident edges.) The same remarks apply to edges, so graphs with labeled edges are called *edge-labeled*. Graphs with labels attached to edges or vertices are more generally designated as *labeled*. Consequently, graphs in which vertices are indistinguishable and edges are indistinguishable are called *unlabeled*. (Note that in the literature, the term *labeled* may apply to other kinds of labeling, besides that which serves only to distinguish different vertices or edges.)

The category of all graphs is the slice category Set $\downarrow D$ where D: Set $\rightarrow$ Set is the functor taking a set s to $s \times s$.

28.4 Examples

- The diagram at right is a graphic representation of the following graph:

  $V = \{1, 2, 3, 4, 5, 6\};$
  $E = \{\{1, 2\}, \{1, 5\}, \{2, 3\}, \{2, 5\}, \{3, 4\}, \{4, 5\}, \{4, 6\}\}.$

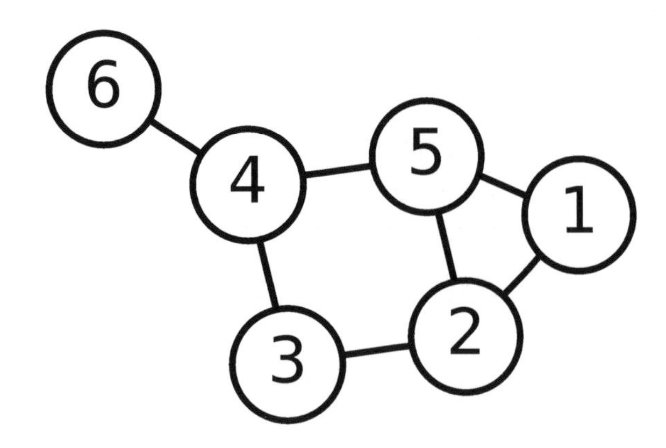

A graph with six nodes.

- In category theory, a small category can be represented by a directed multigraph in which the objects of the category are represented as vertices and the morphisms as directed edges. Then, the functors between categories induce some, but not necessarily all, of the digraph morphisms of the graph.
- In computer science, directed graphs are used to represent knowledge (e.g., conceptual graph), finite state machines, and many other discrete structures.
- A binary relation R on a set X defines a directed graph. An element x of X is a direct predecessor of an element y of X if and only if xRy.
- A directed graph can model information networks such as Twitter, with one user following another.[11][12]

28.5 Graph operations

Main article: Graph operations

There are several operations that produce new graphs from initial ones, which might be classified into the following categories:

- *unary operations*, which create a new graph from an initial one, such as:
  - edge contraction,
  - line graph,
  - dual graph,
  - complement graph,
  - graph rewriting;
- *binary operations*, which create a new graph from two initial ones, such as:

- disjoint union of graphs,
- cartesian product of graphs,
- tensor product of graphs,
- strong product of graphs,
- lexicographic product of graphs,
- series-parallel graphs.

28.6 Generalizations

In a hypergraph, an edge can join more than two vertices.

An undirected graph can be seen as a simplicial complex consisting of 1-simplices (the edges) and 0-simplices (the vertices). As such, complexes are generalizations of graphs since they allow for higher-dimensional simplices.

Every graph gives rise to a matroid.

In model theory, a graph is just a structure. But in that case, there is no limitation on the number of edges: it can be any cardinal number, see continuous graph.

In computational biology, power graph analysis introduces power graphs as an alternative representation of undirected graphs.

In geographic information systems, geometric networks are closely modeled after graphs, and borrow many concepts from graph theory to perform spatial analysis on road networks or utility grids.

28.7 See also

- Conceptual graph
- Dual graph
- Glossary of graph theory
- Graph (abstract data type)
- Graph database
- Graph drawing
- Graph theory
- Hypergraph
- List of graph theory topics
- List of publications in graph theory
- Network theory

28.8 Notes

[1] Trudeau, Richard J. (1993). *Introduction to Graph Theory* (Corrected, enlarged republication. ed.). New York: Dover Pub. p. 19. ISBN 978-0-486-67870-2. Retrieved 8 August 2012. A graph is an object consisting of two sets called its *vertex set* and its *edge set*.

[2] See:

- J. J. Sylvester (February 7, 1878) "Chemistry and algebra," *Nature, 17* : 284. From page 284: "Every invariant and covariant thus becomes expressible by a *graph* precisely identical with a Kekuléan diagram or chemicograph."
- J. J. Sylvester (1878) "On an application of the new atomic theory to the graphical representation of the invariants and covariants of binary quantics, — with three appendices," *American Journal of Mathematics, Pure and Applied, 1* (1) : 64-90. The term "graph" first appears in this paper on page 65.

[3] Gross, Jonathan L.; Yellen, Jay (2004). *Handbook of graph theory*. CRC Press. p. 35. ISBN 978-1-58488-090-5.

[4] See, for instance, Iyanaga and Kawada, *69 J*, p. 234 or Biggs, p. 4.

[5] See, for instance, Graham et al., p. 5.

[6] For example, see. Bollobás, p. 7 and Diestel, p. 25.

[7] Gross (1998), p. 3, Gross (2003), p. 205, Harary, p.10, and Zwillinger, p. 220.

[8] For example, see Balakrishnan, p. 1, Gross (2003), p. 4, and Zwillinger, p. 220.

[9] Fletcher, Peter; Hoyle, Hughes; Patty, C. Wayne (1991). *Foundations of Discrete Mathematics* (International student ed.). Boston: PWS-KENT Pub. Co. p. 463. ISBN 0-53492-373-9. A *weighted graph* is a graph in which a number *w(e)*, called its *weight*, is assigned to each edge *e*.

[10] Strang, Gilbert (2005), *Linear Algebra and Its Applications* (4th ed.), Brooks Cole, ISBN 0-03-010567-6

[11] Grandjean, Martin (2016). "A social network analysis of Twitter: Mapping the digital humanities community". *Cogent Arts & Humanities* **3** (1): 1171458. doi:10.1080/23311983.2016.1171458.

[12] Pankaj Gupta, Ashish Goel, Jimmy Lin, Aneesh Sharma, Dong Wang, and Reza Bosagh Zadeh WTF: The who-to-follow system at Twitter, *Proceedings of the 22nd international conference on World Wide Web*

28.9 References

- Balakrishnan, V. K. (1997-02-01). *Graph Theory* (1st ed.). McGraw-Hill. ISBN 0-07-005489-4.
- Berge, Claude (1958). *Théorie des graphes et ses applications* (in French). Dunod, Paris: Collection Universitaire de Mathématiques, II. pp. viii+277. Translation: -. Dover, New York: Wiley. 2001 [1962].
- Biggs, Norman (1993). *Algebraic Graph Theory* (2nd ed.). Cambridge University Press. ISBN 0-521-45897-8.
- Bollobás, Béla (2002-08-12). *Modern Graph Theory* (1st ed.). Springer. ISBN 0-387-98488-7.
- Bang-Jensen, J.; Gutin, G. (2000). *Digraphs: Theory, Algorithms and Applications*. Springer.
- Diestel, Reinhard (2005). *Graph Theory* (3rd ed.). Berlin, New York: Springer-Verlag. ISBN 978-3-540-26183-4..
- Graham, R.L., Grötschel, M., and Lovász, L, ed. (1995). *Handbook of Combinatorics*. MIT Press. ISBN 0-262-07169-X.
- Gross, Jonathan L.; Yellen, Jay (1998-12-30). *Graph Theory and Its Applications*. CRC Press. ISBN 0-8493-3982-0.
- Gross, Jonathan L., & Yellen, Jay, ed. (2003-12-29). *Handbook of Graph Theory*. CRC. ISBN 1-58488-090-2.
- Harary, Frank (January 1995). *Graph Theory*. Addison Wesley Publishing Company. ISBN 0-201-41033-8.
- Iyanaga, Shôkichi; Kawada, Yukiyosi (1977). *Encyclopedic Dictionary of Mathematics*. MIT Press. ISBN 0-262-09016-3.
- Zwillinger, Daniel (2002-11-27). *CRC Standard Mathematical Tables and Formulae* (31st ed.). Chapman & Hall/CRC. ISBN 1-58488-291-3.

28.10 Further reading

- Trudeau, Richard J. (1993). *Introduction to Graph Theory* (Corrected, enlarged republication. ed.). New York: Dover Publications. ISBN 978-0-486-67870-2. Retrieved 8 August 2012.

28.11 External links

- Weisstein, Eric W., "Graph", *MathWorld*.

Chapter 29

Spin (physics)

This article is about spin in quantum mechanics. For rotation in classical mechanics, see angular momentum.

In quantum mechanics and particle physics, **spin** is an intrinsic form of angular momentum carried by elementary particles, composite particles (hadrons), and atomic nuclei.[1][2]

Spin is one of two types of angular momentum in quantum mechanics, the other being *orbital angular momentum*. The orbital angular momentum operator is the quantum-mechanical counterpart to the classical angular momentum of orbital revolution: it arises when a particle executes a rotating or twisting trajectory (such as when an electron orbits a nucleus).[3][4] The existence of spin angular momentum is inferred from experiments, such as the Stern–Gerlach experiment, in which particles are observed to possess angular momentum that cannot be accounted for by orbital angular momentum alone.[5]

In some ways, spin is like a vector quantity; it has a definite magnitude, and it has a "direction" (but quantization makes this "direction" different from the direction of an ordinary vector). All elementary particles of a given kind have the same magnitude of spin angular momentum, which is indicated by assigning the particle a *spin quantum number*.[2]

The SI unit of spin is the (N·m·s) or ($kg{\cdot}m^2{\cdot}s^{-1}$), just as with classical angular momentum. In practice, spin is given as a dimensionless *spin quantum number* by dividing the spin angular momentum by the reduced Planck constant ħ, which has the same units of angular momentum. Very often, the "spin quantum number" is simply called "spin" leaving its meaning as the unitless "spin quantum number" to be inferred from context.

When combined with the spin-statistics theorem, the spin of electrons results in the Pauli exclusion principle, which in turn underlies the periodic table of chemical elements.

Wolfgang Pauli was the first to propose the concept of spin, but he did not name it. In 1925, Ralph Kronig, George Uhlenbeck and Samuel Goudsmit at Leiden University suggested an (erroneous) physical interpretation of particles spinning around their own axis. The mathematical theory was worked out in depth by Pauli in 1927. When Paul Dirac derived his relativistic quantum mechanics in 1928, electron spin was an essential part of it.

29.1 Quantum number

Main article: Spin quantum number

As the name suggests, spin was originally conceived as the rotation of a particle around some axis. This picture is correct so far as spin obeys the same mathematical laws as quantized angular momenta do. On the other hand, spin has some peculiar properties that distinguish it from orbital angular momenta:

- Spin quantum numbers may take half-integer values.
- Although the direction of its spin can be changed, an elementary particle cannot be made to spin faster or slower.
- The spin of a charged particle is associated with a magnetic dipole moment with a g-factor differing from 1. This could only occur classically if the internal charge of the particle were distributed differently from its mass.

The conventional definition of the **spin quantum number**, s, is $s = n/2$, where n can be any non-negative integer. Hence the allowed values of s are 0, 1/2, 1, 3/2, 2, etc. The value of s for an elementary particle depends only on the type of particle, and cannot be altered in any known way (in contrast to the *spin direction* described below). The spin angular momentum, S, of any physical system is quantized. The allowed values of S are

$$S = \hbar\,\sqrt{s(s+1)} = \frac{h}{4\pi}\,\sqrt{n(n+2)},$$

where h is the Planck constant and $\hbar = h/2\pi$ is the reduced Planck constant. In contrast, orbital angular momentum can only take on integer values of s; i.e., even-numbered values of n.

29.1.1 Fermions and bosons

Those particles with half-integer spins, such as 1/2, 3/2, 5/2, are known as fermions, while those particles with integer spins, such as 0, 1, 2, are known as bosons. The two families of particles obey different rules and *broadly* have different roles in the world around us. A key distinction between the two families is that fermions obey the Pauli exclusion principle; that is, there cannot be two identical fermions simultaneously having the same quantum numbers (meaning, roughly, having the same position, velocity and spin direction). In contrast, bosons obey the rules of Bose–Einstein statistics and have no such restriction, so they may "bunch together" even if in identical states. Also, composite particles can have spins different from the particles which comprise them. For example, a helium atom can have spin 0 and therefore can behave like a boson even though the quarks and electrons which make it up are all fermions.

This has profound practical applications:

- Quarks and leptons (including electrons and neutrinos), which make up what is classically known as matter, are all fermions with spin 1/2. The common idea that "matter takes up space" actually comes from the Pauli exclusion principle acting on these particles to prevent the fermions that make up matter from being in the same quantum state. Further compaction would require electrons to occupy the same energy states, and therefore a kind of pressure (sometimes known as degeneracy pressure of electrons) acts to resist the fermions being overly close. It is also this pressure which prevents stars collapsing inwardly, and which, when it finally gives way under immense gravitational pressure in a dying massive star, triggers inward collapse into a black hole.

  Elementary fermions with other spins (3/2, 5/2, etc.) are not known to exist, as of 2014.

- Elementary particles which are thought of as carrying forces are all bosons with spin 1. They include the photon which carries the electromagnetic force, the gluon (strong force), and the W and Z bosons (weak force). The ability of bosons to occupy the same quantum state is used in the laser, which aligns many photons having the same quantum number (the same direction and frequency), superfluid liquid helium resulting from helium-4 atoms being bosons, and superconductivity where pairs of electrons (which individually are fermions) act as single composite bosons.

  Elementary bosons with other spins (0, 2, 3 etc.) were not historically known to exist, although they have received considerable theoretical treatment and are well established within their respective mainstream theories. In particular, theoreticians have proposed the graviton (predicted to exist by some quantum gravity theories) with spin 2, and the Higgs boson (explaining electroweak symmetry breaking) with spin 0. Since 2013, the Higgs boson with spin 0 has been considered proven to exist.[6] It is the first scalar particle (spin 0) known to exist in nature.

Theoretical and experimental studies have shown that the spin possessed by elementary particles cannot be explained by postulating that they are made up of even smaller particles rotating about a common center of mass analogous to a classical electron radius; as far as can be determined at present, these elementary particles have no inner structure. The spin of an elementary particle is therefore seen as a truly intrinsic physical property, akin to the particle's electric charge and rest mass.

29.1.2 Spin-statistics theorem

The proof that particles with half-integer spin (fermions) obey Fermi–Dirac statistics and the Pauli Exclusion Principle, and particles with integer spin (bosons) obey Bose–Einstein statistics, occupy "symmetric states", and thus can share quantum states, is known as the spin-statistics theorem. The theorem relies on both quantum mechanics and the theory of special relativity, and this connection between spin and statistics has been called "one of the most important applications of the special relativity theory".[7]

29.2 Magnetic moments

Main article: Spin magnetic moment

Particles with spin can possess a magnetic dipole moment, just like a rotating electrically charged body in classical electrodynamics. These magnetic moments can be experimentally observed in several ways, e.g. by the deflection of particles by inhomogeneous magnetic fields in a Stern–Gerlach experiment, or by measuring the magnetic fields generated by the particles themselves.

The intrinsic magnetic moment $\boldsymbol{\mu}$ of a spin 1/2 particle with charge q, mass m, and spin angular momentum $\mathbf{S}$, is[8]

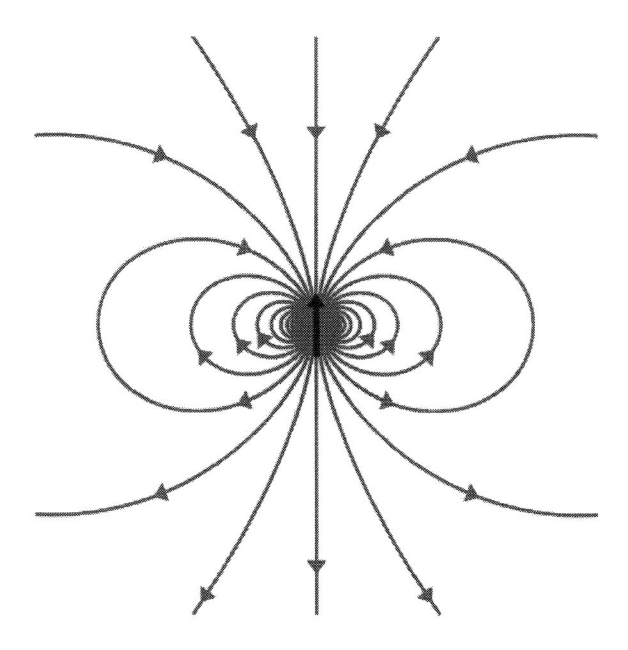

Schematic diagram depicting the spin of the neutron as the black arrow and magnetic field lines associated with the neutron magnetic moment. The neutron has a negative magnetic moment. While the spin of the neutron is upward in this diagram, the magnetic field lines at the center of the dipole are downward.

$$\mu = \frac{g_s q}{2m} \mathbf{S}$$

where the dimensionless quantity g_s is called the spin g-factor. For exclusively orbital rotations it would be 1 (assuming that the mass and the charge occupy spheres of equal radius).

The electron, being a charged elementary particle, possesses a nonzero magnetic moment. One of the triumphs of the theory of quantum electrodynamics is its accurate prediction of the electron g-factor, which has been experimentally determined to have the value −2.0023193043622(15), with the digits in parentheses denoting measurement uncertainty in the last two digits at one standard deviation.[9] The value of 2 arises from the Dirac equation, a fundamental equation connecting the electron's spin with its electromagnetic properties, and the correction of 0.002319304... arises from the electron's interaction with the surrounding electromagnetic field, including its own field.[10] Composite particles also possess magnetic moments associated with their spin. In particular, the neutron possesses a non-zero magnetic moment despite being electrically neutral. This fact was an early indication that the neutron is not an elementary particle. In fact, it is made up of quarks, which are electrically charged particles. The magnetic moment of the neutron comes from the spins of the individual quarks and their orbital motions.

Neutrinos are both elementary and electrically neutral. The minimally extended Standard Model that takes into account non-zero neutrino masses predicts neutrino magnetic moments of:[11][12][13]

$$\mu_\nu \approx 3 \times 10^{-19} \mu_\mathrm{B} \frac{m_\nu}{\mathrm{eV}}$$

where the μv are the neutrino magnetic moments, mv are the neutrino masses, and μB is the Bohr magneton. New physics above the electroweak scale could, however, lead to significantly higher neutrino magnetic moments. It can be shown in a model independent way that neutrino magnetic moments larger than about 10^{-14} μB are unnatural, because they would also lead to large radiative contributions to the neutrino mass. Since the neutrino masses cannot exceed about 1 eV, these radiative corrections must then be assumed to be fine tuned to cancel out to a large degree.[14]

The measurement of neutrino magnetic moments is an active area of research. As of 2001, the latest experimental results have put the neutrino magnetic moment at less than 1.2×10^{-10} times the electron's magnetic moment.

In ordinary materials, the magnetic dipole moments of individual atoms produce magnetic fields that cancel one another, because each dipole points in a random direction. Ferromagnetic materials below their Curie temperature, however, exhibit magnetic domains in which the atomic dipole moments are locally aligned, producing a macroscopic, non-zero magnetic field from the domain. These are the ordinary "magnets" with which we are all familiar.

In paramagnetic materials, the magnetic dipole moments of individual atoms spontaneously align with an externally applied magnetic field. In diamagnetic materials, on the other hand, the magnetic dipole moments of individual atoms spontaneously align oppositely to any externally applied magnetic field, even if it requires energy to do so.

The study of the behavior of such "spin models" is a thriving area of research in condensed matter physics. For instance, the Ising model describes spins (dipoles) that have only two possible states, up and down, whereas in the Heisenberg model the spin vector is allowed to point in any direction. These models have many interesting properties, which have led to interesting results in the theory of phase transitions.

29.3 Direction

Further information: Angular momentum operator

29.3.1 Spin projection quantum number and multiplicity

In classical mechanics, the angular momentum of a particle possesses not only a magnitude (how fast the body is rotating), but also a direction (either up or down on the axis of rotation of the particle). Quantum mechanical spin also contains information about direction, but in a more subtle form. Quantum mechanics states that the component of angular momentum measured along any direction can only take on the values [15]

$$S_i = \hbar s_i, \quad s_i \in \{-s, -(s-1), \dots, s-1, s\}$$

where S_i is the spin component along the i-axis (either x, y, or z), s_i is the spin projection quantum number along the i-axis, and s is the principal spin quantum number (discussed in the previous section). Conventionally the direction chosen is the z-axis:

$$S_z = \hbar s_z, \quad s_z \in \{-s, -(s-1), \dots, s-1, s\}$$

where S_z is the spin component along the z-axis, s_z is the spin projection quantum number along the z-axis.

One can see that there are $2s + 1$ possible values of s_z. The number "$2s + 1$" is the multiplicity of the spin system. For example, there are only two possible values for a spin-1/2 particle: $sz = +1/2$ and $sz = -1/2$. These correspond to quantum states in which the spin is pointing in the +z or −z directions respectively, and are often referred to as "spin up" and "spin down". For a spin-3/2 particle, like a delta baryon, the possible values are +3/2, +1/2, −1/2, −3/2.

29.3.2 Vector

For a given quantum state, one could think of a spin vector $\langle S \rangle$ whose components are the expectation values of the spin components along each axis, i.e., $\langle S \rangle = [\langle S_x \rangle, \langle S_y \rangle, \langle S_z \rangle]$. This vector then would describe the "direction" in which the spin is pointing, corresponding to the classical concept of the axis of rotation. It turns out that the spin vector is not very useful in actual quantum mechanical calculations, because it cannot be measured directly: s_x, s_y and s_z cannot possess simultaneous definite values, because of a quantum uncertainty relation between them. However, for statistically large collections of particles that have been placed in the same pure quantum state, such as through the use of a Stern–Gerlach apparatus, the spin vector does have a well-defined experimental meaning: It specifies the direction in ordinary space in which a subsequent detector must be oriented in order to achieve the maximum possible probability (100%) of detecting every particle in the collection. For spin-1/2 particles, this maximum probability drops off smoothly as the angle between the spin vector and the detector increases, until at an angle of 180 degrees—that is, for detectors oriented in the opposite direction to the spin vector—the expectation of detecting particles from the collection reaches a minimum of 0%.

A single point in space can spin continuously without becoming tangled. Notice that after a 360-degree rotation, the spiral flips between clockwise and counterclockwise orientations. It returns to its original configuration after spinning a full 720 degrees.

As a qualitative concept, the spin vector is often handy because it is easy to picture classically. For instance, quantum mechanical spin can exhibit phenomena analogous to classical gyroscopic effects. For example, one can exert a kind of "torque" on an electron by putting it in a magnetic field (the field acts upon the electron's intrinsic magnetic dipole moment—see the following section). The result is that the spin vector undergoes precession, just like a classical gyroscope. This phenomenon is known as electron spin resonance (ESR). The equivalent behaviour of protons in atomic nuclei is used in nuclear magnetic resonance (NMR) spectroscopy and imaging.

Mathematically, quantum mechanical spin states are described by vector-like objects known as spinors. There are subtle differences between the behavior of spinors and vectors under coordinate rotations. For example, rotating a spin-1/2 particle by 360 degrees does not bring it back to the same quantum state, but to the state with the opposite quantum phase; this is detectable, in principle, with interference experiments. To return the particle to its exact original state, one needs a 720-degree rotation. (The

Plate trick and Mobius strip give non-quantum analogies.) A spin-zero particle can only have a single quantum state, even after torque is applied. Rotating a spin-2 particle 180 degrees can bring it back to the same quantum state and a spin-4 particle should be rotated 90 degrees to bring it back to the same quantum state. The spin 2 particle can be analogous to a straight stick that looks the same even after it is rotated 180 degrees and a spin 0 particle can be imagined as sphere which looks the same after whatever angle it is turned through.

29.4 Mathematical formulation

29.4.1 Operator

Spin obeys commutation relations analogous to those of the orbital angular momentum:

$$[S_i, S_j] = i\hbar\varepsilon_{ijk}S_k$$

where ε_{ijk} is the Levi-Civita symbol, and the expression is summed over the kth index. It follows (as with angular momentum) that the eigenvectors of S^2 and S_z (expressed as kets in the total S basis) are:

$$S^2|s,m_s\rangle = \hbar^2 s(s+1)|s,m_s\rangle$$
$$S_z|s,m_s\rangle = \hbar m_s|s,m_s\rangle.$$

The spin raising and lowering operators acting on these eigenvectors give:

$$S_\pm|s,m_s\rangle = \hbar\sqrt{s(s+1)-m_s(m_s\pm 1)}|s,m_s\pm 1\rangle$$

where $S\pm = Sx \pm i\,Sy$.

But unlike orbital angular momentum the eigenvectors are not spherical harmonics. They are not functions of θ and φ. There is also no reason to exclude half-integer values of s and m_s.

In addition to their other properties, all quantum mechanical particles possess an intrinsic spin (though it may have the intrinsic spin 0, too). The spin is quantized in units of the reduced Planck constant, such that the state function of the particle is, say, not $\psi = \psi(\mathbf{r})$, but $\psi = \psi(\mathbf{r},\sigma)$ where σ is out of the following discrete set of values:

$$\sigma \in \{-s\hbar, -(s-1)\hbar, \cdots, +(s-1)\hbar, +s\hbar\}.$$

One distinguishes bosons (integer spin) and fermions (half-integer spin). The total angular momentum conserved in interaction processes is then the sum of the orbital angular momentum and the spin.

29.4.2 Pauli matrices

Main article: Pauli matrices

The quantum mechanical operators associated with spin-1/2 observables are:

$$\hat{\mathbf{S}} = \frac{\hbar}{2}\boldsymbol{\sigma}$$

where in Cartesian components:

$$S_x = \frac{\hbar}{2}\sigma_x, \quad S_y = \frac{\hbar}{2}\sigma_y, \quad S_z = \frac{\hbar}{2}\sigma_z\,.$$

For the special case of spin-1/2 particles, σ_x, σ_y and σ_z are the three Pauli matrices, given by:

$$\sigma_x = \begin{pmatrix} 0 & 1 \\ 1 & 0 \end{pmatrix} \quad \sigma_y = \begin{pmatrix} 0 & -i \\ i & 0 \end{pmatrix} \quad \sigma_z = \begin{pmatrix} 1 & 0 \\ 0 & -1 \end{pmatrix}.$$

29.4.3 Pauli exclusion principle

For systems of N identical particles this is related to the Pauli exclusion principle, which states that by interchanges of any two of the N particles one must have

$$\psi(\cdots \mathbf{r}_i, \sigma_i \cdots \mathbf{r}_j, \sigma_j \cdots) = (-1)^{2s}\psi(\cdots \mathbf{r}_j, \sigma_j \cdots \mathbf{r}_i, \sigma_i \cdots).$$

Thus, for bosons the prefactor $(-1)^{2s}$ will reduce to +1, for fermions to −1. In quantum mechanics all particles are either bosons or fermions. In some speculative relativistic quantum field theories "supersymmetric" particles also exist, where linear combinations of bosonic and fermionic components appear. In two dimensions, the prefactor $(-1)^{2s}$ can be replaced by any complex number of magnitude 1 such as in the anyon.

The above permutation postulate for N-particle state functions has most-important consequences in daily life, e.g. the periodic table of the chemical elements.

29.4.4 Rotations

See also: symmetries in quantum mechanics

As described above, quantum mechanics states that components of angular momentum measured along any direction can only take a number of discrete values. The most convenient quantum mechanical description of particle's spin is therefore with a set of complex numbers corresponding to amplitudes of finding a given value of projection of its intrinsic angular momentum on a given axis. For instance, for a spin-1/2 particle, we would need two numbers $a_{\pm 1/2}$, giving amplitudes of finding it with projection of angular momentum equal to $\hbar/2$ and $-\hbar/2$, satisfying the requirement

$$\left|a_{\frac{1}{2}}\right|^2 + \left|a_{-\frac{1}{2}}\right|^2 = 1.$$

For a generic particle with spin s, we would need $2s+1$ such parameters. Since these numbers depend on the choice of the axis, they transform into each other non-trivially when this axis is rotated. It's clear that the transformation law must be linear, so we can represent it by associating a matrix with each rotation, and the product of two transformation matrices corresponding to rotations A and B must be equal (up to phase) to the matrix representing rotation AB. Further, rotations preserve the quantum mechanical inner product, and so should our transformation matrices:

$$\sum_{m=-j}^{j} a_m^* b_m = \sum_{m=-j}^{j} \left(\sum_{n=-j}^{j} U_{nm} a_n\right)^* \left(\sum_{k=-j}^{j} U_{km} b_k\right)$$

$$\sum_{n=-j}^{j} \sum_{k=-j}^{j} U_{np}^* U_{kq} = \delta_{pq}.$$

Mathematically speaking, these matrices furnish a unitary projective representation of the rotation group SO(3). Each such representation corresponds to a representation of the covering group of SO(3), which is SU(2).[16] There is one n-dimensional irreducible representation of SU(2) for each dimension, though this representation is n-dimensional real for odd n and n-dimensional complex for even n (hence of real dimension $2n$). For a rotation by angle θ in the plane with normal vector $\hat{\boldsymbol{\theta}}$, U can be written

$$U = e^{-\frac{i}{\hbar}\boldsymbol{\theta}\cdot\mathbf{S}},$$

where $\boldsymbol{\theta} = \theta\hat{\boldsymbol{\theta}}$, and $\mathbf{S}$ is the vector of spin operators.

(Click "show" at right to see a proof or "hide" to hide it.)

Working in the coordinate system where $\hat{\theta} = \hat{z}$, we would like to show that S_x and S_y are rotated into each other by the angle θ. Starting with S_x. Using units where $\hbar = 1$:

$$S_x \to U^\dagger S_x U = e^{i\theta S_z} S_x e^{-i\theta S_z}$$

$$= S_x + (i\theta)[S_z, S_x] + \left(\frac{1}{2!}\right)(i\theta)^2 [S_z, [S_z, S_x]]$$

$$+ \left(\frac{1}{3!}\right)(i\theta)^3 [S_z, [S_z, [S_z, S_x]]] + \cdots$$

Using the spin operator commutation relations, we see that the commutators evaluate to i S_y for the odd terms in the series, and to S_x for all of the even terms. Thus:

$$U^\dagger S_x U = S_x\left[1 - \frac{\theta^2}{2!} + ...\right] - S_y\left[\theta - \frac{\theta^3}{3!}\cdots\right]$$
$$= S_x \cos\theta - S_y \sin\theta$$

as expected. Note that since we only relied on the spin operator commutation relations, this proof holds for any dimension (i.e., for any principal spin quantum number s).[17]

A generic rotation in 3-dimensional space can be built by compounding operators of this type using Euler angles:

$$\mathcal{R}(\alpha,\beta,\gamma) = e^{-i\alpha S_z} e^{-i\beta S_y} e^{-i\gamma S_z}$$

An irreducible representation of this group of operators is furnished by the Wigner D-matrix:

$$D^s_{m'm}(\alpha,\beta,\gamma) \equiv \langle sm'|\mathcal{R}(\alpha,\beta,\gamma)|sm\rangle = e^{-im'\alpha} d^s_{m'm}(\beta) e^{-im\gamma},$$

where

$$d^s_{m'm}(\beta) = \langle sm'|e^{-i\beta s_y}|sm\rangle$$

is Wigner's small d-matrix. Note that for $\gamma = 2\pi$ and $\alpha = \beta = 0$; i.e., a full rotation about the z-axis, the Wigner D-matrix elements become

$$D^s_{m'm}(0,0,2\pi) = d^s_{m'm}(0) e^{-im2\pi} = \delta_{m'm}(-1)^{2m}.$$

Recalling that a generic spin state can be written as a superposition of states with definite m, we see that if s is an integer, the values of m are all integers, and this matrix corresponds to the identity operator. However, if s is a half-integer, the values of m are also all half-integers, giving $(-1)^{2m} = -1$ for all m, and hence upon rotation by 2π the state picks up a minus sign. This fact is a crucial element of the proof of the spin-statistics theorem.

29.4.5 Lorentz transformations

We could try the same approach to determine the behavior of spin under general Lorentz transformations, but we would immediately discover a major obstacle. Unlike SO(3), the group of Lorentz transformations SO(3,1) is non-compact and therefore does not have any faithful, unitary, finite-dimensional representations.

In case of spin-1/2 particles, it is possible to find a construction that includes both a finite-dimensional representation and a scalar product that is preserved by this representation. We associate a 4-component Dirac spinor ψ with each particle. These spinors transform under Lorentz transformations according to the law

$$\psi' = \exp\left(\tfrac{1}{8}\omega_{\mu\nu}[\gamma_\mu, \gamma_\nu]\right)\psi$$

where $\gamma\nu$ are gamma matrices and $\omega\mu\nu$ is an antisymmetric 4 × 4 matrix parametrizing the transformation. It can be shown that the scalar product

$$\langle\psi|\phi\rangle = \bar{\psi}\phi = \psi^\dagger\gamma_0\phi$$

is preserved. It is not, however, positive definite, so the representation is not unitary.

29.4.6 Measurement of spin along the x-, y-, or z-axes

Each of the (Hermitian) Pauli matrices has two eigenvalues, +1 and −1. The corresponding normalized eigenvectors are:

$$\begin{array}{ll} \psi_{x+} = \frac{1}{\sqrt{2}}\begin{pmatrix}1\\1\end{pmatrix}, & \psi_{x-} = \frac{1}{\sqrt{2}}\begin{pmatrix}1\\-1\end{pmatrix}, \\ \psi_{y+} = \frac{1}{\sqrt{2}}\begin{pmatrix}1\\i\end{pmatrix}, & \psi_{y-} = \frac{1}{\sqrt{2}}\begin{pmatrix}1\\-i\end{pmatrix}, \\ \psi_{z+} = \begin{pmatrix}1\\0\end{pmatrix}, & \psi_{z-} = \begin{pmatrix}0\\1\end{pmatrix}. \end{array}$$

By the postulates of quantum mechanics, an experiment designed to measure the electron spin on the x-, y-, or z-axis can only yield an eigenvalue of the corresponding spin operator (S_x, S_y or S_z) on that axis, i.e. $\hbar/2$ or $-\hbar/2$. The quantum state of a particle (with respect to spin), can be represented by a two component spinor:

$$\psi = \begin{pmatrix}a+bi\\c+di\end{pmatrix}.$$

When the spin of this particle is measured with respect to a given axis (in this example, the x-axis), the probability that its spin will be measured as $\hbar/2$ is just $|\langle\psi_{x+}|\psi\rangle|^2$. Correspondingly, the probability that its spin will be measured as $-\hbar/2$ is just $|\langle\psi_{x-}|\psi\rangle|^2$. Following the measurement, the spin state of the particle will collapse into the corresponding eigenstate. As a result, if the particle's spin along a given axis has been measured to have a given eigenvalue, all measurements will yield the same eigenvalue (since $|\langle\psi_{x+}|\psi_{x+}\rangle|^2 = 1$, etc), provided that no measurements of the spin are made along other axes.

29.4.7 Measurement of spin along an arbitrary axis

The operator to measure spin along an arbitrary axis direction is easily obtained from the Pauli spin matrices. Let $u = (ux, uy, uz)$ be an arbitrary unit vector. Then the operator for spin in this direction is simply

$$S_u = \frac{\hbar}{2}(u_x\sigma_x + u_y\sigma_y + u_z\sigma_z)$$

The operator S_u has eigenvalues of $\pm\hbar/2$, just like the usual spin matrices. This method of finding the operator for spin in an arbitrary direction generalizes to higher spin states, one takes the dot product of the direction with a vector of the three operators for the three x-, y-, z-axis directions.

A normalized spinor for spin-1/2 in the (ux, uy, uz) direction (which works for all spin states except spin down where it will give 0/0), is:

$$\frac{1}{\sqrt{2+2u_z}}\begin{pmatrix}1+u_z\\u_x+iu_y\end{pmatrix}.$$

The above spinor is obtained in the usual way by diagonalizing the σ_u matrix and finding the eigenstates corresponding to the eigenvalues. In quantum mechanics, vectors are termed "normalized" when multiplied by a normalizing factor, which results in the vector having a length of unity.

29.4.8 Compatibility of spin measurements

Since the Pauli matrices do not commute, measurements of spin along the different axes are incompatible. This means that if, for example, we know the spin along the x-axis, and we then measure the spin along the y-axis, we have invalidated our previous knowledge of the x-axis spin. This can be seen from the property of the eigenvectors (i.e. eigenstates) of the Pauli matrices that:

$$|\langle\psi_{x\pm} \mid \psi_{y\pm}\rangle|^2 = |\langle\psi_{x\pm} \mid \psi_{z\pm}\rangle|^2 = |\langle\psi_{y\pm} \mid \psi_{z\pm}\rangle|^2 = \tfrac{1}{2}.$$

So when physicists measure the spin of a particle along the x-axis as, for example, ħ/2, the particle's spin state collapses into the eigenstate $| \psi_{x+}\rangle$. When we then subsequently measure the particle's spin along the y-axis, the spin state will now collapse into either $| \psi_{y+}\rangle$ or $| \psi_{y-}\rangle$, each with probability 1/2. Let us say, in our example, that we measure −ħ/2. When we now return to measure the particle's spin along the x-axis again, the probabilities that we will measure ħ/2 or −ħ/2 are each 1/2 (i.e. they are $|\langle\psi_{x+} | \psi_{y-}\rangle|^2$ and $|\langle\psi_{x-} | \psi_{y-}\rangle|^2$ respectively). This implies that the original measurement of the spin along the x-axis is no longer valid, since the spin along the x-axis will now be measured to have either eigenvalue with equal probability.

29.4.9 Higher spins

The spin-1/2 operator **S** = ħ/2**σ** form the fundamental representation of SU(2). By taking Kronecker products of this representation with itself repeatedly, one may construct all higher irreducible representations. That is, the resulting spin operators for higher spin systems in three spatial dimensions, for arbitrarily large s, can be calculated using this spin operator and ladder operators.

The resulting spin matrices for spin 1 are:

$$S_x = \frac{\hbar}{\sqrt{2}}\begin{pmatrix} 0 & 1 & 0 \\ 1 & 0 & 1 \\ 0 & 1 & 0 \end{pmatrix}$$

$$S_y = \frac{\hbar}{\sqrt{2}}\begin{pmatrix} 0 & -i & 0 \\ i & 0 & -i \\ 0 & i & 0 \end{pmatrix}$$

$$S_z = \hbar\begin{pmatrix} 1 & 0 & 0 \\ 0 & 0 & 0 \\ 0 & 0 & -1 \end{pmatrix}$$

for spin 3/2 they are

$$S_x = \frac{\hbar}{2}\begin{pmatrix} 0 & \sqrt{3} & 0 & 0 \\ \sqrt{3} & 0 & 2 & 0 \\ 0 & 2 & 0 & \sqrt{3} \\ 0 & 0 & \sqrt{3} & 0 \end{pmatrix}$$

$$S_y = \frac{\hbar}{2}\begin{pmatrix} 0 & -i\sqrt{3} & 0 & 0 \\ i\sqrt{3} & 0 & -2i & 0 \\ 0 & 2i & 0 & -i\sqrt{3} \\ 0 & 0 & i\sqrt{3} & 0 \end{pmatrix}$$

$$S_z = \frac{\hbar}{2}\begin{pmatrix} 3 & 0 & 0 & 0 \\ 0 & 1 & 0 & 0 \\ 0 & 0 & -1 & 0 \\ 0 & 0 & 0 & -3 \end{pmatrix}$$

and for spin 5/2 they are

$$S_x = \frac{\hbar}{2}\begin{pmatrix} 0 & \sqrt{5} & 0 & 0 & 0 & 0 \\ \sqrt{5} & 0 & 2\sqrt{2} & 0 & 0 & 0 \\ 0 & 2\sqrt{2} & 0 & 3 & 0 & 0 \\ 0 & 0 & 3 & 0 & 2\sqrt{2} & 0 \\ 0 & 0 & 0 & 2\sqrt{2} & 0 & \sqrt{5} \\ 0 & 0 & 0 & 0 & \sqrt{5} & 0 \end{pmatrix}$$

$$S_y = \frac{\hbar}{2}\begin{pmatrix} 0 & -i\sqrt{5} & 0 & 0 & 0 & 0 \\ i\sqrt{5} & 0 & -2i\sqrt{2} & 0 & 0 & 0 \\ 0 & 2i\sqrt{2} & 0 & -3i & 0 & 0 \\ 0 & 0 & 3i & 0 & -2i\sqrt{2} & 0 \\ 0 & 0 & 0 & 2i\sqrt{2} & 0 & -i\sqrt{5} \\ 0 & 0 & 0 & 0 & i\sqrt{5} & 0 \end{pmatrix}$$

$$S_z = \frac{\hbar}{2}\begin{pmatrix} 5 & 0 & 0 & 0 & 0 & 0 \\ 0 & 3 & 0 & 0 & 0 & 0 \\ 0 & 0 & 1 & 0 & 0 & 0 \\ 0 & 0 & 0 & -1 & 0 & 0 \\ 0 & 0 & 0 & 0 & -3 & 0 \\ 0 & 0 & 0 & 0 & 0 & -5 \end{pmatrix}.$$

The generalization of these matrices for arbitrary s is

$$(S_x)_{ab} = \frac{\hbar}{2}(\delta_{a,b+1} + \delta_{a+1,b})\sqrt{(s+1)(a+b-1) - ab}$$

$$(S_y)_{ab} = \frac{\hbar}{2i}(\delta_{a,b+1} - \delta_{a+1,b})\sqrt{(s+1)(a+b-1) - ab}$$

$$1 \le a, b \le 2s+1$$

$$(S_z)_{ab} = \hbar(s+1-a)\delta_{a,b} = \hbar(s+1-b)\delta_{a,b}.$$

Also useful in the quantum mechanics of multiparticle systems, the general Pauli group G_n is defined to consist of all n-fold tensor products of Pauli matrices.

The analog formula of Euler's formula in terms of the Pauli matrices:

$$e^{i\theta(\hat{\mathbf{n}}\cdot\boldsymbol{\sigma})} = I\cos\theta + i(\hat{\mathbf{n}}\cdot\boldsymbol{\sigma})\sin\theta$$

for higher spins is tractable, but less simple.[18]

29.5 Parity

In tables of the spin quantum number s for nuclei or particles, the spin is often followed by a "+" or "−". This refers to the parity with "+" for even parity (wave function unchanged by spatial inversion) and "−" for odd parity (wave function negated by spatial inversion). For example, see the isotopes of bismuth.

29.6 Applications

Spin has important theoretical implications and practical applications. Well-established *direct* applications of spin include:

- Nuclear magnetic resonance (NMR) spectroscopy in chemistry;
- Electron spin resonance spectroscopy in chemistry and physics;
- Magnetic resonance imaging (MRI) in medicine, a type of applied NMR, which relies on proton spin density;
- Giant magnetoresistive (GMR) drive head technology in modern hard disks.

Electron spin plays an important role in magnetism, with applications for instance in computer memories. The manipulation of *nuclear spin* by radiofrequency waves (nuclear magnetic resonance) is important in chemical spectroscopy and medical imaging.

Spin-orbit coupling leads to the fine structure of atomic spectra, which is used in atomic clocks and in the modern definition of the second. Precise measurements of the g-factor of the electron have played an important role in the development and verification of quantum electrodynamics. *Photon spin* is associated with the polarization of light.

An emerging application of spin is as a binary information carrier in spin transistors. The original concept, proposed in 1990, is known as Datta-Das spin transistor.[19] Electronics based on spin transistors are referred to as spintronics. The manipulation of spin in dilute magnetic semiconductor materials, such as metal-doped ZnO or TiO_2 imparts a further degree of freedom and has the potential to facilitate the fabrication of more efficient electronics.[20]

There are many *indirect* applications and manifestations of spin and the associated Pauli exclusion principle, starting with the periodic table of chemistry.

29.7 History

Spin was first discovered in the context of the emission spectrum of alkali metals. In 1924 Wolfgang Pauli introduced what he called a "two-valued quantum degree of freedom" associated with the electron in the outermost shell. This allowed him to formulate the Pauli exclusion principle, stating that no two electrons can share the same quantum state at the same time.

Wolfgang Pauli lecturing

The physical interpretation of Pauli's "degree of freedom" was initially unknown. Ralph Kronig, one of Landé's assistants, suggested in early 1925 that it was produced by the self-rotation of the electron. When Pauli heard about the idea, he criticized it severely, noting that the electron's hypothetical surface would have to be moving faster than the speed of light in order for it to rotate quickly enough to produce the necessary angular momentum. This would violate the theory of relativity. Largely due to Pauli's criticism, Kronig decided not to publish his idea.

In the autumn of 1925, the same thought came to two Dutch physicists, George Uhlenbeck and Samuel Goudsmit at Leiden University. Under the advice of Paul Ehrenfest, they published their results. It met a favorable response, especially after Llewellyn Thomas managed to resolve a factor-of-two discrepancy between experimental results and Uhlenbeck and Goudsmit's calculations (and Kronig's unpublished results). This discrepancy was due to the orientation of the electron's tangent frame, in addition to its position.

Mathematically speaking, a fiber bundle description is needed. The tangent bundle effect is additive and relativistic; that is, it vanishes if c goes to infinity. It is one half of the value obtained without regard for the tangent space orientation, but with opposite sign. Thus the combined effect differs from the latter by a factor two (Thomas precession).

Despite his initial objections, Pauli formalized the theory of spin in 1927, using the modern theory of quantum mechanics invented by Schrödinger and Heisenberg. He pioneered the use of Pauli matrices as a representation of the spin operators, and introduced a two-component spinor wave-function.

Pauli's theory of spin was non-relativistic. However, in 1928, Paul Dirac published the Dirac equation, which described the relativistic electron. In the Dirac equation, a four-component spinor (known as a "Dirac spinor") was used for the electron wave-function. In 1940, Pauli proved the *spin-statistics theorem*, which states that fermions have half-integer spin and bosons have integer spin.

In retrospect, the first direct experimental evidence of the electron spin was the Stern–Gerlach experiment of 1922. However, the correct explanation of this experiment was only given in 1927.[21]

29.8 See also

- Einstein–de Haas effect
- Spin-orbital
- Chirality (physics)
- Dynamic nuclear polarisation
- Helicity (particle physics)
- Holstein–Primakoff transformation
- Pauli equation
- Pauli–Lubanski pseudovector
- Rarita–Schwinger equation
- Representation theory of SU(2)
- Spin-1/2
- Spin-flip
- Spin isomers of hydrogen
- Spin tensor
- Spin wave
- Spin engineering
- Yrast
- Zitterbewegung

29.9 References

[1] Merzbacher, Eugen (1998). *Quantum Mechanics* (3rd ed.). pp. 372–3.

[2] Griffiths, David (2005). *Introduction to Quantum Mechanics* (2nd ed.). pp. 183–4.

[3] "Angular Momentum Operator Algebra", class notes by Michael Fowler

[4] *A modern approach to quantum mechanics*, by Townsend, p. 31 and p. 80

[5] Eisberg, Robert; Resnick, Robert (1985). *Quantum Physics of Atoms, Molecules, Solids, Nuclei, and Particles* (2nd ed.). pp. 272–3.

[6] Information about Higgs Boson in CERN's official website.

[7] Pauli, Wolfgang (1940). "The Connection Between Spin and Statistics" (PDF). *Phys. Rev* **58** (8): 716–722. Bibcode:1940PhRv...58..716P. doi:10.1103/PhysRev.58.716.

[8] Physics of Atoms and Molecules, B.H. Bransden, C.J.Joachain, Longman, 1983, ISBN 0-582-44401-2

[9] "CODATA Value: electron *g* factor". *The NIST Reference on Constants, Units, and Uncertainty*. NIST. 2006. Retrieved 2013-11-15.

[10] R.P. Feynman (1985). "Electrons and Their Interactions". *QED: The Strange Theory of Light and Matter*. Princeton, New Jersey: Princeton University Press. p. 115. ISBN 0-691-08388-6.

> "After some years, it was discovered that this value [−*g*/2] was not exactly 1, but slightly more—something like 1.00116. This correction was worked out for the first time in 1948 by Schwinger as *j**j* divided by 2 pi [*sic*] [where j is the square root of the fine-structure constant], and was due to an alternative way the electron can go from place to place: instead of going directly from one point to another, the electron goes along for a while and suddenly emits a photon; then (horrors!) it absorbs its own photon."

[11] W.J. Marciano, A.I. Sanda (1977). "Exotic decays of the muon and heavy leptons in gauge theories". *Physics Letters* **B67** (3): 303–305. Bibcode:1977PhLB...67..303M. doi:10.1016/0370-2693(77)90377-X.

[12] B.W. Lee, R.E. Shrock (1977). "Natural suppression of symmetry violation in gauge theories: Muon- and electron-lepton-number nonconservation". *Physical Review* **D16** (5): 1444–1473. Bibcode:1977PhRvD..16.1444L. doi:10.1103/PhysRevD.16.1444.

[13] K. Fujikawa, R. E. Shrock (1980). "Magnetic Moment of a Massive Neutrino and Neutrino-Spin Rotation". *Physical Review Letters* **45** (12): 963–966. Bibcode:1980PhRvL..45..963F. doi:10.1103/PhysRevLett.45.963.

[14] N.F. Bell; Cirigliano, V.; Ramsey-Musolf, M.; Vogel, P.; Wise, Mark; et al. (2005). "How Magnetic is the Dirac Neutrino?". *Physical Review Letters* **95** (15): 151802. arXiv:hep-ph/0504134. Bibcode:2005PhRvL..95o1802B. doi:10.1103/PhysRevLett.95.151802. PMID 16241715.

[15] Quanta: A handbook of concepts, P.W. Atkins, Oxford University Press, 1974, ISBN 0-19-855493-1

[16] B.C. Hall (2013). *Quantum Theory for Mathematicians*. Springer. pp. 354–358.

[17] *Modern Quantum Mechanics*, by J. J. Sakurai, p159

[18] Curtright, T L; Fairlie, D B; Zachos, C K (2014). "A compact formula for rotations as spin matrix polynomials". *SIGMA* **10**: 084. arXiv:1402.3541. Bibcode:2014SIGMA..10..084C. doi:10.3842/SIGMA.2014.084.

[19] Datta. S and B. Das (1990). "Electronic analog of the electrooptic modulator". *Applied Physics Letters* **56** (7): 665–667. Bibcode:1990ApPhL..56..665D. doi:10.1063/1.102730.

[20] Assadi, M.H.N; Hanaor, D.A.H (2013). "Theoretical study on copper's energetics and magnetism in TiO_2 polymorphs" (PDF). *Journal of Applied Physics* **113** (23): 233913. arXiv:1304.1854. Bibcode:2013JAP...113w3913A. doi:10.1063/1.4811539.

[21] B. Friedrich, D. Herschbach (2003). "Stern and Gerlach: How a Bad Cigar Helped Reorient Atomic Physics". *Physics Today* **56** (12): 53. Bibcode:2003PhT....56l..53F. doi:10.1063/1.1650229.

29.10 Further reading

- Cohen-Tannoudji, Claude; Diu, Bernard; Laloë, Franck (2006). *Quantum Mechanics* (2 volume set ed.). John Wiley & Sons. ISBN 978-0-471-56952-7.
- Condon, E. U.; Shortley, G. H. (1935). "Especially Chapter 3". *The Theory of Atomic Spectra*. Cambridge University Press. ISBN 0-521-09209-4.
- Hipple, J. A.; Sommer, H.; Thomas, H.A. (1949). *A precise method of determining the faraday by magnetic resonance*. doi:10.1103/PhysRev.76.1877.2.https://www.academia.edu/6483539/John_A._Hipple_1911-1985_technology_as_knowledge
- Edmonds, A. R. (1957). *Angular Momentum in Quantum Mechanics*. Princeton University Press. ISBN 0-691-07912-9.
- Jackson, John David (1998). *Classical Electrodynamics* (3rd ed.). John Wiley & Sons. ISBN 978-0-471-30932-1.
- Serway, Raymond A.; Jewett, John W. (2004). *Physics for Scientists and Engineers* (6th ed.). Brooks/Cole. ISBN 0-534-40842-7.
- Thompson, William J. (1994). *Angular Momentum: An Illustrated Guide to Rotational Symmetries for Physical Systems*. Wiley. ISBN 0-471-55264-X.
- Tipler, Paul (2004). *Physics for Scientists and Engineers: Mechanics, Oscillations and Waves, Thermodynamics* (5th ed.). W. H. Freeman. ISBN 0-7167-0809-4.
- Sin-Itiro Tomonaga, The Story of Spin, 1997

29.11 External links

- "Spintronics. Feature Article" in *Scientific American*, June 2002.
- Goudsmit on the discovery of electron spin.
- *Nature*: "Milestones in 'spin' since 1896."
- ECE 495N Lecture 36: Spin Online lecture by S. Datta

Chapter 30

General covariance

In theoretical physics, **general covariance** (also known as **diffeomorphism covariance** or **general invariance**) is the invariance of the *form* of physical laws under arbitrary differentiable coordinate transformations. The essential idea is that coordinates do not exist *a priori* in nature, but are only artifices used in describing nature, and hence should play no role in the formulation of fundamental physical laws.

A physical law expressed in a generally covariant fashion takes the same mathematical form in all coordinate systems,[1] and is usually expressed in terms of tensor fields. The classical (non-quantum) theory of electrodynamics is one theory that has such a formulation.

Albert Einstein proposed this principle for his special theory of relativity; however, that theory was limited to space-time coordinate systems related to each other by uniform relative motions only , the so-called "inertial frames." Einstein recognized that the general principle of relativity should also apply to accelerated relative motions, and he used the newly developed tool of tensor calculus to extend the special theory's global Lorentz covariance (applying only to inertial frames) to the more general local Lorentz covariance (which applies to all frames), eventually producing his general theory of relativity. The local reduction of the general metric tensor to the Minkowski metric corresponds to free-falling (geodesic) motion, in this theory, thus encompassing the phenomenon of gravitation.

Much of the work on classical unified field theories consisted of attempts to further extend the general theory of relativity to interpret additional physical phenomena, particularly electromagnetism, within the framework of general covariance, and more specifically as purely geometric objects in the space-time continuum.

30.1 Remarks

The relationship between general covariance and general relativity may be summarized by quoting a standard textbook:[2]

> Mathematics was not sufficiently refined in 1917 to cleave apart the demands for "no prior geometry" and for a geometric, coordinate-independent formulation of physics. Einstein described both demands by a single phrase, "general covariance." The "no prior geometry" demand actually fathered general relativity, but by doing so anonymously, disguised as "general covariance", it also fathered half a century of confusion.

A more modern interpretation of the physical content of the original principle of general covariance is that the Lie group $GL4(\mathbf{R})$ is a fundamental "external" symmetry of the world. Other symmetries, including "internal" symmetries based on compact groups, now play a major role in fundamental physical theories.

30.2 See also

- Coordinate conditions
- Coordinate-free
- Covariance and contravariance
- Covariant derivative
- Diffeomorphism
- Fictitious force
- Galilean invariance
- Gauge covariant derivative
- General covariant transformations
- Harmonic coordinate condition
- Inertial frame of reference

- Lorentz covariance
- Principle of covariance
- Special relativity
- Symmetry in physics

30.3 Notes

[1] More precisely, only coordinate systems related through sufficiently differentiable transformations are considered.

[2] Charles W. Misner, Kip S. Thorne, and John Archibald Wheeler (1973). *Gravitation*. Freeman. p. 431. ISBN 0-7167-0344-0.

30.4 References

- O'Hanian, Hans C.; Ruffini, Remo (1994). *Gravitation and Spacetime* (2nd ed.). New York: W. W. Norton. ISBN 0-393-96501-5. See *section 7.1*.

30.5 External links

- General covariance and the foundations of general relativity: eight decades of dispute, by J. D. Norton (file size: 4 MB)
  re-typeset version (file size: 460 KB)

Chapter 31

Background independence

Background independence is a condition in theoretical physics, that requires the defining equations of a theory to be independent of the actual shape of the spacetime and the value of various fields within the spacetime. In particular this means that it must be possible not to refer to a specific coordinate system—the theory must be coordinate-free. In addition, the different spacetime configurations (or backgrounds) should be obtained as different solutions of the underlying equations.

31.1 What is background-independence?

Background-independence is a loosely defined property of a theory of physics. Roughly speaking, it limits the number of mathematical structures used to describe space and time that are put in place "by hand". Instead, these structures are the result of dynamical equations, such as Einstein field equations, so that one can determine from first principles what form they should take. Since the form of the metric determines the result of calculations, a theory with background independence is more predictive than a theory without it, since the theory requires fewer inputs to make its predictions. This is analogous to desiring fewer free parameters in a fundamental theory.

So background-independence can be seen as extending the mathematical objects that should be predicted from theory to include not just the parameters, but also geometrical structures. Summarizing this, Rickles writes: "Background structures are contrasted with dynamical ones, and a background independent theory only possesses the latter type—obviously, background dependent theories are those possessing the former type in addition to the latter type.".[1]

In general relativity, background-independence is identified with the property that the metric of space-time is the solution of a dynamical equation.[2] In classical mechanics, this is not the case, the metric is fixed by the physicist to match experimental observations. This is undesirable, since the form of the metric impacts the physical predictions, but is not itself predicted by the theory.

31.2 Manifest background-independence

Manifest background-independence is primarily an aesthetic rather than a physical requirement. It is analogous, and closely related, to requiring in differential geometry that equations be written in a form that is independent of the choice of charts and coordinate embeddings. If a background-independent formalism is present, it can lead to simpler and more elegant equations. However, there is no physical content in requiring that a theory be **manifestly background-independent** – for example, the equations of general relativity can be rewritten in local coordinates without affecting the physical implications.

Although making a property manifest is only aesthetic, it is a useful tool for making sure the theory actually has that property. For example, if a theory is written in a manifestly Lorentz invariant way, one can check at every step to be sure that Lorentz invariance is preserved. Making a property manifest also makes it clear whether or not the theory actually has that property. The inability to make classical mechanics manifestly Lorentz invariant does not reflect a lack of imagination on the part of the theorist, but rather a physical feature of the theory. The same goes for making classical mechanics, or electromagnetism background independent.

31.3 Theories of quantum gravity

Because of the speculative nature of quantum gravity research, there is much debate as to the correct implementation of background-independence. Ultimately, the answer is to be decided by experiment, but until experiments can probe quantum gravity phenomena, physicists have to settle for debate. Below is a brief summary of the two largest quantum gravity approaches.

Physicists have studied models of 3D quantum gravity, which is a much simpler problem than 4D quantum gravity (this is because in 3D, quantum gravity has no local degrees of freedom). In these models, there are non-zero transition amplitudes between two different topologies,[3] or in other words, the topology changes. This and other similar results lead physicists to believe that any consistent quantum theory of gravity should include topology change as a dynamical process.

31.3.1 String theory

String theory is usually formulated with perturbation theory around a fixed background. While it is possible that the theory defined this way is locally background-invariant, if so it is not manifest, and it is not clear what the exact meaning is. One attempt to formulate string theory in a manifestly background-independent fashion is string field theory, but little progress has been made in understanding it.

Another approach is the conjectured, but yet unproven AdS/CFT duality, which is believed to provide a full, non-perturbative definition of string theory in spacetimes with anti-de Sitter asymptotics. If so, this could describe a kind of superselection sector of the putative background-independent theory. But it would be still restricted to anti-de Sitter space asymptotics, which disagrees with the current observations of our Universe. A full non-perturbative definition of the theory in arbitrary space-time backgrounds is still lacking.

Topology change is an established process in string theory.

31.3.2 Loop quantum gravity

A very different approach to quantum gravity called loop quantum gravity is full non-perturbative, manifest background-independent: Geometric quantities, such as area, are predicted without reference to a background metric or asymptotics (e.g. no need for an background metric or an anti-de Sitter asymptotics), only a given topology.

31.4 See also

- General relativity
- String theory
- Causal dynamical triangulation
- Loop quantum gravity
- Quantum field theory
- Coordinate-free

31.5 References

[1] D. Rickles, Who's Afraid of Background Independence?, p. 4

[2] John Baez, Higher-Dimensional Algebra and Planck-Scale Physics

[3] Hiroshi Ooguri, Partition Functions and Topology-Changing Amplitudes in the 3D Lattice Gravity of Ponzano and Regge

31.6 Further reading

- Rozali, M. (2009). "Comments on Background Independence and Gauge Redundancies". *Advanced Science Letters* **2** (2): 244. arXiv:0809.3962. doi:10.1166/asl.2009.1031.
- Smolin, L. (2005). "The case for background independence". arXiv:hep-th/0507235 [hep-th].
- Colosi, D.; et al. (2005). "Background independence in a nutshell". *Classical and Quantum Gravity* **22** (14): 2971–2989. arXiv:gr-qc/0408079. Bibcode:2005CQGra..22.2971C. doi:10.1088/0264-9381/22/14/008.
- Witten, E. (1993). "Quantum Background Independence in String Theory". arXiv:hep-th/9306122 [hep-th].
- Stachel, J. (1993). "The Meaning of General Covariance: The Hole Story". In J. Earman, A. Janis, G. Massey and N. Rescher. *Philosophical Problems of the Internal and External Worlds: Essays on the Philosophy of Adolf Grünbaum*. University of Pittsburgh Press. pp. 129–160. ISBN 0-8229-3738-7.
- Stachel, J. (1994). "Changes in the Concepts of Space and Time Brought About by Relativity". In C. C. Gould and R. S. Cohen. *Artifacts, Representations and Social Practice*. Kluwer Academic. pp. 141–162. ISBN 0-7923-2481-1.
- Zahar, E. (1989). *Einstein's Revolution: A Study in Heuristic*. Open Court Publishing Company. ISBN 0-8126-9066-4.

Chapter 32

Diffeomorphism

In mathematics, a **diffeomorphism** is an isomorphism of smooth manifolds. It is an invertible function that maps one differentiable manifold to another such that both the function and its inverse are smooth.

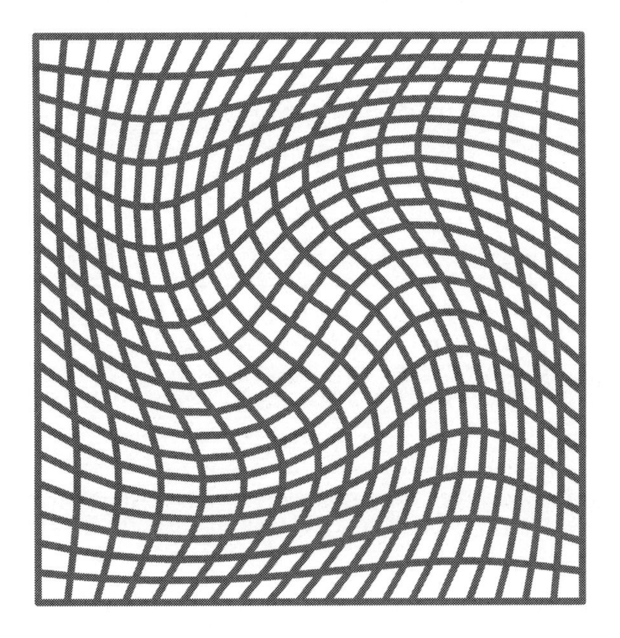

The image of a rectangular grid on a square under a diffeomorphism from the square onto itself.

32.1 Definition

Given two manifolds M and N, a differentiable map $f : M \to N$ is called a **diffeomorphism** if it is a bijection and its inverse $f^{-1} : N \to M$ is differentiable as well. If these functions are r times continuously differentiable, f is called a ***C^r*-diffeomorphism**.

Two manifolds M and N are **diffeomorphic** (symbol usually being $\simeq$) if there is a diffeomorphism f from M to N. They are ***C^r* diffeomorphic** if there is an r times continuously differentiable bijective map between them whose inverse is also r times continuously differentiable.

32.2 Diffeomorphisms of subsets of manifolds

Given a subset X of a manifold M and a subset Y of a manifold N, a function $f : X \to Y$ is said to be smooth if for all p in X there is a neighborhood $U \subset M$ of p and a smooth function $g : U \to N$ such that the restrictions agree $g_{|U\cap X} = f_{|U\cap X}$ (note that g is an extension of f). f is said to be a diffeomorphism if it is bijective, smooth and its inverse is smooth.

32.3 Local description

Model example

If U, V are connected open subsets of $\mathbf{R}^n$ such that V is simply connected, a differentiable map $f : U \to V$ is a **diffeomorphism** if it is proper and if the differential $Df_x : \mathbf{R}^n \to \mathbf{R}^n$ is bijective at each point x in U.

First remark

It is essential for V to be simply connected for the function f to be globally invertible (under the sole condition that its derivative is a bijective map at each point). For example, consider the "realification" of the complex square function

$$\begin{cases} f : \mathbf{R}^2 \setminus \{(0,0)\} \to \mathbf{R}^2 \setminus \{(0,0)\} \\ (x,y) \mapsto (x^2 - y^2, 2xy) \end{cases}$$

Then f is surjective and it satisfies

$$\det Df_x = 4(x^2 + y^2) \neq 0$$

Thus, though *Dfx* is bijective at each point, f is not invertible because it fails to be injective (e.g. $f(1,0) = (1,0) = f(-1,0)$.

Second remark

Since the differential at a point (for a differentiable function)

$$Df_x : T_xU \to T_{f(x)}V$$

is a linear map, it has a well-defined inverse if and only if *Dfx* is a bijection. The matrix representation of *Dfx* is the $n \times n$ matrix of first-order partial derivatives whose entry in the i-th row and j-th column is $\partial f_i / \partial x_j$. This so-called Jacobian matrix is often used for explicit computations.

Third remark

Diffeomorphisms are necessarily between manifolds of the same dimension. Imagine f going from dimension n to dimension k. If $n < k$ then *Dfx* could never be surjective; and if $n > k$ then *Dfx* could never be injective. In both cases, therefore, *Dfx* fails to be a bijection.

Fourth remark

If *Dfx* is a bijection at x then f is said to be a local diffeomorphism (since, by continuity, *Dfy* will also be bijective for all y sufficiently close to x).

Fifth remark

Given a smooth map from dimension n to dimension k, if *Df* (or, locally, *Dfx*) is surjective, f is said to be a submersion (or, locally, a "local submersion"); and if *Df* (or, locally, *Dfx*) is injective, f is said to be an immersion (or, locally, a "local immersion").

Sixth remark

A differentiable bijection is *not* necessarily a diffeomorphism. $f(x) = x^3$, for example, is not a diffeomorphism from $\mathbf{R}$ to itself because its derivative vanishes at 0 (and hence its inverse is not differentiable at 0). This is an example of a homeomorphism that is not a diffeomorphism.

Seventh remark

When f is a map between *differentiable* manifolds, a diffeomorphic f is a stronger condition than a homeomorphic f. For a diffeomorphism, f and its inverse need to be differentiable; for a homeomorphism, f and its inverse need only be continuous. Every diffeomorphism is a homeomorphism, but not every homeomorphism is a diffeomorphism.

$f : M \to N$ is called a **diffeomorphism** if, in coordinate charts, it satisfies the definition above. More precisely: Pick any cover of M by compatible coordinate charts and do the same for N. Let φ and ψ be charts on, respectively, M and N, with U and V as, respectively, the images of φ and ψ. The map $\psi f \varphi^{-1} : U \to V$ is then a diffeomorphism as in the definition above, whenever $f(\varphi^{-1}(U)) \subset \psi^{-1}(V)$.

32.4 Examples

Since any manifold can be locally parametrised, we can consider some explicit maps from $\mathbf{R}^2$ into $\mathbf{R}^2$.

- Let

$$f(x, y) = \left(x^2 + y^3, x^2 - y^3\right).$$

We can calculate the Jacobian matrix:

$$J_f = \begin{pmatrix} 2x & 3y^2 \\ 2x & -3y^2 \end{pmatrix}.$$

The Jacobian matrix has zero determinant if, and only if $xy = 0$. We see that f could only be a diffeomorphism away from the x-axis and the y-axis. However, f is not bijective since $f(x,y)=f(-x,y)$, and thus it cannot be a diffeomorphism.

- Let

$$g(x, y) = (a_0 + a_{1,0}x + a_{0,1}y + \cdots \;,\; b_0 + b_{1,0}x + b_{0,1}y + \cdots)$$

where the $a_{i,j}$ and $b_{i,j}$ are arbitrary real numbers, and the omitted terms are of degree at least two in x and y. We can calculate the Jacobian matrix at $\mathbf{0}$:

$$J_g(0, 0) = \begin{pmatrix} a_{1,0} & a_{0,1} \\ b_{1,0} & b_{0,1} \end{pmatrix}.$$

We see that g is a local diffeomorphism at $\mathbf{0}$ if, and only if,

$$a_{1,0}b_{0,1} - a_{0,1}b_{1,0} \neq 0,$$

i.e. the linear terms in the components of g are linearly independent as polynomials.

- Let

$$h(x, y) = \left(\sin(x^2 + y^2), \cos(x^2 + y^2)\right).$$

We can calculate the Jacobian matrix:

$$J_h = \begin{pmatrix} 2x\cos(x^2 + y^2) & 2y\cos(x^2 + y^2) \\ -2x\sin(x^2 + y^2) & -2y\sin(x^2 + y^2) \end{pmatrix}.$$

The Jacobian matrix has zero determinant everywhere! In fact we see that the image of *h* is the unit circle.

32.4.1 Surface deformations

In mechanics, a stress-induced transformation is called a deformation and may be described by a diffeomorphism. A diffeomorphism $f : U \to V$ between two surfaces *U* and *V* has a Jacobian matrix *Df* that is an invertible matrix. In fact, it is required that for *p* in *U*, there is a neighborhood of *p* in which the Jacobian *Df* stays non-singular. Since the Jacobian is a 2 × 2 real matrix, Df can be read as one of three types of complex number: ordinary complex, split complex number, or dual number. Suppose that in a chart of the surface, $f(x, y) = (u, v)$.

The total differential of *u* is

$$du = \frac{\partial u}{\partial x}dx + \frac{\partial u}{\partial y}dy,$$, and similarly for *v*.

Then the image $(du, dv) = (dx, dy)Df$ is a linear transformation, fixing the origin, and expressible as the action of a complex number of a particular type. When (*dx*, *dy*) is also interpreted as that type of complex number, the action is of complex multiplication in the appropriate complex number plane. As such, there is a type of angle (Euclidean, hyperbolic, or slope) that is preserved in such a multiplication. Due to *Df* being invertible, the type of complex number is uniform over the surface.

Consequently, a surface deformation or diffeomorphism of surfaces has the **conformal property** of preserving (the appropriate type of) angles.

32.5 Diffeomorphism group

Let *M* be a differentiable manifold that is second-countable and Hausdorff. The **diffeomorphism group** of *M* is the group of all C^r diffeomorphisms of *M* to itself, denoted by $\mathrm{Diff}^r(M)$ or, when *r* is understood, Diff(*M*). This is a "large" group, in the sense that – provided *M* is not zero-dimensional – it is not locally compact.

32.5.1 Topology

The diffeomorphism group has two natural topologies: *weak* and *strong* (Hirsch 1997). When the manifold is compact, these two topologies agree. The weak topology is always metrizable. When the manifold is not compact, the strong topology captures the behavior of functions "at infinity" and is not metrizable. It is, however, still Baire.

Fixing a Riemannian metric on *M*, the weak topology is the topology induced by the family of metrics

$$d_K(f, g) = \sup_{x \in K} d(f(x), g(x)) +$$

$$\sum_{1 \le p \le r} \sup_{x \in K} \|D^p f(x) - D^p g(x)\|$$

as *K* varies over compact subsets of *M*. Indeed, since *M* is σ-compact, there is a sequence of compact subsets *Kn* whose union is *M*. Then:

$$d(f, g) = \sum_n 2^{-n} \frac{d_{K_n}(f, g)}{1 + d_{K_n}(f, g)}.$$

The diffeomorphism group equipped with its weak topology is locally homeomorphic to the space of C^r vector fields (Leslie 1967). Over a compact subset of *M*, this follows by fixing a Riemannian metric on *M* and using the exponential map for that metric. If *r* is finite and the manifold is compact, the space of vector fields is a Banach space. Moreover, the transition maps from one chart of this atlas to another are smooth, making the diffeomorphism group into a Banach manifold with smooth right translations; left translations and inversion are only continuous. If $r = \infty$, the space of vector fields is a Fréchet space. Moreover, the transition maps are smooth, making the diffeomorphism group into a Fréchet manifold and even into a regular Fréchet Lie group.

If the manifold is σ-compact and not compact the full diffeomorphism group is not locally contractible for any of the two topologies. One has to restrict the group by controlling the deviation from the identity near infinity to obtain a diffeomorphism group which is a manifold; see (Michor & Mumford 2013).

32.5.2 Lie algebra

The Lie algebra of the diffeomorphism group of *M* consists of all vector fields on *M* equipped with the Lie bracket of vector fields. Somewhat formally, this is seen by making a small change to the coordinate x at each point in space:

$$x^\mu \to x^\mu + \varepsilon h^\mu(x)$$

so the infinitesimal generators are the vector fields

$$L_h = h^\mu(x)\frac{\partial}{\partial x_\mu}.$$

32.5.3 Examples

- When $M = G$ is a Lie group, there is a natural inclusion of G in its own diffeomorphism group via left-translation. Let $\mathrm{Diff}(G)$ denote the diffeomorphism group of G, then there is a splitting $\mathrm{Diff}(G) \simeq G \times \mathrm{Diff}(G, e)$, where $\mathrm{Diff}(G, e)$ is the subgroup of $\mathrm{Diff}(G)$ that fixes the identity element of the group.
- The diffeomorphism group of Euclidean space $\mathbf{R}^n$ consists of two components, consisting of the orientation preserving and orientation reversing diffeomorphisms. In fact, the general linear group is a deformation retract of subgroup $\mathrm{Diff}(\mathbf{R}^n, 0)$ of diffeomorphisms fixing the origin under the map $f(x) \mapsto f(tx)/t$, t ∈& (0,1]. In particular, the general linear group is also a deformation retract of the full diffeomorphism group.
- For a finite set of points, the diffeomorphism group is simply the symmetric group. Similarly, if M is any manifold there is a group extension $0 \to \mathrm{Diff}_0(M) \to \mathrm{Diff}(M) \to \Sigma(\pi_0(M))$. Here $\mathrm{Diff}_0(M)$is the subgroup of $\mathrm{Diff}(M)$ that preserves all the components of M, and $\Sigma(\pi_0(M))$ is the permutation group of the set $\pi_0(M)$ (the components of M). Moreover, the image of the map $\mathrm{Diff}(M) \to \Sigma(\pi_0(M))$ is the bijections of $\pi_0(M)$ that preserve diffeomorphism classes.

32.5.4 Transitivity

For a connected manifold M, the diffeomorphism group acts transitively on M. More generally, the diffeomorphism group acts transitively on the configuration space *CkM*. If M is at least two-dimensional, the diffeomorphism group acts transitively on the configuration space *FkM* and the action on M is multiply transitive (Banyaga 1997, p. 29).

32.5.5 Extensions of diffeomorphisms

In 1926, Tibor Radó asked whether the harmonic extension of any homeomorphism or diffeomorphism of the unit circle to the unit disc yields a diffeomorphism on the open disc. An elegant proof was provided shortly afterwards by Hellmuth Kneser. In 1945, Gustave Choquet, apparently unaware of this result, produced a completely different proof.

The (orientation-preserving) diffeomorphism group of the circle is pathwise connected. This can be seen by noting that any such diffeomorphism can be lifted to a diffeomorphism f of the reals satisfying $[f(x+1) = f(x) + 1]$; this space is convex and hence path-connected. A smooth, eventually constant path to the identity gives a second more elementary way of extending a diffeomorphism from the circle to the open unit disc (a special case of the Alexander trick). Moreover, the diffeomorphism group of the circle has the homotopy-type of the orthogonal group O(2).

The corresponding extension problem for diffeomorphisms of higher-dimensional spheres $\mathbf{S}^{n-1}$ was much studied in the 1950s and 1960s, with notable contributions from René Thom, John Milnor and Stephen Smale. An obstruction to such extensions is given by the finite abelian group Γn, the "group of twisted spheres", defined as the quotient of the abelian component group of the diffeomorphism group by the subgroup of classes extending to diffeomorphisms of the ball B^n.

32.5.6 Connectedness

For manifolds, the diffeomorphism group is usually not connected. Its component group is called the mapping class group. In dimension 2 (i.e. surfaces), the mapping class group is a finitely presented group generated by Dehn twists (Dehn, Lickorish, Hatcher). Max Dehn and Jakob Nielsen showed that it can be identified with the outer automorphism group of the fundamental group of the surface.

William Thurston refined this analysis by classifying elements of the mapping class group into three types: those equivalent to a periodic diffeomorphism; those equivalent to a diffeomorphism leaving a simple closed curve invariant; and those equivalent to pseudo-Anosov diffeomorphisms. In the case of the torus $\mathbf{S}^1 \times \mathbf{S}^1 = \mathbf{R}^2/\mathbf{Z}^2$, the mapping class group is simply the modular group $\mathrm{SL}(2, \mathbf{Z})$ and the classification becomes classical in terms of elliptic, parabolic and hyperbolic matrices. Thurston accomplished his classification by observing that the mapping class group acted naturally on a compactification of Teichmüller space; as this enlarged space was homeomorphic to a closed ball, the Brouwer fixed-point theorem became applicable.

Smale conjectured that if M is an oriented smooth closed manifold, the identity component of the group of orientation-preserving diffeomorphisms is simple. This had first been proved for a product of circles by Michel Herman; it was proved in full generality by Thurston.

32.5.7 Homotopy types

- The diffeomorphism group of $\mathbf{S}^2$ has the homotopy-type of the subgroup O(3). This was proved by Steve Smale.[1]
- The diffeomorphism group of the torus has the homotopy-type of its linear automorphisms: $\mathbf{S}^1 \times \mathbf{S}^1 \times \mathrm{GL}(2, \mathbf{Z})$.
- The diffeomorphism groups of orientable surfaces of genus $g > 1$ have the homotopy-type of their mapping class groups (i.e. the components are contractible).
- The homotopy-type of the diffeomorphism groups of 3-manifolds are fairly well-understood via the work of Ivanov, Hatcher, Gabai and Rubinstein, although there are a few outstanding open cases (primarily 3-manifolds with finite fundamental groups).
- The homotopy-type of diffeomorphism groups of n-manifolds for $n > 3$ are poorly understood. For example, it is an open problem whether or not $\mathrm{Diff}(\mathbf{S}^4)$ has more than two components. Via Milnor, Kahn and Antonelli, however, it is known that provided $n > 6$, $\mathrm{Diff}(\mathbf{S}^n)$ does not have the homotopy-type of a finite CW-complex.

32.6 Homeomorphism and diffeomorphism

Unlike non-diffeomorphic homeomorphisms, it is relatively difficult to find a pair of homeomorphic manifolds that are not diffeomorphic. In dimensions 1, 2, 3, any pair of homeomorphic smooth manifolds are diffeomorphic. In dimension 4 or greater, examples of homeomorphic but not diffeomorphic pairs have been found. The first such example was constructed by John Milnor in dimension 7. He constructed a smooth 7-dimensional manifold (called now Milnor's sphere) that is homeomorphic to the standard 7-sphere but not diffeomorphic to it. There are, in fact, 28 oriented diffeomorphism classes of manifolds homeomorphic to the 7-sphere (each of them is the total space of a fiber bundle over the 4-sphere with the 3-sphere as the fiber).

More unusual phenomena occur for 4-manifolds. In the early 1980s, a combination of results due to Simon Donaldson and Michael Freedman led to the discovery of exotic $\mathbf{R}^4$s: there are uncountably many pairwise non-diffeomorphic open subsets of $\mathbf{R}^4$ each of which is homeomorphic to $\mathbf{R}^4$, and also there are uncountably many pairwise non-diffeomorphic differentiable manifolds homeomorphic to $\mathbf{R}^4$ that do not embed smoothly in $\mathbf{R}^4$.

32.7 See also

- Étale morphism
- Large diffeomorphism
- Local diffeomorphism
- Superdiffeomorphism

32.8 Notes

[1] Smale, "Diffeomorphisms of the 2-sphere", *Proc. Amer. Math. Soc.* 10 (1959), pp. 621–626.

32.9 References

- Chaudhuri, Shyamoli, Hakuru Kawai and S.-H Henry Tye. "Path-integral formulation of closed strings", *Phys. Rev. D*, 36: 1148 (1987).
- Banyaga, Augustin (1997), *The structure of classical diffeomorphism groups*, Mathematics and its Applications, 400, Kluwer Academic, ISBN 0-7923-4475-8
- Duren, Peter L. (2004), *Harmonic Mappings in the Plane*, Cambridge Mathematical Tracts, 156, Cambridge University Press, ISBN 0-521-64121-7
- Hazewinkel, Michiel, ed. (2001), "Diffeomorphism", *Encyclopedia of Mathematics*, Springer, ISBN 978-1-55608-010-4
- Hirsch, Morris (1997), *Differential Topology*, Berlin, New York: Springer-Verlag, ISBN 978-0-387-90148-0
- Kriegl, Andreas; Michor, Peter (1997), *The convenient setting of global analysis*, Mathematical Surveys and Monographs, 53, American Mathematical Society, ISBN 0-8218-0780-3
- Leslie, J. A. (1967), "On a differential structure for the group of diffeomorphisms", *Topology* **6** (2): 263–271, doi:10.1016/0040-9383(67)90038-9, ISSN 0040-9383, MR 0210147
- Michor, Peter W.; Mumford, David (2013), "A zoo of diffeomorphism groups on $\mathbf{R}^n$.", *Annals of Global Analysis and Geometry* **44** (4): 529–540, doi:10.1007/s10455-013-9380-2 (arXiv:1211.5704)
- Milnor, John W. (2007), *Collected Works Vol. III, Differential Topology*, American Mathematical Society, ISBN 0-8218-4230-7

- Omori, Hideki (1997), *Infinite-dimensional Lie groups*, Translations of Mathematical Monographs, 158, American Mathematical Society, ISBN 0-8218-4575-6
- Kneser, Hellmuth (1926), "Lösung der Aufgabe 41.", *Jahresbericht der Deutschen Mathematiker-Vereinigung* (in German) **35** (2): 123

Chapter 33

Poisson bracket

In mathematics and classical mechanics, the **Poisson bracket** is an important binary operation in Hamiltonian mechanics, playing a central role in Hamilton's equations of motion, which govern the time evolution of a Hamiltonian dynamical system. The Poisson bracket also distinguishes a certain class of coordinate transformations, called *canonical transformations*, which map canonical coordinate systems into canonical coordinate systems. A "canonical coordinate system" consists of canonical position and momentum variables (below symbolized by q_i and p_i, respectively) that satisfy canonical Poisson bracket relations. The set of possible canonical transformations is always very rich. For instance, it is often possible to choose the Hamiltonian itself H = H(q, p; t) as one of the new canonical momentum coordinates.

In a more general sense, the Poisson bracket is used to define a Poisson algebra, of which the algebra of functions on a Poisson manifold is a special case. These are all named in honour of Siméon Denis Poisson.

33.1 Properties

For any functions f, g, h of phase space and time:

Anticommutativity $\{f,g\} = -\{g,f\}$

Distributivity $\{f+g,h\} = \{f,h\} + \{g,h\}$

Product rule $\{fg,h\} = \{f,h\}g + f\{g,h\}$

Jacobi identity $\{f,\{g,h\}\}+\{g,\{h,f\}\}+\{h,\{f,g\}\} = 0$

Also, if a function k is time-dependent but constant over phase space, then $\{f,k\}=0$ for any f.

33.2 Definition in canonical coordinates

In canonical coordinates (also known as Darboux coordinates) (q_i, p_i) on the phase space, given two functions $f(p_i, q_i, t)$ and $g(p_i, q_i, t)$,[Note 1] the Poisson bracket takes the form

$$\{f,g\} = \sum_{i=1}^{N} \left(\frac{\partial f}{\partial q_i}\frac{\partial g}{\partial p_i} - \frac{\partial f}{\partial p_i}\frac{\partial g}{\partial q_i} \right).$$

The Poisson brackets of the canonical coordinates are

$$\{q_i, q_j\} = 0$$
$$\{p_i, p_j\} = 0$$
$$\{q_i, p_j\} = \delta_{ij}$$

where δ_{ij} is the Kronecker delta.

33.3 Hamilton's equations of motion

Hamilton's equations of motion have an equivalent expression in terms of the Poisson bracket. This may be most directly demonstrated in an explicit coordinate frame. Suppose that $f(p,q,t)$ is a function on the manifold. Then from the multivariable chain rule, one has

$$\frac{\mathrm{d}}{\mathrm{d}t} f(p,q,t) = \frac{\partial f}{\partial q}\frac{\mathrm{d}q}{\mathrm{d}t} + \frac{\partial f}{\partial p}\frac{\mathrm{d}p}{\mathrm{d}t} + \frac{\partial f}{\partial t}.$$

Further, one may take $p = p(t)$ and $q = q(t)$ to be solutions to Hamilton's equations; that is,

$$\begin{cases} \dot{q} = \frac{\partial H}{\partial p} = \{q,H\} \\ \dot{p} = -\frac{\partial H}{\partial q} = \{p,H\} \end{cases}$$

Then, one has

$$\frac{\mathrm{d}}{\mathrm{d}t}f(p,q,t) = \frac{\partial f}{\partial q}\frac{\partial H}{\partial p} - \frac{\partial f}{\partial p}\frac{\partial H}{\partial q} + \frac{\partial f}{\partial t}$$
$$= \{f, H\} + \frac{\partial f}{\partial t}.$$

Thus, the time evolution of a function f on a symplectic manifold can be given as a one-parameter family of symplectomorphisms (i.e., canonical transformations, area-preserving diffeomorphisms), with the time t being the parameter: Hamiltonian motion is a canonical transformation generated by the Hamiltonian. That is, Poisson brackets are preserved in it, so that *any time t* in the solution to Hamilton's equations, $q(t) = \exp(-t\{H, \bullet\})\, q(0)$, $p(t) = \exp(-t\{H, \bullet\})\, p(0)$, can serve as the bracket coordinates. *Poisson brackets are canonical invariants.*

Dropping the coordinates, one has

$$\frac{\mathrm{d}}{\mathrm{d}t}f = \left(\frac{\partial}{\partial t} - \{H, \cdot\}\right) f.$$

The operator in the convective part of the derivative, $i\hat{L} = -\{H, \bullet\}$, is sometimes referred to as the Liouvillian (see Liouville's theorem (Hamiltonian)).

33.4 Constants of motion

An integrable dynamical system will have constants of motion in addition to the energy. Such constants of motion will commute with the Hamiltonian under the Poisson bracket. Suppose some function *f(p, q)* is a constant of motion. This implies that if *p(t), q(t)* is a trajectory or solution to the Hamilton's equations of motion, then one has that

$$0 = \frac{\mathrm{d}f}{\mathrm{d}t}$$

along that trajectory. Then one has

$$0 = \frac{\mathrm{d}}{\mathrm{d}t}f(p,q) = \{f, H\} + \frac{\partial f}{\partial t}$$

where, as above, the intermediate step follows by applying the equations of motion. This equation is known as the Liouville equation. The content of Liouville's theorem is that the time evolution of a measure (or "distribution function" on the phase space) is given by the above.

If the Poisson bracket of f and g vanishes ($\{f,g\} = 0$), then f and g are said to be **in involution**. In order for a Hamiltonian system to be completely integrable, all of the constants of motion must be in mutual involution.

Furthermore, according to the **Poisson's Theorem**, if two quantities A and B are constants of motion, so is their Poisson bracket $\{A, B\}$. This does not always supply a useful result, however, since the number of possible constants of motion is limited ($2n - 1$ for a system with n degrees of freedom), and so the result may be trivial (a constant, or a function of A and B .)

33.5 The Poisson bracket in coordinate-free language

Let M be symplectic manifold, that is, a manifold equipped with a symplectic form: a 2-form ω which is both **closed** (i.e., its exterior derivative $d\omega = 0$) and **non-degenerate**. For example, in the treatment above, take M to be $\mathbb{R}^{2n}$ and take

$$\omega = \sum_{i=1}^{n} dq_i \wedge dp_i.$$

If $\iota_v\omega$ is the interior product or contraction operation defined by $(\iota_v\omega)(w) = \omega(v, w)$, then non-degeneracy is equivalent to saying that for every one-form α there is a unique vector field Ω_α such that $\iota_{\Omega_\alpha}\omega = \alpha$. Alternatively, $\Omega_{dH} = \omega^{-1}(dH)$. Then if H is a smooth function on M, the Hamiltonian vector field XH can be defined to be Ω_{dH}. It is easy to see that

$$X_{p_i} = \frac{\partial}{\partial q_i}$$
$$X_{q_i} = -\frac{\partial}{\partial p_i}.$$

The **Poisson bracket** $\{\cdot, \cdot\}$ on (M, ω) is a bilinear operation on differentiable functions, defined by $\{f, g\} = \omega(X_f, X_g)$; the Poisson bracket of two functions on M is itself a function on M. The Poisson bracket is antisymmetric because:

$$\{f, g\} = \omega(X_f, X_g) = -\omega(X_g, X_f) = -\{g, f\}$$

Furthermore,

$$\begin{aligned}\{f, g\} &= \omega(X_f, X_g) = \omega(\Omega_{df}, X_g)\\ &= (\iota_{\Omega_{df}}\omega)(X_g) = df(X_g)\\ &= X_g f = \mathcal{L}_{X_g} f\end{aligned}$$

Here Xgf denotes the vector field Xg applied to the function f as a directional derivative, and $\mathcal{L}_{X_g} f$ denotes the (entirely equivalent) Lie derivative of the function f.

If α is an arbitrary one-form on *M*, the vector field $\Omega\alpha$ generates (at least locally) a flow $\phi_x(t)$ satisfying the boundary condition $\phi_x(0) = x$ and the first-order differential equation

$$\frac{d\phi_x}{dt} = \Omega_\alpha|_{\phi_x(t)} \,.$$

The $\phi_x(t)$ will be symplectomorphisms (canonical transformations) for every *t* as a function of *x* if and only if $\mathcal{L}_{\Omega_\alpha}\omega = 0$; when this is true, $\Omega\alpha$ is called a symplectic vector field. Recalling Cartan's identity $\mathcal{L}_X\omega = d(\iota_X\omega) + \iota_X d\omega$ and $d\omega = 0$, it follows that $\mathcal{L}_{\Omega_\alpha}\omega = d(\iota_{\Omega_\alpha}\omega) = d\alpha$. Therefore $\Omega\alpha$ is a symplectic vector field if and only if α is a closed form. Since $d(df) = d^2 f = 0$, it follows that every Hamiltonian vector field *Xf* is a symplectic vector field, and that the Hamiltonian flow consists of canonical transformations. From **(1)** above, under the Hamiltonian flow *XH*,

$$\frac{d}{dt}f(\phi_x(t)) = X_H f = \{f, H\}.$$

This is a fundamental result in Hamiltonian mechanics, governing the time evolution of functions defined on phase space. As noted above, when *{f,H} = 0, f* is a constant of motion of the system. In addition, in canonical coordinates (with $\{p_i, p_j\} = \{q_i, q_j\} = 0$ and $\{q_i, p_j\} = \delta_{ij}$), Hamilton's equations for the time evolution of the system follow immediately from this formula.

It also follows from **(1)** that the Poisson bracket is a derivation; that is, it satisfies a non-commutative version of Leibniz's product rule:

The Poisson bracket is intimately connected to the Lie bracket of the Hamiltonian vector fields. Because the Lie derivative is a derivation,

$$\mathcal{L}_v\iota_w\omega = \iota_{\mathcal{L}_v w}\omega + \iota_w\mathcal{L}_v\omega = \iota_{[v,w]}\omega + \iota_w\mathcal{L}_v\omega$$

Thus if *v* and *w* are symplectic, using $\mathcal{L}_v\omega = 0$, Cartan's identity, and the fact that $\iota_w\omega$ is a closed form,

$$\iota_{[v,w]}\omega = \mathcal{L}_v\iota_w\omega = d(\iota_v\iota_w\omega) + \iota_v d(\iota_w\omega) = d(\iota_v\iota_w\omega) = d(\ \omega(w,v)).$$

It follows that $[v,w] = X_{\omega(w,v)}$, so that

Thus, the Poisson bracket on functions corresponds to the Lie bracket of the associated Hamiltonian vector fields. We have also shown that the Lie bracket of two symplectic vector fields is a Hamiltonian vector field and hence is also symplectic. In the language of abstract algebra, the symplectic vector fields form a subalgebra of the Lie algebra of smooth vector fields on *M*, and the Hamiltonian vector fields form an ideal of this subalgebra. The symplectic vector fields are the Lie algebra of the (infinite-dimensional) Lie group of symplectomorphisms of *M*.

It is widely asserted that the Jacobi identity for the Poisson bracket,

$$\{f, \{g,h\}\} + \{g, \{h,f\}\} + \{h, \{f,g\}\} = 0$$

follows from the corresponding identity for the Lie bracket of vector fields, but this is true only up to a locally constant function. However, to prove the Jacobi identity for the Poisson bracket, it is sufficient to show that:

$$\mathrm{ad}_{\{f,g\}} = [\mathrm{ad}_f, \mathrm{ad}_g]$$

where the operator ad_g on smooth functions on *M* is defined by $\mathrm{ad}_g(\cdot) = \{\cdot, g\}$ and the bracket on the right-hand side is the commutator of operators, $[\mathrm{A}, \mathrm{B}] = \mathrm{AB} - \mathrm{BA}$. By **(1)**, the operator ad_g is equal to the operator *Xg*. The proof of the Jacobi identity follows from **(3)** because the Lie bracket of vector fields is just their commutator as differential operators.

The algebra of smooth functions on M, together with the Poisson bracket forms a Poisson algebra, because it is a Lie algebra under the Poisson bracket, which additionally satisfies Leibniz's rule **(2)**. We have shown that every symplectic manifold is a Poisson manifold, that is a manifold with a "curly-bracket" operator on smooth functions such that the smooth functions form a Poisson algebra. However, not every Poisson manifold arises in this way, because Poisson manifolds allow for degeneracy which cannot arise in the symplectic case.

33.6 A result on conjugate momenta

Given a smooth vector field *X* on the configuration space, let *PX* be its conjugate momentum. The conjugate momentum mapping is a Lie algebra anti-homomorphism from the Poisson bracket to the Lie bracket:

$$\{P_X, P_Y\} = -P_{[X,Y]}.$$

This important result is worth a short proof. Write a vector field X at point q in the configuration space as

$$X_q = \sum_i X^i(q) \frac{\partial}{\partial q^i}$$

where the $\frac{\partial}{\partial q^i}$ is the local coordinate frame. The conjugate momentum to X has the expression

$$P_X(q,p) = \sum_i X^i(q)\, p_i$$

where the *pi* are the momentum functions conjugate to the coordinates. One then has, for a point *(q,p)* in the phase space,

$$\begin{aligned}\{P_X, P_Y\}(q,p) &= \sum_i \sum_j \{X^i(q)\, p_i, Y^j(q)\, p_j\} \\ &= \sum_{ij} p_i Y^j(q) \frac{\partial X^i}{\partial q^j} - p_j X^i(q) \frac{\partial Y^j}{\partial q^i} \\ &= -\sum_i p_i\, [X,Y]^i(q) \\ &= -P_{[X,Y]}(q,p).\end{aligned}$$

The above holds for all *(q, p)*, giving the desired result.

33.7 Quantization

Poisson brackets deform to Moyal brackets upon quantization, that is, they generalize to a different Lie algebra, the Moyal algebra, or, equivalently in Hilbert space, quantum commutators. The Wigner-İnönü group contraction of these (the classical limit, $\hbar \to 0$) yields the above Lie algebra.

33.8 See also

- Poisson algebra
- Phase space
- Lagrange bracket
- Moyal bracket
- Peierls bracket
- Poisson superalgebra
- Poisson superbracket
- Dirac bracket
- Commutator

33.9 References

- Arnold, V. I. (1989). *Mathematical Methods of Classical Mechanics* (2nd ed.). New York: Springer. ISBN 978-0-387-96890-2.
- Landau, L. D.; Lifshitz, E. M. (1982). *Mechanics*. Course of Theoretical Physics. Vol. 1 (3rd ed.). Butterworth-Heinemann. ISBN 978-0-7506-2896-9.
- Karasëv, M. V.; Maslov, V. P.: Nonlinear Poisson brackets. Geometry and quantization. Translated from the Russian by A. Sossinsky [A. B. Sosinskiĭ] and M. Shishkova. Translations of Mathematical Monographs, 119. American Mathematical Society, Providence, RI, 1993.

33.10 External links

- Hazewinkel, Michiel, ed. (2001), "Poisson brackets", *Encyclopedia of Mathematics*, Springer, ISBN 978-1-55608-010-4
- Eric W. Weisstein, "Poisson bracket", *MathWorld*.

33.11 Notes

[1] $f(p_i, q_i, t)$ means f is a function of the 2N + 1 independent variables: momentum, $p_{1...N}$; position, $q_{1...N}$; and time, t

Chapter 34

Wilson loop

"Wilson line" redirects here. For the Wilson Line shipping company, see Thomas Wilson Sons & Co.

In gauge theory, a **Wilson loop** (named after Kenneth G. Wilson) is a gauge-invariant observable obtained from the holonomy of the gauge connection around a given loop. In the classical theory, the collection of all Wilson loops contains sufficient information to reconstruct the gauge connection, up to gauge transformation.[1]

In quantum field theory, the definition of Wilson loop observables as *bona fide* operators on Fock spaces is a mathematically delicate problem and requires regularization, usually by equipping each loop with a *framing*. The action of Wilson loop operators has the interpretation of creating an elementary excitation of the quantum field which is localized on the loop. In this way, Faraday's "flux tubes" become elementary excitations of the quantum electromagnetic field.

Wilson loops were introduced in the 1970s in an attempt at a nonperturbative formulation of quantum chromodynamics (QCD), or at least as a convenient collection of variables for dealing with the strongly interacting regime of QCD.[2] The problem of confinement, which Wilson loops were designed to solve, remains unsolved to this day.

The fact that strongly coupled quantum gauge field theories have elementary nonperturbative excitations which are loops motivated Alexander Polyakov to formulate the first string theories, which described the propagation of an elementary quantum loop in spacetime.

Wilson loops played an important role in the formulation of loop quantum gravity, but there they are superseded by spin networks (and, later, spinfoams), a certain generalization of Wilson loops.

In particle physics and string theory, Wilson loops are often called **Wilson lines**, especially Wilson loops around non-contractible loops of a compact manifold.

34.1 An equation

The **Wilson loop** variable is a quantity defined by the trace of a path-ordered exponential of a gauge field A_μ transported along a closed line C:

$$W_C := \mathrm{Tr}\,(\,\mathcal{P}\exp i \oint_C A_\mu dx^\mu\,)\,.$$

Here, C is a closed curve in space, $\mathcal{P}$ is the path-ordering operator. Under a gauge transformation

$$\mathcal{P}e^{i\oint_C A_\mu dx^\mu} \to g(x)\mathcal{P}e^{i\oint_C A_\mu dx^\mu}g^{-1}(x)$$

where x corresponds to the initial (and end) point of the loop (only initial and end point of a line contribute, whereas gauge transformations in between cancel each other). For SU(2) gauges, for example, one has $g^{\pm 1}(x) \equiv \exp\{\pm i\alpha^j(x)\frac{\sigma^j}{2}\}$; $\alpha^j(x)$ is an arbitrary real function of x , and σ^j are the three Pauli matrices; as usual, a sum over repeated indices is implied.

The invariance of the trace under cyclic permutations guarantees that W_C is invariant under gauge transformations. Note that the quantity being traced over is an element of the gauge Lie group and the trace is really the character of this element with respect to one of the infinitely many irreducible representations, which implies that the operators $A_\mu\,dx^\mu$ don't need to be restricted to the "trace class" (thus with purely discrete spectrum), but can be generally hermitian (or mathematically: self-adjoint) as usual. Precisely because we're finally looking at the trace, it doesn't matter which point on the loop is chosen as the initial point. They all give the same value.

Actually, if A is viewed as a connection over a principal G-bundle, the equation above really ought to be "read" as the parallel transport of the identity around the loop which would give an element of the Lie group G.

Note that a path-ordered exponential is a convenient shorthand notation common in physics which conceals a fair number of mathematical operations. A mathematician would refer to the path-ordered exponential of the connection as "the holonomy of the connection" and characterize it by the parallel-transport differential equation that it satisfies.

At T=0, where T corresponds to temperature, the Wilson loop variable characterizes the confinement or deconfinement of a gauge-invariant quantum-field theory, namely according to whether the variable increases with the *area*, or alternatively with the *circumference* of the loop ("area law", or alternatively "circumferential law" also known as "perimeter law").

In finite-temperature QCD, the thermal expectation value of the Wilson line distinguishes between the confined "hadronic" phase, and the deconfined state of the field, e.g., the quark–gluon plasma.

34.2 See also

- Stochastic vacuum model
- Winding number

34.3 References

[1] Giles, R. (1981). "Reconstruction of Gauge Potentials from Wilson loops". *Physical Review D* **24** (8): 2160. Bibcode:1981PhRvD..24.2160G. doi:10.1103/PhysRevD.24.2160.

[2] Wilson, K. (1974). "Confinement of quarks". *Physical Review D* **10** (8): 2445. Bibcode:1974PhRvD..10.2445W. doi:10.1103/PhysRevD.10.2445.

Chapter 35

Knot invariant

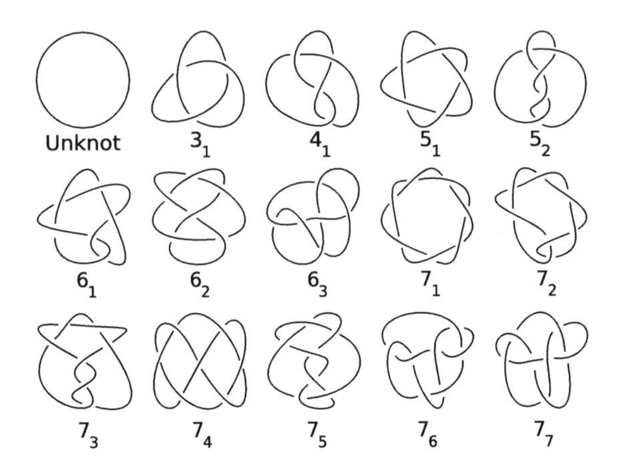

Prime knots are organized by the crossing number invariant.

In the mathematical field of knot theory, a **knot invariant** is a quantity (in a broad sense) defined for each knot which is the same for equivalent knots. The equivalence is often given by ambient isotopy but can be given by homeomorphism. Some invariants are indeed numbers, but invariants can range from the simple, such as a yes/no answer, to those as complex as a homology theory . Research on invariants is not only motivated by the basic problem of distinguishing one knot from another but also to understand fundamental properties of knots and their relations to other branches of mathematics.

From the modern perspective, it is natural to define a knot invariant from a knot diagram. Of course, it must be unchanged (that is to say, invariant) under the Reidemeister moves. Tricolorability is a particularly simple example. Other examples are knot polynomials, such as the Jones polynomial, which are currently among the most useful invariants for distinguishing knots from one another, though currently it is not known whether there exists a knot polynomial which distinguishes all knots from each other, or even which distinguishes just the unknot from all other knots.

Other invariants can be defined by considering some integer-valued function of knot diagrams and taking its minimum value over all possible diagrams of a given knot. This category includes the crossing number, which is the minimum number of crossings for any diagram of the knot, and the bridge number, which is the minimum number of bridges for any diagram of the knot.

Historically, many of the early knot invariants are not defined by first selecting a diagram but defined intrinsically, which can make computing some of these invariants a challenge. For example, knot genus is particularly tricky to compute, but can be effective (for instance, in distinguishing mutants).

The complement of a knot itself (as a topological space) is known to be a "complete invariant" of the knot by the Gordon–Luecke theorem in the sense that it distinguishes the given knot from all other knots up to ambient isotopy and mirror image. Some invariants associated with the knot complement include the knot group which is just the fundamental group of the complement. The knot quandle is also a complete invariant in this sense but it is difficult to determine if two quandles are isomorphic.

By Mostow–Prasad rigidity, the hyperbolic structure on the complement of a hyperbolic link is unique, which means the hyperbolic volume is an invariant for these knots and links. Volume, and other hyperbolic invariants, have proven very effective, utilized in some of the extensive efforts at knot tabulation.

In recent years, there has been much interest in homological invariants of knots which categorify well-known invariants. Heegaard Floer homology is a homology theory whose Euler characteristic is the Alexander polynomial of the knot. It has been proven effective in deducing new results about the classical invariants. Along a different line of study, there is a combinatorially defined cohomology theory of knots called Khovanov homology whose Euler characteristic is the Jones polynomial. This has recently been shown to be useful in obtaining bounds on slice genus whose earlier proofs required gauge theory. Khovanov and Rozansky have since defined several other related cohomology theories whose Euler characteristics recover other classical invariants. Catharina Stroppel gave a representation theoretic interpretation of Khovanov homology by categorifying quantum group invariants.

There is also growing interest from both knot theorists and scientists in understanding "physical" or geometric properties of knots and relating it to topological invariants and knot type. An old result in this direction is the Fary–Milnor theorem states that if the total curvature of a knot K in $\mathbb{R}^3$ satisfies

$$\oint_K \kappa \, ds \leq 4\pi,$$

where $\kappa(p)$ is the curvature at p, then K is an unknot. Therefore, for knotted curves,

$$\oint_K \kappa \, ds > 4\pi.$$

An example of a "physical" invariant is ropelength, which is the amount of 1-inch diameter rope needed to realize a particular knot type.

35.1 Other invariants

- Linking number
- Finite type invariant (or Vassiliev or Vassiliev–Goussarov invariant)
- Stick number

35.2 Further reading

- Rolfsen, Dale (2003). *Knots and Links*. Providence, RI: AMS. ISBN 0-8218-3436-3.
- Adams, Colin Conrad (2004). *The Knot Book: an Elementary Introduction to the Mathematical Theory of Knots* (Repr., with corr ed.). Providence, RI: AMS. ISBN 0-8218-3678-1.
- Burde, Gerhard; Zieschang, Heiner (2002). *Knots* (2nd rev. and extended ed.). New York: De Gruyter. ISBN 3-11-017005-1.

35.3 External links

- J. C. Cha and C. Livingston. "KnotInfo: Table of Knot Invariants", *Indiana.edu*. Accessed: 09:10, 18 April 2013 (UTC)
- "Invariants", *The Knot Atlas*.

35.4 Text and image sources, contributors, and licenses

35.4.1 Text

- **Loop quantum gravity** *Source:* https://en.wikipedia.org/wiki/Loop_quantum_gravity?oldid=721971111 *Contributors:* Bryan Derksen, The Anome, AstroNomer, RK, Toby Bartels, Miguel~enwiki, Schewek, Ewen, Michael Hardy, TakuyaMurata, Islandboy99, GTBacchus, Mcarling, Looxix~enwiki, Ahoerstemeier, Cyp, Kimiko, Palfrey, Jordi Burguet Castell, Mxn, Charles Matthews, Sanxiyn, Maximus Rex, Phys, Omegatron, Finlay McWalter, Dmytro, Sdedeo, Astronautics~enwiki, Peak, Chris Roy, Mirv, Sverdrup, Kn1kda, Hadal, Jheise, Clementi, Connelly, Giftlite, Sj, Fastfission, Herbee, Anville, Dratman, Curps, JeffBobFrank, Jason Quinn, Gzornenplatz, C17GMaster, DÅ,ugosz, PhiloVivero, DefLog~enwiki, Gadfium, HorsePunchKid, Sam Hocevar, Lumidek, Tdent, Joyous!, M1ss1ontomars2k4, Eep^2, Poccil, Rich Farmbrough, Avriette, Pjacobi, Vsmith, MuDavid, Pavel Vozenilek, Bender235, ESkog, Clement Cherlin, Peter M Gerdes, Drhex, John Vandenberg, C S, Cmdrjameson, GTubio, Tweet Tweet, Slicky, Ral315, Lysdexia, Arthena, Xaphan9966, Wtmitchell, Greg Kuperberg, Count Iblis, Egg, Lee-Anne, Kazvorpal, Killing Vector, Linas, Merlinme, HFarmer, Sympleko, Hfarmer, Mpatel, GregorB, J M Rice, Ae7flux, Tjbk tjb, Alienus, BD2412, Fleisher, Sjö, Rjwilmsi, Nightscream, Zbxgscqf, Bubba73, FlaBot, John Baez, Don Gosiewski, Smithbrenon, Chobot, Spasemunki, Bgwhite, Roboto de Ajvol, YurikBot, Wavelength, RobotE, Rt66lt, Hillman, DanMS, Chaos, Salsb, Welsh, Schmock, Crasshopper, Beanyk, Akashmitra, Bota47, JonathanD, Endomion, Modify, Petri Krohn, Ilmari Karonen, Caco de vidro, Benandorsqueaks, SmackBot, Bayardo, FlashSheridan, Unyoyega, Vald, JMiall, Chris the speller, IvanAndreevich, DHN-bot~enwiki, Colonies Chris, Chlewbot, Pepsidrinka, Chrylis, MegaHasher, TriTertButoxy, Lambiam, Vincenzo.romano, Loadmaster, Konklone, K, G-W, Kurtan~enwiki, Harold f, Will314159, Friendly Neighbour, Vyznev Xnebara, Ian Beynon, Myasuda, Gmusser, Rjm656s, Fournax, Headbomb, Marek69, Nick Number, MichaelMaggs, Edokter, Byrgenwulf, Knotwork, Arch dude, Igodard, Yill577, WolfmanSF, Tonyfaull, Skylights76, Rickard Vogelberg, Gwern, AltiusBimm, Melamed katz, Vanished user 47736712, WJBscribe, Izno, KittyHawker, Sheliak, Maxzimet, AlnoktaBOT, Nxavar, Jackfork, Carlorovelli, Anotherak, SieBot, Keskival, AS, Robdunst, Hugh16, Senderista~enwiki, Bnsreenath, Caidh, Oxymoron83, Dcattell, Swiebodzice, Sk8hack, Danthewhale, Martarius, Sfan00 IMG, Shaded0, Djr32, CohesionBot, JavierReynaldo, Arjayay, SchreiberBike, Pqnelson, Mjaniec, DumZiBoT, Ianbay, Neuralwarp, XLinkBot, Fastily, Tenner47, Arthur chos, Avoided, Tenderbuttons, Benplusnumber, Balungifrancis, Addbot, DOI bot, 15lsoucy, Tarosic, Debresser, Favonian, SamatBot, Yobot, Ibayn, 4th-otaku, AnomieBOT, VanishedUser sdu9aya9fasdsopa, Archon 2488, Francois33, Citation bot, Xqbot, Imushfiq, MIRROR, Pra1998, Dumontierc, Omnipaedista, Franco3450, Rr2000, FrescoBot, Paine Ellsworth, Nunc aut numquam, Martlet1215, Citation bot 1, Jonesey95, Tom.Reding, Schiefesfragezeichen, ROMVLVS, Casimir9999, RobinK, Meier99, Orenburg1, Trappist the monk, Dinamik-bot, Bj norge, ElPeste, Afteread, EmausBot, Detogain, John of Reading, Racerx11, GoingBatty, XinaNicole, Ensabah6, Uploadvirus, ZéroBot, Arbnos, Zueignung, WaterCrane, Crown Prince, LaurentRDC, Isocliff, Vodkacannon, Raidr, Helpful Pixie Bot, Titodutta, Bibcode Bot, BG19bot, Spaligo, KateWishing, PhnomPencil, Sylvain.maurin, Kecchina, Halfb1t, JMtB03, Brad7777, Fylbecatulous, Jimw338, MyTuppence, Mogism, LTWoods, Andyhowlett, Jawa0, &reasNink, SomeFreakOnTheInternet, Tentinator, EvergreenFir, DimReg, Pedarkwa, Db9199 24, Anrnusna, Notspelly, Ntomlin1996, Monkbot, Isbromberg, Dsprc, Dakroth, YeOldeGentleman, Tetra quark, Srednuas Lenoroc, Tomuel99, Wulframm, Chemistry1111 and Anonymous: 345

- **History of loop quantum gravity** *Source:* https://en.wikipedia.org/wiki/History_of_loop_quantum_gravity?oldid=599200655 *Contributors:* Miguel~enwiki, Edward, Charles Matthews, Northgrove, Gzornenplatz, Sietse Snel, Hfarmer, Ems57fcva, Journalist, SmackBot, Bayardo, Slashme, Vald, Commander Keane bot, Colonies Chris, Myasuda, Edokter, SchreiberBike, Mjamja, Unara, Skmnet, Rcsprinter123 and Anonymous: 3

- **Loop quantum cosmology** *Source:* https://en.wikipedia.org/wiki/Loop_quantum_cosmology?oldid=701463989 *Contributors:* The Anome, Rursus, Cgmusselman, CWitte, WAS 4.250, Xaxafrad, SmackBot, Bayardo, Bluebot, Kingdon, JorisvS, Gkhanna, Robertinventor, Headbomb, MarshBot, Widefox, Buddachile, Addbot, Jncraton, OlEnglish, Yobot, Ibayn, Dreamer08, Jo3sampl, Citation bot, GrouchoBot, Jonesey95, Uploadvirus, Germanviscuso, Bibcode Bot, PhnomPencil, Kecchina, Qtom.masters, BattyBot, ChrisGualtieri, Monkbot and Anonymous: 13

- **Gravity** *Source:* https://en.wikipedia.org/wiki/Gravity?oldid=722789394 *Contributors:* Bryan Derksen, Danny, Rmhermen, Caltrop, Heron, Montrealais, Stevertigo, Patrick, D, Michael Hardy, Ixfd64, TakuyaMurata, GTBacchus, Wintran, Snarfies, Ahoerstemeier, Mac, Dgaubin, William M. Connolley, Jeff Relf, TonyClarke, Smack, Ec5618, Tpbradbury, Phoebe, Fairandbalanced, BenRG, Pollinator, Lumos3, Jni, Northgrove, Donarreiskoffer, Robbot, Sander123, Jakohn, TimothyPilgrim, Academic Challenger, Caknuck, Bkell, Nerval, Aetheling, Ruakh, Dina, Alan Liefting, Cedars, Ancheta Wis, Giftlite, Mikez, Haeleth, Wolfkeeper, Inkling, Fropuff, Everyking, Frencheigh, Aechols, Bobblewik, Utcursch, Keith Edkins, Antandrus, Phe, Quarl, Elembis, Kiteinthewind, Jossi, Karol Langner, Lindberg G Williams Jr, Demiurge, Trevor MacInnis, Grstain, Mike Rosoft, &Delta, DanielCD, Shipmaster, JimJast, Discospinster, Zaheen, Supercoop, Vsmith, Smyth, Notinasnaid, Dbachmann, Bender235, Rubicon, Loren36, El C, Huntster, Kwamikagami, Aude, Diomidis Spinellis, RoyBoy, Nrbelex, Gershwinrb, Bobo192, Smalljim, Elipongo, I9Q79oL78KiL0QTFHgyc, Larryv, MPerel, Danski14, Alansohn, JYolkowski, Anthony Appleyard, Davetcoleman, Atlant, Maya Levy, Paleorthid, Andrew Gray, Cjthellama, Riana, Goldom, Mac Davis, Wdfarmer, Hdeasy, Snowolf, Mononoke~enwiki, BRW, Yuckfoo, RainbowOfLight, Jesvane, Bookandcoffee, Kazvorpal, Falcorian, Stephen, Feezo, Richard Arthur Norton (1958-), OwenX, Linas, Camw, LOL, Benhocking, JFG, MONGO, Mpatel, Tabletop, GregorB, Eaolson, Isnow, Scm83x, SDC, Philbarker, TheAlphaWolf, Brownsteve, Radiant!, Dysepsion, Sin-man, Ashmoo, Chrispasquale, Chun-hian, Drbogdan, Rjwilmsi, Tangotango, Nichiran, Fel64, Kazrak, Ligulem, Ems57fcva, LjL, Boccobrock, Bhadani, MarnetteD, Matt Deres, Sango123, Syced, Yamamoto Ichiro, Fish and karate, JanSuchy, Magmafox, Titoxd, RobertG, Old Moonraker, Nihiltres, Nivix, RexNL, Gurch, Fresheneesz, Alphachimp, Tardis, Srleffler, King of Hearts, DVdm, Guliolopez, Hall Monitor, Bomb319, Gwernol, EamonnPKeane, Raelx, The Rambling Man, Wavelength, Drdisque, Sceptre, Hairy Dude, Jimp, Hillman, Brandmeister (old), Ohwilleke, Red Slash, Musicpvm, Anonymous editor, Loom91, Bhny, Pigman, Epolk, Philip Hazelden, DanMS, Scott5834, CambridgeBayWeather, Wimt, Ccmccm, NawlinWiki, Nowa, Wiki alf, Ytcracker, SigPig, SCZenz, Apokryltaros, Nick, Brandon, Katrina Graziano, Semperf, Aaron Schulz, Roy Brumback, Addps4cat, Jessemerriman, Rayc, Mgnbar, Dna-webmaster, Eurosong, Heeroyuy135, Enormousdude, 2over0, SFGiants, Chase me ladies, I'm the Cavalry, Lappado, Endomion, E Wing, Fmyhr, Smurrayinchester, Kevin, Geoffrey.landis, Anclation~enwiki, Emc2, Willtron, Kungfuadam, RG2, JDspeeder1, Bo Jacoby, Draicone, FyzixFighter, Mejor Los Indios, Sbyrnes321, DVD R W, Hide&Reason, That Guy, From That Show!, Shenhemu, Luk, TravisTX, Sardanaphalus, SmackBot, Ulterior19802005, Incnis Mrsi, Reedy, KnowledgeOfSelf, Hydrogen Iodide, NaiPiak, David Shear, Kilo-Lima, Jagged 85, Thunderboltz, Delldot, Hardyplants, Cessator, Syckls, BiT, Timotheus Canens, GraemeMcRae, HalfShadow, Typhoonchaser, Yamaguchi??, Gilliam, Algont, Hmains, Ppntori, ERcheck, Andy M. Wang, Kmarinas86, Marc Kupper, The monkeyhate, Saros136, Bluebot, Cush, Keegan, Raymond arritt, Fplay, Silly rabbit, Lehkost, Complexica,

Bbq332,Jeff5102,Sbharris,Hallenrm,CharonM72,Scwlong,Can't sleep,clown will eat me,Scott3,UNHchabo,MJCdetroit,Apostolos Margaritis,Lesnail,Cryocide,Rsm99833,Amazins490,Jmlk17,Cybercobra,Nakon,Steve Pucci,TedE,Redl~enwiki,Jiddisch~enwiki,Dreadstar, Dave-ros,Weregerbil,Cockneyite,Crd721,Bryanmcdonald,Jklin,DMacks,Wizardman,Where,LeoNomis,Risker,Sadi Carnot,Carlosp420,TTE, Yevgeny Kats,Will Beback,Jonpalmer,Nathanael Bar-Aur L.,Bcasterline,Geoffrey Wickham,Rklawton,Djeneba,Sophia,Dsantesteban,Kuru, Thefro552,Titus III,Richard L.Peterson,John,Scientizzle,Stephane Yelle,DocRocksl,Jaganath,Thegathering,Skoobieschnax,JorisvS, LestatdeLioncourt,Coredesat,Accurizer,Minna Sora no Shita,Mgiganteusl,A5y,Nonsuch,Ridersbydelta,Mr. Lefty,AtD,Jess Mars,Ben Moore,JHunterJ,MarkSutton,Slakr,Special-T,Momolee,LuYiSi,Mr Stephen,Samaster1991,Spiel496,Buttle,Novangelis,PSUMark2006, Inquisitus,Dl2000,ShakingSpirit,Hgrobe,Ginkgo100,Vanished user,JMK,Craigboy,Lakers,Newone,MOBle,J Di,StephenBuxton,MattBernius ,Igoldste,Taucetiman,Tofoo,Tawkerbot2,Lsskys,George100,Kurtan~enwiki,Lahiru k,CalebNoble,SkyWalker,JForget,CmdrObot,Tanthalas39 ,Gholson,Porterjoh,Ale jrb,Scohoust,Aherunar,Galo1969X,Picaroon,Shakespeare87,User92361,Zureks,Basawala,Ruslik0,GHe,Dgw, OMGsplosion,ShelfSkewed,WHATaintNOcountryIeverHEARDofDOtheySPEAKenglishINwhat,MarsRover,Hi There,Groosh,Myasuda, Anthony Bradbury,Gregbard,Logicus,Cydebot,Steel,Travelbird,Red Director,Jon Stockton,A Softer Answer,Adolphus79,Nice-sai,Rracecarr, Codingmasters,Ch0rx,Tawkerbot4,Alexnye,Christian75,M a s,DumbBOT,DarkLink,Interwiki gl,FastLizard4,Optimiston the run,Jimip, Waxigloo,SteveMcCluskey,Omicronpersei8,Stoked,Gimmetrow,Sevenaces,Raoul NK,FrancoGG,Thijs!bot,Epbr123,DarlingFriend, Opabinia regalis,Pajz,Ishdarian,Jamesluster,Andyjsmith,24fan24,Gamer007,ClosedEyesSeeing,Headbomb,John254,Bob-blehead,Neil916, Pogogunner,Grayshi,EdJohnston,HistoryMaster1,Zachary,The Hybrid,Nick Number,Lithpiperpilot,CarbonX,Michael-Maggs,Sam42,J.S.B. Anderson,Escarbot,Eleuther,Mentifisto,Hmrox,AntiVandalBot,Yonatan,Luna Santin,CodeWeasel,Themaxeditor,Prolog, Yay unto the Chicken,Dylan Lake,Gdo01,Myanw,Ioeth,JAnDbot,Leuko,Husond,Vorpal blade,ThomasO1989,Roman à clef,MER-C, Nevadacall,Andonic,Hut8.5,100110100,N shaji,Pablothegreat85,Magioladitis,Foobird107,Murgh,Bennybp,Bongwarrior,VoABot II, AuburnPilot,Xn4,Wikidudeman,Hendrixjoseph,Careless hx,Aerographer1981,Crazytonyi,9holdss,ThomasThePolishMan,Bubba hotep,Mi6 agent00g off, BatteryIncluded, Beetfarm Louie, Adrian J.Hunter, Alexei Kojenov, Kane1047, LorenzoB,Mollwollfumble, Scot.parker ,Andykass,Talon Artaine,Chris G,DerHexer,MeEricYay,WLU,Seph Vellius,TheRanger,Patstuart,Seba5618,Oroso,NatureA16,FisherQueen,Hdt83,E.vondarkmoor,MartinBot,Sinfear,Shimwell,Flamingpanda,The Ubik,UnfriendlyFire,APT,Rettetast,Juansidious,Anax-ial, Sm8900,David J Wilson,Mschel,R'n'B,GarrisonGreen,AlexiusHoratius,Pekaje,LittleOldMe old,PrestonH,J.delanoy,Filll,Trusilver, Tonmoy Chowdhury,Bloomingiris,72Dino,MoogleEXE,Lhynard,Ginsengbomb,WarthogDemon,Willow123~enwiki,SubwayEater,Yeti Hunter,Hisagi,James Mead,M C Y1008,Wandering Ghost,Redmotherfive,Rod57,Vertigo900,Mr Rookles,Samtheboy,Gurchzilla,Supuh-star, Pyrospirit,AntiSpamBot,GhostPirate,Belovedfreak,Raichu Trainer,Ohms law,Mitchell is hollywood,SJP,Policron,Touch Of Light, Pwnasaurusrex38,MKoltnow,Mufka,FJPB,Blckavnger,Cmichael,Mohrflies,Stoned Proffesor,Kenneth M Burke,Cosmictinker,RB972,Tiggerjay,U.S.A.U.S.A.U.S.A.,Treisijs,Mike V,Redrocket,Gtg204y,MtyQuinn,Darkfrog24,Jxzj,Annax3,Smartman10,Ronbo76,Micmic28, Yecril,Missphysics,GoldenGolem,Xiahou,Lorax835,Steel1943,Washboard6,Sheliak,Funandtrvl,Gravityc,Tecsup,Black Kite,Chinneeb,Deor, VolkovBot, TreasuryTag, TJ Elliot Scott, Meaningful Username, Danwills, DSRH,Mtesm, Indubitably, Gseletko, Cullaloe, Veddan, Boaex,Dominics Fire,Heerojyuy,Philip Trueman,Childhoodsend,TXiKiBoT,Oshwah,GVIlleneuve27,Davecas0,Adamwang,BuickCenturyDriver,99DBSIMLR,MeStevo,Lolerballer~enwiki,Brocql8,Andrius.v,Z.E.R.O.,Anonymous Dissident,Jonnymagic,Ask123,Trentc, MattCarterSurrealist,IPSOS,Sarthella,Seraphim,Fizzackerly,Wvogeler,The One Cause,Tallcreek,Markp93,Drappel,PDFbot,Freak104, Manticore55,Cremepuff222,Blackdragon1002,Wiae,Vrsixfire,Liberal Classic,Witchzilla,Noel rebeira,Inductiveload,Gladiator2155,Sydweighz,Spiral5800,Larklight,RobertFritzius,Dirkbb,SQL,SallyBoseman,Frag1983,Synthebot,ChillDeity,Speria,Heroesrule17,Enviroboy, Hollop09,Sylent,Kaseman519,Ballsucker22,Sesshomaru,Brianga,Skins88,Chickyfuzz14,AlleborgoBot,GavinTing,Shaidar cuebiyar,Happyhacker101,FlyingLeopard2014,Steven Weston,D.Recorder,Kastrel,Al.Glitch,Xarr,The Random Editor,Mr. dick008,SieBot,Tosun, Crunchedfor6,High2lowo,K.Annoyomous,Thong123456789,Work permit,Scarian,Invmog,Gerakibot,Mazza uk04,Mickeyd24,LealandA, Caltas,BreakfastTom,Beethoven314,Pieman123456789,RJaguar3,Triwbe,Rangutan,Chs³,Tubular bells83,The way,the truth,and the light, Garrett gagne,Keilana,Elliott Fontain,CaptainIraq,RadicalOne,Flyer22Reborn,Tiptoety,CombatCraig,Bhahn0125,Oda Mari,Sunayanaa, Cnormansen,Thin joe,M keshe,Jirachipokemon,Kcin213,Chives4life,Luskj,Oxymoron83,Faradayplank,Smilesfozwood,AngelOfSadness, Nuttycoconut,Edwardwittenstein,Zharradan.angelfire,John fromer,Lightmouse,AMCKen,Iain99,Techman224,Skateboardsfly,Mr mimises, Blobbucket,BenoniBot~enwiki,Dillard421,Panicum,Fedosin,C'est moi,Hamiltondaniel,ShexRox,Dolphin51,M2Ys4U,Into The Fray,Canglesea,C0nanPayne,Albin&dani,Quinling,Muhends,Moony1000,Atif.t2,Tomasz Prochownik,ClueBot,Thejman123456789,Artichoker, PipepBot,The Thing That Should Not Be,JavaJesus,Mr.Pinklesworth,Meekywiki,Artist7337,Drmies,Firth m,Mild Bill Hiccup,Boing! said Zebedee,CounterVandalismBot,Kitty9992,VandalCruncher,Harlandl,Otolemur crassicaudatus,Jackey0105,Andwor9,AikBkj,Another Matt,RogerEllman,Puchiko,Shocky95,Gakusha,DragonBot,Joshisamazing,Chris earl89,Alexbot,Mccann tom,Icreaser,Robbie098,Sct 5333,Shinkolobwe,Fsunka,Itel94,Ola Hansen,Ice Cold Beer,Aurora2698,Jotterbot,PhySusie,Glacialvortex,Razorflame,Handcannon-beast, Applejacks47,Chaosdruid,Thingg,Venera7,Wnt,Myagooshki666,Deproduction,MasterOfHisOwnDomain,Jaykay0424,Ioannes93, TimothyRias,Misterbeal,InternetMeme,Mhamhamha,XLinkBot,Royboturso,Gwandoya,Crustillicus,BodhisattvaBot,Rror,Ance.cdas,Baconlover13,Ost316,Srossman07,Noctibus,JinJian,Truthnlove,Ttimespan,Airplaneman,Infonation101,EEng,Freestyle-69,Kbdankbot,Xerbaycom,Addbot,Nyw195,Willking1979,NateDres23,Bobocheese,Betterusername,Non-dropframe,Olli Niemitalo,Cre84u,TutterMouse, Presentabsent,Eedlee,Veraptor,WFPM,Ashanda,MrOllie,Lost on Belmont,FerrousTigrus,Ld100,Delaszk,Glass Sword,Maddox1,JasperDeng ,Harvardstudent,Kicka,Tide rolls,EugeneKantarovich,Cesiumfrog,Krano,WikiDreamer Bot,Hartz,Narutolovehinata5,Legobot,Dr-pickem, Luckas-bot,Dov Henis,Senator Palpatine,II MusLiM HyBRiD II,JustWong,Becky Sayles,THEN WHO WAS PHONE?,Anakn-gAraw, Solo Zone,Azcolvin429,Atqueamemus,MacTire02,AnomieBOT,Jdiyef,Ahmediq152,Jt16733,Gatoradeparade,Kanwaraj,Pianonon troppo, Chaosmaker39,Gsd65,LlywelynII,Rhlowe,Unicornlad,VCleemput,Kanat Abildinov,Typesships,Are you ready for IPv6?,Dje8,Citation bot, Oddball.bfi,Nyrox395,Maxis ftw,Persistent76,Chase4813,ArthurBot,SnorlaxMonster,Gravityforce,ChococatR,Xqbot,Tin-ucherianBot II, Wikidushyant,TechBot,Millahnna,Gap9551,Markell West,GrouchoBot,Celebration1981,Gsard,Amaury,Charvest,DoulosChristos,A.di M., Acannas,CES1596,Ninjainventor,LucienBOT,Paine Ellsworth,Tobby72,TedlyW,Lookang,Lipsquid,,SławomirBiały,ArkianNWM, Allen Jesus,Parvons,Cannolis,Orion8,Citation bot1,Pinethicket,Tom.Reding,RedBot,IceBlade710,SpaceFlight89,Rohitphy,Jauhienij, RockSolidCosmo,FoxBot,TobeBot,Yunshui,Gafferuk,Comet Tuttle,Le Docteur,Schwede66,MitchLay,Oms22,Earth-andmoon,Tbhotch, RobertMfromLI,Brambleclawx,RjwilmsiBot,Androstachys,DASHBot,EmausBot,John of Reading,Mnkyman,Atwar-withthem,Qurq,Gfoley4 ,Kueller1,Dewritech,Baguettes,Syncategoremata,Jmencisom,Dcirovic,Sheeana,The Blade of the Northern Lights,Solomonfromfinland, JSquish,Ryan.vilbig,ZéroBot,Crua9,GoldRenet,Dffgd,Everard Proudfoot,Quondum,Wikfr,Confession0791,JosJuice,Ocaasi,Wtsbeynon, BrokenAnchorBot,Brandmeister, L Kensington,Zayzya, Ally1604,Checkmark56, Fluctuating metric,RockMagnetist,Teapeat,Laned130, Rememberway,ClueBot NG,Nebulosus,CocuBot,PoqVaUSA,Jj1236,Lilptrsn,Megalobingosaurus,MerlIwBot,Helpful Pixie Bot, BG19bot, Negativecharge, Furkhaocean, GKFX, Bryanpiczon, Ninney, Cadiomals, Mr.viktor.stepanov, RGloucester, BattyBot, Tutelary, David.moreno72, GravityForce, Padenton, Khazar2, JYBot, Davidlwinkler, LightandDark2000, Mogism, Rudrene, Reatlas, CsDix, Everymorning, Vanderoops, Yuan Jullian Morales, Gacman67, DavidLeighEllis, Prokaryotes, Jwratner1, My name is not dave, Mfb, Konveyor Belt, Mproncace, Mahusha, Monkbot, Opencooper, Filedelinkerbot, Gronk Oz, 3primetime3, WaryLouka, Oiyarbepsy, Stefania.deluca, Loraof, Danmcz, Ps20231131, Gladamas, Pishcal, Freedom2003, Tetra quark, James 123234, Inorout, Supdiop, KasparBot, MusikBot, The oracle 2015, Sweepy, Jeffryan123, ImperatorRomanorvm, Addycrisp, Sir Cumference, The golden colten, CAPTAIN RAJU, J.A.Witt (Tony), Hiimemilylol, Dan6233, Sunshine2night, H.dryad, Dr Peter Donald Rodgers, L3erdnik, Worldandhistory, Rekzy FFA, New Speech Killer, BIGMANFI17, Crazyguys123, Ahmadhammo2, Dat nuke tho, Apokryltaroes and Anonymous: 1274

- **Quantum gravity** *Source:* https://en.wikipedia.org/wiki/Quantum_gravity?oldid=723545634 *Contributors:* AstroNomer, Matusz, Miguel~enwiki, Roadrunner, Stevertigo, Ubiquity, Bobby D. Bryant, Mcarling, NuclearWinner, Anders Feder, Susurrus, Coren, Charles Matthews, Timwi, Reddi, Tpbradbury, Phys, Bevo, Raul654, BenRG, Frazzydee, Jeffq, Sdedeo, Rholton, Wereon, Ilya (usurped), Seth Ilys, Ancheta Wis, Giftlite, Herbee, Fropuff, Endlessnameless, Malyctenar, Jason Quinn, Finn-Zoltan, YapaTi~enwiki, Lumidek, Marcus2, Joyous!, TJSwoboda, Vitaleyes, Davidclifford, JimJast, Rich Farmbrough, Guanabot, FT2, Masudr, Pjacobi, Pie4all88, David Schaich, Bender235, Clement Cherlin, El C, PhilHibbs, Army1987, Apyule, VBGFscJUn3, PWilkinson, Daniel Arteaga~enwiki, Keenan Pepper, Cjthellama, DonJStevens, Velella, Dabbler, Tycho, Cal 1234, RJFJR, Count Iblis, ThomasWinwood, Anarchimede, Scarykitty, Woohookitty, Igny, ToddFincannon, Mpatel, GregorB, Joke137, Christopher Thomas, Marudubshinki, Graham87, Yurik, Kroggz, Rjwilmsi, Eoghanacht, Jrasowsky, JHMM13, Smithfarm, Ems57fcva, FayssalF, Itinerant1, Lmatt, Chobot, Hmonroe, YurikBot, Hillman, ErkDemon, JocK, SCZenz, Roy Brumback, Bota47, Zunaid, JonathanD, 2over0, Arthur Rubin, Modify, LeonardoRob0t, Caco de vidro, RG2, KasugaHuang, Resolute, SmackBot, Samdutton, Vald, Eskimbot, Hbackman, Onebravemonkey, Gilliam, Chris the speller, Ben.c.roberts, Cthuljew, Silly rabbit, Complexica, Colonies Chris, Scwlong, QFT, Soosed, Theanphibian, Shushruth, Ck lostsword, Yevgeny Kats, DJIndica, Lambiam, Mike1901, Vampus, Vincenzo.romano, Jaganath, JorisvS, Bjankuloski06, RoboDick~enwiki, IronGargoyle, Dicklyon, SirFozzie, Treyp, Twunchy, Iridescent, Piccor, Kurtan~enwiki, Harold f, CalebNoble, Duduong, Paulmlieberman, TVC 15, Phatom87, UncleBubba, TAz69x, Sam Staton, ST47, B, Patrick O'Leary, Epbr123, Koeplinger, Klasovsky, Markus Pössel, Keraunos, Headbomb, Marek69, MichaelMaggs, Tim Shuba, MER-C, ParadiZio, Clementvidal, Perlygatekeeper, VoABot II, Alvatros~enwiki, Bdalevin, SHCarter, Jpod2, DAGwyn, Nucleophilic, LorenzoB, Rickard Vogelberg, DancingPenguin, Rettetast, Victor Blacus, AstroHurricane001, Yonidebot, Acalamari, Mstuomel, Fullmetal2887, NewEnglandYankee, DorganBot, CardinalDan, Idioma-bot, Sheliak, VolkovBot, Pleasantville, Seattle Skier, AlnoktaBOT, TXiKiBoT, Dllahr, Red Act, Rdekleer, Saibod, Cyberchip, Wikiwikimoore, Carlorovelli, LoreMiles, StevenJohnston, SieBot, LeadSongDog, Bentogoa, Coldcreation, ReluctantPhilosopher, StaticG, Randy Kryn, GarbagEcol, ClueBot, The Thing That Should Not Be, EoGuy, Polyamorph, Andwor9, Notburnt, Tms9, Alexbot, Resoru, Eeekster, Tamaratrouts, Brews ohare, SchreiberBike, Askahrc, BOTarate, Lambtron, DumZiBoT, XLinkBot, Rror, Facts707, SilvonenBot, Theonlydavewilliams, Mhsb, Truthnlove, Ttimespan, Trifonov~enwiki, Addbot, Mortense, Grayfell, Eric Drexler, Gravitophoton, DOI bot, AkhtaBot, CanadianLinuxUser, Frosty726, LaaknorBot, Delaszk, Tassedethe, Tide rolls, Taketa, Titan1129, Krano, Luckas-bot, Yobot, WikiDan61, Pigetrational, Wireader, Allowgolf~enwiki, Wiki Roxor, Jim1138, IRP, Sz-iwbot, Quantity, Materialscientist, Citation bot, ArthurBot, LilHelpa, Amareto2, Ekwos, KrisBogdanov, Rolfguthmann, StealthCopyEditor, ??, Dan6hell66, Rabsmith, Hep thinker, Paine Ellsworth, DrArthurRubinPHD, Lagelspeil, Nunc aut numquam, Vacuunaut, Van Speijk, Knowandgive, Craig Pemberton, Udifuchs, Citation bot 2, Citation bot 1, Citation bot 4, Jonesey95, Hirvenkürpa, Tom.Reding, Pmokeefe, Serols, Casimir9999, Dac04, Dude1818, Valeriy Pischenko, Follyland, TrueTeargem, N0814444, Earthandmoon, Korepin, DARTH SIDIOUS 2, Musictime4me, RjwilmsiBot, EmausBot, Francophile124, Octaazacubane, Fotoni, Slightsmile, Garfield Salazar, Hhhippo, JSquish, John Cline, Fæ, LostAlone, Brazmyth, Throwmeaway, Arbnos, Ebrambot, Kusername, DanielBurnstein, TonyMath, L Kensington, Maschen, Donner60, Parusaro, Apratim07, Terra Novus, Isocliff, Googledin!, ClueBot NG, SpikeTorontoRCP, Science writer, Preon, Raidr, Jhmmok, 336, Widr, Helpful Pixie Bot, Bibcode Bot, BG19bot, Bardsley Rides a Segway, Apelikedawg, FiveColourMap, Trevayne08, Mr.viktor.stepanov, Brainssturm, BattyBot, Jimw338, Ryanr666, Kryomaxim, Garuda0001, CuriousMind01, Saehry, TwoTwoHello, Sanathdevalapurkar, Andyhowlett, GabeIglesia, Sanathlab, Roiwallace, Spencer.mccormick, Spencerfjase, MrShlongNo1, Marc D. Garrett, D00d00ballz, Gigantmozg, Susan.grayeff, Polytope24, Frinthruit, Anrnusna, Dfyytj, Monkbot, Umut Alihan Dikel, Negative24, Amortias, Klj1234, Pfpguy, Egarcitenre, KasparBot, Christos Theopoulos, Schidan, Jespergrimstrup, DiscreteEditor, Peter SamFan, Xerxeese, VehementurInhorrui and Anonymous: 308
- **Spin network** *Source* :https://en.wikipedia.org/wiki/Spin_network?oldid=719879627 *Contributors* :Miguel~enwiki ,Gabbe ,Mcarling ,Smack, Jake Nelson ,Phys ,Topbanana ,Robbot ,Anthony ,Giftlite ,Lumidek ,Jag123 ,PWilkinson ,Oleg Alexandrov ,Linas ,Rjwilmsi ,Itinerant 1,Russ-Bot ,Jpbowen ,SmackBot ,Bayardo ,Deadlyvices ,Colonies Chris ,BWDuncan ,Princehahaha ,JRSpriggs ,Thijs!bot ,Nagarjunag ,David Eppstein , Gwern ,TXiKiBoT ,PolarBot ,PixelBot ,SchreiberBike ,MilesAgain ,Triathematician ,Addbot ,Mesopelagicity ,Lightbot ,Baxxterr ,AnomieBOT , Scienceking 5,Xqbot ,Topological defect ,The Wiki ghost ,D'ohBot ,Sławomir Biały ,Gil987 ,Gluino ,GoingBatty ,ZéroBot ,Maschen ,Bibcode Bot,F=q(E+v^B)and Anonymous:23
- **Spin foam** *Source:* https://en.wikipedia.org/wiki/Spin_foam?oldid=719879488 *Contributors:* The Anome, Palfrey, Charles Matthews, Centrx, Lumidek, Vsmith, Brian0918, Grnch, Bluemoose, BD2412, Rjwilmsi, Mansari, Hamidifar, Conscious, Bondegezou, Roy Fultun, Princehahaha, Knotwork, MegX, David Eppstein, Gwern, Camrn86, Carlorovelli, PerryTachett, Tomas e, MystBot, Addbot, Haasfelix, SpBot, Yobot, Anypodetos, AnomieBOT, NorTalf, FrescoBot, Paine Ellsworth, Sławomir Biały, Citation bot 2, Jonesey95, Casimir9999, Dr. LaFave, EmausBot, ZéroBot, Bibcode Bot, Kecchina and Anonymous: 28
- **Planck length** *Source:* https://en.wikipedia.org/wiki/Planck_length?oldid=714552835 *Contributors:* AxelBoldt, Mav, Uriyan, Roadrunner, GrahamN, Bdesham, Michael Hardy, Dominus, Dcljr, Glenn, Alf, Dying, Dysprosia, Doradus, Nnh, BenRG, Digizen, Blainster, DHN, Netjeff, Mattflaschen, Giftlite, JamesMLane, Python eggs, Pne, Mckaysalisbury, YapaTi~enwiki, Utcursch, Sonjaaa, Superborsuk, JimWae, Icairns, Tdent, Yuriz, Chris Howard, Sysy, Hidaspal, Luqui, Florian Blaschke, Alistair1978, Bender235, Pt, El C, Rbj, Obradovic Goran, Wendell, Jhertel, DanielLC, Keenan Pepper, Axl, Sligocki, PAR, Dominic, Oleg Alexandrov, GVOLTT, StradivariusTV, GregorB, Christopher Thomas, Radiant!, JIP, Zbxgscqf, FlaBot, John Baez, DevastatorIIC, QuicksilverJohny, LeCire~enwiki, DVdm, Jimp, RussBot, Amakuha, Elizabeyth, NorsemanII, JonathanD, Ayeomans, Gtdp, Neoliten, TLSuda, Groyolo, KasugaHuang, SmackBot, Gold333, DLH, Skizzik, Amux, Bluebot, Scwlong, Voyajer, Mr.Z-man, Valenciano, Sadi Carnot, Tesseran, Lambiam, Petr Kopač, Adj08, Korean alpha for knowledge, JH-man, Loadmaster, JHunterJ, Slakr, Ashaver, Krispos42, Achoo5000, Roland.barrat, Zureks, Zginder, Quibik, Agony, Nonagonal Spider, Greg L, Oreo Priest, Defaultdotxbe, Tyco.skinner, Spartaz, Blaine Steinert, 100110100, Magioladitis, Nakkisormi, Harelx, Mtiffany71, Tanvirzaman, Stuver, Meam5555, Brian.ellis, BJ Axel, Peter Chastain, Melamed katz, Maurice Carbonaro, Foober, Grshiplett, Sigmundur, Rotemdanzig, VolkovBot,

Thurth, Philip Trueman, TXiKiBoT, UnitedStatesian, Cojones893, Newbyguesses, Likebox, Ellohir, Fhburton, Danthewhale, ClueBot, Cygnis insignis, Unitfreak, Brews ohare, NuclearWarfare, BOTarate, Faramarz.M, PL290, Truthnlove, Addbot, Luckas-bot, Yobot, Crispmuncher, Linktex, AnomieBOT, DemocraticLuntz, The High Fin Sperm Whale, Xqbot, Brufydsy, Br77rino, Srich32977, Champlax, حامد میرزاحسینی, Brycee, Sewblon, Constructive editor, FrescoBot, Jonesey95, Tom.Reding, Serols, RobinK, DixonDBot, Fama Clamosa, Javierito92, Is it Protagoras?, Autumnalmonk, EmausBot, John of Reading, Bookalign, GoingBatty, 1howardsr1, Medeis, Quondum, Rocketrod1960, ClueBot NG, ClaudeDes, Widr, StanS, Helpful Pixie Bot, CrannySmith, Bibcode Bot, BG19bot, Chess, Mynameisnoted, Mark Arsten, Phil.gagner, Klilidiplomus, Penguinstorm300, Pinkie Pie, Timothy Gu, Dtm1234, 069952497a, Jiawhein, ReconditeRodent, Zane6324, JustBerry, Ginsuloft, FDMS4, Crabbyhamster, Coreyemotela, Tobywheeer, Tobyepic, Monkbot, Alexander Klimets, Martijnvans, Gamebuster19901, Isambard Kingdom, Dtrumpnairobanbirthcertificate, Nertuop and Anonymous: 212

- **Big Bounce** *Source:* https://en.wikipedia.org/wiki/Big_Bounce?oldid=722905799 *Contributors:* Bryan Derksen, Edward, Furrykef, Omegatron, Korath, Peak, Academic Challenger, GreatWhiteNortherner, Jason Quinn, Junuxx, LucasVB, Antandrus, FT2, Bender235, I9Q79oL78KiL0QTFHgyc, Pearle, Knucmo2, Sade, Velella, Gpvos, Vashti, Sin-man, Drbogdan, Rjwilmsi, Linuxbeak, Phoenix2~enwiki, Bgwhite, Stephenb, Gaius Cornelius, JonathanD, CWenger, SmackBot, Armeria, Dane Sorensen, J 1982, *Zzzzzzzzzzz*, TPIRFanSteve, Robertinventor, Thijs!bot, Headbomb, Peter Gulutzan, BlytheG, Mentifisto, DagosNavy, VoABot II, Skylights76, Rob Lindsey, Ildus58, MooresLaw, Fullmetal2887, KylieTastic, Mike V, CardinalDan, Andyvphil, Knightshield, PaddyLeahy, MarcelloBarnaba, Filos96, Hatster301, Atif.t2, ClueBot, Eetvartti, Unbuttered Parsnip, CohesionBot, Simon Villeneuve, Nonunitary, NonvocalScream, Addbot, Luckas-bot, Fraggle81, Backslash Forwardslash, AnomieBOT, Ipatrol, Rtyq2, Citation bot, StrontiumDogs, Adrianilias, Theone567hunter, Finncarey, Citation bot 1, Seryo93, Trappist the monk, Aiurdin, Slightsmile, Dpieski, Crux007, Terraflorin, ClueBot NG, Helpful Pixie Bot, Bibcode Bot, BG19bot, NUMB3RN7NE, Dannie996, Lianatajo, Anthony Felizardo, Monkbot, Cyrej, Tetra quark, Safderg, Equinox and Anonymous: 88
- **Accelerating expansion of the universe** *Source:* https://en.wikipedia.org/wiki/Accelerating_expansion_of_the_universe?oldid=723518856 *Contributors:* AxelBoldt, Bth, Dbundy, Tim Starling, EddEdmondson, Alfio, Looxix~enwiki, William M. Connolley, Julesd, Evercat, Hashar, Timwi, Dysprosia, Mw66, The Anomebot, DW40, Dragons flight, Phys, Peak, Lowellian, Rursus, JerryFriedman, Giftlite, Graeme Bartlett, DavidCary, Barbara Shack, Joe Kress, ConradPino, Bbbl67, Burschik, Eep², JimJast, Guanabot, Vsmith, LeeHunter, Dbachmann, Bender235, AdamSolomon, RJHall, Mr. Billion, Sietse Snel, Bobo192, Foobaz, I9Q79oL78KiL0QTFHgyc, Hackwrench, Eric Kvaalen, Vuo, Falcorian, Bobrayner, Zanaq, Richard Arthur Norton (1958-), Woohookitty, OCNative, Joke137, Wisq, Driftwoodzebulin, Drbogdan, Rjwilmsi, Seandop, Kolbasz, ThunderPeel2001, Marcperkel, Chase me ladies, I'm the Cavalry, Closedmouth, Dutch-Bostonian, SmackBot, Melchoir, Vald, Nickst, AnOddName, Dreadstar, Pulu, Wkerney, Ckatz, Hypnosifl, Xxxiv34, Geral Corasjo, CRGreathouse, Kjknohw, Meno25, Michael C Price, Doug Weller, DumbBOT, Peter Gulutzan, Dawnseeker2000, KrakatoaKatie, Seaphoto, Obeattie, Steelpillow, Arch dude, Bpmullins, Cardamon, Cgingold, Robin S, NatureA16, Hedwig in Washington, McSly, Mannhoodd, Tarotcards, VolkovBot, UnitedStatesian, Insanity Incarnate, Nihil novi, Puzhok, Lightmouse, Coldcreation, Gevgiorbran, Aggel000, Dr. Leif Rongved, Dmyersturnbull, Kentgen1, El bot de la dieta, DanielPharos, DumZiBoT, XLinkBot, Ladsgroup, Ost316, Addbot, Ridgepg, Simonm223, Ilketitz93, AkhtaBot, Marx01, Verbal, Legobot, Cosmos72, Luckas-bot, Yobot, Systemizer, Fraggle81, Amirobot, Amble, Synchronism, AnomieBOT, Jim1138, Citation bot, Xqbot, Gap9551, GrouchoBot, RibotBOT, Waleswatcher, Bellerophon, Michael93555, DivineAlpha, Citation bot 4, Jonesey95, Tom.Reding, Issuesixty soulsgreat, Trappist the monk, Lotje, RjwilmsiBot, John of Reading, Primefac, Tinss, Vanjka-ivanych, Solomonfromfinland, ZéroBot, Brandmeister, ChuispastonBot, MrChandmari, Mechachomp, ClueBot NG, Astrocog, Bibcode Bot, BG19bot, Yizlpku, Anukool.rajoriya, Minsbot, Pheng13, RiseUpAgain, Mediran, Kozmokonstans, Makecat-bot, Wjs64, Illuusio, Rfassbind, Yheyma, Blackbombchu, The Herald, BDwinds, Leegrc, Jsaur, Igby Kollektiv, Garfield Garfield, HannahFord428, Tetra quark, Isambard Kingdom, MusikBot, Sir Cumference, Youknowwhatimsayin, ICameHereToEditNotToFeel, Xx Cool Guy7202 xX and Anonymous: 93
- **Inflation (cosmology)** *Source:* https://en.wikipedia.org/wiki/Inflation_(cosmology)?oldid=723096289 *Contributors:* Bryan Derksen, The Anome, Diatarn iv~enwiki, Roadrunner, David spector, Hephaestos, Stevertigo, Edward, Nealmcb, Boud, Michael Hardy, Tim Starling, Dcljr, Cyde, Ellywa, William M. Connolley, Theresa knott, Jeff Relf, Mxn, Timwi, Rednblu, Bartosz, Pierre Boreal, Raul654, Chuunen Baka, Robbot, Gandalf61, Rursus, Ancheta Wis, Giftlite, Barbara Shack, Mikez, Lethe, Dratman, Curps, Jcobb, Gracefool, Just Another Dan, Andycjp, HorsePunchKid, Beland, Elroch, JDoolin, Burschik, Shadypalm88, Eep², Mike Rosoft, DanielCD, Noisy, Rich Farmbrough, FT2, Pjacobi, Luxdormiens, Dbachmann, Bender235, AdamSolomon, Pt, Worldtraveller, Art LaPella, Orlady, Drhex, Guettarda, I9Q79oL78KiL0QTFHgyc, Jeodesic, Rsholmes, Anthony Appleyard, Plumbago, JHG, Schaefer, EmmetCaulfield, Cgmusselman, Dirac1933, Oleg Alexandrov, Matevzk, Yeastbeast, StradivariusTV, BillC, Bluemoose, Wdanwatts, Joke137, Rnt20, Malangthon, Ketiltrout, Drbogdan, Rjwilmsi, Zbxgscqf, Mattmartin, Strait, Eyu100, Jehochman, Ems57fcva, Bubba73, FlaBot, Nihiltres, Itinerant1, Phoenix2~enwiki, Chobot, Hermitage, Bgwhite, YurikBot, Wavelength, Supasheep, Ytrottier, Gaius Cornelius, Anomalocaris, NawlinWiki, LiamE, Davemck, JonathanD, Enormousdude, 2over0, Arthur Rubin, Argo Navis, Physicsdavid, Protero, Luk, SmackBot, Haza-w, KnowledgeOfSelf, Lawrencekhoo, Onsly, Jdthood, Salmar, Jefffire, Hve, QFT, Vanished User 0001, Stevenmitchell, BIL, Lostart, Ligulembot, Yevgeny Kats, Byelf2007, Lambiam, J 1982, Rcapone, JorisvS, Heliogabulus, Dan Gluck, Spebudmak, JoeBot, UncleDouggie, Fsotrain09, Oshah, JRSpriggs, Chetvorno, Friendly Neighbour, Drinibot, Vanished user 2345, Brownlee, SuperMidget, Cydebot, BobQQ, Mortus Est, Cyhawk, Ttiotsw, Julian Mendez, Dr.enh, Michael C Price, Kozuch, LilDice, Thijs!bot, Headbomb, Z10x, Jklumker, Alfredr, Dawnseeker2000, Pollira, Rico402, Lfstevens, Gmarsden, JAnDbot, Olaf, LinkinPark, GurchBot, Magioladitis, Jpod2, Vanished user ty12kl89jq10, Rickard Vogelberg, Dr. Morbius, Bhenderson, TomS TDotO, Tarotcards, Wesino, Student7, Potatoswatter, Ollie 9045, Ja 62, Useight, Idioma-bot, Sheliak, Tokenhost, VolkovBot, ABF, ColdCase, Philip Trueman, TXiKiBoT, Calwiki, Thrawn562, Gobofro, SwordSmurf, Northfox, PaddyLeahy, SieBot, Wing gundam, OpenLoop, Likebox, Flyer22 Reborn, Maynard-Clark, Mimihitam, Hockeyboi34, Lightmouse, Sunrise, Southtown, Hamiltondaniel, Epistemion, ClueBot, Niceguyedc, ChandlerMapBot, Jusdafax, ResidueOfDesign, Ploft, Scog, SchreiberBike, Carlroddam, TimothyRias, Katsushi, MidwestGeek, Addbot, Roentgenium111, DOI bot, Blethering Scot, Ronhjones, Glane23, Deamon138, TStein, Barak Sh, Tassedethe, Zorrobot, Ben Ben, Legobot, Yinweichen, Luckas-bot, Yobot, Amirobot, Aldebaran66, Amble, Isotelesis, Magog the Ogre, AnomieBOT, Pyrrhon8, Rubinbot, Piano non troppo, Collieuk, Ulric1313, Citation bot, Xqbot, Plastadity, Capricorn42, P14nic997, False vacuum, Waleswatcher, Ignoranteconomist, Bigger digger, Chatul, ??, CES1596, FrescoBot, Mesterhd, Paine Ellsworth, Schnufflus, Charles Edwin Shipp, Bbhustles, Ahnoneemoos, Pinethicket, Tom.Reding, Ganondolf, Σ, Aknochel, Mercy11, Trappist the monk, Jordgette, Wdanbae, Aabaakawad, Michael9422, CobraBot, Deathflyer, Mathewsyriac, EmausBot, Thucyd, GoingBatty, Wikipelli, Dcirovic, Kiatdd, Italia2006, Werieth, ZéroBot, Chasrob, Wackywace, Chharvey, Bamyers99, Suslindisambiguator, AManWithNoPlan, RaptureBot, Maschen, HCPotter, Crux007, RockMagnetist, Whoop whoop pull up, ClueBot NG, J kay831, Law of Entropy, Supermint, Helpful Pixie Bot, Bibcode Bot, Lowercase sigmabot, BG19bot, Negativecharge, MSgtpotter, Badon, BML0309, Zedshort, Hamish59, Minsbot, BattyBot, SupernovaExplosion, ChrisGualtieri, JYBot, Rfassbind, Ikjyotsingh, Astroali, Lepton01, Pkanella, Chwon,

Rolf h nelson, Comp.arch, Kogge, Hilmer B, Anrnusna, Stamptrader, Dodi 8238, Epaminondas of Thebes, Man of Steel 85, Abitslow, Monkbot, Accnln, BradNorton1979, YeOldeGentleman, Waters.Justin, Tetra quark, Isambard Kingdom, Sleepy Geek, Anand2202, Jmc76, Quasiopinionated, EnigmaLord515, Phseek, Trekkiepanda and Anonymous: 212

- **String theory** *Source:*https://en.wikipedia.org/wiki/String_theory?oldid=723494177*Contributors:* AxelBoldt,Sodium,Mav,Bryan Derksen, Zundark, The Anome, Tarquin, Taw, Eean, Malcolm Farmer, Hephaestos, Olivier, Drseudo, Stevertigo, Spiff~enwiki, Edward, PhilipMW, Michael Hardy,Bewildebeast,Dante Alighieri,Gabbe,Graue,Tgeorgescu,Mcarling,CesarB,Looxix~enwiki,Ahoerstemeier,Theresa knott, Suisui,Angela,Den fjättrade ankan~enwiki,Jdforrester,Julesd,Salsa Shark,Schneelocke,Charles Matthews,Timwi,Bemoeial,Jitse Niesen, 4lex,Greenrd,ErikStewart,Furrykef,Saltine,Phys,Omegatron,Bevo,Topbanana,Trent,Nufy8,Robbot,Craig Stuntz,Fredrik,Chris73, R3m0t,COGDEN,Mirv,Wjhonson,Sverdrup,Academic Challenger,DHN,Hadal,Khlo,ElBenevolente,HaeB,Xanzzibar,Tea2min,Giftlite, DocWatson42,Christopher Parham,Awolf002,Mporter,Amorim Parga,Mikez,Harp,Kim Bruning,Tom harrison,Ferkelparade,Leflyman, Fropuff, No Guru, Anville, Moyogo, Curps, Pashute, Nomad~enwiki, Mboverload, Solipsist, SWAdair, DemonThing, Wmahan, Btphelps, MSTCrow,Decoy,Chowbok,Gadfium,Steuard,Pgan002,Quadell,Carandol~enwiki,Antandrus,Beland,JoJan,Khaosworks,Tothebarricades.tk,Thincat,Tomruen,Shidobu,Icairns,Lumidek,NoPetrol,Avihu,Fanghong~enwiki,Trevor MacInnis,Lacrimosus,Zro,Mike Rosoft, D6,Urvabara,Felix Wan,Jkl,Discospinster,ElTyrant,Rich Farmbrough,Rhobite,Pjacobi,Alien life form,Vapour,Silence,Kzzl,LindsayH, Manil,Pavel Vozenilek,Paul August,Bender235,Kjoonlee,Mashford,Kelvinc,Perlman10s,Panu~enwiki,Brian0918,Dpotter,Livajo,El C, Laurascudder,Shanes,Zegoma beach,RoyBoy,Causa sui,Bobo192,Directorstratton,Janna Isabot,Smalljim,John Vandenberg,Flxmghvgvk, I9Q79oL78KiL0QTFHgyc, Physicistjedi, Bongoo, 4v4l0n42, Merope, Geschichte, Linuxlad, Phils, Merenta, Alansohn, Gary, JYolkowski, Enirac Sum,Ryanmcdaniel,Arthena,Borisblue,Rd232,Plumbago,Axl,R Calvete,Lightdarkness,Kocio,Bartl33,Wtmitchell,Isaac,Tycho, Call234,Fadereu,CloudNine,Sciurinæ,Computerjoe,Kusma,DV8 2XL,Pwqn,Gene Nygaard,Ringbang,Ceyockey,Falcorian,Bobrayner, Joriki,Mel Etitis,Linas,BillC,Jacobolus,HFarmer,Before My Ken,Netdragon,MONGO,GeorgeOrr,Mpatel,Bbatsell,GregorB,,Joke137, Christopher Thomas,Dysepsion,GSlicer,Jan.bannister,Graham87,Magister Mathematicae,Hillbrand,BD2412,Elvey,Galwhaa, Raymond Hill,JIP,RxS,Athelwulf,Edison,Sjakkalle,Rjwilmsi,Xgamer4,Jake Wartenberg,Arabani,MarSch,TheRingess,Jmcc150,Aero66, Crazynas, Juan Marquez, R.e.b., Bubba73, DoubleBlue, Zelos, AlisonW,Asafavi, Lionelbrits, Conorific, Zunz, Mathbot, Crazycomputers,RexNL,Gurch,Algri,TeaDrinker,Zifnabxar,XAXISx,Erik4,Phoenix2~enwiki,Antimatter15,Ggb667,Chobot,Visor, DVdm,Mhking,VolatileChemical,Bgwhite,Algebraist,Ben Tibbetts,YurikBot,Ugha,Wavelength,Borgx,NuclearFusion~enwiki, Angus Lepper,Hairy Dude,Jimp,Hillman,Cyferx,Wolfmankurd,Pip2andahalf,RussBot,Moronoman,Crazytales,Pippo2001,Bhny,Pigman, SpuriousQ,Branman515,Stephenb,Gaius Cornelius,Eleassar,Rsrikanth05,Bovineone,Cheesus,Shanel,NawlinWiki,Tong~enwiki,Mikel8 xx,SCZenz,Cleared asfiled,Bdiah,Pym98,SColombo,Haemo,FF2010,Closedmouth,Reyk,Brina700,Chris Brennan,Vicarious,Brianlucas, Geoffrey.landis,Hitch-hiker89,Spliffy,Pred,ArielGold,Roy Fultun,Ilmari Karonen,Katieh5584,Pentasyllabic,Lunch,DVD R W,WikiFew, That Guy,From ThatShow!,Street Scholar,AndrewWTaylor,QSquared,Sardanaphalus,Vanka5,MacsBug,Hvitlys,SmackBot,Kurochka, Zazaban,Tom Lougheed,Prodego,KnowledgeOfSelf,Hydrogen Iodide,Melchoir,Vald,Skrewtape,Atomota,Canthusus,GaeusOctavius,Cool 3,Andyvn22,Gilliam,Skizzik,RobertM525,Dauto,Bluebot,SSJ5,Keegan,Aidan Croft,Thumperward,Oli Filth,Silly rabbit,Timneu22, SchfiftyThree,MosheConstantine Hassan Al-Silverburg,Complexica,Rediahs,RayAYang,Aero77,Adamstevenson,Ikiroid,Epastore,Baronnet ,Ned Scott,Sbhar-ris, Colonies Chris, Konstable, Sct72, Scwlong, Can't sleep, clown will eat me, Timothy Clemans, Onorem, Neilanderson, EvelinaB,TKD,KerathFreeman,Addshore,UU,The tooth,Pepsidrinka,Somebody2292, --=The Doctor=--,Fuhghettaboutit, Cybercobra,Irish Souffle,Nakon,Jdlambert,James McNally,MichaelBillington,Lostart,Insineratehymn,Drphilharmonic,SpiderJon,DMacks, Ihatetoregister,Where,MichaelIFA,Yevgeny Kats,Vasiliy Faronov,Byelf2007,Angela26,Visium,Rory096,Zymurgy,Harryboyles,Mdl 53711,T-dot,Titus III,Ergativerlt,MagnaMopus,UberCryxic,Vgy7ujm,Lazylaces,Linnell,Mgiganteus1,Nonsuch,IronGargoyle,Ckatz, DoItAgain,AstroGod,Kirbytime,Jimbo Mahoney,FredrickS,Invisifan,Ryulong,Ryanjunk,MathStuf,Mike Doughney,Norm mit,Hindol, Dan Gluck,Huntscorpio,Irides-cent,K,Sunoco,You? Me? Us?,CzarB,Rabinzkaman,JoeBot,Lottamiata,Tony Fox,Vrkaul,Torrazzo, Gil Gamesh,Areldyb,Courcelles,Tawkerbot2,Gebrah,Shamvil,Fdssdf,DKqwerty,Lbrl23,Harold f,Heqs,Devourer09,Duduong,Sarvagnya, Dewayne76,JForget,Cg-realms,InvisibleK,CRGreathouse,CmdrObot,Earthlyreason,Van helsing,Olaf Davis,CBM,Rawling,Jibal, Witten Is God,Nunquam Dormio,Giko,KnightLago,Thubsch,Leujohn,SlashDot,TheTito,Karenjc,Myasuda,Emarv,Cydebot,Gmusser, Gogo Dodo,Jkokavec,Kahananite,Qua-jafrie,Michael C Price,Doug Weller,DumbBOT,Narayanese,AlphaNumeric,SRoughsedge, Vanished User jdksfajlasd,Woland37,Zalgo,Daniel Olsen,UberScienceNerd,Bkazaz,DJBullfish,Thijs!bot,Epbr123,Rwmnau,Babemachine, Pimpin101,Mbell,O,Faigl.ladislav,Ucan-lookitup,Andyjsmith,Headbomb,Tcturner2002,Marek69,Brahmajnani,Arthurcprado~enwiki,Y.t., D3gtrd,Babemonkey,Dark dude,DuncanMcB,EdJohnston,MichaelMaggs,Ancientanubis,Natalie Erin,Hempfel,Jomoal99,Mmortal03,Mentifisto,Geekdom04,AntiVandalBot,LunaSantin,Seaphoto,Ed270791,Opelio,Doc Tropics,David136a,NithinBekal,Dotdotdotdash,Helicoptor, Poshzombie,MontanNito,Dylan Lake,Maximilian77,Shlomi Hillel,Db63376,SamIAmNot,Knotwork,Res2216firestar,Superior IQ Genius, MER-C,Andonic,Sitethief,100110100,TallulahBelle,Nestamachine,Savant13,Daynightrader,Goldenglove,Charibdis,Acroterion,Ophion, Aigisthos,Editmyhandman,Aruben537,Magioladitis,WolfmanSF,Bongwarrior,VoABot II,Yandman,JamesBWatson,بلال,Qutt,Jespinos, Kevinmon,Aka042,Froid,DAGwyn,Catgut,Panser Born,Ensign beedrill,Perspectival,JJ Harrison,Dirac66,Justanother,Azizl005, Cpl Syx,ChazBeckett,Teardrop onthefire,WLU,Stephen shenker,Robin S,SkepticVK,Joshua Davis,Mkroh,B9hummingbird hovering,S3000, Hdt83,MartinBot,FlieGerFaUstMe262,Ytomem,Shimwell,Arjun01,KrishSundaresan,Anaxial,Jay Litman,Alexcalamaro,Andrej.westermann,Smokizzy,LedgendGamer,CyrusAndiron,Peteryoungl44,Tgeairn,Artaxiad,HEL,AlphaEta,J.delanoy,AstroHurricane001, Maurice Carbonaro,Yonidebot,Morris729,MC Y1008,69gangsta420,It Is Me Here,Shawn in Montreal,Janus Shadowsong,Bailo26, Fredsie,Madagaskar07,Duchesserin,AntiSpam-Bot,CHIAGEHYANG,Chiswick Chap,Watsupl313,Belovedfreak,HaloInverse, NewEnglandYankee,Scott1329m,Thesis4Eva,Policron,Jrcla2,KylieTastic,WJBscribe,Rnricklefs,Jamesofur,Eyelidlessness,Jonnyk aus, Kvdveer,JavierMC,Izno,Xiahou,CardinalDan,Sheliak,HamatoKameko,Malik Shabazz,Concertmusic,JohnBlackburne,JustinHagstrom, Fences and windows,Wooba doob,Philip Trueman,Door-sAjar,HowardFrampton,TXiKiBoT,Oshwah,Zidonuke,Red Act,Kriak,Calwiki, Technopat,Hqb,Andrius.v,Anonymous Dissident,Crohnie,AlysTarr,Qxz,Vanished user ikijeirw34iuaeolaseriffic,Impunv,Seraphim,Martin451 ,Don4of4,ABigGreenHippo,Huperphuff,LeaveSleaves,Kaenneth,StringyGuy,Maxim,Erth64net,Meters,Lamro,Rickstauduhar,Enviroboy, Turgan,Anna512,PhysPhD,Northfox,NPguy,MatthewSanders,Luke Walkerson,Newbyguesses,MissMJ,SieBot,Escher26,J.A.Ireland,BA(IHPST),4wajzkd02,Robdunst,Dreamafter,Pallabl234,Dbelange,MTHarden,Lemonflash,Kylemew,Yintan,GlassCobra,Discrete,Bentogoa, Likebox,Flyer22Reborn,Exert,ProGeek314,Arbor toSJ,Babawhitemoose,Caidh,Dhatfield,Audree,Oxymoron83,Pretty Green, Weaselstomp,Manway,Alex.muller,Taco Manipulator,Tschach,Manheat84,Anchor Link Bot,Mikebernstein,ImperialismGo,Nergaal,Ionfield,Ayleuss,Sh4wz0r,Naturespace,ImageRemovalBot,Martar-ius,Phyte,ClueBot,The Thing That Should Not Be,String4d,Illusion96, Polyamorph,Mpd1989,Alexdeburca18,Wiggl3sLimited,Excirial,Kjramesh,Jusdafax,Resoru,WikiZorro,Eeekster,Verum~enwiki, Tamaratrouts,Gtstricky,Humanino,Brews ohare,NuclearWarfare,Cenarium, Arjayay, Razorflame, Scoobey, BOTarate, Sideswiper, Thingg, Capudo, BVBede, Versus22, Introductory adverb clause, MelonBot, SoxBot III, Egmontaz, Notpayingthepsychiatrist, DumZiBoT, BahTab, TimothyRias, Aj00200, Reaperfromhell, Dunkaroo207, XLinkBot, AlexGWU, Impshum, Saeed.Veradi, Little Mountain 5, Guy392, David424, Truthnlove, Qweeveen, Tayste, Addbot, Steven66s, Denali134, Elemented9, Varrey280303, Eric Drexler, Some jerk on the Internet, Fizzycyst, Uruk2008, DOI bot, Jojhutton, AngryBacon, Non-dropframe, Captain-tucker, Auspex1729, Kongr43gpen, Fgnievinski, Rhetoric Of A Sophist, Ronhjones, CanadianLinuxUser, Cst17, Download, Glane23, Bassbone-rocks, Chzz, Favonian, Kronix35, LinkFA-Bot, Udugunit, Aktsu, Tassedethe, Numbo3-bot, Anpecota, Tide rolls, HerpesVirus, SDJ, OlEnglish, Scourge of God, Davidmedlar, Couldbenoway66, Yobot, Maxdamantus, Terrisknickers, Kartano, TaBOT-zerem, Julia W, Unique and proud of it, FireMouseHQ, Terrifictriffid, ArchonMagnus, CinchBug, Synchronism, AnomieBOT, Cleeseheb, 1exec1, Charlesvi, Bigdaddy4x4, Gitman4, Jim1138, IRP, Mintrick, Drweetmola, Ornamentalone, M00npirate, Gautam10, Csigabi, Poli-Psy, Materialscientist, 90 Auto, Citation bot, Teleprinter Sleuth, Vuerqex, Twri, Frankenpuppy, Fuzzy Bob Saget, DirlBot, Georgepowell2008, Heidisql, Cureden, Ekwos, Capricorn42, Gen-sanders, NFD9001, Anna Frodesiak, Tomwsulcer, A23649, Pra1998, Coretheapple, RadiX, Jagbag2, Vandalism destroyer, Ab1, Omnipaedista, Bandit5005, Shirik, RibotBOT, Waleswatcher, Saalstin, Amaury, Aaron35510, Caz34, Doulos Christos, Sewblon, Born Gay, Capricorn24, SchnitzelMannGreek, A. di M., SpacePyjamas, Kierkkadon, A.amitkumar, Dougofborg, StringLove, Nobelprizewinner, Astiburg, FrescoBot, Fortdj33, Paine Ellsworth, Goodbye Galaxy, HJ Mitchell, Steve Quinn, Vhann, Kwiki, Xhaoz, Citation bot 1, Batong, Gil987, Pinethicket, I dream of horses, Tallboyhoops1991, Three887, Steveo27five, RedBot, Sardinita, Serols, Vhsatheeshkumar, Swisstingle, DeletionUK, Recon-sider the static, IVAN3MAN, Remingtonhill1, Orenburg1, Coltonhs, Angus Guilherme, Smamaret, Bethovenn, Dinamik-bot, Dc987, Oswaldo Zapata, Egemont, Syebo, Alaithiran, Reaper Eternal, Seahorseruler, Ybungalobill, Quaker phil, Specs112, Dr. Aakash Patel, Bj norge, Tb-hotch, StormbringerUK, Minimac, Mathgenius3141592, Keegscee, Omgwaffels, Mick le pick, Solancel, Aznhero3793, Dwielark, Afteread, Enauspeaker, EmausBot, MaooaM, Immunize, Az29, Milkocookie, Faolin42, Fotoni, RA0808, RenamedUser01302013, 8digits, Yukisea-side, Slightsmile, Tommy2010, Winner 42, Wikipelli, Dcirovic, JonezyKiDx, Joe Gazz84, ZéroBot, Timeitsways, John Cline, Cogiati, Quaqa, Chrispaps2413, Nasulikid, Vollrath2323, Benjamin14141414141414141414, Arbnos, Green Lane, A930913, Bamyers99, Azeraphale, H3llBot, En-cyclopadia, Danga1988, Ollainen, PoisonGM, Wayne Slam, OnePt618, Knome335, L Kensington, Lulzprotuns, Kranix, Rpcappello, Maschen, Vastly~enwiki, Donner60, CatFiggy, CountMacula, Orange Suede Sofa, Etov, M1k3 101, Bill william compton, Wakabaloola, TERBAFAN, Nickslspride34, NeuralLotus, Isocliff, Brechbill123, Xanchester, ClueBot NG, Martti Muukkonen, KagakuKyouju, Jeff Song, This lousy T-shirt, Satellizer, Name Omitted, Marcdean123, Wiki incorp, Frietjes, O.Koslowski, Alexdamaino9, Dream of Nyx, Blackhall616, Widr, Sash-here, WikiPuppies, Stu181, T00g00d96, Pluma, Storm.sarup, Helpful Pixie Bot, Manzeet, Waffleboy36, HMSSolent, Mikeshelton1, Bibcode Bot, 2001:db8, Phillip.phillipson, Hoaxinator, Lowercase sigmabot, Thor cherubim, BG19bot, Mrshabam, Nishch, Flowerhat15, AvocatoBot, Housegeek224, MahRanch, Benzband, Altaïr, Benhenchdickthomas, Shreyakstring, Sweaty maori sphincter, DaFalk, Dsabo74, Ratanmaitra, MM4EVAH, Steven.w.kowalski, Minsbot, JGallardo2600, Dylanlatham, Myfriendganesha, OCCullens, Likeaboss189, Sean271293, LinusE8, BattyBot, Several Pending, Aldrich2122, CommanderMoka, Cyberbot II, The Illusive Man, ChrisGualtieri, KoalamaN2, Trevorkid45, Cat-sloveit07, Alex Modzz, Rustyjamsen, Goh ryangoh, Dexbot, Exolius, Hilander316, Alman1234321, SuperCalzer, LightandDark2000, MeekMe-lange, BQND, Cdarrai1, Kephir, TheMonkeyboy524, Michael Anon, TwoTwoHello, Mattfat8, Lugia2453, Anruy, Rachel weld, Jamesx12345, AHusain314, BossEditors, Hillbillyholiday, Joeinwiki, Mattninja, Theshadow444, Asaa82, Jakemarz197, Kzhang1025, Epicgenius, Spong-bob456789, ?, TestMaster, Ianreisterariola, GrapperJ, Makeitnasty, Moemajdi, I am One of Many, Nualalvy, BAZINGASS, St3fanPC, Eye-snore, Isaac grozd, Jordanissexyaf1999, Baruch6525, Mosbruckercj, Ihatedirac2k13, Jonamithy121314, 123physicsquantum, Jt198, RaphaelQS, HeyJude70, AParker628, DimReg, A.k.blaze1, Joshuk, Zenibus, Nianoobasik, Ihelpapplen, Gamo To Apoel, SacredLabyrinth, Ginsuloft, Vam-pre1122, Dimension10, Howard Wolowitz, AddWittyNameHere, Polytope24, Elysion, Tutun12$, Longerboats5, SimonWombat8, Konveyor Belt, Vtank54, Micheal545, Hck24, Caliae19, Hexafish, Simpick, TheRealTheKoi, Bballbro62, Monkbot, ArmyPath, Gabero.88, TheQ Editor, Jtsmith098, Joshmiller1, Hanseer360, XXvPIEvXx, Dbennett 24, Ghikpenos, Nick65633, Saundra03, Thehippothatknows, Sewwgers, Tee-laskeletor, Cirksena, Balockaye1234, PloppyDoo, Yesufu29, Lumpy2k14, Podayeruma, Abstract92, Sbenfiel, Monkman2k4, Swegwegdgfyetk-foffkkfkfkv, John95541234, Poopman224, ScrapIronIV, Tetra quark, GeneralizationsAreBad, Shivansh2014n, KasparBot, SHUCKYLUCKY, Fabiotheoto, FartGoblin, Joca potato, Joshcool246, Theoretical Physisist4444, JanetTom55, Reg7d88, CHANDLER MERRILL, Baking Soda, FklfjDKFd bfl, Rajputclann, Entranced98, Jahziahk, Mjhog, Strong81, WikiTikiDude007, ILoveShukli, Qwerty2345ß, Irene000, A1D1A2D2 and Anonymous: 1601

- **Loop representation in gauge theories and quantum gravity** *Source:* https://en.wikipedia.org/wiki/Loop_representation_in_gauge_theories_and_quantum_gravity?oldid=712144868 *Contributors:* Michael Hardy, Bearcat, DragonflySixtyseven, Rich Farmbrough, Sjö, Kateshortforbob, Wiae, David Condrey, Yobot, Ibayn, I dream of horses, Orenburg1, KateWishing, Mogism, OccultZone, Robevans123 and Anonymous: 3
- **Hamiltonian constraint of LQG** *Source:* https://en.wikipedia.org/wiki/Hamiltonian_constraint_of_LQG?oldid=712144130 *Contributors:* Guy Macon, Carriearchdale, Pqnelson, Yobot, Ibayn, 4th-otaku, AnomieBOT, Omnipaedista, Orenburg1 and Anonymous: 4
- **Lorentz invariance in loop quantum gravity** *Source:* https://en.wikipedia.org/wiki/Lorentz_invariance_in_loop_quantum_gravity?oldid=716399680 *Contributors:* Charles Matthews, Phys, NetBot, Itinerant1, Chris the speller, Colonies Chris, Phuzion, Colonel Warden, Nick Number, Biscuittin, Niceguyedc, SchreiberBike, Boleyn, Addbot, Yobot, AnomieBOT, PowerUserPCDude, JimVC3, Omnipaedista, Paine Ellsworth, Maschen, Helpful Pixie Bot, BG19bot, Cerabot~enwiki, Chemistry1111 and Anonymous: 4
- **Quantum configuration space** *Source:* https://en.wikipedia.org/wiki/Quantum_configuration_space?oldid=693178559 *Contributors:* Yobot, Ibayn, Carturo222 and BG19bot
- **Classical limit** *Source:* https://en.wikipedia.org/wiki/Classical_limit?oldid=722241029 *Contributors:* Patrick, Michael Hardy, Cyan, Charles Matthews, Chris Howard, Mpatel, YurikBot, Salsb, SmackBot, Jpvinall, Complexica, Radagast83, Cydebot, Thijs!bot, Headbomb, Alphachimp-bot, Cuzkatzimhut, Lisatwo, Dthomsen8, MystBot, Addbot, Yobot, Erik9bot, RedBot, Nora lives, GoingBatty, Support.and.Defend, BG19bot, Dexbot and Anonymous: 7
- **Quantum mechanics** *Source:* https://en.wikipedia.org/wiki/Quantum_mechanics?oldid=723354617 *Contributors:* AxelBoldt, Paul Drye, Chenyu, Derek Ross, CYD, Eloquence, Mav, The Anome, AstroNomer, Taral, Ap, Magnus~enwiki, Ed Poor, XJaM, Rgamble, Christian List, William Avery, Roadrunner, Ellmist, Mjb, Olivier, Stevertigo, Bdesham, Michael Hardy, Tim Starling, JakeVortex, Vudujava, Owl, Norm, Gabbe, Menchi, Ixfd64, Axlrosen, TakuyaMurata, Shanemac, Alfio, Looxix~enwiki, Mdebets, Ahoerstemeier, Cyp, Stevenj, J-Wiki, Theresa knott, Snoyes, Gyan, Nanobug, Cipapuc, Jebba, Александър, Glenn, Kyokpae~enwiki, Nikai, Dod1, Jouster, Mxn, Charles Matthews, Tantalate, Timwi, Stone, Jitse Niesen, Rednblu, Wik, Dtgm, Patrick0Moran, Tpbradbury, Nv8200pa, Phys, Bevo, Jecar, Fvw, Stormie, Sokane, Optim, Bcorr, Johnleemk, Jni, Rogper~enwiki, Robbot, Ke4roh, Midom, MrJones, Jaleho, Astronautics~enwiki, Fredrik, Chris 73, Moncrief, Goethean,

Bkalafut, Lowellian, Centic, Gandalf61, StefanPernar, Academic Challenger, Rursus, Texture, Matty j, Moink, Hadal, Papadopc, Johnstone, Fuelbottle, Lupo, HaeB, Mcdutchie, Xanzzibar, Tea2min, David Gerard, Enochlau, Ancheta Wis, Decumanus, Giftlite, Donvinzk, DocWatson42, ScudLee, Awolf002, Barbara Shack, Harp, Fudoreaper, Lethe, Fastfission, Zigger, Monedula, Wwoods, Anville, Alison, Bensaccount, Tromer, Sukael, Andris, Jason Quinn, Gracefool, Solipsist, Nathan Hamblen, Foobar, SWAdair, Mckaysalisbury, AdamJacobMuller, Utcursch, CryptoDerk, Knutux, Yath, Amarvc, Pcarbonn, Stephan Leclercq, Antandrus, JoJan, Savant1984, Jossi, Karol Langner, CSTAR, Rdsmith4, APH, Anythingyouwant, Thincat, Aaron Einstein, Edsanville, Robin klein, Muijz, Zondor, Guybrush, Grunt, Lacrimosus, Chris Howard, L-H, Ta bu shi da yu, Freakofnurture, Sfngan, Venu62, Spiffy sperry, CALR, Ultratomio, KeyStroke, Noisy, Discospinster, Caroline Thompson, Rich Farmbrough, H0riz0n, FT2, Pj.de.bruin, Hidaspal, Pjacobi, Vsmith, Wk muriithi, Silence, Smyth, Phil179, Moogoo, WarEagleTH, Smear~enwiki, Paul August, Dmr2, Bender235, ESkog, Nabla, Dataphile, Dpotter, Floorsheim, El C, Lankiveil, Kross, Laurascudder, Edward Z. Yang, Shanes, Spearhead, RoyBoy, Femto, MPS, Bobo192, Army1987, John Vandenberg, AugustinMa, Geek84, GTubio, Clarkbhm, SpaceMonkey, Sjoerd visscher, I9Q79oL78KiL0QTFHgyc, Sriram sh, Matt McIrvin, Sasquatch, BM, Firewheel, MtB, Nsaa, Storm Rider, Alansohn, Gary, ChristopherWillis, Tek022, ZiggyZig, Keenan Pepper, La hapalo, Gpgarrettboast, Pippu d'Angelo, PAR, Batmanand, Hdeasy, Bart133, Snowolf, Wtmitchell, Tycho, Leoadec, Jon Cates, Mikeo, Dominic, Bsadowski1, W7KyzmJt, GabrielF, DV8 2XL, Alai, Nick Mks, KTC, Dan100, Chughtai, Falcorian, Oleg Alexandrov, Ashujo, Ott, Feezo, OwenX, Woohookitty, Linas, Superstring, Tripodics, Shoyer, StradivariusTV, Kzollman, Kosher Fan, JeremyA, Tylerni7, Pchov, GeorgeOrr, Mpatel, Adhalanay, Firien, Wikiklrsc, GregorB, AndriyK, SeventyThree, Wayward, Prashanthns, DL5MDA, Palica, Pfalstad, Graham87, Magister Mathematicae, Chun-hian, FreplySpang, Baker APS, JIP, RxS, Search4Lancer, Canderson7, Sjö, Saperaud~enwiki, Rjwilmsi, Jake Wartenberg, Linuxbeak, Tangotango, Brucelee, Darguz Parsilvan, Mike Peel, Pasky, HappyCamper, Ligulem, The wub, Ttwaring, Reinis, Hermione1980, Sango123, Oo64eva, St33lbird, Kevmitch, Titoxd, Das Nerd, Alejo2083, FlaBot, Moskvax, RobertG, Urbansky~enwiki, Arnero, Latka, Nihiltres, Pathoschild, Quuxplusone, Srleffler, Kri, Cpcheung, Phoenix2~enwiki, Chobot, DVdm, Gwernol, Niz, YurikBot, Wavelength, Paulraine, Arado, Loom91, Xihr, GLaDOS, Khatharr, Firas@user, Gaius Cornelius, Chaos, Rsrikanth05, Rodier, Wimt, Anomalocaris, Royalbroil, David R. Ingham, NawlinWiki, Grafen, NickBush24, RazorICE, Stephen e nelson, JocK, SCZenz, Randolf Richardson, Vb, E2mb0t~enwiki, Tony1, Syrthiss, SFC9394, DeadEyeArrow, Bota47, Kkmurray, Werdna, Bmju, Wknight94, WAS 4.250, FF2010, Donbert, Light current, Enormousdude, 21655, Zzuuzz, TheKoG, Lt-wiki-bot, Nielad, Closedmouth, Ketsuekigata, E Wing, Brina700, Modify, Dspradau, Netrapt, Petri Krohn, Badgettrg, Peter, Willtron, Mebden, RG2, GrinBot~enwiki, Mejor Los Indios, Sbyrnes321, CIreland, Luk, Itub, Hvitlys, SmackBot, Paulc1001, Moeron, Rex the first, InverseHypercube, KnowledgeOfSelf, Royalguard11, K-UNIT, Lagalag, Pgk, Jagged 85, Clpo13, Chairman S., Pxfbird, Grey Shadow, Delldot, Petgraveyard, Weiguxp, David Woolley, Lithium412, Philmurray, Yamaguchi??, Robbjedi, Gilliam, Slaniel, Betacommand, Skizzik, Dauto, Holy Ganga, JSpudeman, Modusoperandi, Amatulic, Stevenwagner, DetlevSchm, MK8, Jprg1966, MalafayaBot, Marks87, Silly rabbit, Complexica, Colonies Chris, Darth Panda, Sajendra, Warbirdadmiral, El Chupacabra, Zhinz, Can't sleep, clown will eat me, Physika~enwiki, Scott3, Scray, ApolloCreed, Ackbeet, Le fantome de l'opera, Onorem, Surfcuba, Voyajer, Addshore, Stiangk, Paul E T, Huon, Khoikhoi, Kingdon, DenisDiderot, Cybercobra, Nakon, Nick125, SnappingTurtle, Dreadstar, Richard001, Akriasas, Freemarket, Weregerbil, DeFoaBuSe, DMacks, Salamurai, LeoNomis, Sadi Carnot, Pilotguy, Byelf2007, Xezlec, DJIndica, Akubra, Rory096, Bcasterline, Harryboyles, JzG, Richard L. Peterson, RTejedor, AmiDaniel, UberCryxic, Wtwilson3, Zslevi, LWF, Gobonobo, Jaganath, JorisvS, Evan Robidoux, Mgiganteus1, Zarniwoot, Goodnightmush, Jordan M, Ex nihil, SirFozzie, Waggers, MarphyBlack, Caiaffa, Asyndeton, Dan Gluck, BranStark, Iridescent, JMK, Dreftymac, Joseph Solis in Australia, UncleDouggie, Rnb, Hikui87~enwiki, Cain47, Mbenzdabest, Nturton, Civil Engineer III, Cleric12121, Tawkerbot2, Chetvorno, Carborn1, Mustbe, SkyWalker, JForget, Frovingslosh, Ale jrb, Peace love and feminism, Wafulz, Sir Vicious, Asmackey, Dycedarg, Lavateraguy, Van helsing, The ed17, Bad2101, Jayunderscorezero, BeenAroundAWhile, JohnCD, Nunquam Dormio, Harriemkali, Swwright, Wquester, N2e, Melicans, Smallpond, Myasuda, Gregbard, Xana's Servant, Dragon's Blood, Cydebot, Wrwrwr, Beek man, Meznaric, Jack O'Lantern, Peterdjones, Meno25, Gogo Dodo, Islander, DangApricot, NijaMunki, Pascal.Tesson, Hughgr, Benvogel, Michael C Price, Doug Weller, Christian75, DumbBOT, FastLizard4, Waxigloo, Amit Moscovich, FrancoGG, CieloEstrellado, Thijs!bot, Epbr123, Derval Sloan, Koeplinger, Mbell, N5iln, Headbomb, Marek69, Ujm, Second Quantization, Martin Hedegaard, Philippe, CharlotteWebb, Nick Number, MichaelMaggs, Sbandrews, Mentifisto, Austin Maxwell, Cyclonenim, AntiVandalBot, Luna Santin, Widefox, Tkirkman, Eveross, Lontax, Grafnita, Rakniz, Prolog, Gnixon, CStar, TimVickers, Dylan Lake, Casomerville, Danger, Farosdaughter, Tim Shuba, North Shoreman, Yellowdesk, Glennwells, Byrgenwulf, GaaraMsg, Figma, JAnDbot, Leuko, Husond, Superior IQ Genius, MER-C, CosineKitty, Matthew Fennell, Eurobas, IJMacD, Andonic, Dcooper, Hut 8.5, 100110100, Skewwhiffy, Four Dog Night, Acroterion, Magioladitis, Connormah, Mattb112885, Bongwarrior, VoABot II, AtticusX, JamesBWatson, SHCarter, FagChops, Bfiene, Rivertorch, Michele123, Zooloo, Jmartinsson, Thunderhead~enwiki, Couki, Catgut, Indon, ClovisPt, Dirac66, 28421u2232nfenfcenc, Joe hill, Schumi555, Adventurer, Cpl Syx, Robb37, Quantummotion, DerHexer, Chaujie328, Khalid Mahmood, Teardrop onthefire, Guitarspecs, Info D, Seba5618, Gjd001, CiA10386, MartinBot, Arjun01, Rettetast, Mike6271, Fpaiano~enwiki, CommonsDelinker, AlexiusHoratius, Andrej.westermann, Tgeairn, Dinkytown, J.delanoy, DrKay, Trusilver, Kaesle, Numbo3, NightFalcon90909, Uncle Dick, Maurice Carbonaro, Kevin aylward, 5Q5, StonedChipmunk, Foober, Acalamari, Metaldev, Bot-Schafter, Katalaveno, DarkFalls, McSly, Bustamonkey2003, Ignatzmice, Tarotcards, JayJasper, Gcad92, Detah, LucianLachance, Midnight Madness, NewEnglandYankee, Rwessel, Nin0rz4u 2nv, SJP, MKoltnow, KCinDC, Han Solar de Harmonics, Cmichael, Juliancolton, Cometstyles, MoForce, Chao129, Elenseel, Wfaze, Samlyn.josfyn, Martial75, GrahamHardy, CardinalDan, Sheliak, Spellcast, Signalhead, Pgb23, Zakuragi, MBlue2020, Pleasantville, LokiClock, Lear's Fool, Soliloquial, Philip Trueman, TXiKiBoT, Oshwah, Maximillion Pegasus, SanfordEsq, RyanB88, SCriBu, Nxavar, Sean D Martin, Sankalpdravid, ChooseAnother, Qxz, Someguy1221, Liko81, Bsharvy, Olly150, XeniaKon, Clarince63, Seraphim, Saibod, Fizzackerly, Zolot, Raymondwinn, David in DC, Handsome Pete, Geometry guy, Ilyushka88, Leavage, Krazywrath, V81, Sodicadl, RandomXYZb, Lerdthenerd, Andy Dingley, Enigmaman, Meters, Lindsaiv, Synthebot, Antixt, Falcon8765, Enviroboy, Spinningspark, H1nomaru senshi, The Devil's Advocate, Monty845, Allebor-goBot, Nagy, The Mad Genius, Logan, PGWG, DarthBotto, Vitalikk, Belsazar, Katzmik, EmxBot, Givegains, Kbrose, Mk2rhino, YohanN7, SieBot, Ivan Štambuk, Nibbleboob, WereSpielChequers, Dawn Bard, AdevarTruth, RJaguar3, Hekoshi, Yintan, 4RM0~enwiki, Ujjwol, Bentogoa, Jc-S0CO, JSpung, Arjen Dijksman, Oxymoron83, Antonio Lopez, Henry Delforn (old), Hello71, AnonGuy, Lightmouse, Radzewicz, Hobartimus, Jaquesthehunter, Michael Courtney, Macy, Hatster301, Swegei, Curlymeatball38, Quackbumper, Coldcreation, Zenbullets, StaticGull, Heptarchy of teh Anglo-Saxons, baby, Mygerardromance, Fishnet37222, Stentor7, Mouselb, Randy Kryn, Velvetron, ElectronicsEnthusiast, Darrellpenta, Soporaeternus, Martarius, ClueBot, NickCT, Mod.torrentrealm, Scottstensland, Yeahyeahkickball, The Thing That Should Not Be, EMC125, Zero over zero, Infrasonik, MichaelVernonDavis, Herakles01, Drmies, Cp111, Diafanakrina, Mackafi92, Mrsastrochicken, VandalCruncher, Agge1000, Otolemur crassicaudatus, Ridge Runner, Neverquick, Asdf1990, DragonBot, Djr32, Ondon, Excirial, HounsGut, Welsh-girl-Lowri, Quercus basaseachicensis, Jusdafax, Krackenback, Winston365, Brews ohare, Sukaj, Viduoke, NuclearWarfare, Ice Cold Beer, Arjayay, Terra Xin, PhySusie, Kding, Imalad, The Red, Mikaey, SchreiberBike, Vlatkovedral, Jfioeawfjdls453, Thingg, Russel Mcpigmin, Aitias, Scalhotrod, Versus22, Mafiaparty303, SoxBot III, Apparition11, Mrvanner, Crowsnest, Vanished user uih38riiw4hjlsd, DumZiBoT,

Finalnight, CBMIBM, Javafreakin, X41, XLinkBot, Megankerr, Yokabozeez, Arthur chos, Odenluna, Matthewsassel, Feinoha, Ajcheema, AndreNatas, Paul bunion, WikHead, Loopism, NellieBly, Mifter, JinJian, Truthnlove, Airplaneman, Billcosbyislonelypart2, Mojska, Stephen Poppitt, Willieru18, Tayste, Addbot, Ryan ley, 11341134a, Willking1979, Manuel Trujillo Berges, Kadski, TylerM37, Wareagles18, XTRENCHARD29x, 11341134b, Tcncv, Betterusername, Captain-tucker, Robertd514, Fgnievinski, Mjamja, Harrytipper, SunDragon34, Blethering Scot, Ronhjones, PandaSaver, WMdeMuynck, Aboctok, JoshTW, CanadianLinuxUser, Looie496, Cst17, MrOllie, BuffaloBill90, Mitchellsims08, Chzz, AnnaFrance, Favonian, LinkFA-Bot, Adolfman, Brufnus, Barak Sh, AgadaUrbanit, Ehrenkater, Tide rolls, Lightbot, NoEdward, Romaioi, Jan eissfeldt, Teles, Jarble, Csdavis1, Ttasterul, Luckas-bot, Yobot, OrgasGirl, Tohd8BohaithuGh1, TaBOT-zerem, Niout, II MusLiM HyBRiD II, Kan8eDie, Nallimbot, Brougham96, KamikazeBot, Fearingfearitself, Positivetruthintent, IW.HG, Solo Zone, Jackthegrape, Eric-Wester, Magog the Ogre, Armegdon, N1RK4UDSK714, Octavianvs, AnomieBOT, Captain Quirk, Jim1138, IRP, Rnpg1014, Piano non troppo, AdjustShift, Giants27, Materialscientist, Gierens22, Suppplersmart, The High Fin Sperm Whale, Citation bot, Bci2, Frankenpuppy, LilHelpa, The Firewall, Joshuafilmer, Rightly, Mollymop, Xqbot, Nxtid, Sionus, Raaziq, Amareto2, Melmann, Capricorn42, Jostylr, Dbroesch, Mark Swiggle, TripLikeIDo, Benvirg89, Sokratesinabasket, Gilo1969, Physprof, Grim23, P99am, Qwertyuio 132, Gap9551, Almabot, Polgo, GrouchoBot, Abce2, Jagbag2, Frosted14, Toofy mcjack34, Richard.decal, Qzd800, Trurle, Omnipaedista, Mind my edits, WilliunWeales, Kesaloma, Charvest, The Spam-a-nata, Dale Ritter, Shipunits, FaTony, Gr33k b0i, Shadowjams, Adrignola, Dingoatscritch, Spakwee, A. di M., Naturelles, Dougofborg, Cigarettizer, ⁇, C.c. hopper, JoshC306, Chjoaygame, GliderMaven, Bboydill, Magnagr, Kroflin, Tobby72, Pepper, Commander zander, Guy82914, PhysicsExplorer, Kenneth Dawson, Colinue, Steve Quinn, N4tur4le, Pratik.mallya, Razataza, Machine Elf 1735, 06twalke, TTGL, Izodman2012, Xenfreak, Iquseruniv, HamburgerRadio, Citation bot 1, Cheryledbernard, Greg HWWOK Shaw, WQUlrich, Brettwats, Pinethicket, I dream of horses, Pink Bull, Tom.Reding, Lithium cyanide, DanielGlazer, Deaddogwalking, FloridaSpaceCowboy, RobinK, Liarliar2009, JeffreyVest, Seattle Jörg, Reconsider the static, Fredkinfollower, Superlions123, GreenReflections, Roseohioresident, Tjlafave, FoxBot, Chris5858, Anonwhymus, Trappist the monk, Buddy23Lee, 3peasants, Beta Orionis, Train2104, Hickorybark, Creativethought20, Lotje, PorkHeart, Michael9422, Lmp883, Bowlofknowledge, Leesy1106, Doc Quintana, Reaper Eternal, Azatos, SeriousGrinz, Pokemon274, Specs112, Vera.tetrix, Earthandmoon, MicioGeremia, Tbhotch, Jesse V., Sideways713, Dannideak, Factosis, MR87, Borki0, Taylo9487, Updatehelper, Seawolf1111, Onesmoothlefty, Carowinds, Bento00, Beyond My Ken, Andy chase, WildBot, Deadlyops, Phyguy03, EmausBot, John of Reading, Davejohnsan, Orphan Wiki, Bookalign, WikitanvirBot, Mahommed alpac, Dr Aaij, Gfoley4, Roxbreak, Word2need, Beatnik8983, Alamadte, Racerx11, Dickwet89, GoingBatty, Minimac's Clone, NotAnonymous0, Dmblub, KHamsun, Wham Bam Rock II, Solarra, Stevenganzburg, Elee, Tommy2010, Uploadvirus, Wikipelli, Dcirovic, Elitedarklord dragonslyer 3.14159, AsceticRose, JSquish, AlexBG72, White Trillium, Harddk, Checkingfax, Angelsages, NickJRocks95, Fæ, Josve05a, Stanford96, MithrandirAgain, Imperial Monarch, 1howardsr1, Plotfeat, User10 5, Brazmyth, Raggot, Alvindclopez, Dalma112211221122, Wayne Slam, Tolly4bolly, EricWesBrown, Mattedia, Jacksccsi, L Kensington, Qmtead, Lemony123, Final00123, Maschen, Donner60, HCPotter, Scientific29, Notolder, Pat walls1, ChuispastonBot, Roberts Ken, RockMagnetist, TYelliot, Llightex, DJDunsie, DASHBotAV, The beings, Whoop whoop pull up, Isocliff, ClueBot NG, KagakuKyouju, Professormeowington, CocuBot, MelbourneStar, This lousy T-shirt, Satellizer, MC ShAdYzOnE, Baseball Watcher, Sabri Al-Safi, Arespectablecitizen, Jj1236, Braincricket, ScottSteiner, Wikishotaro, Widr, Machdeep, Ciro.santilli, Mikeiysnake, Dorje108, Anupmehra, Theopolisme, MerlIwBot, BlooddRose, Helpful Pixie Bot, Novusuna, Olaniyob, Billybobjow, Leo3232, Elochai26, Jubobroff, Ieditpagesincorrectly, Bibcode Bot, Psaup09, Lowercase sigmabot, Saurabhagat, BG19bot, Physics1717171, Brannan.brouse, ThisLaughingGuyRightHere, Happyboy2011, Hashem sfarim, The Mark of the Beast, Northamerica1000, Declan12321, Cyberpower678, BobTheBuilder1997, Yowhatsupdude, Metricopolus, Solomon7968, Mark Arsten, Bigsean0300, Chander, Guythundar, Joydeep, Trevayne08, Roopydoop55, Aranea Mortem, Jamessweeen, F=q(E+v^B), Vagigi, Snow Blizzard, Hipsupful, Laye Mehta, Glacialfox, Winston Trechane, In11Chaudri, Achowat, Bfong2828, PinkShinyRose, Tm14, Lieutenant of Melkor, Penguinstorm300, Pkj61, Williamxu26, Jnracv, Samwalton9, Lbkt, Kisokj, Bakkedal, Cyberbot II, StopTheCrackpots, Callum Inglis, Davidwhite18, Macven, Khazar2, Adwaele, Gdrg22, BuzyBody, BrightStarSky, Dexbot, Webclient101, AutisticCatnip, Garuda0001, William.winkworth, Belief action, Harrycol123, Saehry, Jamesx12345, Josophie, Brirush, Athomeinkobe, Thepalerider2012, JustAMuggle, Reatlas, Joeinwiki, Mmcev106, Darvii, Loganfalco, Everymorning, Jakec, Rod Pierce, Backendgaming, DavidLeighEllis, Geometriccentaur, Rauledc, Eapbar, Ryomaiinsai12345, Pr.malek, LieutenantLatvia, Quadratic formula, Desswarrior, Ray brock, The Herald, Shawny J, DrYusuf786, Bubblynoah, Asherkirschbaum, W. P. Uzer, Cfunkera, SJ Defender, Melquiades Babilonia, Atticus Finch28, Dfranz1012, PhuongAlex, JaconaFrere, 15petedc, AspaasBekkelund, QuantumMatt101, Htp0020, Derenek, Russainbiaed na, Internucleotide, Emmaellix, Renegade469, Nikrulz07, HiYahhFriend, Johntrollston1233, BethNaught, HolLak456, Castielsbloodyface, Trackteur, Black789Green456, Kinetic37, Theskruff, DaleReese1962, Zazzi01, Garfield Garfield, Potatomuncher2000, 3primetime3, 420noscopekills, HMSLavender, The Original Bob, EvilLair, 427454LSX, ChamithN, Suman Chatterjee DHEP, HelloFriendsOfPlanetEarth, Zppix, Audiorew, Trentln1852, CheeseButterfly, Blackbeast75, Justdausualtf, Whijus19, Govindaharihari, Dubsir, Virophage, Lanzdsey, Tetra quark, Isambard Kingdom, Rohin2002, Bloodorange1234, Harsh mahesheka, Skipfortyfour, Username1234567890101112131415161718 1920, Camisboss5, WebdriverHead, SamuelFey666, Cnbr15, Amccann421, Jerodlycett, KasparBot, MintyTurtle01, Peter Richard Obama, Sweepy, Pengyulong7, TheDoctor07, Tropicalkitty, Ipskycak, Iavathunderman, Seventhorbitday, 420BlazeItPhaggot, Jeffjef, Matthewadinatajapati, Fthatshiit, Zackwright07, RedExplosiveswiki, Urmomisdumb69, JonahSpars, GreenC bot, Eisengetribe13, PlayGatered, Eep03, PANDA12346, Konic004, Mindopener420, Imsarvesh18, Ecnomercy and Anonymous: 1784

- **Quantum field theory** *Source:* https://en.wikipedia.org/wiki/Quantum_field_theory?oldid=723590432 *Contributors:* AxelBoldt, CYD, Mav, The Anome, XJaM, Roadrunner, Stevertigo, Michael Hardy, Tim Starling, IZAK, TakuyaMurata, SebastianHelm, Looxix~enwiki, Ahoerstemeier, Cyp, Glenn, Rotem Dan, Stupidmoron, Charles Matthews, Timwi, Jitse Niesen, Kbk, Rudminjd, Wik, Phys, Bevo, BenRG, Northgrove, Robbot, Bkalafut, Gandalf61, Rursus, Fuelbottle, Tea2min, Ancheta Wis, Giftlite, Lethe, Dratman, Alison, St3vo, Mboverload, DefLog~enwiki, ConradPino, Amarvc, Pcarbonn, Karol Langner, APH, AmarChandra, D6, CALR, Urvabara, Discospinster, Guanabot, Igorivanov~enwiki, Masudr, Pjacobi, Vsmith, Nvj, MuDavid, Bender235, Pt, El C, Shanes, Sietse Snel, Physicistjedi, KarlHallowell, PWilkinson, Helix84, Thialfi, Varuna, Gcbirzan, Docboat, Count Iblis, Egg, Tripodics, Mpatel, Marudubshinki, Graham87, Opie, Vanderdecken, Rjwilmsi, MarSch, Earin, R.e.b., RE, Strobilomyces, Arnero, Itinerant1, Alfred Centauri, Srleffler, Chobot, UkPaolo, Wavelength, Bambaiah, Hairy Dude, RussBot, TimNelson, Archelon, CambridgeBayWeather, SCZenz, Odddmonster, E2mb0t~enwiki, Semperf, Tetracube, Garion96, Erik J, Robert L, Banus, RG2, SmackBot, Stephan Schneider, Tom Lougheed, Melchoir, Rentier, KocjoBot~enwiki, Mcld, Dauto, Chris the speller, Complexica, Threepounds, RuudVisser, QFT, Jmnbatista, Cybercobra, Rebooted, Victor Eremita, DJIndica, Lambiam, Mgiganteus1, Zarniwoot, Jim.belk, Stwalkerster, SirFozzie, Hu12, Dan Gluck, Iridescent, Joseph Solis in Australia, Albertod4, Van helsing, BeenAroundAWhile, Witten Is God, Cydebot, Jamie Lokier, Meno25, Michael C Price, The 80s chick, Mendicus~enwiki, AstroPig7, Msebast~enwiki, Mbell, Headbomb, Nick Number, Mentifisto, AntiVandalBot, Bt414, Bananan~enwiki, Martin Kostner, Moltrix, Kasimann, Kromatol, Puksik, Lerman, LLHolm, RogueNinja, Tlabshier, JEH, Nikolas Karalis, Storkk, JAnDbot, Igodard, Four Dog Night, N shaji, Bongwarrior, Andrea Allais, Soulbot, Etale, Maliz, Custos0, HEL, J.delanoy, Maurice Carbonaro, Acalamari, Jeepday, Policron, Blckavnger, Juliancolton, Skou, Telecomtom,

GrahamHardy, Sheliak, Cuzkatzimhut, VolkovBot, Pleasantville, Bktennis2006, Marksr, HowardFrampton, Oshwah, The Original Wildbear, Dj thegreat, Markisgreen, TBond, Lejarrag, Moose-32, Raphtee, Sue Rangell, Neparis, Drschawrz, YohanN7, SieBot, TCO, Yintan, Likebox, Paolo.dL, Tugjob, Henry Delforn (old), Jecht (Final Fantasy X), OKBot, Randy Kryn, StewartMH, ClueBot, EoGuy, Wwheaton, Niceguyedc, The Wild West guy, Shvav~enwiki, Bob108, Brews ohare, Thingg, Count Truthstein, XLinkBot, PSimeon, SilvonenBot, Truthnlove, Hexa-Chord, Addbot, ConCompS, Pinkgoanna, Leapold~enwiki, Dmhowarth26, Glane23, Hanish.polavarapu, Lightbot, Scientryst, R.ductor, Ettrig, Yndurain, Legobot, Luckas-bot, Yobot, Ht686rg90, Niout, Tamtamar, AnomieBOT, Ciphers, Palpher, IRP, Gjsreejith, Materialscientist, Citation bot, Bci2, ArthurBot, Northryde, LilHelpa, Caracolillo, Amareto2, MIRROR, Professor J Lawrence, Plasmon1248, Omnipaedista, RibotBOT, Spellage, JayJay, FrescoBot, Kenneth Dawson, D'ohBot, Knowandgive, N4tur4le, Hyqeom, Newt Winkler, Hickorybark, Lotje, Dinamik-bot, LilyKitty, Fortesque666, Reaper Eternal, Minimac, Marie Poise, Yaush, Dylan1946, EmausBot, Racerx11, GoingBatty, Carbosi, Thecheesykid, ZéroBot, Cogiati, Jjspinorfield1, Suslindisambiguator, Quondum, Maschen, Zueignung, Davidaedwards, RockMagnetist, Lom Konkreta, ClueBot NG, Gilderien, Iloveandrea, Vacation9, Heyheyheyhohoho, Fortune432, The ubik, Zak.estrada, Widr, Helpful Pixie Bot, Bibcode Bot, Guy vandegrift, Evanescent7, Ykentluo, Martin.uecker, Walterpfeifer, Pfeiferwalter, Klilidiplomus, David.moreno72, W.D., CarrieVS, Khazar2, Momo1381, Dexbot, Cerabot~enwiki, Garuda0001, AHusain314, Thepalerider2012, A.entropy, Mark viking, Faizan, Aj7s6, संजीव कुमार, Lemnaminor, BerFinelli, Axel.P.Hedstrom, Kclongstocking, Mutley1989, I art a troler, Liquidityinsta, Prokaryotes, DemonThuum, Dingdong2680, Asherkirschbaum, Monkbot, Gjbayes, Thedarkcheese, BradNorton1979, UareNumber6, Teelaskeletor, YeOldeGentleman, Mret81, KasparBot, Ami.bangali, CAPTAIN RAJU, Koitus~nlwiki, Rico deutsch and Anonymous: 310

- **Hamiltonian constraint** *Source:* https://en.wikipedia.org/wiki/Hamiltonian_constraint?oldid=722755857 *Contributors:* Phys, Joy, Meelar, HorsePunchKid, Mindspillage, Rich Farmbrough, Prsephone1674, Linas, SeventyThree, Conscious, Salsb, Dialectric, SmackBot, Cuzkatzimhut, Flyer22 Reborn, Flaming, Carriearchdale, Yobot, Ibayn, 4th-otaku, AnomieBOT, Omnipaedista, Bj norge, John of Reading and Anonymous: 20
- **Hamiltonian mechanics** *Source:* https://en.wikipedia.org/wiki/Hamiltonian_mechanics?oldid=718047665 *Contributors:* CYD, Zundark, Mjb, David spector, Michael Hardy, Pit~enwiki, Cyde, Looxix~enwiki, Stevan White, AugPi, Charles Matthews, Reddi, Jitse Niesen, Phys, Bevo, Chuunen Baka, Robbot, Sverdrup, Tea2min, Snobot, Giftlite, Lethe, Dratman, Zhen Lin, Jason Quinn, HorsePunchKid, Chris Howard, D6, Bender235, Laurascudder, Army1987, Rephorm, Linuxlad, Jheald, Gene Nygaard, Linas, PhoenixPinion, Isnow, K3wq, Nanite, RE, Rbeas, Mathbot, Srleffler, Kri, Chobot, Sanpaz, YurikBot, Borgx, RobotE, RussBot, KSmrq, Archelon, David R. Ingham, Bachrach44, Hyandat, Crasshopper, Reyk, Gbmaizol, Darrel francis, Mebden, Teply, Samuel Blanning, SmackBot, Errarel, 7segment, Frédérick Lacasse, TimBentley, Movementarian, MK8, Complexica, Akriasas, Wybot, Atoll, Xenure, JRSpriggs, OS2Warp, Mct mht, Cydebot, Rwmcgwier, BobQQ, Bb vb, Dchristle, Ebyabe, Mbell, Headbomb, Paquitotrek, Sbandrews, Rico402, JAnDbot, Felix116, Epq, Andrej.westermann, Maurice Carbonaro, LordAnubisBOT, Hessammehr, STBotD, Borat fan, Sheliak, Cuzkatzimhut, Gazok, VolkovBot, JohnBlackburne, Barbacana, Thurth, BertSen, Red Act, Voorlandt, Cgwaldman, Geometry guy, YohanN7, SieBot, JerrySteal, JerroldPease-Atlanta, Commutator, Anchor Link Bot, PerryTachett, StewartMH, UrsusArctosL71, AstroMark, Razimantv, HHHEB3, Alexey Muranov, DS1000, Crowsnest, Addbot, EjsBot, SPat, Zorrobot, Luckas-bot, Ptbotgourou, KamikazeBot, Freeskyman, AnomieBOT, Stefansquintet, Citation bot, Frederic Y Bois, Omnipaedista, RibotBOT, Craig Pemberton, DrilBot, Vrenator, Doctor Zook, EmausBot, John of Reading, Helptry, Netheril96, Gerasime, SporkBot, AManWithNoPlan, Maschen, Zueignung, RockMagnetist, Jorgecarleitao, Helpful Pixie Bot, Jcc2011, BG19bot, Dzustin, Jamontaldi, F=q(E+v^B), ChrisGualtieri, Kylarnys, Hublolly, Dimoroi, Elenceq, Epic Wink, Zmilne, CAPTAIN RAJU, Fgnv052 and Anonymous: 117
- **Lie algebra** *Source:* https://en.wikipedia.org/wiki/Lie_algebra?oldid=722830756 *Contributors:* AxelBoldt, Zundark, Miguel~enwiki, Michael Hardy, Wshun, Joel Koerwer, TakuyaMurata, Suisui, Kragen, Rossami, Iorsh, Loren Rosen, Charles Matthews, Dysprosia, Michael Larsen, Grendelkhan, Phys, BenRG, Tea2min, David Gerard, Weialawaga~enwiki, Tosha, Giftlite, BenFrantzDale, Lethe, Fropuff, Curps, Jeremy Henty, Jason Quinn, Python eggs, Chameleon, DefLog~enwiki, CryptoDerk, CSTAR, Pyrop, Guanabot, Pj.de.bruin, Vsmith, Gauge, Pt, Kwamikagami, Wood Thrush, Reinyday, Foobaz, Msh210, Arthena, Spangineer, Dirac1933, Drbreznjev, Oleg Alexandrov, Linas, Isnow, BD2412, NatusRoma, MarSch, Mathbot, Margosbot~enwiki, RexNL, Masnevets, YurikBot, Wavelength, Hairy Dude, Michael Slone, Lenthe, Stephenb, Grubber, Trovatore, Asimy, Crasshopper, Curpsbot-unicodify, Sbyrnes321, SmackBot, Incnis Mrsi, Grokmoo, Kmarinas86, Bluebot, Silly rabbit, Nbarth, Thomas Bliem, Chlewbot, BlackFingolfin, Noegenesis, Rschwieb, AlainD, Harold f, CmdrObot, Shirulashem, Headbomb, Second Quantization, Dachande, RobHar, B-80, Jrw@pobox.com, Deflective, Englebert, Vanish2, David Eppstein, R'n'B, Bogey97, Maurice Carbonaro, Supermanifold, Policron, Fylwind, Cuzkatzimhut, VolkovBot, JohnBlackburne, LokiClock, Ndbrian1, Hesam7, Geometry guy, Drorata, Arcfrk, StevenJohnston, YohanN7, SieBot, Stca74, Jenny Lam, Paolo.dL, JackSchmidt, Mr. Stradivarius, Fatchat, Veromies, JP.Martin-Flatin, Count Truthstein, Addbot, Roentgenium111, Lightbot, Legobot, Luckas-bot, Yobot, Niout, Jason Recliner, Esq., DutchCanadian, Delilahblue, AnomieBOT, Twri, SassoBot, Kaoru Itou, D'ohBot, Darij, Juniuswikiae, Prtmrz, Rausch, Jkock, Adam cohenus, TobeBot, Lotje, Doctor Zook, Slawekb, Quondum, Mikhail Ryazanov, ClueBot NG, Dd314, BG19bot, Teika kazura, Walterpfeifer, Pfeiferwalter, IkamusumeFan, Stefan.Groote, Flbsimas, Deltahedron, Saung Tadashi, Mark L MacDonald, Danielbrice, Enyokoyama, CsDix, Icarot, 314Username, Forgetfulfunctor00, CaptainLama, KasparBot, Texnico, Ryanexler, Dougal37 and Anonymous: 95
- **Lie group** *Source:* https://en.wikipedia.org/wiki/Lie_group?oldid=709075836 *Contributors:* AxelBoldt, Zundark, Josh Grosse, XJaM, Miguel~enwiki, Stevertigo, Xavic69, Michael Hardy, TakuyaMurata, GTBacchus, Looxix~enwiki, Barak~enwiki, Charles Matthews, Dysprosia, Jitse Niesen, Zoicon5, David Shay, Itai, Phys, Josh Cherry, Saaska, Tea2min, Weialawaga~enwiki, Tosha, Giftlite, JamesMLane, BenFrantzDale, Lethe, Fropuff, Wgmccallum, Jason Quinn, Bobblewik, DefLog~enwiki, Lockeownzj00, Beland, Pmanderson, Abdull, Dablaze, MuDavid, Paul August, ChrisJ, Bender235, Tompw, Rgdboer, Kwamikagami, Shanes, Cherlin, Msh210, PAR, Alex Varghese, Oleg Alexandrov, Zntrip, Joriki, Linas, Dzordzm, Isnow, SDC, AnmaFinotera, Frankie1969, Graham87, Porcher, Rjwilmsi, NatusRoma, MarSch, Salix alba, HappyCamper, R.e.b., VKokielov, BMF81, Masnevets, Chobot, Algebraist, Wavelength, Hillman, RussBot, Michael Slone, KSmrq, Archelon, Buster79, Arkapravo, Smaines, Orthografer, Ekeb, Kier07, Pred, RodVance, JDspeeder1, SmackBot, Incnis Mrsi, Tom Lougheed, FlashSheridan, Davewild, Mhss, Kmarinas86, Bluebot, Badger014, Silly rabbit, DHN-bot~enwiki, Bears16, Akriasas, KeithB, Lambiam, Ninte, Siva1979, John, Ulner, Jim.belk, Michael Kinyon, Inquisitus, Mathchem271828, Rschwieb, Krasnoludek, Yggdrasil014, CRGreathouse, CBM, Logical2u, Myasuda, Kupirijo, MotherFunctor, The real dan, Dr.enh, Xantharius, Thijs!bot, Headbomb, Marek69, JustAGal, Richard-Veryard, RobHar, Salgueiro~enwiki, Dougher, Len Raymond, JAnDbot, Deflective, Unifey~enwiki, Homeworlds, Magioladitis, Bongwarrior, Cmelby, WhatamIdoing, Sullivan.t.j, David Eppstein, The Real Marauder, Benjamin.friedrich, David J Wilson, Jesper Carlstrom, Maproom, TomyDuby, Rocket71048576, Pidara, Fylwind, Dorftrottel, Lseixas, Borat fan, Cuzkatzimhut, Trevorgoodchild, JohnBlackburne, Ndbrian1, James.r.a.gray, Hesam7, Geometry guy, Jmath666, Eubulides, Brian Huffman, Genuine0legend, Drorata, Arcfrk, Smylei, Oscarbaltazar, YohanN7, JackSchmidt, S2000magician, Beastinwith, Mr. Stradivarius, Deciwill, Sidiropo, Leontios, Heckledpie, Cacadril, SchreiberBike,

Marc van Leeuwen, MystBot, Addbot, Topology Expert, LaaknorBot, Ozob, Tanath, Tide rolls, Luckas-bot, Yobot, Ht686rg90, Niout, Amirobot, AnomieBOT, Citation bot, ArthurBot, Br77rino, Kaoru Itou, FrescoBot, Anterior1, Sławomir Biały, RedBot, Tinfoilcat, EmausBot, KbReZiE 12, Darkfight, Slawekb, Suslindisambiguator, Maschen, Zueignung, Anita5192, ClueBot NG, Mgvongoeden, Kasirbot, Helpful Pixie Bot, Daviddwd, BG19bot, CitationCleanerBot, Fraisière, NotWith, MathKnight-at-TAU, Hemlisp, Suhagja, Brirush, CsDix, Sol1, Blackbombchu, Pwm86, Mathphysman, Abitslow, Cbartondock, Victoryhuy, KasparBot, Egdunne, Referencing, Chemistry1111 and Anonymous: 114

- **Lie derivative** *Source:* https://en.wikipedia.org/wiki/Lie_derivative?oldid=711576420 *Contributors:* Michael Hardy, TakuyaMurata, Karada, Delirium, Docu, Charles Matthews, Dysprosia, Phys, Tea2min, Tosha, Giftlite, Donvinzk, BenFrantzDale, Fropuff, CryptoDerk, AmarChandra, Rich Farmbrough, TedPavlic, Mat cross, Bender235, Gauge, CanisRufus, EmilJ, Pearle, SteinbDJ, Alai, Oleg Alexandrov, Japanese Searobin, JoeRiel, Linas, Mpatel, Isnow, Ryan Reich, BD2412, Grammarbot, MarSch, Mathbot, Itinerant1, YurikBot, Gaius Cornelius, Orthografer, SmackBot, Melchoir, Eskimbot, KHarbaugh, Silly rabbit, Ewjw, Henning Makholm, Leo C Stein, Dl2000, JRSpriggs, Alaibot, Dharma6662000, Deflective, Enlil2, RogierBrussee, Sullivan.t.j, Alu042, Lantonov, Policron, Cuzkatzimhut, 28bytes, Trevorgoodchild, AlnoktaBOT, Saibod, Geometry guy, Wbrenna36, Perturbationist, Gulmammad, ChrisHodgesUK, Panzieri, AnonyScientist, Addbot, Dabsent, Cesiumfrog, Yobot, Ht686rg90, AnomieBOT, Teleprinter Sleuth, Point-set topologist, Hep thinker, Night Jaguar, Tcnuk, Rausch, TobeBot, Sayantan m, Fly by Night, GoingBatty, Stephan Spahn, Slawekb, ZéroBot, Quondum, Innumerate1979, Maschen, Bomazi, Dylan Moreland, Mgvongoeden, Tbennert, Helpful Pixie Bot, Jimw338, Manstein1942, Mark viking and Anonymous: 67
- **Gauge theory** *Source:* https://en.wikipedia.org/wiki/Gauge_theory?oldid=718602614 *Contributors:* The Anome, Michael Hardy, Grendelkhan, Tea2min, Ancheta Wis, Beland, TedPavlic, Xezbeth, MuDavid, Bender235, Pt, Phils, BD2412, Rjwilmsi, JocK, Modify, Teply, SmackBot, RD-Bury, Henning Makholm, Byelf2007, Michael C Price, Biblbroks, Headbomb, Nick Number, Fashionslide, VectorPosse, Magioladitis, Bakken, Email4mobile, JaGa, Policron, Squids and Chips, Cuzkatzimhut, VolkovBot, Red Act, Michael H 34, Setreset, Jwpitts, Tcamps42, Moonriddengirl, ClueBot, Mastertek, TimothyRias, XLinkBot, Addbot, Mortense, Eric Drexler, Bte99, Zorrobot, Luckas-bot, Maximilian Reininghaus, AnomieBOT, Christopher.Gordon3, Citation bot, Northryde, Xqbot, Pra1998, Omnipaedista, Gsard, A. di M., Erik9bot, FrescoBot, Fortdj33, Citation bot 1, Ganondolf, RedBot, RobinK, Mary at CERN, EmausBot, Brent Perreault, Slawekb, Cogiati, Maschen, Isocliff, ClueBot NG, Helpful Pixie Bot, Bibcode Bot, Dzustin, Brendan.Oz, ChrisGualtieri, SD5bot, Dexbot, Enyokoyama, Joeinwiki, Dath Thou Even Lift, Dhm4444, Dbw1976, KasparBot, Marianna251 and Anonymous: 46
- **Holonomy** *Source:* https://en.wikipedia.org/wiki/Holonomy?oldid=707205764 *Contributors:* Zundark, Charles Matthews, Dysprosia, Phys, Tosha, Giftlite, Fropuff, Mennucc, Lumidek, Serenus~enwiki, Zaslav, Rgdboer, Firsfron, Rjwilmsi, MarSch, Salix alba, R.e.b., Mathbot, Chobot, YurikBot, Wavelength, Archelon, Orthografer, Yaco, RUZA, SmackBot, Moocowpong1, Silly rabbit, Tesseran, Mets501, Dan Gluck, Comech, Myasuda, Headbomb, David Eppstein, Lantonov, Squids and Chips, Camulogene, Aaport, Mathphd1989, Nilradical, SchreiberBike, MystBot, Addbot, DOI bot, Lightbot, Citation bot, Point-set topologist, Jatosado, FrescoBot, Citation bot 1, Ordnascrazy, Julian Birdbath, Fly by Night, Klbrain, Just granpa, Helpful Pixie Bot, ????, Brad7777, Wolfgang42, BattyBot, IkamusumeFan, Zalcberg, Monkbot, Luciaernaga and Anonymous: 29
- **Wheeler–DeWitt equation** *Source:* https://en.wikipedia.org/wiki/Wheeler%E2%80%93DeWitt_equation?oldid=709499327 *Contributors:* The Anome, Charles Matthews, Phys, Giftlite, Lethe, Herbee, Micru, Lumidek, Chris Howard, CALR, Eb.hoop, Dmr2, Bender235, John Vandenberg, Jag123, Physicistjedi, Weyes, Mpatel, Zzyzx11, Rjwilmsi, Strobilomyces, Itinerant1, Sbove, Chobot, Conscious, Archelon, EEMIV, Larsobrien, Teply, KasugaHuang, SmackBot, Lambiam, Harryboyles, Mets501, Michael C Price, Karl-H, Marek69, .anacondabot, Geniac, Rhwawn, GreenJoe, Alro, Tarotcards, Hulten, Pamputt, AlleborgoBot, Reinderien, Henry Delforn (old), WurmWoode, Debsuvra, DragonBot, Djr32, SchreiberBike, Pqnelson, Addbot, EdgeNavidad, LaaknorBot, Yobot, Nallimbot, AnomieBOT, Novel Zephyr, Xqbot, FrescoBot, HRoest-Bot, Σ, Abromwell, Bookalign, Throwmeaway, Maschen, Cristi Stoica, Kartasto, Raidr, Ben morphett, Helpful Pixie Bot, Bibcode Bot, BattyBot, Sonarjetlens, ChrisGualtieri, EvergreenFir, Db9199 24, Kanawishi, Sofia Koutsouveli and Anonymous: 31
- **Graph (discrete mathematics)** *Source:* https://en.wikipedia.org/wiki/Graph_(discrete_mathematics)?oldid=720258080 *Contributors:* The Anome, Manning Bartlett, XJaM, Tomo, Stevertigo, Patrick, Michael Hardy, W~enwiki, Zocky, Wshun, Chris-martin, Karada, Ahoerstemeier, Den fjättrade ankan~enwiki, Jiang, Dcoetzee, Dysprosia, Doradus, Zero0000, McKay, BenRG, Robbot, LuckyWizard, Mountain, Altenmann, Mayooranathan, Gandalf61, MathMartin, Timrollpickering, Bkell, Tea2min, Tosha, Giftlite, Dbenbenn, Harp, Tom harrison, Chinasaur, Jason Quinn, Matt Crypto, Neilc, Erhudy, Knutux, Yath, Joeblakesley, Tomruen, Peter Kwok, Aknorals, Chmod007, Abdull, Corti, PhotoBox, Discospinster, Rich Farmbrough, Andros 1337, Paul August, Bender235, Zaslav, Gauge, Tompw, Crisófilax, Yitzhak, Kine, Bobo192, Jpiw~enwiki, Mdd, Jumbuck, Zachlipton, Sswn, Liao, Rgclegg, Paleorthid, Super-Magician, Mahanga, Joriki, Mindmatrix, Wesley Moy, Oliphaunt, Brentdax, Jwanders, Tbc2, Cbdorsett, Ch'marr, Davidfstr, Xiong, Marudubshinki, Tslocum, Magister Mathematicae, Ilya, BD2412, SixWingedSeraph, Sjo, Rjwilmsi, Salix alba, Bhadani, FlaBot, Nowhither, Mathbot, Gurch, MikeBorkowski~enwiki, Chronist~enwiki, Silversmith, Chobot, Peterl, Siddhant, Borgx, Karlscherer3, Hairy Dude, Gene.arboit, Michael Slone, Gaius Cornelius, Rsrikanth05, Shanel, Gwaihir, Dtrebbien, Dureo, Doetoe, Wknight94, Arthur Rubin, Netrapt, RobertBorgersen, Cjfsyntropy, RonnieBrown, Burnin1134, SmackBot, Nihonjoe, Stux, McGeddon, BiT, Algont, Ohnoitsjamie, Chris the speller, Bluebot, TimBentley, Theone256, Cornflake pirate, Zven, Anabus, Can't sleep, clown will eat me, Tamfang, Cybercobra, Jon Awbrey, Kuru, Nat2, Tomhubbard, Dicklyon, Cbuckley, Quaeler, BranStark, Wandrer2, George100, Ylloh, Vaughan Pratt, Repied, CRGreathouse, Citrus538, Jokes Free4Me, Requestion, Myasuda, Danrah, Robertsteadman, Eric Lengyel, Headbomb, Urdutext, AntiVandalBot, Hannes Eder, JAnDbot, MER-C, Dreamster, Struthious Bandersnatch, JNW, Catgut, David Eppstein, JoergenB, MartinBot, Rettetast, R'n'B, J.delanoy, Hans Dunkelberg, Yecril, Pafcu, Ijdejter, Deor, ABF, Maghnus, TXiKiBoT, Sdrucker, Someguy1221, PaulTanenbaum, Lambyte, Ilia Kr., Jpeeling, Falcon8765, RaseaC, Insanity Incarnate, Zenek.k, Radagast3, Debamf, Debeolaurus, SieBot, Minder2k, Dawn Bard, Cwkmail, Jon har, SophomoricPedant, Oxymoron83, Henry Delforn (old), Ddxc, Svick, Phegyi81, Anchor Link Bot, Jarauh, ClueBot, Vacio, Nsk92, JuPitEer, Huynl, JP.Martin-Flatin, Xavexgoem, UKoch, Mitmaro, Editor70, Watchduck, Hans Adler, Suchap, Wikidsp, Muro Bot, 3ICE, Aitias, Versus22, Djk3, Kruusamägi, SoxBot III, XLinkBot, Marc van Leeuwen, Libcub, WikiDao, Tangi-tamma, Addbot, Gutin, Athenray, Willking1979, Royerloic, West.andrew.g, Tyw7, Zorrobot, LuK3, Luckas-bot, Yobot, TaBOT-zerem, THEN WHO WAS PHONE?, E mraedarab, Tempodivalse, Пика Пика, Ulric1313, RandomAct, Materialscientist, Twri, Dockfish, Anand jeyahar, Miym, Prunesqualer, Andyman100, VictorPorton, JonDePlume, Shadowjams, A.amitkumar, Kracekumar, Edgars2007, Citation bot 1, Maggyero, DrilBot, Amintora, Pinethicket, Calmer Waters, RobinK, Barras, Tgv8925, DARTH SIDIOUS 2, Powerthirst123, DRAGON BOOSTER, Mymyhoward16, Kerrick Staley, Ajraddatz, Wgunther, Bethnim, Akutagawa10, White Trillium, Josve05a, D.Lazard, L Kensington, Maschen, Inka 888, Chewings72, ClueBot NG, Wcherowi, MelbourneStar, Kingmash, O.Koslowski, Joel B. Lewis, Andrewsky00, Timflutre, Helpful Pixie Bot, HMSSolent, BG19bot, Grolmusz, John Cummings, Stevetihi, Канеюку, Void-995, MRG90, Vanischenu, Tman159, BattyBot, Ekren,

Lugia2453, Jeff Erickson, CentroBabbage, Nina Cerutti, Chip Wildon Forster, Yloreander, Manul, JaconaFrere, Monkbot, Hou710, Anon124, Aryan5496, Dr.basheer09, SlvrKy, Fmadd and Anonymous: 361

- **Spin (physics)** *Source:* https://en.wikipedia.org/wiki/Spin_(physics)?oldid=722063311 *Contributors:* AxelBoldt, CYD, The Anome, Larry_Sanger, Andre Engels, XJaM, David spector, Stevertigo, Xavic69, Michael Hardy, Tim Starling, Dominus, Cyp, Stevenj, Glenn, AugPi, Rossami, Nikai, Andres, Med, Mxn, Charles Matthews, Timwi, Kbk, 4lex, Reina riemann, E23~enwiki, Phys, Wtrmute, Bevo, Elwoz, Robbot, Gandalf61, Blainster, DHN, Hadal, Papadopc, Jheise, Anthony, Diberri, Xanzzibar, Giftlite, Smjg, Lethe, Lupin, MathKnight, Xerxes314, Average Earthman, AlistairMcMillan, Ato, Andycjp, Gzuckier, Beland, Karol Langner, Spiralhighway, Elroch, B.d.mills, Tsemii, Frau Holle, Mike Rosoft, Igorivanov~enwiki, FT2, MuDavid, Paul August, Pt, Susvolans, Army1987, Wood Thrush, SpeedyGonsales, Physicistjedi, Obradovic Goran, Neonumbers, Keenan Pepper, Count Iblis, Egg, Linas, Palica, Torquil~enwiki, Ashmoo, Grammarbot, Zoz, Rjwilmsi, Zbxgscqf, Drrngrvy, FlaBot, Mathbot, TheMidnighters, Itinerant1, Ewlyahoocom, Gurch, Fresheneesz, Srleffler, Kri, Chobot, DVdm, YurikBot, Bambaiah, Hairy Dude, JabberWok, Rsrikanth05, NawlinWiki, Buster79, Hwasungmars, Kkmurray, Werdna, Djdaedalus, Simen, Netrapt, Mpjohans, KSevcik, GrinBot~enwiki, Joshronsen, Bo Jacoby, Sbyrnes321, DVD R W, Shanesan, KasugaHuang, That Guy, From That Show!, SmackBot, Unyoyega, Bluebot, Complexica, DHN-bot~enwiki, Sergio.ballestrero, V1adis1av, QFT, Voyajer, Terryeo, Ryanluck, Radagast83, Jgrahamc, MichaelBillington, Richard001, DMacks, Daniel.Cardenas, Bidabadi~enwiki, Sadi Carnot, Bdushaw, Andrei Stroe, Tesseran, SashatoBot, Leo C Stein, Vanished user 9i39j3, UberCryxic, Jonas Ferry, Vgy7ujm, Loodog, Jaganath, Terry Bollinger, Wierdw123, Inquisitus, Beefyt, Jc37, Dreftymac, Newone, RokasT~enwiki, Jaksmata, Aepryus, JRSpriggs, Joostvandeputte~enwiki, CRGreathouse, David s graff, Ahmes, Jason-Hise, Eric Le Bigot, Bmk, Myasuda, Mct mht, FilipeS, Cydebot, A876, Thijs!bot, Barticus88, Headbomb, Brichcja, Davidhorman, Oreo Priest, Widefox, Orionus, Tlabshier, Accordionman, Astavats, JAnDbot, Em3ryguy, MER-C, Igodard, PhilKnight, .anacondabot, Sangak, Magioladitis, Swpb, Dirac66, LorenzoB, Monurkar~enwiki, TechnoFaye, Brilliand, R'n'B, CommonsDelinker, Victor Blacus, J.delanoy, Numbo3, Sackm, Maurice Carbonaro, Klatkinson, Cmichael, Uberdude85, Craklyn, CardinalDan, Cuzkatzimhut, VolkovBot, Error9312, JohnBlackburne, Bolzano~enwiki, TXiKiBoT, Hqb, Anonymous Dissident, Costela, BotKung, Lamro, Kganjam, Petergans, Kbrose, SieBot, BotMultichill, The way, the truth, and the light, RadicalOne, Flyer22 Reborn, Jasondet, Paolo.dL, R J Sutherland, Lightmouse, JackSchmidt, Martarius, ClueBot, JonnybrotherJr, Warbler271, Mild Bill Hiccup, David Trochos, Outerrealm, Sbian, Peachypoh, SchreiberBike, Ant59, Crowsnest, XLinkBot, Addbot, Mathieu Perrin, Narayansg, Imeriki al-Shimoni, Sriharsha.karnati, Numbo3-bot, Tide rolls, Lightbot, Luckas-bot, Yobot, Nallimbot, AnomieBOT, Lendtuffz, Citation bot, Nepahwin, ArthurBot, Obersachsebot, Xqbot, Sionus, WandringMinstrel, Francine Rogers, Pradameinhoff, Tom1936, Ernsts, A. di M., NoldorinElf, Daleang, Baz.77.243.99.32, LucienBOT, Paine Ellsworth, Tobby72, Freddy78, Craig Pemberton, C.Bluck, Jondn, Pokyrek, Citation bot 1, I dream of horses, Adlerbot, Casimir9999, Kallikanzarid, Jkforde, Trappist the monk, Michael9422, Miracle Pen, Sgravn, 8af4bf06611c, Garuh knight, EmausBot, Beatnik8983, GoingBatty, JustinTime55, Zhenyok 1, Atomicann, JSquish, ZéroBot, Harddk, Neh0000, Quondum, Jacksccsi, Maschen, Zueignung, Rasinj, RockMagnetist, Eg-T2g, ClueBot NG, Paolo328, Gilderien, Frietjes, Widr, PhiMAP, Helpful Pixie Bot, BroOkWiki, Bibcode Bot, BG19bot, PUECH P.-F., Mark Arsten, Yudem, F=q(E+v^B), Blaspie55, Halfb1t, Robertwilliams2011, Dexbot, Foreverascone, ScitDei, Mark viking, Pedantchemist, YiFeiBot, W. P. Uzer, Francois-Pier, Mathphysman, Aidan Clark, Brotter121, Cpt Wise, KasparBot, Solobear34 and Anonymous: 247
- **General covariance** *Source:* https://en.wikipedia.org/wiki/General_covariance?oldid=722951852 *Contributors:* The Anome, Charles Matthews, Aenar, Sdedeo, Isopropyl, Giftlite, Anythingyouwant, Lumidek, PhotoBox, Pearle, Oleg Alexandrov, Cleonis, Mpatel, MarSch, Ligulem, Ems57fcva, Reedbeta, Mathbot, Hillman, Salsb, Petri Krohn, SmackBot, Bytesmythe, Polonium, Ligulembot, JRSpriggs, Storm63640, Michael C Price, Thijs!bot, Headbomb, DAGwyn, BatteryIncluded, J Hill, JCarlos, Kevin aylward, Lantonov, Tarotcards, Henry Delforn (old), Michel421, Brews ohare, MystBot, Addbot, Ozob, AnomieBOT, Citation bot, MauritsBot, Xqbot, Point-set topologist, Gsard, Anterior1, HRoestBot, RobinK, Rausch, Dcirovic, Gilderien, Helpful Pixie Bot, BattyBot, I3roly and Anonymous: 19
- **Background independence** *Source:* https://en.wikipedia.org/wiki/Background_independence?oldid=708783336 *Contributors:* The Anome, Timc, Phil Boswell, Sdedeo, Merovingian, Jason Quinn, Poccil, Jkl, Pjacobi, Cmdrjameson, Hooperbloob, Lysdexia, Danski14, Axl, Woohookitty, Linas, Opie, Ligulem, Itinerant1, Kjlewis, McGinnis, Jpeob, JonathanD, Caco de vidro, SmackBot, Reedy, AndyZ, Lambiam, Joe-Bot, CmdrObot, Ian Beynon, Unclenuclear, Hebrides, Thijs!bot, Dogaroon, Headbomb, Marek69, JustAGal, Lself, Igodard, Autotheist, Markov Chain, Tonyfaull, MujiKha, Katharineamy, Wesino, DorganBot, Ross Fraser, Venny85, MvL1234, Bobathon71, Environnement2100, Addbot, Yobot, Ibayn, AnomieBOT, Citation bot, NOrbeck, Omnipaedista, Waleswatcher, Citation bot 1, Minimac, Mathewsyriac, Cogiati, AvicAWB, Quondum, Rezabot, Helpful Pixie Bot, ChrisGualtieri, Enyokoyama, DimReg, Monkbot, Gjbayes, Loraof, Chemistry1111 and Anonymous: 30
- **Diffeomorphism** *Source:* https://en.wikipedia.org/wiki/Diffeomorphism?oldid=723300256 *Contributors:* JeLuF, Maury Markowitz, Michael Hardy, TakuyaMurata, AugPi, Poor Yorick, Med, Charles Matthews, Dysprosia, Kuszi, MathMartin, Pascalromon, Tosha, Connelly, Giftlite, Lethe, Fropuff, CryptoDerk, Paul August, Rgdboer, Physicistjedi, Msh210, BRW, Oleg Alexandrov, R.e.b., Mathbot, Lmatt, Bgwhite, Mhwu, YurikBot, Wavelength, RussBot, Woseph, Gaius Cornelius, SmackBot, Silly rabbit, Nakon, Dreadstar, Mathsci, Myasuda, Dharma6662000, Thijs!bot, Headbomb, LachlanA, Nosirrom, Ensign beedrill, Policron, LokiClock, Oshwah, Geometry guy, Rybu, BotMultichill, JerroldPease-Atlanta, He7d3r, Topology Expert, PV=nRT, Legobot, Yobot, AnomieBOT, Point-set topologist, RibotBOT, Sławomir Biały, Citation bot 1, Åkebråke, Redrose64, MondalorBot, Cgqyyflz, Fly by Night, Slawekb, ZéroBot, Quondum, Chester Markel, Uni.Liu, Helpful Pixie Bot, BG19bot, Muses' house, Herve.lombaert, Jeremy112233, Hillbillyholiday, CsDix, Pwm86, BD2412bot and Anonymous: 47
- **Poisson bracket** *Source:* https://en.wikipedia.org/wiki/Poisson_bracket?oldid=723398517 *Contributors:* Michael Hardy, Charles Matthews, Jitse Niesen, Phys, Chuunen Baka, Tea2min, Ancheta Wis, Giftlite, Lethe, Fropuff, DefLog~enwiki, CryptoDerk, Rpchase, Chris Howard, Gauge, Linuxlad, Keenan Pepper, RoySmith, RJFJR, Linas, Sympleko, Mandarax, BD2412, Nanite, HappyCamper, Lionelbrits, Mathbot, Natkuhn, Chobot, Roboto de Ajvol, YurikBot, Archelon, Kier07, SmackBot, RDBury, Melchoir, Complexica, Nick Levine, Sammy1339, Geminatea, Owlbuster, Dicklyon, Andreas Rejbrand, PetaRZ, ShelfSkewed, Cydebot, Skittleys, Ebyabe, Headbomb, JAnDbot, .anacondabot, Michael K. Edwards, B. Wolterding, Joshua Davis, Policron, Lseixas, Cuzkatzimhut, VolkovBot, Red Act, Geometry guy, Pamputt, Rdengler, AlleborgoBot, YohanN7, SophomoricPedant, Hadrianheugh, Addbot, Download, StarLight, عس‌ی, Legobot, Luckas-bot, Rubinbot, Citation bot, GrouchoBot, FrescoBot, Sławomir Biały, Darij, Tkuvho, Night Jaguar, Qmechanic, Zfeinst, Zueignung, Bbbbbbbbba, Helpful Pixie Bot, BendelacBOT, ChrisGualtieri, Mark viking, JudgeDeadd, L2boyer and Anonymous: 47
- **Wilson loop** *Source:* https://en.wikipedia.org/wiki/Wilson_loop?oldid=693540804 *Contributors:* Bryan Derksen, The Anome, Miguel~enwiki, Stevenj, Charles Matthews, Phys, Finlay McWalter, Robbot, Giftlite, AmarChandra, Lumidek, Pjacobi, Paul August, Pt, Guy Harris, Joke137, Ketiltrout, Lionelbrits, Closedmouth, SmackBot, YellowMonkey, Chris the speller, Colonies Chris, Hongooi, Myasuda, Headbomb, Maliz, Racepacket, Basicfiend, VolkovBot, Calwiki, Spitfire8520, DragonBot, Addbot, DOI bot, Tassedethe, Feelinabsurd, Lightbot, ZéroBot, Bibcode Bot, Makecat-bot, Ruby Murray and Anonymous: 25

- **Knot invariant** *Source:* https://en.wikipedia.org/wiki/Knot_invariant?oldid=720134568 *Contributors:* Matthew Woodcraft, Michael Hardy, Bagpuss, Charles Matthews, Hyacinth, Kwantus, Fropuff, Bender235, Gauge, C S, Roger Fenn, Linas, Reetep, Iswyn, SmackBot, PJTraill, Chris the speller, ChuckHG, Sciyoshi~enwiki, Cronholm144, Lrudolphlrudolph, Shlomi Hillel, Hermel, David Eppstein, DavidCBryant, Rybu, Addbot, Prijutme4ty, WikitanvirBot, Hitchcock007, Helpful Pixie Bot and Anonymous: 12

35.4.2 Images

- **File:6n-graf.svg** *Source:* https://upload.wikimedia.org/wikipedia/commons/5/5b/6n-graf.svg *License:* Public domain *Contributors:* Image: 6n-graf.png simlar input data *Original artist:* User:AzaToth
- **File:A_Swarm_of_Ancient_Stars_-_GPN-2000-000930.jpg** *Source:* https://upload.wikimedia.org/wikipedia/commons/6/6a/A_Swarm_of_Ancient_Stars_-_GPN-2000-000930.jpg *License:* Public domain *Contributors:* Great Images in NASA Description *Original artist:* NASA, The Hubble Heritage Team, STScI, AURA
- **File:AdS3.svg** *Source:* https://upload.wikimedia.org/wikipedia/commons/4/47/AdS3.svg *License:* CC BY-SA 3.0 *Contributors:* This file was derived from: AdS3 (new).png
Original artist:
- derivative work: Alex Dunkel (Maky)
- **File:Ambox_important.svg** *Source:* https://upload.wikimedia.org/wikipedia/commons/b/b4/Ambox_important.svg *License:* Public domain *Contributors:* Own work, based off of Image:Ambox scales.svg *Original artist:* Dsmurat (talk · contribs)
- **File:Apollo_15_feather_and_hammer_drop.ogg** *Source:* https://upload.wikimedia.org/wikipedia/commons/3/3c/Apollo_15_feather_and_hammer_drop.ogg *License:* Public domain *Contributors:* Taken from Spacecraftfilms.com DVD "Apollo 15: The Great Explorations Begin" *Original artist:* NASA
- **File:Asymmetric_Ashes_(artist'{}s_impression).jpg** *Source:* https://upload.wikimedia.org/wikipedia/commons/d/db/Asymmetric_Ashes_%28artist%27s_impression%29.jpg *License:* CC BY 4.0 *Contributors:* http://www.eso.org/public/images/eso0644a/ *Original artist:* ESO
- **File:Black_Hole_Merger.jpg** *Source:* https://upload.wikimedia.org/wikipedia/commons/d/d1/Black_Hole_Merger.jpg *License:* Public domain *Contributors:* http://chandra.harvard.edu/resources/illustrations/quasar.html (image link) *Original artist:* NASA/CXC/A.Hobart
- **File:Bundesarchiv_Bild183-R57262,_Werner_Heisenberg.jpg** *Source:* https://upload.wikimedia.org/wikipedia/commons/f/f8/Bundesarchiv_Bild183-R57262%2C_Werner_Heisenberg.jpg *License:* CC BY-SA 3.0 de *Contributors:* This image was provided to Wikimedia Commons by the German Federal Archive (Deutsches Bundesarchiv) as part of a cooperation project. The German Federal Archive guarantees an authentic representation only using the originals (negative and/or positive), resp. the digitalization of the originals as provided by the Digital Image Archive. *Original artist:* Unknown<a href='//www.wikidata.org/wiki/Q4233718' title='wikidata:Q4233718'><img alt='wikidata:Q4233718' src='https://upload.wikimedia.org/wikipedia/commons/thumb/f/ff/Wikidata-logo.svg/20px-Wikidata-logo.svg.png' width='20' height='11' srcset='https://upload.wikimedia.org/wikipedia/commons/thumb/f/ff/Wikidata-logo.svg/30px-Wikidata-logo.svg.png 1.5x, https://upload.wikimedia.org/wikipedia/commons/thumb/f/ff/Wikidata-logo.svg/40px-Wikidata-logo.svg.png 2x' data-file-width='1050' data-file-height='590' /></a>
- **File:CERN_LHC_Tunnel1.jpg** *Source:* https://upload.wikimedia.org/wikipedia/commons/f/fc/CERN_LHC_Tunnel1.jpg *License:* CC BY-SA 3.0 *Contributors:* Own work *Original artist:* Julian Herzog (Website)
- **File:Calabi-Yau-alternate.png** *Source:* https://upload.wikimedia.org/wikipedia/commons/5/55/Calabi-Yau-alternate.png *License:* CC BY-SA 2.5 *Contributors:* Transferred from en.wikipedia to Commons by Lunch. *Original artist:* The original uploader was Lunch at English Wikipedia
- **File:Calabi-Yau.png** *Source:* https://upload.wikimedia.org/wikipedia/commons/d/d4/Calabi-Yau.png *License:* CC BY-SA 2.5 *Contributors:* own work by Lunch
http://en.wikipedia.org/wiki/Image:Calabi-Yau.png (english Wikipedia) *Original artist:* Lunch
- **File:Calabi_yau.jpg** *Source:* https://upload.wikimedia.org/wikipedia/commons/f/f3/Calabi_yau.jpg *License:* Public domain *Contributors:* Mathematica output, created by author *Original artist:* Jbourjai
- **File:Circle_as_Lie_group.svg** *Source:* https://upload.wikimedia.org/wikipedia/commons/8/82/Circle_as_Lie_group.svg *License:* Public domain *Contributors:* self-made with en:Inkscape *Original artist:* Oleg Alexandrov
- **File:Clebsch_Cublic.png** *Source:* https://upload.wikimedia.org/wikipedia/commons/7/7c/Clebsch_Cublic.png *License:* CC BY-SA 3.0 *Contributors:* I created this on my own computer using the free software Surfer *Original artist:* Fly by Night
- **File:Commons-logo.svg** *Source:* https://upload.wikimedia.org/wikipedia/en/4/4a/Commons-logo.svg *License:* CC-BY-SA-3.0 *Contributors:* ? *Original artist:* ?
- **File:Compactification_example.svg** *Source:* https://upload.wikimedia.org/wikipedia/commons/f/f5/Compactification_example.svg *License:* CC BY-SA 4.0 *Contributors:* Brian Greene (2004). The Elegant Universe (DVD). Part II (String's the thing): WGBH Boston Video. Event occurs at 43:55. OCLC 54019786 *Original artist:* Alex Dunkel (Maky)
- **File:Complete_graph_K5.svg** *Source:* https://upload.wikimedia.org/wikipedia/commons/c/cf/Complete_graph_K5.svg *License:* Public domain *Contributors:* Own work *Original artist:* David Benbennick wrote this file.
- **File:Crab_Nebula.jpg** *Source:* https://upload.wikimedia.org/wikipedia/commons/0/00/Crab_Nebula.jpg *License:* Public domain *Contributors:* HubbleSite: gallery, release. *Original artist:* NASA, ESA, J. Hester and A. Loll (Arizona State University)
- **File:Cyclic_group.svg** *Source:* https://upload.wikimedia.org/wikipedia/commons/5/5f/Cyclic_group.svg *License:* CC BY-SA 3.0 *Contributors:*
- Cyclic_group.png *Original artist:*

- derivative work: Pbroks13 (talk)
- **File:D3-brane_et_D2-brane.PNG** *Source:* https://upload.wikimedia.org/wikipedia/commons/8/88/D3-brane_et_D2-brane.PNG *License:* Public domain *Contributors:* Image:D-brane.PNG, oeuvre personnelle. *Original artist:* <a href='//commons.wikimedia.org/w/index.php?title=User:Rogilbert,<span>,&,</span>,action=edit,<span>,&,</span>,redlink=1' class='new' title='User:Rogilbert (page does not exist)'>Rogilbert</a>
- **File:Dark_Energy.jpg** *Source:* https://upload.wikimedia.org/wikipedia/commons/c/ce/Dark_Energy.jpg *License:* Public domain *Contributors:* http://hubblesite.org/newscenter/archive/releases/2001/09/image/g/ OR http://science.nasa.gov/astrophysics/focus-areas/what-is-dark-energy/ *Original artist:* Ann Feild (STScI)
- **File:Diffeomorphism_of_a_square.svg** *Source:* https://upload.wikimedia.org/wikipedia/commons/5/51/Diffeomorphism_of_a_square.svg *License:* Public domain *Contributors:* self-made with MATLAB *Original artist:* Oleg Alexandrov
- **File:Directed.svg** *Source:* https://upload.wikimedia.org/wikipedia/commons/a/a2/Directed.svg *License:* Public domain *Contributors:* ? *Original artist:* ?
- **File:Earth-moon.jpg** *Source:* https://upload.wikimedia.org/wikipedia/commons/5/5c/Earth-moon.jpg *License:* Public domain *Contributors:* NASA [1] *Original artist:* Apollo 8 crewmember Bill Anders
- **File:Edit-clear.svg** *Source:* https://upload.wikimedia.org/wikipedia/en/f/f2/Edit-clear.svg *License:* Public domain *Contributors:* The *Tango! Desktop Project*. *Original artist:*

  The people from the Tango! project. And according to the meta-data in the file, specifically: "Andreas Nilsson, and Jakub Steiner (although minimally)."
- **File:Edward_Witten.jpg** *Source:* https://upload.wikimedia.org/wikipedia/commons/9/97/Edward_Witten.jpg *License:* Public domain *Contributors:* Own work *Original artist:* Ojan
- **File:Energia_template.svg** *Source:* https://upload.wikimedia.org/wikipedia/commons/0/00/Energia_template.svg *License:* CC-BY-SA-3.0 *Contributors:* Own work *Original artist:* user:Urutseg
- **File:ExponentialMap-01.png** *Source:* https://upload.wikimedia.org/wikipedia/commons/0/06/ExponentialMap-01.png *License:* Public domain *Contributors:* ? *Original artist:* ?
- **File:Falling_ball.jpg** *Source:* https://upload.wikimedia.org/wikipedia/commons/0/02/Falling_ball.jpg *License:* CC BY-SA 3.0 *Contributors:* Own work *Original artist:* MichaelMaggs
- **File:Feynman-Diagram.svg** *Source:* https://upload.wikimedia.org/wikipedia/commons/e/e3/Feynman-Diagram.svg *License:* Public domain *Contributors:* own work, based on Image:Feynman-Diagram.jpg *Original artist:* helix84
- **File:Folder_Hexagonal_Icon.svg** *Source:* https://upload.wikimedia.org/wikipedia/en/4/48/Folder_Hexagonal_Icon.svg *License:* Cc-by-sa-3.0 *Contributors:* ? *Original artist:* ?
- **File:GPB_circling_earth.jpg** *Source:* https://upload.wikimedia.org/wikipedia/commons/d/d1/GPB_circling_earth.jpg *License:* Public domain *Contributors:* http://www.nasa.gov/mission_pages/gpb/gpb_012.html *Original artist:* NASA
- **File:GabrieleVeneziano.jpg** *Source:* https://upload.wikimedia.org/wikipedia/commons/9/95/GabrieleVeneziano.jpg *License:* CC BY-SA 2.5 *Contributors:* Taken by Betsythedevine *Original artist:* The original uploader was Betsythedevine at English Wikipedia
- **File:GalacticRotation2.svg** *Source:* https://upload.wikimedia.org/wikipedia/commons/b/b9/GalacticRotation2.svg *License:* CC-BY-SA-3.0 *Contributors:* Own work in Inkscape 0.42 *Original artist:* PhilHibbs
- **File:Gravity_Probe_B.jpg** *Source:* https://upload.wikimedia.org/wikipedia/commons/5/51/Gravity_Probe_B.jpg *License:* Public domain *Contributors:* ? *Original artist:* ?
- **File:Gravity_action-reaction.gif** *Source:* https://upload.wikimedia.org/wikipedia/commons/7/7d/Gravity_action-reaction.gif *License:* CC BY-SA 3.0 *Contributors:* Own work *Original artist:* Orion 8
- **File:HAtomOrbitals.png** *Source:* https://upload.wikimedia.org/wikipedia/commons/c/cf/HAtomOrbitals.png *License:* CC-BY-SA-3.0 *Contributors:* ? *Original artist:* ?
- **File:History_of_the_Universe.svg** *Source:* https://upload.wikimedia.org/wikipedia/commons/d/db/History_of_the_Universe.svg *License:* CC BY-SA 3.0 *Contributors:* Own work *Original artist:* Yinweichen
- **File:Horizonte_inflacionario.svg** *Source:* https://upload.wikimedia.org/wikipedia/commons/b/b4/Horizonte_inflacionario.svg *License:* CC-BY-SA-3.0 *Contributors:* Transferred from en.wikipedia to Commons.; original: *I created this work in Adobe Illustrator. Original artist:* Joke137 at English Wikipedia
- **File:Hydrogen_Density_Plots.png** *Source:* https://upload.wikimedia.org/wikipedia/commons/e/e7/Hydrogen_Density_Plots.png *License:* Public domain *Contributors:* the English language Wikipedia (log).

  Original artist: PoorLeno (talk)
- **File:Ilc_9yr_moll4096.png** *Source:* https://upload.wikimedia.org/wikipedia/commons/3/3c/Ilc_9yr_moll4096.png *License:* Public domain *Contributors:* http://map.gsfc.nasa.gov/media/121238/ilc_9yr_moll4096.png *Original artist:* NASA / WMAP Science Team
- **File:Infinite_potential_well.svg** *Source:* https://upload.wikimedia.org/wikipedia/commons/2/27/Infinite_potential_well.svg *License:* Public domain *Contributors:* Created by bdesham in Inkscape. *Original artist:* Benjamin D. Esham (bdesham)
- **File:Joseph_Polchinski.jpg** *Source:* https://upload.wikimedia.org/wikipedia/commons/c/cf/Joseph_Polchinski.jpg *License:* Public domain *Contributors:* Transferred from en.wikipedia to Commons by Magnus Manske using CommonsHelper. *Original artist:* The original uploader was Lumidek at English Wikipedia
- **File:JuanMaldacena.jpg** *Source:* https://upload.wikimedia.org/wikipedia/commons/b/bc/JuanMaldacena.jpg *License:* CC BY 3.0 *Contributors:* Own work by the original uploader *Original artist:* Lumidek at English Wikipedia

- **File:KleinInvariantJ.jpg** *Source:* https://upload.wikimedia.org/wikipedia/commons/3/37/KleinInvariantJ.jpg *License:* Public domain *Contributors:* made with mathematica, own work *Original artist:* Jan Homann
- **File:Knot_table.svg** *Source:* https://upload.wikimedia.org/wikipedia/commons/1/12/Knot_table.svg *License:* Public domain *Contributors:* Own work *Original artist:* Jkasd
- **File:LQG_black_hole_Horizon.jpg** *Source:* https://upload.wikimedia.org/wikipedia/en/9/9d/LQG_black_hole_Horizon.jpg *License:* CC-BY-SA-3.0 *Contributors:*

  created on xfig
  Previously published: 2007-09-01
  Original artist:
  Ibayn
- **File:Labeled_Triangle_Reflections.svg** *Source:* https://upload.wikimedia.org/wikipedia/commons/3/38/Labeled_Triangle_Reflections.svg *License:* Public domain *Contributors:* Own work *Original artist:* Jim.belk
- **File:LeonardSusskindStanford2009_cropped.jpg** *Source:* https://upload.wikimedia.org/wikipedia/commons/f/f8/LeonardSusskindStanford2009_cropped.jpg *License:* CC BY-SA 3.0 *Contributors:* File:LeonardSusskindStanford2009.jpg *Original artist:* Jonathan Maltz
- **File:Liealgebra.png** *Source:* https://upload.wikimedia.org/wikipedia/commons/d/d2/Liealgebra.png *License:* Public domain *Contributors:* http://en.wikipedia.org/wiki/File:Liealgebra.png *Original artist:* Phys
- **File:Limits_of_M-theory.svg** *Source:* https://upload.wikimedia.org/wikipedia/commons/b/b8/Limits_of_M-theory.svg *License:* CC BY-SA 3.0 *Contributors:*

  Limits of M-theory.png

  Original artist:
- derivative work: Alex Dunkel (Maky)
- **File:Max_Born.jpg** *Source:* https://upload.wikimedia.org/wikipedia/commons/f/f7/Max_Born.jpg *License:* Public domain *Contributors:* ? *Original artist:* ?
- **File:Max_Planck_(1858-1947).jpg** *Source:* https://upload.wikimedia.org/wikipedia/commons/a/a7/Max_Planck_%281858-1947%29.jpg *License:* Public domain *Contributors:* http://www.sil.si.edu/digitalcollections/hst/scientific-identity/CF/display_results.cfm?alpha_sort=p *Original artist:* Unknown<a href='//www.wikidata.org/wiki/Q4233718' title='wikidata:Q4233718'><img alt='wikidata:Q4233718' src='https://upload.wikimedia.org/wikipedia/commons/thumb/f/ff/Wikidata-logo.svg/20px-Wikidata-logo.svg.png' width='20' height='11' srcset='https://upload.wikimedia.org/wikipedia/commons/thumb/f/ff/Wikidata-logo.svg/30px-Wikidata-logo.svg.png 1.5x, https://upload.wikimedia.org/wikipedia/commons/thumb/f/ff/Wikidata-logo.svg/40px-Wikidata-logo.svg.png 2x' data-file-width='1050' data-file-height='590' /></a>
- **File:Meissner_effect_p1390048.jpg** *Source:* https://upload.wikimedia.org/wikipedia/commons/5/55/Meissner_effect_p1390048.jpg *License:* CC-BY-SA-3.0 *Contributors:* self photo *Original artist:* Mai-Linh Doan
- **File:Milky_Way_Emerges_as_Sun_Sets_over_Paranal.jpg** *Source:* https://upload.wikimedia.org/wikipedia/commons/9/90/Milky_Way_Emerges_as_Sun_Sets_over_Paranal.jpg *License:* CC BY 4.0 *Contributors:* http://www.eso.org/public/images/potw1517a/ *Original artist:* ESO/J. Colosimo
- **File:Neutron_spin_dipole_field.jpg** *Source:* https://upload.wikimedia.org/wikipedia/commons/1/15/Neutron_spin_dipole_field.jpg *License:* CC BY-SA 4.0 *Contributors:* Own work *Original artist:* Bdushaw
- **File:Nuvola_apps_edu_mathematics_blue-p.svg** *Source:* https://upload.wikimedia.org/wikipedia/commons/3/3e/Nuvola_apps_edu_mathematics_blue-p.svg *License:* GPL *Contributors:* Derivative work from Image:Nuvola apps edu mathematics.png and Image:Nuvola apps edu mathematics-p.svg *Original artist:* David Vignoni (original icon); Flamurai (SVG convertion); bayo (color)
- **File:Open_and_closed_strings.svg** *Source:* https://upload.wikimedia.org/wikipedia/commons/5/56/Open_and_closed_strings.svg *License:* Public domain *Contributors:* Own work *Original artist:* Xoneca
- **File:Parallel_Transport.svg** *Source:* https://upload.wikimedia.org/wikipedia/commons/7/7d/Parallel_Transport.svg *License:* CC BY-SA 4.0 *Contributors:* <a href='//commons.wikimedia.org/wiki/File:Parallel_transport.png' class='image'><img alt='Parallel transport.png' src='https://upload.wikimedia.org/wikipedia/commons/thumb/6/6d/Parallel_transport.png/100px-Parallel_transport.png' width='100' height='112' srcset='https://upload.wikimedia.org/wikipedia/commons/thumb/6/6d/Parallel_transport.png/150px-Parallel_transport.png 1.5x, https://upload.wikimedia.org/wikipedia/commons/thumb/6/6d/Parallel_transport.png/200px-Parallel_transport.png 2x' data-file-width='510' data-file-height='570' /></a> *Original artist:* Fred the Oyster
- **File:Parametrized_Harmonic_Oscillator.jpg** *Source:* https://upload.wikimedia.org/wikipedia/en/5/5c/Parametrized_Harmonic_Oscillator.jpg *License:* CC-BY-SA-3.0 *Contributors:*

  On xfig package.
  Previously published: 2006-06-06
  Original artist:
  Ibayn
- **File:Pascual_Jordan_1920s.jpg** *Source:* https://upload.wikimedia.org/wikipedia/commons/a/a6/Pascual_Jordan_1920s.jpg *License:* Public domain *Contributors:* http://www.gettyimages.co.uk/detail/news-photo/the-austrian-physicist-paul-ehrenfest-posing-in-front-of-a-news-photo/141551561 *Original artist:* Unknown (Mondadori Publishers)
- **File:Point&string.png** *Source:* https://upload.wikimedia.org/wikipedia/commons/4/47/Point%26string.png *License:* Public domain *Contributors:* ? *Original artist:* ?

- **File:Qm_step_pot_temp.png** *Source:* https://upload.wikimedia.org/wikipedia/commons/8/87/Qm_step_pot_temp.png *License:* Public domain *Contributors:* Own work *Original artist:* F=q(E+v^B)
- **File:Qm_template_pic_4.svg** *Source:* https://upload.wikimedia.org/wikipedia/commons/f/fe/Qm_template_pic_4.svg *License:* CC0 *Contributors:* Own work *Original artist:* Maschen
- **File:QuantumHarmonicOscillatorAnimation.gif** *Source:* https://upload.wikimedia.org/wikipedia/commons/9/90/QuantumHarmonicOscillatorAnimation.gif *License:* CC0 *Contributors:* Own work *Original artist:* Sbyrnes321
- **File:Quantum_gravity.svg** *Source:* https://upload.wikimedia.org/wikipedia/commons/6/64/Quantum_gravity.svg *License:* CC0 *Contributors:* File:Quantum gravity.png *Original artist:* original by Raidr
- **File:Question_book-new.svg** *Source:* https://upload.wikimedia.org/wikipedia/en/9/99/Question_book-new.svg *License:* Cc-by-sa-3.0 *Contributors:*

  Created from scratch in Adobe Illustrator. Based on Image:Question book.png created by User:Equazcion *Original artist:*

  Tkgd2007
- **File:Question_dropshade.png** *Source:* https://upload.wikimedia.org/wikipedia/commons/d/dd/Question_dropshade.png *License:* Public domain *Contributors:* Image created by JRM *Original artist:* JRM
- **File:Richard_Feynman_Nobel.jpg** *Source:* https://upload.wikimedia.org/wikipedia/en/4/42/Richard_Feynman_Nobel.jpg *License:* ? *Contributors:*

  http://www.nobelprize.org/nobel_prizes/physics/laureates/1965/feynman-bio.html *Original artist:*

  The Nobel Foundation
- **File:Rtd_seq_v3.gif** *Source:* https://upload.wikimedia.org/wikipedia/commons/5/51/Rtd_seq_v3.gif *License:* CC BY 3.0 *Contributors:* Tool: Resonant Tunneling Diode Simulation with NEGF simulator on www.nanoHUB.org. Link: http://nanohub.org/resources/8799 *Original artist:* Saumitra R Mehrotra & Gerhard Klimeck
- **File:Sir_Isaac_Newton_(1643-1727).jpg** *Source:* https://upload.wikimedia.org/wikipedia/commons/8/83/Sir_Isaac_Newton_%281643-1727%29.jpg *License:* Public domain *Contributors:* http://www.phys.uu.nl/~{ }vgent/astrology/images/newton1689.jpg] *Original artist:* Sir Godfrey Kneller
- **File:Solar_system.jpg** *Source:* https://upload.wikimedia.org/wikipedia/commons/8/83/Solar_system.jpg *License:* Public domain *Contributors:* http://photojournal.jpl.nasa.gov/catalog/PIA03153 *Original artist:* NASA/JPL
- **File:Solvay_conference_1927.jpg** *Source:* https://upload.wikimedia.org/wikipedia/commons/6/6e/Solvay_conference_1927.jpg *License:* Public domain *Contributors:* http://w3.pppl.gov/ *Original artist:* Benjamin Couprie
- **File:Spacetime_curvature.png** *Source:* https://upload.wikimedia.org/wikipedia/commons/2/22/Spacetime_curvature.png *License:* CC-BY-SA-3.0 *Contributors:* ? *Original artist:* ?
- **File:Spacetime_lattice_analogy.svg** *Source:* https://upload.wikimedia.org/wikipedia/commons/6/63/Spacetime_lattice_analogy.svg *License:* CC BY-SA 3.0 *Contributors:* Own work. Self -made in Blender & Inkscape. *Original artist:* Mysid
- **File:Spin_One-Half_(Slow).gif** *Source:* https://upload.wikimedia.org/wikipedia/commons/6/6e/Spin_One-Half_%28Slow%29.gif *License:* CC0 *Contributors:* Own work *Original artist:* JasonHise
- **File:Spin_foam_from_Hamiltonian_constraint.jpg** *Source:* https://upload.wikimedia.org/wikipedia/en/3/34/Spin_foam_from_Hamiltonian_constraint.jpg *License:* CC-BY-SA-3.0 *Contributors:*

  On a graphics tool.
  Previously published: 2013-04-13

  Original artist:

  Ibayn
- **File:Spin_network.svg** *Source:* https://upload.wikimedia.org/wikipedia/commons/5/52/Spin_network.svg *License:* CC BY-SA 3.0 *Contributors:* Own work *Original artist:* Markus Poessel (Mapos)
- **File:StringTheoryDualities.svg** *Source:* https://upload.wikimedia.org/wikipedia/commons/8/8a/StringTheoryDualities.svg *License:* CC BY-SA 3.0 *Contributors:*
- StringTheoryDualities.jpg *Original artist:*
- derivative work: Alex Dunkel (Maky)
- **File:Stylised_Lithium_Atom.svg** *Source:* https://upload.wikimedia.org/wikipedia/commons/e/e1/Stylised_Lithium_Atom.svg *License:* CC-BY-SA-3.0 *Contributors:* based off of Image:Stylised Lithium Atom.png by Halfdan. *Original artist:* SVG by Indolences. Recoloring and ironing out some glitches done by Rainer Klute.
- **File:Text_document_with_red_question_mark.svg** *Source:* https://upload.wikimedia.org/wikipedia/commons/a/a4/Text_document_with_red_question_mark.svg *License:* Public domain *Contributors:* Created by bdesham with Inkscape; based upon Text-x-generic.svg from the Tango project. *Original artist:* Benjamin D. Esham (bdesham)
- **File:The_Mandelstam_identity.jpg** *Source:* https://upload.wikimedia.org/wikipedia/en/5/5f/The_Mandelstam_identity.jpg *License:* CC-BY-SA-3.0 *Contributors:*

  On graphics package
  Previously published: 2005-06-25

  Original artist:

  Ibayn

- **File:Undirected.svg** *Source:* https://upload.wikimedia.org/wikipedia/commons/b/bf/Undirected.svg *License:* Public domain *Contributors:* ? *Original artist:* ?
- **File:Uniform_tiling_433-t0_(formatted).svg** *Source:* https://upload.wikimedia.org/wikipedia/commons/2/21/Uniform_tiling_433-t0_%28formatted%29.svg *License:* CC BY-SA 4.0 *Contributors:* This file was derived from: Uniform tiling 433-t0.png *Original artist:*
- derivative work: Polytope24
- **File:Wikibooks-logo.svg** *Source:* https://upload.wikimedia.org/wikipedia/commons/f/fa/Wikibooks-logo.svg *License:* CC BY-SA 3.0 *Contributors:* Own work *Original artist:* User:Bastique, User:Ramac et al.
- **File:Wikinews-logo.svg** *Source:* https://upload.wikimedia.org/wikipedia/commons/2/24/Wikinews-logo.svg *License:* CC BY-SA 3.0 *Contributors:* This is a cropped version of Image:Wikinews-logo-en.png. *Original artist:* Vectorized by Simon 01:05, 2 August 2006 (UTC) Updated by Time3000 17 April 2007 to use official Wikinews colours and appear correctly on dark backgrounds. Originally uploaded by Simon.
- **File:Wikiquote-logo.svg** *Source:* https://upload.wikimedia.org/wikipedia/commons/f/fa/Wikiquote-logo.svg *License:* Public domain *Contributors:* ? *Original artist:* ?
- **File:Wikisource-logo.svg** *Source:* https://upload.wikimedia.org/wikipedia/commons/4/4c/Wikisource-logo.svg *License:* CC BY-SA 3.0 *Contributors:* Rei-artur *Original artist:* Nicholas Moreau
- **File:Wikiversity-logo-Snorky.svg** *Source:* https://upload.wikimedia.org/wikipedia/commons/1/1b/Wikiversity-logo-en.svg *License:* CC BY-SA 3.0 *Contributors:* Own work *Original artist:* Snorky
- **File:Wikiversity-logo.svg** *Source:* https://upload.wikimedia.org/wikipedia/commons/9/91/Wikiversity-logo.svg *License:* CC BY-SA 3.0 *Contributors:* Snorky (optimized and cleaned up by verdy_p) *Original artist:* Snorky (optimized and cleaned up by verdy_p)
- **File:Wiktionary-logo-en.svg** *Source:* https://upload.wikimedia.org/wikipedia/commons/f/f8/Wiktionary-logo-en.svg *License:* Public domain *Contributors:* Vector version of Image:Wiktionary-logo-en.png. *Original artist:* Vectorized by Fvasconcellos (talk · contribs), based on original logo tossed together by Brion Vibber
- **File:Wolfgang_Pauli_young.jpg** *Source:* https://upload.wikimedia.org/wikipedia/commons/4/43/Wolfgang_Pauli_young.jpg *License:* Public domain *Contributors:* ? *Original artist:* ?
- **File:World_lines_and_world_sheet.svg** *Source:* https://upload.wikimedia.org/wikipedia/commons/2/25/World_lines_and_world_sheet.svg *License:* Public domain *Contributors:* Point&string.png *Original artist:* Kurochka, svg version by Actam

35.4.3 Content license

Made in the USA
Middletown, DE
19 December 2017